Odontodes

The odontode system, which encompasses teeth and other dentine-based structures, is ancient. Odontodes are present in the oldest vertebrate fossils, dating back 500 million years, and still play an important role in the anatomy and function of living jawed vertebrates. Fossils preserve odontode tissues with remarkable nanoscale fidelity, allowing the evolution and diversification of the odontode system to be studied in deep time as well as across the diversity of living vertebrates. This synthetic volume presents an overview of odontode research by internationally leading researchers from different fields of biology.

Key features

- Summarizes classic and cutting-edge research devoted to dental development and evolution
- Focuses on the cellular basis of odontogenesis
- Documents the structural and functional diversity of odontode tissues
- Describes the patterning mechanisms of dentitions in various vertebrate groups
- Provides a thorough index for students

Evolutionary Cell Biology

Published Titles

Evolutionary Cell Processes in Primates: Genes, Skin, Energetics, Breathing, and Feeding, Volume II
Edited by M. Kathleen Pitirri and Joan T. Richtsmeier

The Notochord: Development, Evolution and contributions to the vertebral column
Eckhard P. Witten and Brian K. Hall

Evolution of Neurosensory Cells and Systems: Gene regulation and cellular networks and processes
Edited by Bernd Fritzsch and Karen L. E. Thompson

The Evolution of Multicellularity
Edited by Matthew D. Herron, Peter Conlin, and William C. Ratcliff

Phenotypic Plasticity & Evolution: Causes, Consequences, Controversies
Edited By David W. Pfennig

Hox Modules in Evolution and Development
Edited By David E. K. Ferrier

Odontodes: The Developmental and Evolutionary Building Blocks of Dentitions
Edited By Donglei Chen

For more information about this series, please visit: www.crcpress.com/Evolutionary-Cell-Biology/book-series/CRCEVOCELBIO

Odontodes

The Developmental and Evolutionary Building Blocks of Dentitions

Edited by Donglei Chen

CRC Press
Taylor & Francis Group
Boca Raton London New York

CRC Press is an imprint of the
Taylor & Francis Group, an **informa** business

Designed cover image: Virtual section through the upper and lower jaws of a bichir larva (*Polypterus senegalus*, TL 5cm)
Cover designed by Donglei Chen
3D model produced by Donglei Chen
Synchrotron scan performed by Jake Leyhr and Sophie Sanchez
Specimen prepared by Jan Stundl

First edition published 2024
by CRC Press
6000 Broken Sound Parkway NW, Suite 300, Boca Raton, FL 33487–2742

and by CRC Press
4 Park Square, Milton Park, Abingdon, Oxon, OX14 4RN

CRC Press is an imprint of Taylor & Francis Group, LLC

Library of Congress Cataloging-in-Publication Data
Names: Chen, Donglei, editor.
Title: Odontodes : the developmental and evolutionary building blocks of dentitions / edited by Donglei Chen.
Description: First edition. | Boca Raton, FL : CRC Press, 2024. | Includes bibliographical references and index.
Identifiers: LCCN 2023015543 (print) | LCCN 2023015544 (ebook) | ISBN 9781032065151 (hardback) | ISBN 9781032574967 (paperback) | ISBN 9781003439653 (ebook)
Subjects: LCSH: Dentition. | Teeth. | Evolution (Biology)
Classification: LCC QL858 .O36 2024 (print) | LCC QL858 (ebook) | DDC 612.3/11—dc23/eng/20230822
LC record available at https://lccn.loc.gov/2023015543
LC ebook record available at https://lccn.loc.gov/2023015544

ISBN: 978-1-032-06515-1 (hbk)
ISBN: 978-1-032-57496-7 (pbk)
ISBN: 978-1-003-43965-3 (ebk)

DOI: 10.1201/9781003439653

Typeset in Times
by Apex CoVantage, LLC

Contents

Series Preface

In recent decades, evolutionary principles have been integrated into biological disciplines such as developmental biology, ecology, and genetics. As a result, major new fields emerged, chief among which are Evolutionary Developmental Biology (or Evo-Devo) and Ecological Developmental Biology (or Eco-Devo). Inspired by the integration of knowledge of change over single life spans (ontogenetic history) and change over evolutionary time (phylogenetic history), Evo-Devo produced a unification of developmental and evolutionary biology that generated unanticipated synergies: Molecular biologists employ computational and conceptual tools generated by developmental biologists and by systematists, while evolutionary biologists use detailed analysis of molecules in their studies. These integrations have shifted paradigms and enabled us to answer questions that were once thought to be intractable.

Major highlights in the development of modern Evo-Devo are a comparison of the evolutionary behavior of cells, evidenced in Stephen J. Gould's 1979 proposal of changes in the timing of the activity of cells during development—heterochrony—as a major force in evolutionary change, and numerous studies demonstrating how conserved gene families across numerous cell types "explain" development and evolution. Advances in technology and in instrumentation now allow cell biologists to make ever more detailed observations of the structure of cells and the processes by which cells arise, divide, differentiate, and die. In recent years, cell biologists have increasingly asked questions whose answers require insights from evolutionary history. As just one example: How many cell types are there, and how are they related? Given this conceptual basis, cell biology—a rich field in biology with history going back centuries—is poised to be reintegrated with evolution to provide a means of organizing and explaining diverse empirical observations and testing fundamental hypotheses about the cellular basis of life. Integrating evolutionary and cellular biology has the potential to generate new theories of cellular function and to create a new field, "*Evolutionary Cell Biology*."

Mechanistically, cells provide the link between the genotype and the phenotype, both during development and in evolution. Hence the proposal for a series of books under the general theme of "*Evolutionary Cell Biology: Translating Genotypes into Phenotypes*" is to document, demonstrate, and establish the central role played by cellular mechanisms in the evolution of all forms of life.

Brian K. Hall
Sally A. Moody

Preface

Odontodes is the collective term for oral and skin teeth and their derivatives. They are mainly composed of dentine, usually coated by enamel/enameloid, and attached by fibrous tissues to bone or cartilage. An odontode is thus a developmental module integrating various hard tissues—which are themselves among the major evolutionary innovations of vertebrates. Thanks to their hardness, odontodes are well-preserved in the fossil record and thus provide a window on the deep-time history of an organ system and its tissues that cannot be matched by other systems only based on soft anatomy. As a feeding apparatus or body armor, albeit under strong selection, all odontodes share a deep homology with a conserved gene regulatory network. Their 500-million-year evolutionary history of morphological plasticity and experimentation has been built upon a relatively simple suite of cell–cell interactions between epithelium and mesenchyme. All these make the odontode skeleton a Complex Adaptive System that can be interrogated effectively in order to test evolutionary developmental phenomena and hypotheses, such as co-option and heterochrony. This volume brings together the most recent perspectives from world-class biologists of different realms, based on investigations with modern techniques, to understand the odontode system from multiple dimensions.

From Chapter 1, we learn how a single set of cell types can attain new functions, like generating novel dental tissues at different parts of an odontode and at different stages of the dental ontogeny. For instance, odontoblasts at the labial and lingual sides of the mouse incisor express different genes and produce different dentines. There thus emerge a variety of cell subtypes or subpopulations by borrowing and modifying preexisting gene expression programs from other cell lineages. However, the concept of discrete cell types and the demarcation between the progenitor and the differentiated cell states might be artificial. An example of the continuum is the evolution of the hard-tissue-depositing ameloblasts from the epithelial progenitors that serve as the enamel knots or signaling centers for tooth initiation.

While pharyngeal teeth gradually disappeared through the fish–tetrapod transition, oral teeth are occasionally suppressed within the teleost radiation. In some teleosts, like zebrafish, teeth are only distributed on certain posterior pharyngeal arches. To initiate teeth in the endodermal territory, a persisting physical link with the ectoderm is needed. Chapter 2 uncovers how the odontogenic competence is transferred from the external to the internal epithelia by the periderm, a transient germ layer. An odontogenic epithelium invariably becomes stratified before tooth morphogenesis, with the surface layer ectodermalized, so that the basal layer, regardless of its germ layer origin, can produce the enamel organ.

Once it has undergone an epithelial-to-mesenchymal transition, the neural crest (NC), "the fourth germ layer" of a vertebrate, gains mesenchymal traits. By reviewing historical and recent experiments, Chapter 3 proposes that NC-derived mesenchyme alone, even without a competent epithelium, has the potential to form odontodes and dermal bones, not only within the head, but also all over the body. Osteoblasts can respond to mechanical stimuli, while odontoblasts can additionally detect temperature and chemical signals, and both cell types are of a possible NC neuroglial origin. To protect the sensory transduction, collagen matrix is secreted around the odontoblast processes and mineralized. Hence, biomineralization might first take place at the dermis, as the dermal odontodes that provided a protective armor for the body of early vertebrates and evolved from the neurosensory organs scattered throughout the skin of their soft-bodied ancestors.

Chapter 4 introduces the shared ancestral processes of the biomineralization of dental tissues and goes through the main categories of protein families involved in these processes, which, however, display considerable taxonomic variation. Gene loss is thought to be associated with the loss or reduction of teeth/scales and enamel/enameloid. However, there is substantial case-by-case variation in which genes are lost or remain but appear as pseudogenes, making it difficult to identify the genes responsible for particular types of odontodes or hypermineralized tissues. The gene that is linked to the loss of dentine in amniotes, *Dspp*, lacks orthologs in other vertebrates, even though the presence of dentine is conserved. The presence of type I collagen fibers distinguishes enameloid from enamel, and *col1a1* is expressed by ameloblasts in amphibians and teleosts during the secretion of enameloid matrix, but surprisingly not in the ameloblasts of sharks.

In chondrichthyans, although most of the known odontode tissue matrix components appear missing, odontoblasts exhibit a tremendously pluripotent repertoire, even in dental plates that have lost the form of teeth. Dentine can also be hypermineralized. Chapter 5 describes exhaustively the cytology of osteodentine, trabecular dentine, tubate dentine, and whitlockin, illustrating microstructural details like the spatial relationship between the branching dentine tubules and fiber bundles. The deposition of the dental batteries is compared between the chondrichthyan and osteichthyan durophagous dentitions. The characteristic infilling nature of pleromin brought by the migration of odontoblasts deep into vascular spaces allows dentine to perform responsive wear repair in early vertebrates by invading the broken bone without resorption.

Like dentine, periodontal tissues are formed by NC-derived mesenchymal cells. Chapter 6 summarizes the modes of tooth attachment across vertebrate groups and concludes that cementum, periodontal ligament, and alveolar bone are plesiomorphic. The gomphosis attachment, in which teeth are suspended by non-mineralized fibers, is not unique to mammals and crocodylians, but a paedomorphic truncation of the mineralization of the Sharpey's fibers in ankylosed teeth by maintaining a gap between cementum and alveolar bone. The epithelium shaping the teeth is also an intervening wall between the odontoblasts inside the dental papilla and the cementoblasts outside in the dental follicle. Attachment of the periodontal tissues to the dentine requires either that this physical barrier breaks down, or that the dentine grows down below the edge of the epithelium. This creates a variability in the geometry of tooth attachment. The extreme asymmetry of pleurodont implantation is the result of a long epithelial boundary along the lingual side of the teeth with a short one on the labial side. The ever-growing and self-sharpening incisors of rodents and lagomorphs are achieved by the shift of the crown–root boundary so that the epithelium is only fragmented on the lingual face to allow the formation of cementum, but continues producing enamel on the labial face. This was also the developmental mechanism of the dental batteries in hadrosaurid and ceratopsid dinosaurs.

Just as amending when and where a dental tissue is deposited can give rise to various types of teeth, when and where a tooth is initiated can bring about diverse arrangements of dentitions as well. Analyzing the multi-row dentitions that grow in either an x/y-mode or a splay-mode, Chapter 7 explores the periodic dental patterning established through the reaction–diffusion system, with the molecular players identified. Unlike hair development, which has multiple signaling centers across a large competent field arising at the same time, tooth development starts with a single signaling center, followed by a directionally incremental addition of new teeth at the propagating front as the competent field is progressively expanding. The first tooth or first generation of teeth at the first signaling center is the initiator of tooth patterning, but can sometimes be evaginated, avascularized, non-mineralized, non-erupted, or non-renewed. Vestigial decidual tooth buds are often seen in mammals, whose role is to initiate the permanent teeth of its tooth class. Having lost some terminal "tooth-producing" genes, rudimentary tooth germs are still initiated in chick embryo before the internal invasion of the rhamphotheca, and odontogenesis can be reactivated if the epithelial signaling center at the oral–aboral (non-keratinizing/keratinizing) boundary is reposited lingually to overlie the competent mesenchyme. It is consistent with the initiator teeth of axolotl, which are always generated at the ectoderm–endoderm boundary or from the ectodermal region.

Chapter 8 reconstructs the 3D dental ontogeny in the earliest bony fish: The initiator odontodes at the original oral–dermal boundary diversify into teeth lingually and dermal odontodes labially, multiple tooth rows are added in the splay mode, non-shedding gradually become partial shedding and complete shedding, and at last the labial tooth rows are overgrown as the dermal odontodes invade the oral domain. During this developmental process, insertion and suppression of tooth position and splitting and merging of tooth families occur frequently, which are also often noticed in the detailed spatiotemporal mapping of dentitions in other vertebrates. Therefore, it is hypothesized that tooth addition is a gap-filling process, arranged by the first principle of close packing. If the expansion of the competent field is too little to accommodate additional teeth, site-specific resorption will be induced to create a gap. For example, when the jaw is elongated, larger teeth are initiated, but the jaw width barely increases enough to house another tooth row, and the previous tooth row will be replaced. This is probably how the in-situ tooth replacement originated, simply by adjusting the relative growth rate between bone growth and tooth initiation.

The coordination between the development of teeth and dental arches, two Complex Adaptive Systems, is explored in humans in Chapter 9 with multi-model thinking based on rich clinical data. Variations in tooth number, size, and shape, from an evolutionary aspect not considered as anomalies, can arise not only from genetic, epigenetic, and environmental factors, but also from interactions between cells, matrix, and the developing tooth germs. The interactions can be compensatory, for instance, the absence of the maxillary lateral incisor leads to significantly larger central incisors. The more teeth are missing, the fewer cusps are borne by the formed teeth, and conversely in the case of supernumerary teeth, reflecting the down- and up-regulation of *Shh*, respectively. However, missing and supernumerary teeth can happen in the same dentition and even in the same tooth class, suggesting no single model can incorporate all situations. During the transition stages of primary-mixed-permanent dentition, the alignment of teeth constantly changes through occlusal development, which affects alveolar growth and hence influences the growing direction and extent of the dental arches.

Each chapter, on the one hand, provides vital information and, on the other hand, questions the conventional knowledge that is narrowly based on the specified conditions of model organisms. The aim of this volume is not only to provide insights into the state-of-the-art in odontode research, but also to inspire readers the potential directions for future evolutionary developmental studies both related and unrelated to odontodes.

Donglei Chen
March 2023

About the Editor

Born in Guangzhou, China, Dr. Donglei Chen has studied and worked at Uppsala University in Sweden for 15 years. She studied for her PhD under the supervision of Prof. Per Ahlberg, focusing on the dentitions of the Silurian (approximately 420 to 425 million years old) osteichthyans *Andreolepis* and *Lophosteus*. This research was based on microtomographic data sets with sub-micrometer resolution produced at the European Synchrotron Radiation Facility (ESRF) in Grenoble, France, and was the first use of such data to fully reconstruct the three-dimensional architecture of early vertebrate dentitions. She received the Photon Science Award 2018, from the Centre for Photon Science at Uppsala University, for showing the most innovative use of photon science in her PhD thesis.

Dr. Chen's research has provided profound mechanistic insights allowing her to address evolutionary and developmental questions, such as the origin of teeth in jawed vertebrates, the origin of in-situ tooth replacement in bony fish, and the relationship between teeth and dermal odontodes. Based on synchrotron microtomography, she uses three-dimensional virtual histology to reconstruct the ontogenetic histories of early vertebrate dermoskeletons. This three-dimensional "reverse engineering" of ontogeny allows cell behaviors, such as the spatial regulation of odontoclast and osteoclast activity within the resorption zones, to be inferred in fossils and compared with those of living animals. It opens the door to integrating different strands of cutting-edge research, from developmental genomics to paleohistology, in order to explore current topics about the evolution, development, and patterning of teeth.

Contributors

Igor Adameyko
Karolinska Institutet
Stockholm, Sweden

Marianne E. Bronner
California Institute of Technology
Pasadena, California, USA

Alan Henry Brook
University of Adelaide
Adelaide, Australia

Robert Cerny
Charles University
Prague, Czech Republic

Donglei Chen
Uppsala University
Uppsala, Sweden

Mélanie Debiais-Thibaud
University of Montpellier
Montpellier, France

Kaj Fried
Karolinska Institutet
Stockholm, Sweden

Tatjana Haitina
Uppsala University
Uppsala, Sweden

Ann Huysseune
Ghent University
Ghent, Belgium

Zerina Johanson
Natural History Museum
London, UK

Jan Krivanek
Masaryk University
Brno, Czech Republic

Aaron R.H. LeBlanc
King's College
London, UK

Matthew Brook O'Donnell
University of Pennsylvania
Philadelphia, Pennsylvania, USA

Alexa Sadier
University of California
Los Angeles, California, USA

Moya Meredith Smith
King's College
London, UK

Vladimír Soukup
Charles University
Prague, Czech Republic

Jan Stundl
California Institute of Technology
Pasadena, California, USA

Charlie Underwood
University of London
London, UK

P. Eckhard Witten
Ghent University
Ghent, Belgium

1 On Dental Cell Types and Cell Populations, Also in Light of Evolution

Jan Krivanek, Kaj Fried and Igor Adameyko

1.1 INTRODUCTION

The evolutionary diversity of odontodes in extant and extinct animals raises questions about how the ancient cell types evolved to build the first proto-odontode, and what was the most archetypical design expanding into a diversity of shapes and matrix arrangements. Thus, the emergence of odontodes is closely related to the evolution of hard matrix-producing cell types and specific cells forming signaling centers guiding the morphogenetic events. For instance, the evolution of enamel suggests a historical transition from the epithelial cells performing morphogenetic function into the cells secreting the hardest mineralized material.

Although odontodes show quite a diversity of shapes and mineralized matrixes, it seems that the major cell types and their positions are similar either due to a deep homology or, alternatively, due to a convergent derivation. Indeed, the diversity of odontode shape and structure can be generated by varying signaling mechanisms and morphogenetic progressions instead of differences in a cell-type composition. Therefore, we might expect to find the fundamentally homologous cell types in shark odontodes and gecko teeth, independently of a shape, self-renewal capacity, and other details. Still, we cannot exclude that some odontodes from different animal lineages will reveal non-homologous cell types or drastic deviations from the expected repertoire. The recent breakthroughs in methodology of cell-type identification and classification will undoubtfully resolve this question in the near future. However, today, we do not possess enough molecular information about cell identities and roles of cell types, inhabiting all kinds of odontodes found in extant animals. Therefore, in this chapter, we will focus our attention on the examples studied to a greater depth.

Teeth represent the most studied odontodes. This is mainly because teeth are essential to the quality of our daily life and often make unexpected troubles requiring visiting a dentist. This painful connection brought funds and bright minds to the field of dental research, and the results of the corresponding profound investigations yielded the most fundamental basis for understanding odontodes in general.

On the other hand, our recent progress was technology-driven. The emergence of different single-cell methods, such as single-cell transcriptomics and spatial transcriptomics (Aldridge and Teichmann, 2020; Longo et al., 2021), solved a historical problem of identifying and defining cell types in an unbiased way (Shekhar and

DOI: 10.1201/9781003439653-1

Menon, 2019). Previously, the criteria for defining cell types were based on cell morphology, location, and histological or molecular marker staining (Thesleff, 2003; Wegner et al., 2017). Similar or evolutionary close cell types were nearly impossible to distinguish, and the researchers could not assign a clear role to many cell types (Liu et al., 2016). Now that the times when the cell types were defined solely based on a cellular shape, location, or a couple of molecular markers are gone, we base our diagnostic conclusions on the entire transcriptomes reflecting cell specializations and functions. At the beginning of the 21st century, we finally mastered the cellular composition of tissues and organs in any species with a mathematical precision.

The majority of single-cell techniques are based on prepping individual cells, encasing them into droplets or micro-wells, and sequencing the individual mRNA pools for the downstream analysis (Hagemann-Jensen et al., 2020; Salomon et al., 2019). Different protocols range in their sensitivity and costs, revealing from few hundreds and up to 12,000 expressed genes per cell (Hagemann-Jensen et al., 2020). This allows defining the entire transcriptomes of cell types with the minimal requirement of having several cells of a sufficient quality per cell type or cell state. The consequences of applying the single-cell techniques are vast and include the identification of a complete transcriptional factor code defining cell identities, defining uninterrupted differentiation trajectories, or predicting regulatory interactions between neighboring and distant cell types (Krivanek et al., 2020; Soldatov et al., 2019). Finally, these methods provide unbiased mathematical means for classifying cell types and discovering new cell types in any multicellular system (Adil et al., 2021; Kharchenko, 2021). The evolutionary comparisons of individual transcriptomes from the homologous cell types might help to establish the deviations of functions under selective pressure and molecular mechanisms responsible for their control. Furthermore, the recent advancements in single-cell epigenetic analysis (chromatin accessibility and DNA methylation profiles) improve our understanding of gene expression control and causality of gene regulation at the individual cell level (Ahn et al., 2021; Baek and Lee, 2020; Yan et al., 2020). The spatial transcriptomics and spatial multi-omics approaches, unlike the approaches relying on cell dissociation, preserve the tissue integrity and positional information and help to establish the cell–cell interaction modes, as well as the micro-environmental contexts for any given tissue (Asp et al., 2020; Liao et al., 2021).

Although the "classic" eras of dental and odontode-oriented studies clearly revealed the compendium of general and readily apparent cell types (ameloblasts, cementoblasts, odontoblasts, pulp and immune cells) (Krivanek et al., 2017), the single-cell methods uncovered additional cell types and enabled observing uninterrupted developmental- and growth-related differentiation trajectories of all lineages building, maintaining, and regenerating teeth (Chiba et al., 2020; Fresia et al., 2021; Krivanek et al., 2020; Pagella et al., 2021; Sharir et al., 2019). Definitive cell types, forming at the tips of such differentiation trajectories, appeared to be well-discernable and did not raise tricky questions. However, the progenitors revealed smooth transitions from the upstream to downstream states, which raised the question of how to split the smooth sequence of progenitor states into more discrete progenitor cell types (Krivanek et al., 2020; Sharir et al., 2019). For instance, the classical approach splits amelogenesis into four stages: The pre-secretory, secretory, transitory, and maturation stages (Bartlett, 2013). Contrary to this discrete picture, the single-cell trajectory shows a smooth transition of amelogenesis with less discrete phases (Krivanek et al., 2020). This is in line with the notion that the production of a hard matrix and corresponding scaffolding proteins is not restricted to any discrete stage, although specific stages demonstrate higher mRNA levels of relevant programs (Krivanek et al., 2020). Of note, the levels of mRNA might not reflect the dynamics of the matrix allocation and protein synthesis, as those are often regulated separately. Still, the previous subdivisions might appear artificial during further research and could be revised in the future.

Overall, many of the intermediate differentiation steps during odontogenic process were not detected by the previous methods and became apparent only in recent single-cell transcriptomics studies. Beyond revealing novel progenitor states and transitions during developmental and growth dynamics, the single-cell methods helped to identify rare definitive subtypes of dental cells arising from the epithelial and mesenchymal lineages and also highlighted high degree of plasticity and cell–cell communication. The identification of rare dental cell types based on information-enriched individual transcriptomes helped to predict their function on the basis of specific gene-expression modules corresponding to stem, metabolite transfer of sensory roles (Krivanek et al., 2020).

In the first part of this chapter, we will discuss in detail how single-cell methods applied on mammalian tooth transformed our understanding of dental cell types from developing, continuously growing, and stable mature teeth. In the second part of the chapter, we will connect these novel results to the ideas concerning the evolution of odontodes.

1.1.1 Types of Teeth

During the evolution of vertebrates, teeth underwent several major modifications, which were caused by the pressure to adapt to a specific lifestyle. Because all these adaptations are associated with the particular cell type heterogeneity, we will briefly discuss some of these aspects in this paragraph. For more detailed information, see the other chapters of this book.

Some adaptations enabled an extension of the abrasion period, which allowed longer time for tooth growth. In addition to the brachyodont (short-crowned) type of teeth, found for example in humans, there are hypsodont and hypselodont types. Hypsodont teeth are defined as high-crowned teeth with short roots, whereas hypselodont teeth are referred to as constantly growing teeth. However, these definitions often appear oversimplified. The characterization of the crown and root, as we know them from brachyodont teeth, is a way more complex and unclear task in hypsodont and hypselodont teeth. For instance, most typically, cementum covers the roots, and enamel covers the crown. In hypsodont teeth, the cementum is a portion of the "crown" part with both structural and functional roles (Sahara, 2014; Thesleff and Tummers, 2008). Hypselodont teeth are defined as teeth that never finish their growth. In different species, the continuously growing teeth acquired different shape and morphology corresponding to their diverse functions. In some of these teeth, the identification of crown and root parts might be non-intuitive. For example, in continuously-growing rodent incisor, the crown and root analogues are not positioned along the rostral (coronal)–caudal (apical) axis. These features and capacities evolved multiple times in different animal lineages and subtypes of teeth, including incisors, canines, premolars, and molars (Renvoise and Michon, 2014). Hypsodont and hypselodont teeth are often, but not always, associated with highly abrasive eating habits. Besides, hypselodont teeth are sometimes used beyond food processing and play a role in fighting or as a sign of dominance (e.g., elephant, wild boar) or even as a sensory organ (in narwhals) (Nweeia et al., 2014; Renvoise and Michon, 2014).

Next, according to the modes of implantation into a jaw, teeth are divided into three groups: Acrodont, pleurodont, and thecodont. Thecodont teeth are characterized by the presence of roots, which are anchored inside of deep sockets of a jawbone. In contrast, acrodont teeth are fixed right at the apexes of the jawbone without being submerged into the bone. Pleurodont teeth are positioned against the lingual wall of a jawbone (Gaengler and Metzler, 1992; Jenkins and Shaw, 2020).

Based on the modes of attachment to the jaw and the presence or absence of periodontium, we can distinguish between gomphosis (presence of flexible periodontal tissue) and ankylosis (teeth are firmly fused with the bone). For more details, see Chapter 6 of this book. Depending on how many times teeth are replaced in the lifetime of an animal, we distinguish among monophyodont (no replacement), diphyodont (there is an additional generation of teeth), or polyphyodont (multiple generations of teeth) types of dentition (Whitlock and Richman, 2013). Finally, homodont dentition refers to a setting, where all teeth have the same or very similar shape. In cases of a heterodont dentition, evolution resulted in different shapes of teeth with specialization based on their function (Stock, 2001).

1.1.2 Developmental Origin of Dental Cell Types

The cells comprising future teeth originate from facial neural crest-derived ectomesenchyme, ectoderm (and sometimes endoderm), and mesoderm. Ectomesenchyme of the face will contribute to dentin-producing odontoblasts and pulp cells; ecto- and endoderm will contribute to the epithelial structure of future teeth, including enamel producing ameloblasts; and mesoderm will give rise to the infrastructural components such as blood vessels and immune cells (Figure 1.1) (summed up in (Krivanek et al., 2017)).

Odontogenesis is commonly divided into six subsequent stages: The initiation stage, the bud stage, the cap stage, the bell stage, root formation, and final maturation (Figure 1.2). Odontogenic signaling is initiated early in the oral ectoderm and will progress as interplay between the primitive oral epithelium and the neural crest-derived mesenchyme (or ectomesenchyme) below. Many signaling pathways are involved in the initial stages of odontogenesis. The epithelial cells proliferate, while the epithelium thickens into a placode, and then grow deep into the ectomesenchyme to form the dental lamina (DL). From the DL, cells that will form teeth bud off. In species with only one set of teeth, DL disintegrates, while it remains in diphyodont species to allow for a second set of teeth originating from the free end of the lamina and then disappears. In species with continuous tooth replacement, various mechanisms preserving odontogenic epithelium have evolved to secure repetitive

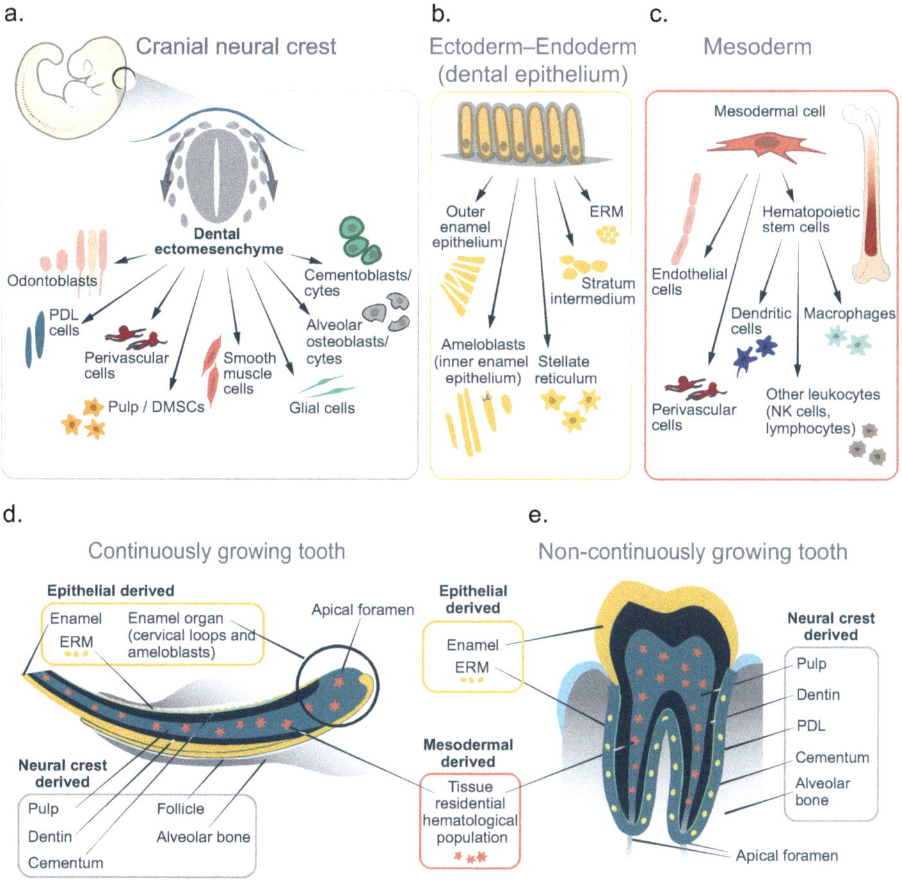

FIGURE 1.1 The developmental origin of different cell populations in teeth. (a) Cranial neural crest cells give rise to ectomesenchymal embryonic facial tissue, which contributes to the mesenchymal compartment of a tooth including different pulp cells, odontoblasts, periodontal ligament (PDL), cementum, and alveolar osteoblasts/osteocytes. (b) Ectoderm (and endoderm in some types of teeth) give rise to epithelial compartment represented by dental inner and outer epithelium, ameloblasts, stellate reticulum cells, stratum intermedium, epithelial cell rests of Malassez (ERM), etc. (c) Mesoderm contributes to endothelial, immune, and some perivascular cells in the tooth. Long bone with bone marrow shows the source of blood lineage cells in the tooth. (d, e) The continuously growing mouse tooth (d) and non-continuously growing e.g. human tooth (e) reveal similarities in their structure and cellular origins. After Krivanek et al. (2017).

teeth renewal. For more details, see Buchtova et al. (2012), Tucker and Fraser (2014), and Westergaard and Ferguson (1987, 1990).

During embryonic growth in the mandible and maxilla, the dental epithelium rapidly invaginates into the underlying mesenchyme. From here on, odontogenic signaling relocates from the epithelium to the mesenchyme, which condenses into

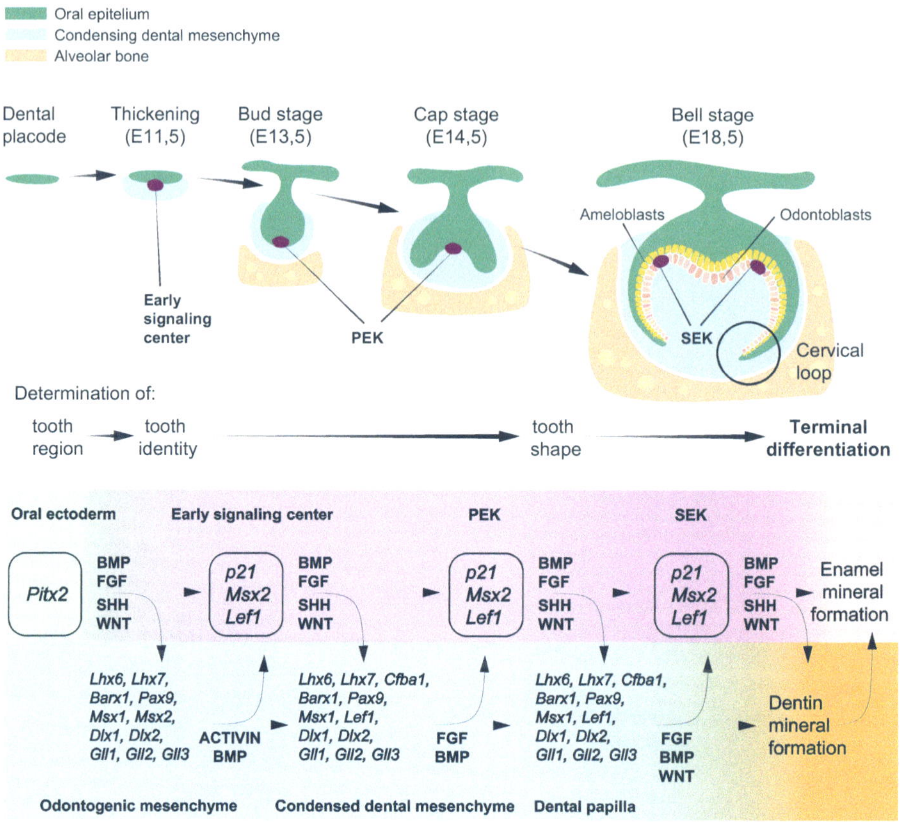

FIGURE 1.2 Stages of mammalian molar odontogenesis with signaling mechanisms coordinating epithelial and mesenchymal compartments. *Abbreviations*: PEK, primary enamel knot; SEK, secondary enamel knot. After Thesleff (2003).

the dental papilla, the future dental pulp—this is the way of how a tooth bud forms.

As cell proliferation proceeds, the tooth bud develops into the cap stage, and at that point the odontogenic potential resides solely in the mesenchyme (Kollar and Baird, 1969). At this stage, a more elaborate structural arrangement is discernible. The epithelium-derived cells form the stratified enamel organ, which makes up for the largest part of the tooth bud. It consists of an outer layer of cells called the outer enamel epithelium (OEE), the inner enamel epithelium (IEE) and a large space in between these layers, the stellate reticulum. This space is loosely populated with star-shaped cells. Furthermore, at the border of the IEE, the stellate reticulum is flanked by a layer of cells denoted as the stratum intermedium, the layer, which supports ameloblasts differentiating from IEE. The entire developing tooth organ is by now surrounded by a vascularized layer of mesenchymal cells, the dental follicle (DF). This cell layer will eventually contribute to the formation of the root and the periodontium.

Within the epithelium at the tip of the tooth, a collection of cells has emerged to form the primary enamel knot, a signaling center, which is instrumental in determining the structure and outline of the future tooth crown. It arises through an influence of dental papilla-derived factors, especially BMP4 (Jernvall et al., 1998), and is fully differentiated by the cap stage. The enamel knot cells themselves do not divide but orchestrate tissue–tissue interaction by means of at least 10 signaling molecules, belonging to the BMP, FGF, HH, and WNT families. Among other things, they seem to trigger proliferation of adjacent epithelial cells, which form cervical loops where the OEE and IEE join. They also have an influence on mesenchymal cells of the dental papilla (Thesleff et al., 2001). Primary enamel knot finally disappears through apoptosis.

In multicuspid teeth, secondary enamel knots appear during the late cap stage. Secondary enamel knots then guide the dental epithelium to produce folds, which set the pattern for the cusps of the teeth (Jernvall and Thesleff, 2000). Through an incremented growth, the tooth anlage attains a bell-shaped morphology and enters the bell stage of development. As a result, from epithelial–mesenchymal interactions, cell types specific for teeth such as ameloblasts and odontoblasts will now start to differentiate into states that eventually will produce enamel and dentin, respectively. Ameloblasts arise from the IEE and as development proceeds, they mature through influences from odontoblast- and/or predentin-derived molecular factors (Zeichner-David et al., 1995). Odontoblasts, in turn, differentiate from mesenchymal cells situated at the border of the dental papilla and are induced by signals from the dental epithelium, including FGFs and BMP2. The fastest differentiation of odontoblasts and ameloblasts occurs on the tips of the cusps, and in this way the cusp patterns determined by the secondary enamel knot are eventually finalized, as odontoblasts begin to lay down the mineralized dentin matrix (Thesleff et al., 2001).

Epithelial cells of the enamel organ, which form the circle at the apical part of tooth (region known as cervical loop) proliferate to form an epithelial root sheath, also denoted Hertwig's epithelial root sheath (HERS). These transient cells induce odontoblast differentiation in adjacent ectomesenchymal cells, and in this way tooth root formation/elongation sets off. A wide range of growth and transcription factors are involved in this process, especially BMP/SHH signaling cascades, but also others (see Li et al. (2017) and Figure 1.3). The HERS disintegrates almost completely, when the root formation is done, although some cell clusters remain as a thin mesh-like structure around the adult roots, called the ERM (the Epithelial cell Rests of Malassez) (Xiong et al., 2013). The ERM is the only living epithelial structure, which remains after complete tooth development and eruption. The role of ERM is not clearly understood, although there is an evidence about ERM's contribution to periodontal regeneration (Rincon et al., 2006). The fully formed root consists of a pulpal root canal surrounded by dentin, which is covered by cementum. Root dentinogenesis follows the same pattern as seen in the crown. Cementogenesis occurs through the formation of cementoblasts, which differentiate from ectomesenchymal progenitors after the formation of a dentinous root. To anchor the tooth in the dental socket, cementoblasts form an organic matrix, where hydroxyapatite crystallizes to create mineralized cementum—the primary, acellular, type of cementum. In some species, there can be observed also a second type of cementum—the secondary,

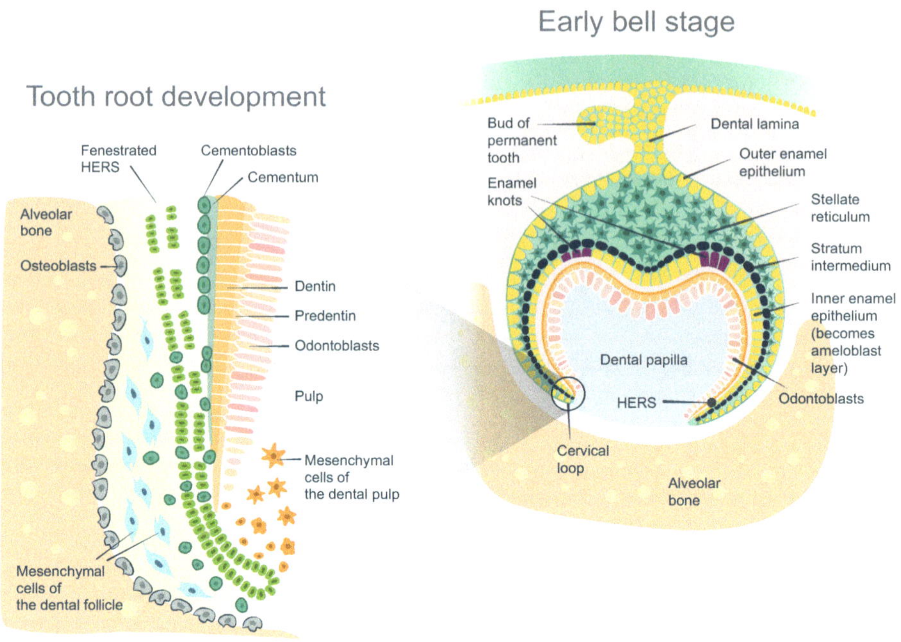

FIGURE 1.3 The development of a dental root in a brachyodont tooth. *Abbreviation*: HERS, Hertwig's epithelial root sheath cells.

cellular, type cementum. This is usually developed only in the apical part of the tooth on the basis of mechanical stimuli. Cementoblasts here continue to deposit the extracellular matrix and become entrapped in newly formed cementum. They form cellular cementum and become cementocytes, similar to bone-residing osteocytes. As such, cementoblasts are similar in function and structure to osteoblasts. However, unlike bone, cementum is not remodeled during the normal appositional growth (Zhao et al., 2016).

Additionally, as the root formation begins, dental follicle (DF) cells of ectomesenchymal origin are remodeled to differentiate into cells of a periodontal ligament. This process commences at the cemento-enamel junction, where roots start to develop, and proceeds further toward the apical direction forming periodontium around the roots. The periodontal ligament attaches to the tooth via the bundles of collagen fibers, the ends of which at either cementum or bone surface are denoted as Sharpey's fibers (Yao et al., 2008).

As the root development proceeds, a number of factors cooperate to enable the eruption of the tooth. This event follows the same pattern in all teeth, but the timescale may vary considerably between types of teeth. For eruption, the bone above the crown of the growing tooth organ needs to undergo resorption through osteoclast activity, and cellular/molecular machinery has to be in place that permits movement of the tooth. Interestingly, in di- and polyphyodont species, cementoclasts and odontoclasts contribute to the root resorption during the tooth replacement process. These cells are similar in the shape, function, and developmental origin to osteoclasts. A series of experiments have demonstrated that the cells of the DF are responsible for the initiation and progress of tooth eruption. Specifically, the coronal region of the DF regulates the osteoclastogenesis necessary for the process, while the basal one-half synchronizes the osteogenesis. Monocytes that will differentiate into osteoclasts seem to be recruited to the future alveolar sites through colony-stimulating factor-one (CSF-1) and monocyte chemotactic protein-1 (MCP-1) signaling from the DF. Osteoclast differentiation then proceeds via DF-derived CSF-1 and RANKL, which promote osteoclastogenesis, and downregulation of osteoprotegerin (OPG). OPG binds the RANKL, preventing RANKL binding to RANK, which ultimately results in the inhibition of bone resorption. For osteogenesis and bone formation below the tooth, which likely is needed for eruption, BMP2 signaling from the DF appears to be a major driver (Que and Wise, 1997). At later stages, the periodontal ligament might help to elevate the tooth to the final level, at least in teeth with continuous eruption (Moxham and Berkovitz, 1974).

1.1.3 Cell Types and Continuously Growing Teeth

During the evolution of vertebrates, many different types of teeth emerged for animals adapting to all sorts of environments. Although they share the same evolutionary origin, and the similar developmental pathways drive their development, these teeth significantly differ in size, shape, number, or even the style of growth and renewal.

The continuously growing (hypselodont) teeth, also known as tusks, convergently evolved in several mammalian and some non-mammalian species to fulfil diverse roles (Renvoisé and Michon, 2014; Whitney et al., 2021). These teeth serve as a sign of dominancy or are used for fighting (e.g., elephants or wild boar), they can be an outer-environment-sensing organs (narwhals), or be heavily abraded during gnawing (e.g., rodents or lagomorphs) (Nweeia et al., 2014). In the latter case, the hypselodont teeth grow enormously fast and are fully renewed multiple times during the lifespan of an animal. Thus, studying the self-renewing teeth is key for discovering features of stem cell niches, differentiation processes, and other aspects connected to tissue dynamics.

The development of an odontode starts with the mutual epithelial and mesenchymal cross-interaction. Such interaction initiates dental development and, in the crown, directs the shape of each tooth via the activity of enamel knots. Further development varies significantly between different tooth types. In brachyodont teeth, once the tooth crown is predetermined and dentin and enamel matrices start being deposited, the dental epithelium and mesenchyme continue to set the base of the roots. When roots reach their intended length, the interaction between dental epithelium and mesenchyme is reduced, and the epithelium is disintegrated into ERM. In hypsodont and hypselodont teeth, it is much harder and counter-intuitive to draw a clear border between the root and the crown. For example, in hypsodont teeth, once the enamel deposition is finished and enamel epithelium becomes disintegrated, the "crown" part is covered by cementum.

Eventually, tooth wear reveals alternating cementum, enamel, and dentin parts throughout the whole "crown-part", altogether having a key role in food processing (Sahara, 2014; Thesleff and Tummers, 2008). Significant variations occur in hypselodont teeth. For example, in rodents and some other species, we distinguish the root-analogue and crown-analogue (at the lingual-labial axis of the tooth), both parts being continuously regenerating and erupting, thus, enabling the tip of the tooth to be sharpened during non-homogenous shortening. However, in some other species, for instance, in elephants, only the tips of the tusks are covered by enamel, whereas the rest of the tooth is covered by cementum. In special cases, enamel is missing completely, for example, in narwhal tusks (Nweeia et al., 2014; Virág, 2012). Therefore, despite the shift of the boundary between the crown and the root, the "root-like" parts of the teeth can also erupt.

The continuous growth of hypselodont teeth is conditioned by the fact that during the late development, the odontogenic dental epithelium is not fully degraded and keeps contributing to the tooth growth postnatally (Figures 1.1–1.3) (Krivanek et al., 2017). Both epithelial and mesenchymal compartments, thus, maintain functional stem cell niches (Figures 1.4 and 1.5). This ensures that transient cellular populations (which are present only during development in non-continuously growing teeth (e.g., human dentition)) are still active and contribute to the continuous tooth growth. The permanent replenishment of the worn tissue by keeping active stem cell niches renders a continuously growing tooth an important model organ for studying stem cells, growth, and mutual interactions of cell subtypes during differentiation.

1.2 CURRENT PERSPECTIVE ON DENTAL CELL TYPES

Recent advances in the single-cell transcriptomics and similar methods enabled looking into the developmental pathways and cellular composition of continuously growing teeth at unprecedented level (Krivanek et al., 2020; Pagella et al., 2021; Seidel et al., 2017; Sharir et al., 2019; Yianni and Sharpe, 2020) (Figures 1.6–1.7). The single-cell approach relieved the burdens of preconceived attitude and opened new era of discovering previously unknown transitions, cell types, cell identity codes, and interactions at single-cell level in dental pulp, epithelium, or dental follicle.

1.2.1 Epithelium-Derived Cell Types

Dental epithelium has two main roles. It controls differentiation of adjacent mesenchyme via mutual interactions, and it forms the enamel matrix. In brachyodont (e.g., human) teeth, enamel organ is present only during the development. In contrast to this, hypselodont (e.g. mouse incisors) teeth have an enamel organ active throughout their lives. It consists of several different cell types, all having their specific histologic position and function.

Previously, it has been shown that dental epithelium consists of stellate reticulum, stratum intermedium, outer enamel epithelium, and inner enamel epithelium consisting of different stages of ameloblasts (Nanci, 2016). Stellate reticulum is a pseudo-mesenchymal reticular tissue of dental epithelium positioned between the inner and outer epithelium (Figure 1.3). It holds the shape (and provides space to

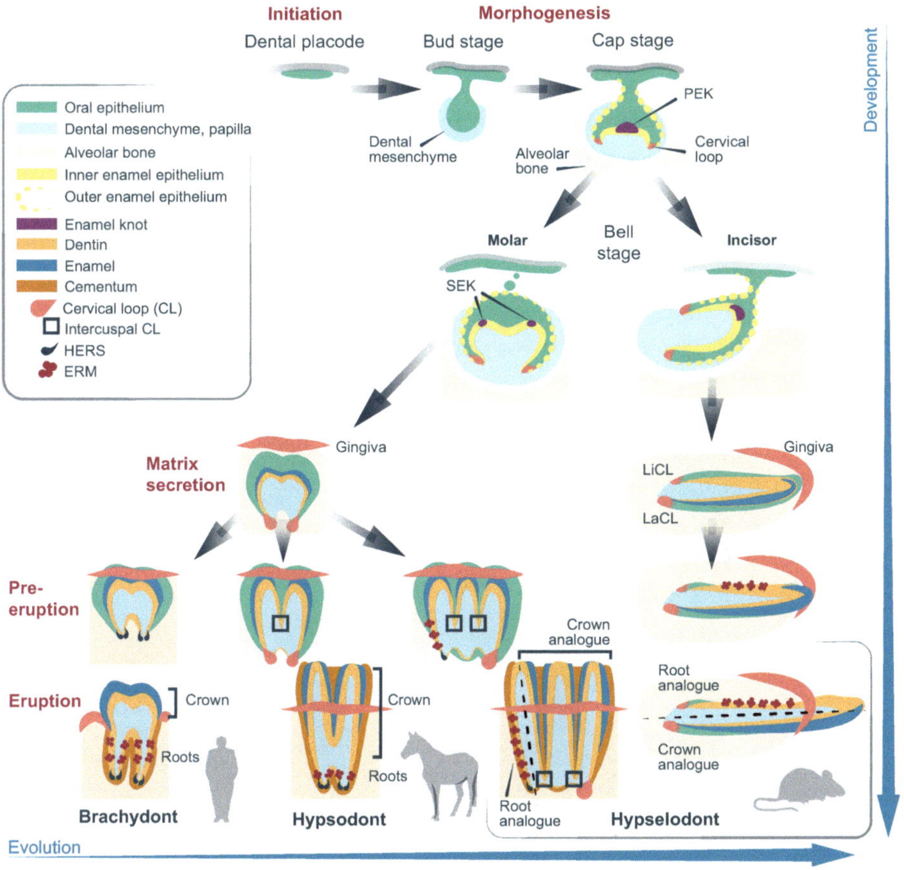

FIGURE 1.4 Different types of teeth with and without preserved stem cell niches and corresponding self-renewal capacity. *Abbreviations*: HERS, Hertwig's epithelial root sheath cells; ERM, epithelial cell rests of Malassez; PEK, primary enamel knot; LiCL, lingual cervical loop; LaCL, labial cervical loop. After Renvoise and Michon (2014).

growth) of enamel organ and is source of epithelial stem cells (Juuri et al., 2012; Sun et al., 2016). Outer enamel epithelium, a layer of dental epithelium, which does not give rise to ameloblasts is positioned on the outermost part of enamel organ. Stratum intermedium is a supportive cell layer for ameloblasts positioned below, the inner enamel epithelium. Inner enamel epithelium gradually differentiates into enamel-producing ameloblasts. The first stage of differentiation are preameloblasts, which still do not produce enamel matrix. They further differentiate into secretory ameloblasts forming the early protein-rich enamel, which, in turn, become mature ameloblasts responsible for further calcification of the enamel matrix. At the last stage, some mature ameloblasts undergo apoptosis, whereas the remaining population ends up in post-maturation (protective) stage to cover the newly formed enamel until tooth eruption (Nanci, 2016).

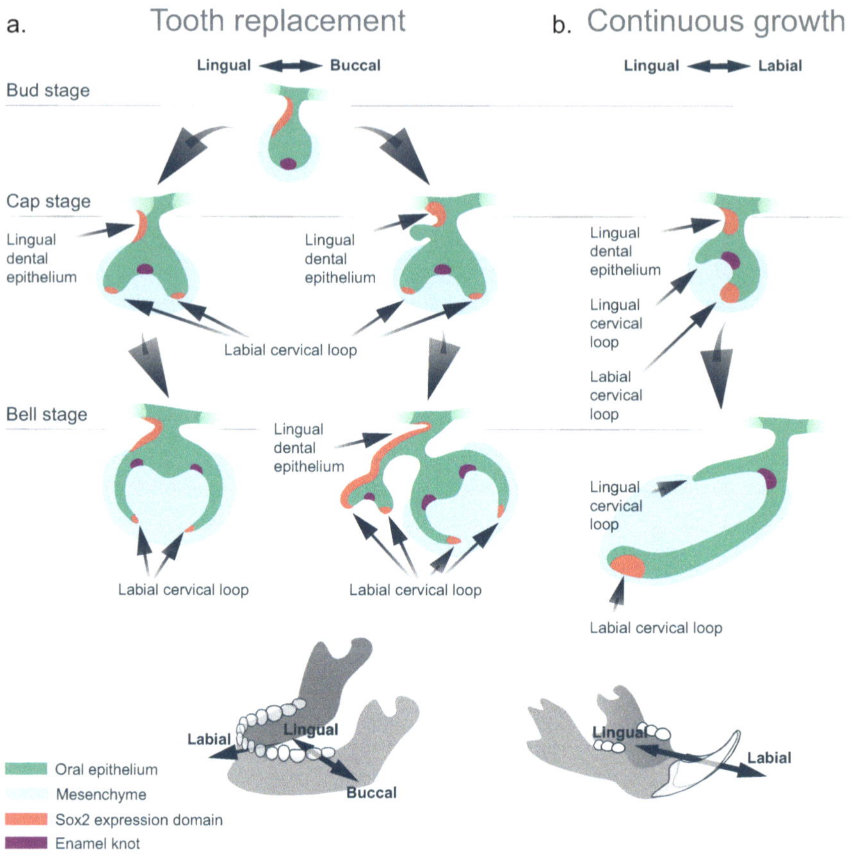

FIGURE 1.5 Teeth can be replaced via induction of the secondary tooth buds as compared to the strategy of maintaining the existing teeth via continuous growth. Sox2 is a marker of dental epithelial stem cells. After Balic (2019).

In addition to this classical knowledge, the latest single-cell studies showed that the adult dental epithelium of a continuously growing tooth consists of at least 13 (Krivanek et al., 2020) or 12 (Sharir et al., 2019) cell subtypes, uncovering several of previously unknown clusters with new transcriptional programs and cell identities in the dental epithelial region of a growing tooth. One of these recent studies by Krivanek et al. (2020) revealed further subdivision of a stratified stellate reticulum into cells of stellate reticulum itself (characterized by *Gjb3* expression) and developmentally more advanced stratum intermedium progenitors and pre-ameloblasts. Furthermore, Krivanek and co-authors (2020) identified a new cell layer within the stratum intermedium. This layer, named “cuboidal layer” (Figure 1.7), is made of cuboidal cells positioned between the layer of developing and maturing ameloblasts and other two subtypes of stratum intermedium. Expression data suggest that this new cuboidal layer, characterized by the expression of *Thbd* and other genes (*Cygb*,

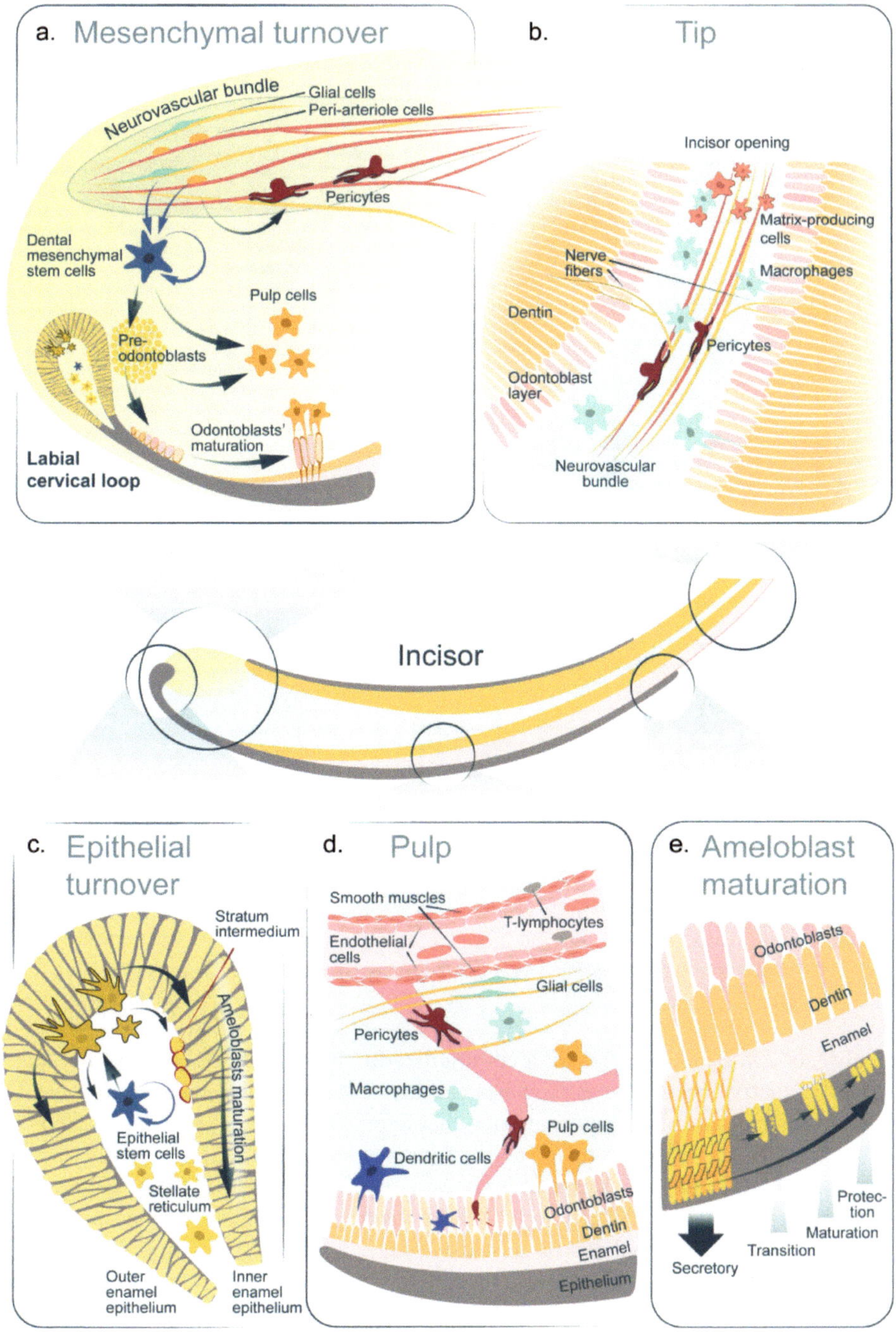

FIGURE 1.6 The outline of adult cell types in a continuously growing hypselodont tooth. With the advent of single-cell RNA-seq studies, the hierarchy of cell types is being further refined and shifted from the crude identification of major cell types to the discrimination of fine cell states and transitions between them. After Krivanek et al. (2017).

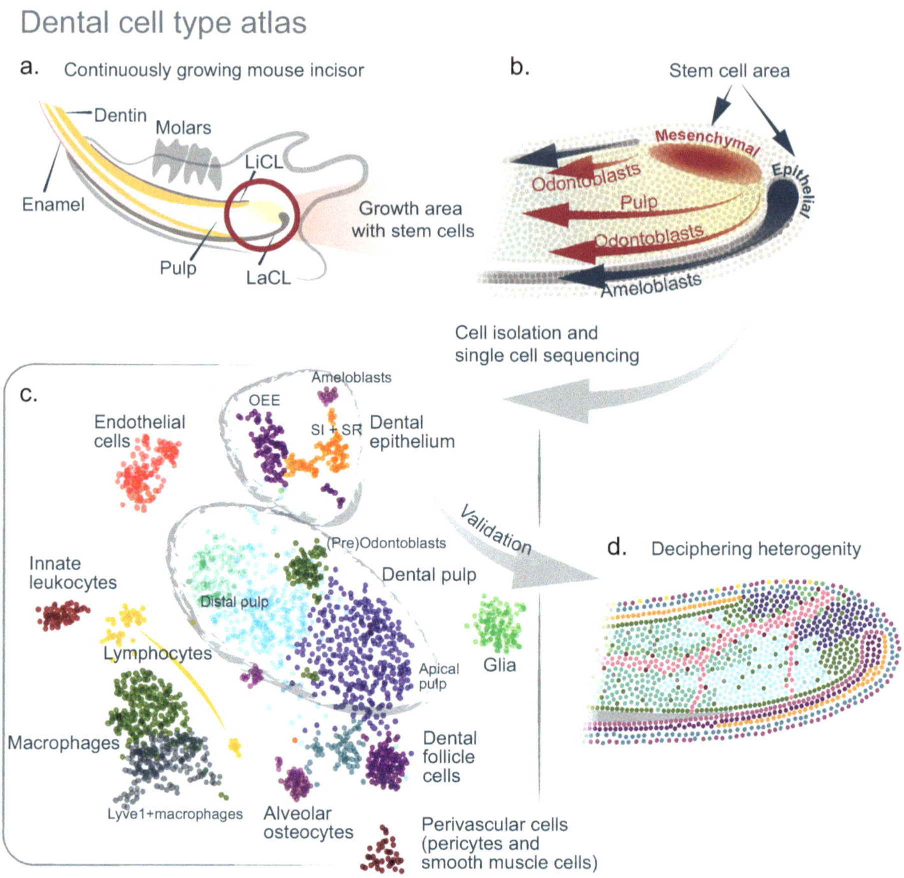

FIGURE 1.7 Example of a single-cell atlas of a continuously growing hypselodont tooth. *Abbreviations*: LiCL, lingual cervical loop; LaCL, labial cervical loop; SI, stratum intermedium; SR, stellate reticulum; OEE, outer enamel epithelium. After Krivanek et al. (2020).

Nphs1, or *Rhcg*), is essential for a communication between blood vessels and metabolically active secretory ameloblasts (Figure 1.8). Next, the same study revealed new transient short-living progenitors characterized by the expression of *Egr1* or *Fos* and giving rise solely to the cells of outer enamel epithelium. Finally, the differentiation trajectory of ameloblasts showed unprecedented level of smoothness and detail (Krivanek et al., 2020) including transition from stellate reticulum, through pre-ameloblasts, ameloblasts in secretory stage, ameloblasts in maturation stage, to the ameloblasts in post-maturation stage. Each step of ameloblast differentiation was characterized by the gradual change of expression transcription factors. This knowledge of sequential transcriptional activation might help to develop new tissue engineering techniques enabling dental regeneration. Single-cell transcriptomics approach also uncovered a new putative mechanosensory cell type characterized by expression of *Ryr2*, *Mylk*, and *Sox5* (Krivanek et al., 2020). Their grouped

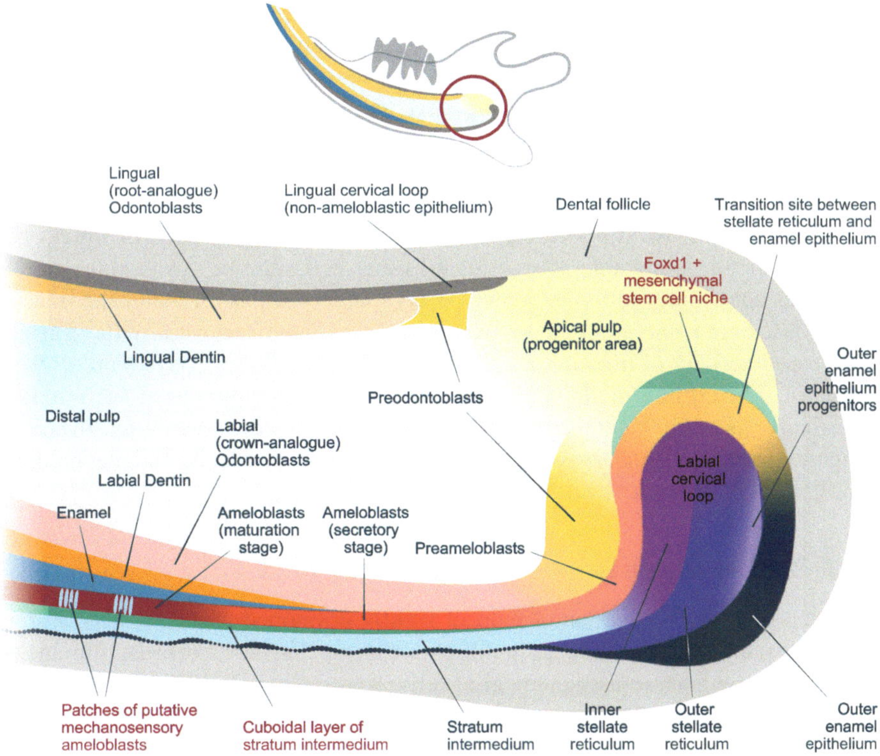

FIGURE 1.8 Integration of recently identified and previously known cell types in a single-cell atlas of a continuously growing hypselodont tooth. The recent discoveries include cuboidal and putative mechanosensory epithelial cells, *Foxd1*⁺-regionalized mesenchymal stem cells, stratification of dental pulp and stellate reticulum, and diversification of labial and lingual odontoblasts together with corresponding difference in dentin structure (Krivanek et al., 2020; Lavicky et al., 2021).

(patches-forming) position within the maturation-stage ameloblasts was confirmed by immunohistochemistry. The question about the exact role of these cells remains to be answered.

The question about the identity, multipotency, and a niche of stem cells residing in the dental epithelium resonates in the dental research community for many decades. Historically, the most widely accepted marker of dental epithelial stem cells in mouse incisor is SOX2 (Juuri et al., 2012). The lineage tracing experiments revealed that *Sox2*⁺ cells, positioned in the stellate reticulum, have a capacity to renew all parts of the dental epithelium (Juuri et al., 2012). Later on, the expression of other genes (*Gli1*, *Lgr5*, *Bmi1*, *Lrig1*, *Meis1* or *Igfbp5*) comes up and traces dental epithelial cells (Chiba et al., 2020; Krivanek et al., 2020; Sanz-Navarro et al., 2019; Yoshizaki et al., 2020). The analysis by Krivanek et al. (2020) uncovered a compact subcluster of epithelial cells, which co-expressed most of these established stem cell markers and showed the presence of other specific transcripts. To confirm the stem cell nature of this *Acta2*$^+$ cluster, Krivanek and co-authors performed *Acta2*-based lineage tracing and proved that *Acta2*$^+$ population, represented only by a very limited number of cells in LaCL, gives rise to all cell types of dental epithelium similar to epithelial cells expressing *Sox2* and other stem cell markers (Krivanek et al., 2020). Sharir et al. (2019) also applied the single-cell approach and showed scattered expression of previously characterized dental epithelial stem cell markers (*Sox2*, *Bmi1*, *Gli1*, *Lrig1*), which was confirmed by the *in situ* hybridization. This study suggested that a pool of actively cycling and diverse cells of inner enamel epithelium forms a stable source for all other cell subtypes of dental epithelium and thus contributes to tissue homeostasis (Sharir et al., 2019). After the injury, the epithelial organ (including ameloblasts) is being faster renewed by the transiently increased activity of progenitor cells as well as via a direct conversion of *Notch1*$^+$ cells of stratum intermedium into ameloblasts. This points out at the remarkable plasticity inside the dental epithelial niche. Further studies involving lineage tracing will explain the diversity of dental mesenchymal stem cells and mechanisms of the epithelial plasticity and renewal.

1.2.2 Mesenchyme-Derived Cell Types

In a classical view, the dental pulp of a fully developed tooth consists of dentin-forming odontoblasts, dental pulp (fibroblast-like) cells, blood vessels, immune cells, and nerves with nerve-associated glial cells (Krivanek et al., 2017). Dentin, the most abundant hard matrix of the tooth, gives each tooth its shape and mechanical properties. In contrast to enamel, dentin is a living tissue penetrated by dental tubules with cell processes and keeps forming throughout life. Formation of dentin is dependent on the function of pseudoepithelial cell layer of odontoblasts. These cells line the entire inner surface of the dental pulp cavity and firmly attach to the dentin (Khatibi Shahidi et al., 2015). Odontoblasts have a highly polarized and often branching tree of processes extending toward dentinal–enamel junction and projecting within dentinal tubules, being accompanied by fine nerve processes.

Human dental pulp is also often referred as a source of dental pulp mesenchymal stem cells (Ledesma-Martinez et al., 2016; Paes et al., 2021). Although some cells isolated from human dental pulp show multipotency *in vitro* and are able to differentiate into many different tooth-unrelated cell types (Chan et al., 2021; Nuti et al., 2016), the origin and behavior of such cells in living teeth still stay enigmatic.

The size and cellular composition of dental pulp significantly vary in different species. One of the extreme examples of tooth minimalism is the dental mesenchyme of the first generation of teeth in small teleost medaka fish (*Oryzias latipes*) and newts (*Pleurodeles waltl*). The whole "dental pulp" is composed of a single odontoblast (Davit-Béal et al., 2007; Larionova et al., 2021). The specification of this unique cell, enwrapped by a bunch of epithelial cells, occurs early during tooth development, and, later on, this single cell builds the enameloid and dentin matrix. This ultimate miniaturization does not allow the presence of other cell types, and the entire mesenchymal part consists of a single-cell type.

On the other side of a spectrum, there are mammals, often possessing exceptionally large teeth. Those include the tusks of elephants and marine mammals weighting kilograms with adequately large dental pulps. These teeth, as well as the teeth of rodents and humans, have dental pulps consisting of dozens of cell types.

The most numerous cell type of a mammalian dental pulp is represented by the fibroblast-like pulp cells. For a long time, we believed that these cells form a uniform population, which fills the dental cavity and produces the inner-pulp extracellular matrix. This soft matrix forms a structural support for intra-pulpal blood vessels and nerves. Several recent studies utilized single-cell RNA-sequencing (scRNA-seq) and unbiased clustering of cell populations from mouse and human teeth to uncover that the heterogeneity of dental pulp cells is much more intricate and complex (Krivanek et al., 2020; Pagella et al., 2021). For instance, the two studies described six (Pagella et al., 2021) or four (Krivanek et al., 2020) transcriptionally different populations of dental pulp cells in adult human wisdom tooth. These populations shared key molecular identifiers, which grouped them in one dental pulp cluster. However, specific molecular identifiers enabled to distinguish and group them into fine subclusters with rather unknown functional specializations. In parallel to the analysis of dental pulp of an adult molar, Krivanek et al. (2020) performed scRNA-seq analysis of mesenchymal cells of the apical papilla isolated from the human molars with growing roots. Grouping the data of growing and non-growing human teeth provided additional information about the complexity of dental pulp in connection with tooth development, which lead to the identification of six distinct cell populations. The two populations unique for the growing human molar teeth were characterized by the expression of e.g. *Sfrp2* and *Bcl11a* (first cluster) and *Sfrp1* and *Fbn2* (second cluster). On the other hand, the other two populations unique for adult-only dental pulp were identified by expression of e.g. *Apoe* and *Igfbp3* (fifth cluster) and *Kif5c* and *Coch* (sixth cluster). The remaining two pulp subpopulations appeared to be common for both adult and growing human molar teeth. Further research will clarify the role of such pulpal cell type heterogeneity and the breadth of its variation in different teeth and animals. Next, the heterogeneity of dental pulp cells from human molars was compared to the cell populations identified in a continuously growing mouse incisor. Importantly, the clusters found only in human apical papilla (growing roots) demonstrated a specific similarity to the corresponding apical growing part of a mouse incisor.

Furthermore, the single-cell approach to the mouse incisor uncovered at least seven subpopulations of dental mesenchyme (see Figure 1.7 for an example). These subpopulations were grouped into three main differentiation states: The apical pulp (characterized by the expression of *Smoc2* or *Fgf10*), the distal pulp (characterized by the expression of *Igfbp5* or *Syt6*), and odontoblast sub-lineage (characterized by the expression of *Dspp* or *Dmp1*). The gradual differentiation trajectories smoothly progressed through the transitory states as supported by the RNA velocity approach (Krivanek et al., 2020; La Manno et al., 2018).

The question about the in-vivo identity of dental mesenchymal stem cells represents a long-standing attraction for dental researchers. Single-cell analysis in human teeth done by Pagella et al. (2021) uncovered one cell subtype (further subdivided into three subclusters) referred as a human dental pulp mesenchymal stem cells (Pagella et al., 2021). This cluster, characterized by the increased expression of *Frzb*, *Notch3*, *Thy1*, and *Myh11* appeared quite abundant and formed 12% of all pulp cells.

Immunohistochemistry for FRZB revealed that these cells surround blood vessels. On the other hand, Krivanek et al. (2020) observed a similar cell type in single-cell data, which was identified as a perivascular cell cluster. This perivascular cell cluster (characterized by the expression of *Rgs5*, *Notch3*, and *Thy1*) further branched into three subclusters characterized by the expression of: *Cd24a* and *Stc1* (first cluster), *Dcn* and *Lum* (second cluster), and *Acta2* and *Myl9* (third cluster) (Krivanek et al., 2020). The use of experimental lineage tracing revealed that perivascular cells can be the source of dental mesenchymal stem cells for newly emerging pulp cells and functional odontoblasts after the damage of adult mouse molar (used molecular marker: *Acta 2*) (Vidovic et al., 2017) or incisor (used molecular marker: *Cspg4* (also known as *Ng2*) (Feng et al., 2011). In line with this, *Tagln*$^+$ pericytes found in mouse molars give rise to mature odontoblasts during molar development (Yianni and Sharpe, 2020). Together, these findings support that perivascular cells represent one of the sources for dental mesenchyme and functional odontoblasts. The high degree of dental mesenchymal stem cell niche heterogeneity proves also recent discovery of very specific and long-lasting type of *Foxd1*$^+$ stem cells in mouse incisor (Krivanek et al., 2020). The existence of *Foxd1*$^+$ stem cells was predicted using bioinformatic analysis of in-depth single-cell sequenced data from mouse incisor. The existence of these stem cells was further confirmed in vivo. They were observed specifically in the mesenchymal region in the closest proximity to the labial cervical loop, and functional lineage tracing experiments proved their long-lasting stemness and the contribution to pulp cells and odontoblasts.

When it comes to heterogeneity of a population of mature odontoblasts, the scRNA-seq studies did not provide much of the insight. One of the causes could include methodological difficulties pertaining to the physical isolation of odontoblasts attached to the dentin matrix. At the same time, Lavicky et al. described some structural, developmental, functional, and molecular heterogeneity of odontoblasts residing in the crown and root parts as well as in labial and lingual sides of mouse incisor (Lavicky et al., 2021). Odontoblasts differ in their morphology (most importantly, in the shape of their processes), as well as in the gene expression depending on the location. The odontoblasts at the labial side of incisor (crown-analogue) express some *Wnt*-related genes including *Dkk1*, *Wisp1*, or *Sall1*, which are not expressed in odontoblasts at the lingual side (root-analogue) of the same tooth. These two identified subtypes of odontoblasts also produce dentin of different microstructure, density, and elemental composition, which might be important for mechanical and biological properties of the tooth (Figure 1.8) (Lavicky et al., 2021).

1.2.3 Blood Vessels

Blood supply of teeth is extensive and complex. In case of a thecodont dentition of mammals, the dental pulp is accessible for blood and nerve supply only via the apical foramen, and, in some teeth, via accessory foramen on the roots. At the entrance to a dental pulp, there are few arterioles, which produce smaller lateral branches and end up in a dense vascular mesh in the subodontoblastic and odontoblastic areas (Nanci, 2016). The main blood vessels enter and run through the pulp in adjacency with the major nerves forming a neurovascular bundle (Franca et al., 2019). Endothelial cells, the key building blocks of the dental vasculature, revealed their heterogeneity in two independent scRNA-seq studies focusing on human molars. Both teams identified three transcriptionally different populations of endothelial cells (Krivanek et al., 2020; Pagella et al., 2021). Pagella et al. (2021) further defined these populations as arterial endothelial cells (expressing *Edn1* and *Cldn5*), postcapillary cells and collecting venule cells (expressing *Ackr1* and *Cd234*), and the third population by expression of *Insr* and Rgcc (Pagella et al., 2021). In the mouse incisor, three endothelial populations showed up as well. These were characterized by the expression of *Esm1* and *Ramp3* (first cluster), *Mafb* and *Lbp* (second cluster), and *Gja3* and *Bmx* (third cluster) (Krivanek et al., 2020). Most tissues are drained by lymphatic vessels. The existence of lymphatic vessels in teeth remains controversial. However, most of the studies incline toward the conclusion that the healthy dental pulp does not host lymphatic vessels (Gerli et al., 2010; Lohrberg and Wilting, 2016; Takahashi et al., 2012).

1.2.4 Dental Innervation and Associated Cell Types

The soft interiors of teeth are among the most densely innervated tissues of the mammalian body. Although even the Greeks of the Antiquity realized that teeth were innervated, this knowledge was lost, and it was not until Vesalius' descriptions in the 16th century that the pulpal nerves were "rediscovered." When activated, pulpal nerves convey signals that are perceived as pain only in a conscious nervous system. This is regardless of the nature of the stimuli, whether hot, cold, mechanical, acidic, etc. Meanwhile, those nerves that subserve the periodontal apparatus transmit invaluable mechanosensory information related to obtaining, chewing, and processing of food. The question of why teeth are so extremely sensitive to external stimuli has been the subject of discussions for centuries (reviewed by (Hildebrand et al., 1995). From an evolutionary point of view, in mammals, teeth are obviously necessary not only for hunting and finding food, for chewing, but also as weapons for aggression and defense. Consequently, the loss of teeth will lead to weakening of the animal and eventually to death.

During evolution, the phenotypic change from jawless to jawed vertebrates signaled a major shift. Perhaps concomitant with a shift to predatory behavior, innervated oral odontodes, which were providing mechanosensory innervation important for feeding, may have undergone an adaptation to provide other types of sensory information. Obviously, the increasingly vital function of teeth would greatly benefit from a protective sensory alarm system. Structurally, this could have taken place through an expansion of periodontal mechanosensory nerves into pulpal territories to gradually extend the target field. If so, it seems that a majority of the pulp-innervating nerve fibers maintained their mechanosensory nature. This explains why a large number of dentinal afferents respond to weak mechanical stimuli, which in other body places would be perceived as harmless. However, but by virtue of what must be an altered central connectivity and neurotransmitter content, they activate pain-signaling neurons in the trigeminal brain stem complex and thus evoke pain (Fried et al., 2011).

Adaptations of denticle or odontode innervation to the environment and behavior are not unusual. An interesting example is provided by innervated skin denticles in an Ecuadorian cave fish, *Astroblepus pholeter* (Haspel et al., 2012). This fish lives in fast turbulent waters in complete darkness. Fish generally receive mechanosensory information through specialized sense organs, neuromasts. In a species that cannot rely on visual information, a hypertrophy of neuromasts would be expected. However, *A. pholeter* has surprisingly few of these, but, instead, highly sensitive skin denticles have evolved. These are innervated at their base and seem to provide the fish with necessary information in order to navigate in its habitat (Haspel et al., 2012). This case of adaptation highlights an ancestral sensory role for denticles and illustrates an aspect of odontode and tooth nerve dynamics during evolution.

Overall, there is no experimental evidence to suggest that neuronal processes are necessary for the initiation of odontogenesis in mammals, although this could be the case in teleosts as well as in amphibians (Makanae et al., 2020; Tuisku and Hildebrand, 1994). Specifically, in the salamander, axons in the oral epithelium seem to partly control the invagination of the dental lamina by secreting signaling molecules (Makanae et al., 2020). Of note, though, in human, tooth agenesis occurs in certain dental arch positions which share a developmentally set pattern of innervation. This has led to the proposal that neuro-osteological malformations might lead to a loss of tooth formation in affected areas (Kjaer, 1998).

Ingrowth of nerves into the dental papilla (i.e., the future dental pulp) is a comparatively late developmental event and occurs *pari passu* with the onset of mineralization (Luukko and Kettunen, 2014). Nonetheless, the dental follicle of early cap stage tooth buds rapidly becomes entangled in a web of trigeminal sensory axons originating from jaw nerve trunks and remains there for a considerable developmental period. Recently, genetic tracing in mice revealed that this network of nerves actually partakes in the continued development of the tooth germ (Kaukua et al., 2014). Thus, although it had for long been common knowledge that dental mesenchyme and odontoblasts are derived directly from migrating cranial neural crest cells, an additional important source turned out to be the glial cells of the axonal arborization around the dental follicle. Here, Schwann Cell Precursors leave their nerve branches and position themselves as dental mesenchymal stem cells, together with those that have arrived directly through neural crest migration. In this way, they contribute with pulp cells as well as odontoblasts in clonal configurations (Kaukua et al., 2014).

There is by now data to suggest that although the tooth anlage-associated neural formations continue to branch and eventually also contribute to the final periodontal innervation, neuro-repelling molecular factors initially prevent nerve fibers from invading the dental papilla. Major such factors include members of the semaphorin family, in particular SEMA3A. SEMA3A signaling, in turn, is controlled by other local signals, including the TGF-β superfamily members, FGF, and WNT, which of course also are essential for tooth organogenesis. Thus, SEMA3A influence on axons is integrated into key odontogenic signaling networks (Kettunen et al., 2005). However, during the late bell stage, there is a shift in expression of factors with effects on axons, from repelling to neurotrophic agents. Major such molecules belong to the NGF and GDNF families, especially NGF and GDNF themselves, acting on appropriate TRK and GFRa axonal receptors. The pioneer axons use specific routes to enter the dental papilla and navigate toward their target fields and establish an initial crude final pattern of innervation (Hildebrand et al., 1995). The specificity of this process suggests that it is regulated by an array of neuroregulatory factors expressed in the ectomesenchyme. Among those are probably ephrins, netrins and laminins, and locally expressed *Sema3a*, outlining borders between outgrowing nerve fiber bundles (Fried et al., 2007, 2000; Luukko and Kettunen, 2016).

As with nerve fibers, blood vessels are present in the mesenchyme that surrounds the developing tooth anlagen during the early fetal stage. However, the growth of vessels into the dental papilla, which likely involves target-derived expression of *Vegf*, precedes ingrowth of neurites. At the embryonic bell stage, the first blood vessels acquire the coverage of smooth muscle cells (Shadad et al., 2019). Although this indicates that the establishment and patterning of blood vessels and nerve fibers are governed by differential molecular processes, it cannot be excluded that some at successive stages actually are shared. Furthermore, once both tissues are present, neurovascular bundles are formed in the apical tooth. The neurovascular bundles play an integral role in developmental growth as well as in tissue regeneration after injury, providing niches for pericytes and dental mesenchymal stem cells. At least in the rodent incisor, dental MSCs reside in periarteriolar tissue in the neurovascular bundle and are regulated here through local interactions. The sensory axons of this structure secrete SHH, which activates GLI1 in periarterial cells contributing to all mesenchymal derivatives and forming the basis for growth and homeostasis (Zhao et al., 2014). Within the niche, MSCs progress into transient amplifying cells through an IGF-WNT communicating system. The MSC fate then depends on tissue-autonomous canonical WNT signaling, involving a transient amplifying cells-MSC feedback signaling (Jing et al., 2021). As for pericytes, they seem mostly to be involved in repair after damage (Zhao et al., 2014).

Throughout pre-, peri-, and postnatal periods, the process of axon ingrowth and terminal maturation proceeds, until the final young mature nerve supply is established (Hildebrand et al., 1995). A limited axonal branching occurs in the root region, where 70–80% are unmyelinated, as seen in the electron microscope, and the myelinated fibers are within the Aδ size range (~1–5 μm), with occasional examples reaching Aβ class diameters (~5–8 μm). As the fibers approach the pulpal chamber, the nerves undergo an increased arborization, and some axons form terminals within this region. However, what appears to be the majority will create the impressive terminal nerve pattern that characterizes the coronal pulp–dentin border, in close association with odontoblast cell bodies and processes as well as with local cells with numerous intertwined extensions (Hildebrand et al., 1995; Khatibi Shahidi et al., 2015) (Figure 1.9). Throughout life, the plasticity of neurite trees is maintained, and the intrapulpal nerve fiber sprouting may occur, sometimes profusely, as a response to e.g. trauma or infection. In diphydont and polyodont species, nerve trunks leading into teeth that are about to be shed emit new branches that will supply the replacing tooth. In older life and senescence, degeneration of dental nerves is common, and this together with age-related obliteration of dental pulps may leave old teeth devoid of a pulpal sensory apparatus (Fried et al., 2000).

Intrapulpal axons are largely Aδ- and C-fibers, but the parent axons of most pulp afferents are myelinated and have a large diameter and rapid conduction velocity at their point of entry into the tooth. A majority of dental afferents, thus, seems to branch, taper, and lose their myelin after they have entered the pulp. Fibers that are

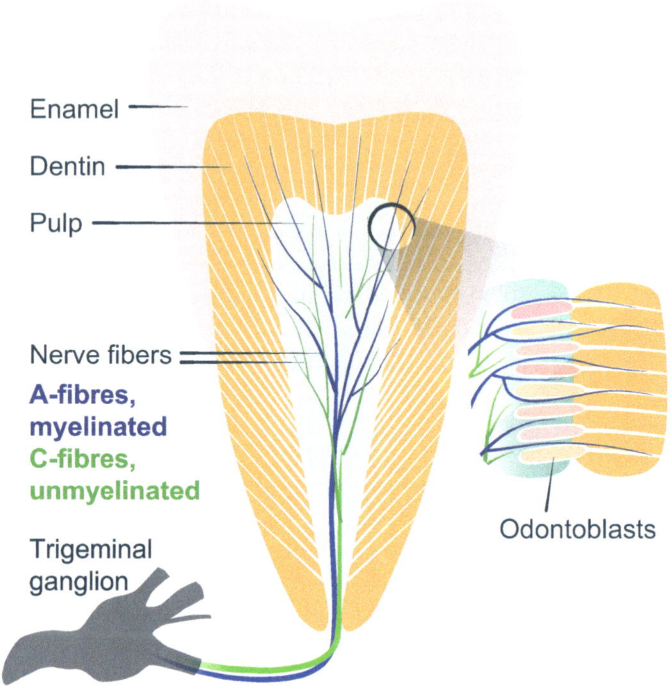

FIGURE 1.9 Innervation of the mammalian tooth with different types of sensory fibers.

unmyelinated along their entire length belong partly to the sympathetic nerve supply, which regulates the local blood flow. Other unmyelinated axons are most likely true classical C-fibers. That many intradental afferents are tapering Aβ-fibers is consistent with the observation that most (70–90%) trigeminal ganglion tooth pulp innervating cells are large or medium in diameter and produce cytochemical markers common in low threshold mechanosensory neurons. These include RT97 (NF200), carbonic anhydrase, calretinin, parvalbumin, calbindin, epithelial Na^+ channels (ENaCs), ASIC3, TREK1, and TREK2. Pulp injury causes the upregulation of NPY, another characteristic of LTM afferents (see Fried et al. (2011), for references). This is a paradox: Neurons of this category are generally not associated with pain transmission after subtle stimulation. Recently, detailed scRNA-seq has been utilized to classify primary sensory cells into altogether 11 distinct neuronal types, including LTM, proprioceptive, thermosensitive, itch-sensitive, and nociceptive neurons with different molecular properties (Usoskin et al., 2015). It is at present unclear where tooth-pulp-related trigeminal neurons fit into this classification, and this issue will have to await individual RNA sequencing.

Although it seems clear that an overwhelming majority of the extremely dense and complex nerve arborizations at the pulp–dentin border are peripheral terminals of mechanosensory nerve cells, the transduction mechanism that converts external stimuli into electrical signals remains unknown. Some arguments brought forward have claimed that nerves at the pulp–dentin border could be activated directly. This might be the case as far as temperature concerns, since dental nerve endings host thermo-sensitive TRP receptors such as TRPV1, TRPV2, TRPV3, TRPV4, and TRPM3 (see Chung et al., 2013; Lee et al., 2019) and perhaps TRPM8 and TRPA1 (Lee et al., 2019). However, in light of the structural arrangements with nerves inside or below mineralized matrix, it seems unlikely that many other pain-provoking stimuli of the tooth are elicited via direct activation of receptors/ion channels on axonal membranes. Instead, based on observations already in the early 20th century and eventually conceived in the 1960s (Brannstrom and Astrom, 1972), focus has been on hydrodynamic interactions between axonal endings and cellular processes related to dentinal tubules. Initially, it was proposed that various stimuli such as heat and cold or exposure of dentin to a wide range of mechanical, chemical, or physiological agents, including even light air puffs, would initiate an either inward or outward movement of the fluid that occupies the space in the dental tubuli. Odontoblast processes in the tubuli would be brought into movement by this and mechanically activate closely associated nerve endings to set off electrical signals. The cilia of odontoblast might play an additional role in stimuli transduction according to Cox and co-authors (Cox et al., 2017). However, these general proposals are weakened by the fact that specialized odontoblast-nerve membrane contacts, which could form a substrate for this interaction, have never been found, despite decades of imaging attempts at ultrastructural levels. Nonetheless, the basic concept of this proposal/explanation has not been ruled out.

In this context, it may be of relevance to point at a remarkable example of tooth–nerve interactions in the tusk of the narwhal *Monodon monoceros*, which apparently do not involve pain. This tooth lacks enamel but contains cementum, dentin, and a pulp with associated blood vessels and nerves (Nweeia et al., 2014). It has a highly unusual anatomy, with a spiral form and a pronounced left–right asymmetry, with, in male specimens, the left tooth fully erupted and the right one embedded. The microscopic architecture of the pulp–dentin border is structurally very similar to that of other mammalian teeth. Odontoblasts with processes extend into dentinal tubules, closely associated with axonal branches. The patent dentin tubules are connected to channels in the cementum, which face seawater. Through this hydrodynamic system, pulpal nerves are able to sense gradients in the water, from high-salt to fresh-water solutions, involving changes in temperature, pressure, hydration status, and/or electrochemical and osmotic gradients (Nweeia et al., 2014). This is likely the normal mechanism for narwhals to sense environmental stimuli and apparently does not result in pain (the mechanisms are not fully resolved), as in other mammalian teeth. Overall, the mechanism might be seen as a particular evolutionary tooth adaptation to environmental demands.

With regard to the generation of pain signals from teeth, another theoretical approach to explain dental nerve transduction at the terminal involves a role for the odontoblast as a specialized receptor cell, analogous to e.g. Merkel cells or other mechanically sensitive cells. Odontoblasts are excitable (Lundquist et al., 2000; Shibukawa and Suzuki, 2003), and possess voltage-gated Na^+ channels, voltage-gated K^+ channels, calcium-activated K^+ channels, store-operated calcium channels, Na^+/Ca^{2+} exchanger, and mechanosensitive ion channels such as PIEZO2, TRPM7, and TREK-1 channels (Khatibi Shahidi et al., 2015). Thus, when exposed to appropriate dental external stimuli, odontoblast processes become mechanically impinged directly or by the movements of the fluid in the dentinal tubuli and become depolarized. However, again, functional evidence for how an electrically excited odontoblast would activate nerve endings is lacking, although some studies have suggested signaling pathways utilizing ATP via P2X3 receptors (Sato et al., 2018).

In a quite recent study, which may bear relevance to transducer mechanisms in dental nerves, the authors revealed that in a mammalian skin, a specialized glial cell type with extensive processes in the subepidermal border conveys noxious thermal and mechanical sensitivity. This glia–nerve complex has an essential physiological role in sensing noxious dermal stimuli (Abdo et al., 2019). The pulp–dentin border is richly furnished with profusely branching cells with glial characteristics (Krivanek et al., 2020), and it cannot be excluded that they are involved in sensory sensing in the dental pulp.

The fact that any stimulus that activates dental LTMs, regardless of its nature, will result in a pain experience remains an enigma. We suggest that this is due to a stepwise rearrangement of synapses at the brainstem level. Thus, during evolution, peridental nerve fibers may gradually expand their target area to include pulps, primarily to increase mechanosensitivity for feeding. Of note, it was recently shown that the target area that mechanoreceptors innervate controls the developmental organization of its central synapses (Lehnert et al., 2021). Hence, during an evolutionary process of pulpal expansion, mechanoreceptors might come under the influence of local cell- or mineralized tissue-derived target molecule(s) or, alternatively, by loss of some. As a result, pulpal neurons would gradually reorganize their brainstem pattern of termination. Through novel synaptic contacts (or strengthening of existing and weakening of others), pulpal signals would gradually be exclusively conveyed by second-order neurons in the pain system in the spinal trigeminal nucleus, which also receives classical mechanosensitive afferents (Tabata and Karita, 1991). This would be evolutionary advantageous as teeth more and more attained their role in fight and feeding.

1.2.5 Tissue-Residential Immune Cells of the Pulp

Different types of leukocytes transiently populate teeth, usually in reaction to inflammatory stimuli, or permanently reside in the dental pulp tissue. Although most of the research of dental tissue-residential immune cells focuses on their role in tooth protection, the recent findings point at their diversity in composition and function in teeth (Krivanek et al., 2017, 2020; Neves et al., 2020; Pagella et al., 2021).

The resident immune cells in the healthy dental pulp are represented by macrophages, dendritic cells, and mast cells, all originating from the myeloid lineage (Cooper et al., 2017; Ferenbach and Hughes, 2008; Krystel-Whittemore et al., 2015; Walsh, 2003). These cells populate tooth via passaging through the walls of blood vessels or expand from the immune cells already residing in the dental pulp (Iwasaki et al., 2011).

The key role of tissue residential immune cells (represented mostly by macrophages and dendritic cells) is to act as sentinels by signaling inflammation, actively fighting infection, and attracting other immune cells, such as neutrophils and T and B cells, through chemokine and cytokine signaling (Cooper et al., 2017). However, the immune cells are also present in the dental pulp during an early tooth development and in the differentiating parts of continuously growing rodent incisors. An interesting hypothesis suggests that these immune cells take a role in the activation of dental pulp stem cells during tertiary dentinogenesis (Krivanek et al., 2020; Neves et al., 2020). This suggests that these cells can have also other non-immune roles in teeth.

The recent single-cell studies revealed the immune cell type heterogeneity in addition to dental pulp and biomineralized matrix-producing cells. For instance, Pagella et al. described five different clusters of immune cells in the human dental pulp using scRNA-seq (Pagella et al., 2021). When it comes to the mouse incisor, Krivanek et al. reported eight different clusters of immune cells, which included four types of macrophages, natural killer cells, lymphocytes, and two types of neutrophils (Krivanek et al., 2020). Notably, these immune cell subtypes might contain cells from the blood contamination. The experimental validations confirmed the presence of two types of M2 macrophages ($LYVE1^+$ and $LYVE1^-$) with different spatial localization and $DPP4^+$ natural killer cells in the healthy dental pulp tissue (outside the blood vessels) via immunohistochemistry (Krivanek et al., 2020).

Overall, further studies are necessary to better define dental-specific subpopulations of immune cells and to understand their roles in tooth protection and beyond protecting teeth.

1.2.6 Cellular Composition of Structures Anchoring Teeth in Jaws

Although the structures responsible for tooth attachment in jaws are structurally and functionally different from the inner dental pulp, they are of the same ectomesenchymal origin. Types of anchoring tissues significantly differ among animal species. A tooth is surrounded by the alveolar bone, being linked to the bone via nonmineralized periodontal ligaments. This type of attachment called gomphosis offers several advantages such as flexibility in tooth micromovement causing higher resistance to a mechanical stress (Beertsen et al., 1997; Bertin et al., 2018). Gomphosis is characterized by the development of associated tissues (cementum, periodontal ligaments, alveolar bone, and gingiva), which are formed together with the dental follicle by precisely orchestrated epithelial–mesenchymal interactions (Li et al., 2017). The second type of tooth attachment in bone is ankylosis where the tooth is firmly attached to the bone without any presence of periodontium (Landova Sulcova et al., 2020). For detailed overview, see Bertin et al. (2018).

At the cellular level, the alveolar bone is formed by osteoblasts, maintained by osteocytes, and, importantly, is continually remodeled by the lytic activity of osteoclasts. This ability is essential for clearing the way during teeth eruption and for the teeth movements followed by the changing pressure or tension acting on teeth. This behavior is largely used during orthodontic procedures. On the surface of tooth roots is deposited a thin layer of cementum formed by cementoblasts. There exist two basic types of cementum: Acellular and cellular. The cellular cementum also contains cells similar to osteocytes, which are entrapped inside this hard tissue. In contrast to the bone, cementum is not remodeled. It serves as a glue layer, which anchors periodontal ligament via Sharpey's fibers. The soft tissue between cementum and bone is highly vascularized and innervated and consists mainly of periodontal ligaments-forming fibroblasts. This space also contains various immune cells (Li et al., 2020) and a previously mentioned ERM and serves as a source of stem cells for periodontal ligaments and cementoblasts (Zhao et al., 2021). At the stem cell level, there are several different sources of stem cells in this area. The residues of dental epithelium inside the periodontal space, known as ERM, can undergo epithelial-to-mesenchymal transition and become a source of stem cells (Xiong et al., 2013). Recently, it was shown that perivascular-derived and perivascular-associated cells are the source of cementoblasts during homeostasis and after the induction of periodontal disease (Zhao et al., 2021).

1.3 EVOLUTION OF CELL TYPES BUILDING ODONTODES

1.3.1 Evolution of Odontoblasts and Osteocytes

Ameloblasts, odontoblasts, and osteocytes are the cell types producing hydroxyapatite-based hard matrixes in a multitude of shapes, functions, and structures. Only exceptional non-chordate animals use hydroxyapatite as a hard matrix for strengthening their cuticles (Bentov et al., 2016), whereas the majority of non-chordates prefers calcium carbonate, silica, or other chemistry (Boskey and Villarreal-Ramirez, 2016; Murdock, 2020). Tracking the origin of such biomineralizing capacity in the vertebrate paleontological record does not seem feasible now, which we will discuss later. On the other hand, the recent study by Sorrentino and co-authors revealed that crystalline calcite serves as a precursor of later hydroxyapatite deposits in biomineralizing vertebrate cells. The authors detected both calcite and hydroxyapatite in early osteogenic mammalian cells, which suggests the temporal shift from calcium carbonate to calcium phosphate production during a continual process of matrix allocation (Sorrentino et al., 2021). From the evolutionary point of view, this means that the same enzymatic and matrix scaffolding systems can work with both calcium carbonate and phosphate, which might imply that calcium-carbonate-producing cells eventually switched to preferential phosphate crystal production during the evolution of chordates. The capacity to generate the massive calcite-based matrixes did not disappear from the vertebrate lineage or, alternatively, was elaborated *de novo* (Boskey and Villarreal-Ramirez, 2016), as we know from the examples of reptilian and bird eggshells (Gautron et al., 2021; Le Roy et al., 2021). If we switch to other related deuterostomes, such as echinoderms, we will see that they have a dermal skeleton made of calcium-carbonate-based calcareous plates or ossicles (Heatfield, 1970). Those somehow resemble the integumentary skeleton of vertebrates and are produced by dermal cell type called "sclerocyte." The sclerocytes express *Alx* transcription factors, which are driving their biomineralizing capacity (Khor and Ettensohn, 2020). Interestingly, *Alx* genes in vertebrates are also associated with skeletogenic cell types and bone development by being expressed in the skeletogenic populations of the cranial neural crest. There are multiple craniofacial malformations associated with misexpression or loss of *Alx* genes in vertebrates (Beverdam et al., 2001; Khor and Ettensohn, 2020). Both echinoderm and vertebrate *Alx* genes drive the downstream expression of *Fam20c*, which is a protein kinase essential for biomineralization (phosphorylates matrix-associated siblings) (Khor et al., 2019; Tagliabracci et al., 2012). Therefore, we hypothesize that the common ancestor of all deuterostomes possessed "dermal" biomineralizing cell types, which were capable of allocating calcium carbonate crystals and shared the fundamental cell identity core with echinoderms and chordates. Later on, this archetypical cell type switched to calcium phosphate deposits in vertebrates.

There are several hypotheses attempting to explain the predominant use of hydroxyapatite crystals instead of carbonate (calcite) in vertebrates. One hypothesis suggests that the evolutionary-early hard matrix-producing cells were the storage cells responsible for ion homeostasis and, most of all, uninterrupted Ca^{2+} supply. According to Ruben and Bennett (1981) and Sire and Kawasaki (2012), the pH of the inner body fluids could fluctuate in metabolically active early vertebrates during their activity and rest. This could lead to the intermittent phases of low pH, which, in turn, could cause the instability of calcium carbonate (Ruben and Bennett, 1981; Sire and Kawasaki, 2012). Indeed, calcite can be extra-sensitive even to slight changes of pH, which is causing troubles to organisms with calcite-based skeleton due to current ocean acidification (Andersson et al., 2008). Thus, hydroxyapatite was a better choice for a stable storage and release of Ca^{2+}, as it is more stable under acidic pH conditions, which could help to minimize hypercalcinemia at certain metabolic rates (Sire and Kawasaki, 2012). This might explain why the vertebrate skeleton relies on calcium phosphate instead of calcium carbonate.

Another hypothesis suggests that the electroconductive hard matrix emerged to protect and stabilize the electro-, mechano-, and other receptors near the body surface. This suggestion is supported by the fact the odontodes are sensory organs, and, in line with this, dentin represents a highly innervated tissue. Indeed, modern dentin is a sensory tissue, and the alignment of nerves during evolution could drive the necessity to form oriented dentinal tubules with nerve projections leading to the external surface of an odontode. This could stimulate the formation of a spatial alignment of odontoblasts organizing them into a pseudo-epithelial mode. This specific hypothesis also might explain why the dermal mineralized skeleton appeared before the internal mineralized skeleton according to Sire and Kawasaki (2012). According to this scenario, mesenchymal cell types surrounding the nerve endings and sensory skin end organs could form condensations and capsules of ensheathing hard matrix to stabilize and protect the fine sensory structures. Besides, calcium apatite shows a directional component, which could aid the electroreception as mentioned in Sire and Kawasaki (2012). In other, more general scenario of the odontode origin, the authors suggest the evolutionary conversion of sensory placodes into odontodes, which retained some essential sensory functions (Fraser et al., 2010). In line with this, the animals with a body surface covered by odontodes (*Placodermi*, for instance) must have had channels for sensory nerves traversing the body armor to provide sensory information arriving from the body surface similar to the innervation of modern teeth and scales.

The true odontodes, including teeth, emerged in the paleontological record in a quite advanced and diverse spectrum, and since the earliest finds of jawless vertebrates from the early Ordovician from around 460 Ma, they already include clearly distinguishable dentin and alveolar bone. These paleontological data support the early co-presence of odontoblasts and osteocytes in the same animals (jawless heterostracans) and anatomical structures according to Keating et al. (2018) and Sire and Kawasaki (2012). Therefore, so far we cannot conclude what came first—odontogenic or osteogenic process, and what was the ancestral cell type, which for the very first time managed the extracellular allocation of hydroxyapatite crystals in dermis or elsewhere. According to Haridy and coauthors, the phosphorus homeostasis and its control were the primary drivers of the initial evolution of osteocytes and bone (interpreted as a storage of minerals). Indeed, the signs of osteocytic osteolysis playing a role in mineral homeostasis became evident in an osteostracan bone (Haridy et al., 2021).

Enameloid, being a capping tissue formed by the acellular dentin- and enamel-like matrix, occupies the superficial position typically characterizing the enamel. The production of enameloid requires a joint activity of odontoblasts and ameloblasts, which might highlight the early steps of evolution of both cell types (Sire et al., 2009). In addition, the evolution of enameloid could require a more specific and tunable interaction between odontoblasts and ameloblasts for the production of this integrated tissue layer.

The evolutionary variation in odontoblasts and odontoblast-like cells enabled the expansion of dentin-related matrices. For instance, lamellin represents an acellular and atubular dentinous matrix found in the scales of the putative early chondrichthyan *Mongolepidida* (Karatajùtė-Talimaa, 1995). The other fossil and some extant animals reveal a series of different and, yet, more modern-type dentinous matrixes entrapping cell bodies and cell processes. For instance, modern bony fishes show the presence of superficial dentin and cell-entrapping elasmodin with plywood-like organization of collagens and matrix. On the other hand, orthodentin, the true dentin, is generated by a variation of modern type well-polarized and aligned odontoblasts and their processes. Mesodentin, being found mostly in the osteostracan and acanthodian dermal armor, differs from orthodentin by lower density of odontoblasts, with some odontoblasts being trapped in the matrix similar to osteocytes. The branching pattern of odontoblasts in mesodentin is much less polarized as compared to orthodentin (Ørvig, 1967; Sire and Kawasaki, 2012). Semidentin (found in placoderms) differs from mesodentin in one important aspect: It shows a much higher degree of polarization of odontoblast processes. Osteodentin found in teeth of some vertebrates is vascularized and resembles alveolar bone. Similarly, vasodentin represents heavily vascularized atubular dentin.

The diversity of evolutionary ancient dentin types and odontode structures raises a question about co-evolution of pulp cells and pulp cavity, as in many cases, the proper pulp cavity does not form, and the pulp cells might mingle within smaller lacunas next to odontoblasts, nerves, and blood vessels.

The later steps of odontoblast evolution in tetrapods highlight examples like plicidentin (commonly found in fish, as well as in varanids and stegocephalians) and tubulidentin (extinct tubulidentates and modern aardvark). Plicidentin shows the presence of abundant collagen lamellae and otherwise is similar to the true dentin.

However, it might be that the formation of plicidentin depends on the folding of epithelium, not the tissue composition.

Tubulidentin is found in continuously growing enamel-free teeth and is penetrated by dentinal tubules surrounding larger pulp channels (Sire et al., 2009; Sire and Kawasaki, 2012). All these types of dentinous matrix show the presence of associated odontoblasts projecting into dentinal tubules or being submerged into the matrix or a combination of both.

The evolution of these subtypes of dentin with differentially oriented dentinal tubules and other properties required evolving specific cellular adaptations in dentin-allocating cells, which could result in phylogenetic variation of odontoblasts found in different animals (see Chapter 5 for extensive discussion). Overall, it seems that the evolution of odontodes could start with the acellular mineralized matrices and only then moved on to cellular hard matrices with non-polarized or partly-polarized submerged cells. Later, or in parallel, the odontodes evolved non-submerged polarized cells with well-organized processes. Thus, after the acellular phase, further evolution could proceed from less polarized to more polarized and finally pseudo-epithelial arrangements of odontoblasts. This logic is suggested by the odontoblast positions and shapes in mesodentin, semidentin, vasodentin, and osteodentin, all of which show some degree of similarity to vascularized bone, where osteocytes possess a non-polarized tree of processes and are trapped in the matrix. Thus, odontoblasts show a great variation of cell polarization, structure of their processes, spatial localization, and characteristics of cell–cell contacts (see Sire et al. (2009) and Sire and Kawasaki (2012) for extended discussion).

Notably, osteocytes also vary in different animal groups—starting from those, where we observe the presence of acellular bone (with cells allocating bony matrix only at the peripheral borders of the bone) (Keating et al., 2018) and ending up with the endochondral ossification, where some chondrocytes directly transdifferentiate into bone-embedded osteocytes (Sire et al., 2009). Little is known about the evolution of osteocytes and their diversity in various animal groups. The majority of current single-cell methods rely on the analysis of dissociated cells, which are hard to obtain in case of a bone. Still, we anticipate that technical issues will resolve, and individual osteocytes from different animals and types of bone will be sequenced and compared.

Finally, the evolution seriously touched upon the cell types and signaling pathways responsible for patterning the odontodes. The way in which the odontode complexes changed during the evolution suggests a significant plasticity of patterning signals across different animal lineages. The primary patterning unit could shift from a single odontode primordium to polyodontode scales, and the anchoring bony base could become one the major patterning units of such dermal armor (Qu et al., 2013). Therefore, the patterning and morphogenetic processes intertwine odontogenic and osteogenic cell types by building complex and integrated functional units. In line with this reasoning, dermal bone becomes the major contributor to the odontode development and evolution (Qu et al., 2013). Thus, co-evolution of cell types serving as signaling centers was a key for generating a diversity of individual odontodes and polyodontodes integrated into the rest of the integumentary skeleton. However, we know very little about the evolutionary steps forming signaling centers orchestrating odontogenic and osteogenic process (see Chapter 8). Even more, we can hardly imagine how the first odontogenic developmental processes evolved or where does the dermis and dermal cell types take their origin in pre-chordate ancestral organisms. The analysis of evolving molecular pathways can provide a better reference for a future research in this direction.

Although the paleontological samples do not help to understand the origin of the initial hard matrix-laying capacity, the interrogations of the molecular toolkit suggest that hard-matrix producing dermal cell types evolved after SPARC (secreted protein acidic and rich in cysteine) proteins diversified and gave rise to the small calcium-binding phosphoproteins (SCPP) family (Bertrand et al., 2013). This family includes EMPs (Enamel Matrix Proteins, for instance ENAM, AMBN, AMEL), as well as dentin- and bone-related proteins DSPP1, DSPP2, DMP1, IBSP, MEPE, SPP1, SPARCL1, and SCPPPQ1 (Kawasaki, 2011). The members of this family became employed to generate various types of hard matrix in varieties of bone, dentin, and enamel (Kawasaki, 2011). Before diversifying and acquiring biomineralization capacity, the ancient SPARC proteins were likely associated and co-expressed with collagens, which suggests the early scaffolding role of SPARC in the soft extracellular matrix (see Chapter 4). In line with this, the early paralogues of SPARC and SPARCB are still preserved in an amphioxus genome and are associated with collagen-synthesizing cells located in soft tissues of mesenchymal origin (Bertrand et al., 2013).

Theoretically, based on a set of molecular similarities, it might be possible to reconstruct the archetypical mineralizing cell, which arose in a dermis of early agnathans (or even earlier in a stem deuterostome) and later diversified into osteocytes and odontoblasts during further evolution of osteogenic and odontogenic developmental processes.

Despite being similar and evolutionary related cell types, odontoblasts and osteocytes evolved different molecular identity codes and regulation of matrix proteins. For instance, odontoblasts rely on the expressions of *Trps1*, *Klf5*, *Nfic*, *Atf6*, *Lef1*, and *Sall1* transcription factors (Balic and Thesleff, 2015; Chen et al., 2017; de Sousa-Romero and Moreno-Fernández, 2016; Krivanek et al., 2020; Kuzynski et al., 2014; Lee et al., 2011; Lim et al., 2021) switching sequentially for their differentiation, whereas the forming osteocytes use different regulation relying on the expressions of *Dlx5*, *Egr2*, *Atf4*, *Jun*, *Ets*, *Men1*, *Hey1*, and *Stat1* (Franceschi et al., 2009; Huang et al., 2007; Kobayashi and Kronenberg, 2005; Komori, 2006).

In the literature, we observe an ongoing debate about the embryonic cell lineage where the odontogenic competence evolved (see Chapter 3), and most of the opinions lean toward the neural crest cells known as the fourth germ layer (Eames, 2020; Gillis et al., 2017; Hall and Gillis, 2013). The alternative point of view suggests that the odontogenic competence evolved in the dermal cells of mesodermal origin.

Today, it does not look possible to resolve this conundrum using embryological or genetic evidence for the reasons we explain here. First, we do not see strong arguments for the odontogenic process is unique to the neural crest lineage despite the ongoing attempts to lineage-trace trunk odontodes for revealing skeletogenic contribution from the neural crest in the trunk (Gillis et al., 2017). Second, the integumentary skeleton of echinoderms is not derived from the crest, as those animals presumably lack cell types homologous to the crest of vertebrates. Next, the border between the neural crest- and mesoderm-derived tissues is not conserved in the cranial domain and drifts between species (Eames, 2020). At the same time, the resulting skeletal and other structures with fluctuating crest-mesoderm border stay unperturbed and evolutionary stable (Hirasawa and Kuratani, 2015). The fate convergence of mesenchymal and skeletogenic cells types originating from the crest and from the mesoderm is near perfect to the extent that nobody could identify any morphologic or function-related distinction. Furthermore, the cranial neural crest cells seem to co-opt the mesodermal gene expression program to achieve such convergence of resulting cell types (chondrocytes, osteocytes, etc.) (Eames, 2020; Ivashkin and Adameyko, 2013). In any case, if we assume for the moment that the initial capacity to form odontodes could originate in the neural crest lineage, this capacity could be co-opted into the mesodermal sub-lineages and *vice versa*.

1.3.2 Evolution of Pulp Cells

Dental pulp represents a conglomerate of highly specialized mesenchymal populations, secreting non-calcified extracellular matrix and exchanging signals with other cell types. Dental pulp is not uniform, as we already discussed before, and often hosts dental mesenchymal stem cells, which are competent to give rise to odontoblasts as well. Thus, some pulp populations are the reservoir of odontogenic and, possibly, osteogenic multipotent stem cells (Krivanek et al., 2020, 2017).

The evolutionary origin of dental pulp is enigmatic and presumably begins with dermal mesenchymal populations, which were competent to exchange signals with the overlaying ectoderm. Thus, the origin of odontode pulp is key for the entire process of the odontode induction, as it starts as an interplay between localized and patterned epithelial signaling centers and competent underlying mesenchyme (Balic and Thesleff, 2015). Furthermore, the signaling role of the developing dental papilla and pulp is vast and major, mainly because the developmental competence of the dental mesenchyme influences dental shaping and self-renewal properties (Jernvall and Thesleff, 2012; Thesleff, 2018; Venugopalan et al., 2011).

If we ponder about the possible evolutionary scenario, the cell types of primitive mesenchymal condensations in the dermal layer eventually acquired the properties to generate and support the hard matrix-producing cells of odontodes. Furthermore, they developed a capacity to host the extrinsically arriving cell types such as immune cells, vasculature, and nervous system-associated cell types, that is, projections of neurons covered by different glial cells. For instance, pulp cells in mammalian teeth secrete homing factors for macrophages and dendritic cells (Krivanek et al., 2020). Next, the pulp cells might have evolved the capacity to mediate the ingression of neuronal projections into the pulp (by secreting neurotrophic factors), which is necessary for further interaction with odontoblasts and dentinal tubules (Nosrat et al., 2001). The mechanosensory and pain-related innervation improved the shielding and biting role of odontodes by bringing the capacity to sense the touch, mechanical pressure, and to detect trauma, which obviously aided the overall evolutionary success of odontodes. Thus, the signaling, homing, and integrating roles of different pulp cell types were the major evolutionary successes, which converted odontodes into sensing, self-protecting, and highly dynamic structures.

The extrinsically arriving cells also must have evolved additional properties, which contributed to the pulp integrity and homeostasis. The methods of single-cell biology will help to understand if there are odontode-specific evolutionary adaptations in immune cell types, especially if there are specific odontode-residential populations of macrophages, dendritic, and B-cells possessing locally tuned protective and signaling properties. Finally, the evolving plasticity of glial and perivascular cell types must have aided dental regeneration and repair by their capacity to transdifferentiate and respond to local cues signifying trauma or infection (Kaukua et al., 2014; Yianni and Sharpe, 2019). Indeed, as the odontodes are at the frontier of physical interaction with the outer world, the damage-resistance and capacity to regenerate become major evolutionary benefits.

1.3.3 Evolution of Cell Types Forming Tooth Attachment—Cementoblasts and Periodontal Ligamentum

The attachment of teeth to a jaw is not uniform across different animals. The most basic attachment system represents a fusion of the dentin to the alveolar and then jawbone, whereas some more recent and advanced types include the attachment via layers of cementum anchoring the periodontal ligamentum (see Chapter 6). The ligamentum, in turn, is attached to the alveolar bone. The latter type of attachment represents a characteristic of crocodilian and mammalian teeth, together called "thecodont teeth", which is currently suggested as a plesiomorphic feature of amniotes (LeBlanc and Reisz, 2013). The roots of the thecodont teeth are coated with cementum, which is produced by the specialized cells called cementoblasts (Krivanek et al., 2017). The evolutionary origin of cementoblasts is unclear. Based on their external position, they are unlikely related to the lineage of odontoblast and dental pulp and most likely represent some evolutionary branch derived from the osteocytes. The fact that cementocytes are routinely found entrapped in a hard matrix might support their relatedness to bone rather than dentin.

The cell types of periodontal ligamentum and their evolutionary cell sources are not well understood either. The periodontal ligamentum has a complex structure with many cell types being involved or located nearby, such as dense innervation with associated glial cells, resident immune cells, and blood vessels. The single-cell analysis suggested the existence of a couple of subpopulations forming the ligamentum itself, and those show the presence of classical markers of a connective tissue (Krivanek et al., 2020).

1.3.4 Evolution of Ameloblasts

Ameloblasts produce extraordinary hard capping tissues such as enamel and homologous matrices, which cover teeth and other odontodes. Ameloblasts are transient cells originating from the epithelial lineage during embryonic development and dying upon the tooth eruption due to their external position (Krivanek et al., 2017). It seems that dentin and dermal bone preceded the enamel in evolution, which is supported by the fact that enamel and homologous matrices are complementary and can undergo secondary loss in different animal lineages that acquire alternative adaptations to feeding (Davit-Beal et al., 2009). At the same time, dentin and especially dermal bone stay rather preserved during the evolution of odontodes, despite the fact that the tooth loss (when enamel and dentine are lost together) happens more often as compared to the selective loss of enamel. Furthermore, some stem osteichthyans (e.g., *Lophosteus*) or stem gnathostomes show the presence of odontodes not covered by enamel (Qu et al., 2015), which might support the absence of enamel-producing cell types in these animals. However, some other stem gnathostomes revealed the presence of enameloid coating the odontodes, including the most primitive teeth (Vaskaninova et al., 2020). Probably, these stem animals lacking enamel already had proto-ameloblasts, which, however, did not specialize yet into the population expressing the enamel-related genes. When these proto-ameloblast cells secreted mineralizing ions into the predentin at earlier stage, they produced enameloid (Davit-Beal et al., 2007).

The hydroxyapatite represents the universal building material for bone, dentin, and enamel. Ameloblasts, being a later evolutionary acquisition, hypothetically might have co-opted some molecular machinery necessary for allocating the crystals of hydroxyapatite from other cell types. However, the analysis of modern enamel and homologous matrixes shows that ameloblasts evolved their own molecular toolkit to scaffold hydroxyapatite with their original matrix proteins, including AMEL, AMBN, ENAM, and SCPP5. These enamel-specific proteins evolved via gene duplications events in basic SCPP cluster resulting in several P/Q-rich SCPPs that specifically engaged into biomineralization of enamel and homologous matrixes (Kawasaki, 2011). Notably, the acidic SCPP genes are associated with bone and dentin, whereas numerous P/Q-rich SCPP genes ensure the formation of enamel and homologous matrixes (Kawasaki, 2011, 2018). This suggests that ameloblasts underwent a number of independent evolutionary steps toward developing biomineralization, which resulted in a competitive alternative to bone and dentin.

Of course, the regulatory mechanisms for such relatively independent molecular toolkit also evolved independently from odontoblast and osteocyte gene expression programs. Indeed, the differentiation trajectory of ameloblasts revealed by several studies relies on *Pitx2*, *Dlx1*, *Dlx2*, *Foxo1*, *Tbx1*, *Klf4*, *Nr1d1*, or *Sox21* (Athanassiou-Papaefthymiou et al., 2011; Babajko et al., 2014; Bei, 2009; de Sousa-Romero and Moreno-Fernández, 2016; Krivanek et al., 2020; Lezot et al., 2008; Tao et al., 2019).

Despite these differences in transcription factor codes, there are major similarities shared by biomineralizing cell types. For example, the development of osteoblast and odontoblasts shares the expression of *Klf4* (Tao et al., 2019; Yu et al., 2021); ameloblast and odontoblast share the expression of *Sp7*, *Klf5*, *cMyb*, and *Pax6* (Bae et al., 2018; Chen et al., 2017; de Sousa-Romero and Moreno-Fernández, 2016; Krivanek et al., 2020; Lei et al., 2014); and ameloblasts and osteoblast share *Satb2* and *Sp3* (Dobreva et al., 2006; Gollner et al., 2001; Huang et al., 2007; Krivanek et al., 2020). Interestingly, the expression of transcription factors *Runx2* and *Dlx3* is shared by all three cell types (Chen et al., 2009, 2005; Chu et al., 2018; de Sousa-Romero and Moreno-Fernández, 2016; Duverger et al., 2017; Hassan et al., 2004; Krivanek et al., 2020; Li et al., 2021; Zhang and Li, 2015).

What were the ancient cell types from which proto-ameloblasts evolved? To answer this question, we need to attend the complexities of dental organogenesis.

Ameloblasts stem from the epithelial progenitors forming bending epithelial layers responsible for the shape of future teeth. These epithelial progenitors incorporate populations that serve as signaling centers called enamel knots directing the shaping of a dental crown and, even before that, the epithelial placodes, which interact with underlying mesenchyme to initiate dental development (Balic and Thesleff, 2015; Jernvall and Thesleff, 2012). Therefore, the progenitors of ameloblasts represent epithelial cells responsible for signaling and morphogenesis. Without the epithelial signals, dental induction is impossible, which suggests that the odontodes lacking enamel relied on the epithelial signals for their induction and shaping (see (Jernvall and Thesleff, 2012) for developmental mechanisms). Thus, the evolution of regulatory centers and signaling cell types in the epithelium and epithelial placodes is at the core of the evolutionary emergence of odontodes. The evolutionary and developmental progenitors of ameloblasts seem to be responsible for the earliest steps of the odontode emergence, patterning, distribution, shaping, and mechanical properties. In line with this, the "morphogenetic" role of the dental epithelium is much more essential for the integrity and formation of teeth, as compared to the role of hard matrix deposition (see Chapter 2). Ablation of enamel-related genes and abrogation of enamel production do not lead to dental loss or severe shaping defects. This suggests that the production of enamel developed on top of other, rather signaling roles, during the evolution of odontodes.

Beyond the grand steps of ameloblast evolution, the smaller steps also played an important role in adapting the enamel to the challenges of different feeding strategies. Indeed, the microstructure of the true enamel varies in different species, even within compact phylogenetic groups such as mammals (Filippo et al., 2020).

Ameloblast use their Tomes' process to allocate differentially oriented enamel prisms, which form perplexed mesh in various species-specific arrangements to provide oriented extra strength to the resulting enamel layer. This ends up in an anisotropic nano-material, resistant to directional pressure beyond what is expected from mechanical properties of amorphous or crystalline hydroxyapatite (Borrero-Lopez et al., 2020; Wilmers and Bargmann, 2020). The evolutionary adaptations of enamel mesh depend on the shape and operation of Thome's process, which might be species-specific. In the future, the comparative single-cell transcriptomics might help to elucidate the genetic programs responsible for the differences in Thome's processes in various groups of animals.

1.4 PERSPECTIVES OF SINGLE-CELL OMICS METHODS IN THE EVOLUTION OF CELL TYPES BUILDING ODONTODES

The odontodes, being quite diverse across phylogenetic groups and even within a single organism, contain evolutionary and adaptive variations of multiple cell types that require new comparative framework. Today, this is technically achieved by using single-cell omics approaches. The progress in single-cell transcriptomics enabled the direct comparison of gene expression programs in similar or presumably related cell types (Tanay and Sebe-Pedros, 2021; Wang et al., 2021) (Figure 1.10). Furthermore, the total gene expression in a given cell can be subdivided into co-regulated modules corresponding to specific functions, or even into the modules,

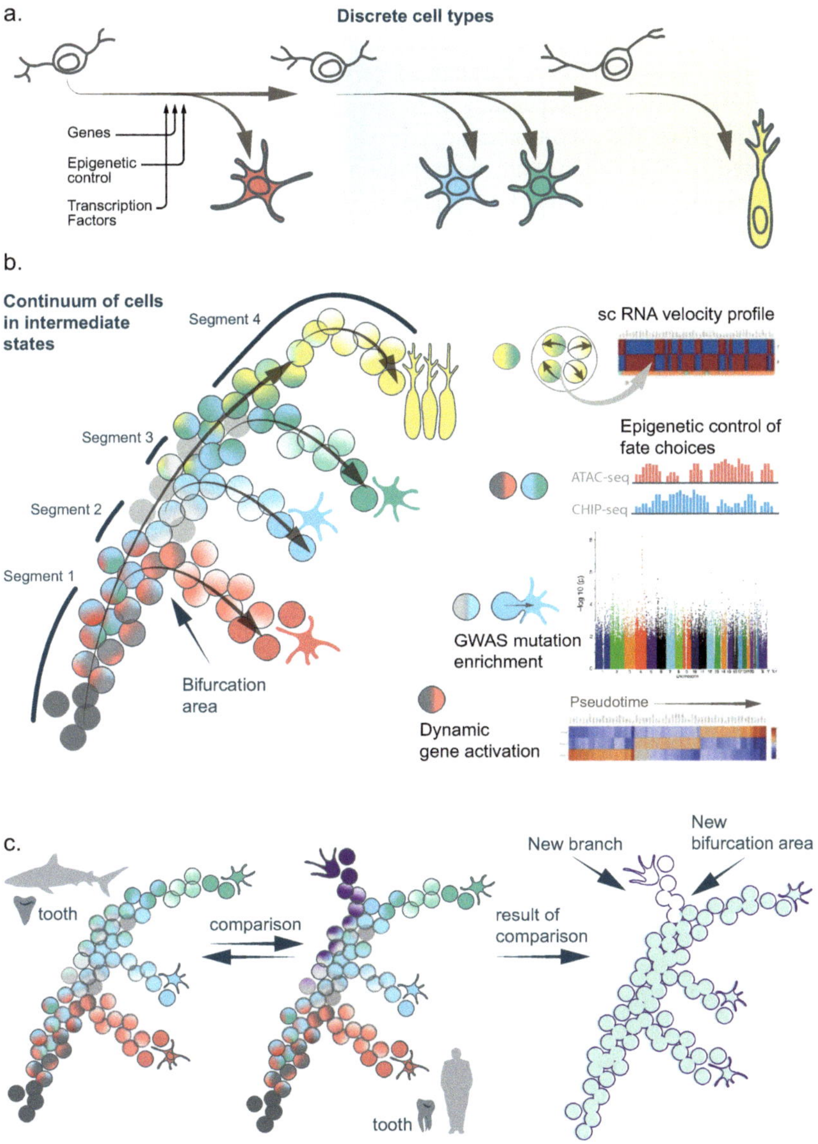

FIGURE 1.10 Single-cell analysis is transforming dental research and our understanding of cell types and their evolution. (a) Former views on cell lineage development. The differentiation steps are viewed as discrete switches between progenitor types. (b) The use of single-cell transcriptomics increased the resolution of differentiation steps and revealed the continual nature of differentiation trajectories. RNA velocity approach helps to pinpoint the orientation of cell differentiation, whereas the integration of single transcriptomics data with epigenomic and GWAS studies enables mapping the origin of diseases and predicting plasticity of cell phenotypes. (c) The comparative single-cell approach allows addressing the evolution of cell types at a new level, including better understanding of evolving fate decisions and lineage tree structures. After Kameneva and Adameyko (2019).

where individual transcripts and their regulatory regions co-evolved in an orchestrated way (Cope et al., 2020; Harris et al., 2021; Ovens et al., 2021; Wang et al., 2016; Zrimec et al., 2020). Such analysis of similar cell types might compare the gene expression modules separately or in any specific context depending on a study design. Therefore, the evolution of specific cellular functions and properties can be investigated in connection with the evolution of cell identity codes. Overall, such comparisons highlight transcriptional similarities and differences, which allow to debate homology or analogy of cell types or their specific cell functions.

Notably, the concept of homology has issues when applied to cell types, as the cell functions depend on the underlying genetic programs. These genetic programs can be inherited from the transcriptional repertoire of a given cell lineage, or, alternatively, such programs can be co-opted from clearly non-homologous cells. Thus, cell types can emerge via diversifying or transforming the preexisting gene expression programs available for the given cell lineage (Hodge et al., 2019) or via blending and pooling together gene expression programs from other unrelated cell lineages, which might lead to tremendous innovations (Eames, 2020; Ivashkin and Adameyko, 2013). For instance, epithelial ameloblasts could emerge by recruiting the expression of P/Q-rich SCPPs, which were previously expressed broader in other biomineralizing cell types (ancient odontoblasts and osteoblasts) from mesodermal and neural crest lineages. Later, the expression of these SCPPs segregated selectively into the ameloblast differentiation trajectory, whereas the expression of other SCPP genes segregated into odontoblasts and osteocytes. Similarly, cranial neural crest cells, which give rise to the craniofacial skeletogenic and odontogenic mesenchyme, might have emerged by pooling the transcriptional programs of the ectoderm-derived neural plate border and mesenchymogenic mesoderm (Eames, 2020). Indeed, many cell types including chondrocytes, osteocytes and, presumably, odontoblasts are convergently produced by the neural crest and mesoderm lineages in vertebrates.

Despite the fact that there is an issue with the homology concept when it comes to the comparisons of specific cell types, the concept of homology can be surely applied to a cell lineage, where external co-options can be tracked and ruled out when we take into the account broader landscapes of rather conserved multi-step modules underlying cell differentiation. What does this mean for the evolution of the cell types building odontodes? For instance, the single-cell transcriptomics analysis of ameloblasts involved in developing enamel, enameloid, and ganoin can rule out if the corresponding cell types are truly homologous or not, and if they rely on the same basic identity programs and belong to the same cell lineages with common developmental origin. Contrary to this, some most recent comparisons were based on the expression of a small number of enamel-specific genes (AMEL, AMBN, ENAM, and SCPP5) (Kawasaki et al., 2021), which we discussed in the previous paragraph. Despite being an extraordinary valuable, the comparisons based on few genes are incomplete, as they cannot rule out the co-option of gene expression or radical divergence of a function in unexpectedly homologous cell types.

One of the key ideas postulates the increase in a number of cell types in an animal lineage during evolutionary timing (Marquez-Zacarias et al., 2021). Such an increase in cell type repertoires is rather explained by "the labor division model", which suggests that a multifunctional "generalist" cell eventually splits into a spectrum of cell types with more segregated subsets of functions (Arendt et al., 2016). Therefore, with time, cells tend to become more specialized and diverse in terms of specializations. This increasing diversity of cell types might be reflected in the diversity of pulp cells or matrix-producing cells building a single odontode or a spectrum of different odontodes within one organism. For instance, we observe the evolution of a tooth attachment system with corresponding emergence of cell types, producing dental cementum and periodontal ligamentum in thecodont teeth (LeBlanc and Reisz, 2013), as compared to the more simple and ancient attachment scheme, where the tooth is ankylosed to the jaw bone. Overall, the detailed comparisons of transcriptional programs are necessary for further studies to delineate the evolution of cell types building the odontodes and defining their diversity in extinct and extant animal lineages (Tanay and Sebe-Pedros, 2021).

Next, because this chapter is dedicated to cell types building odontodes, we find it useful to mention the complexity of a debate related to what is a cell type and how to define it, especially in a context of recent studies addressing dental heterogeneity at a single-cell level. Similar to the debates of the past, which focused on how to define a concept of species, we slowly arrive to a debate about the concept of a cell type. The developmental and evolutionary continuum of cell states is uninterrupted, and, by living in an ideal world, we would observe this continuum to a full extent, which would preclude the intuitive discretization. However, because we historically dealt with definitive adult cell types observed in a limited number of organisms, a discrete picture emerges on its own. Today, when we analyze the development of cell lineages at a single-cell level by building uninterrupted trajectories of transcriptional states (Wagner et al., 2018), the concept of a lineage segment between two adjacent fate splits (upstream and downstream of a segment) might substitute a concept of a progenitor cell type. The fate split points represent special parts of the transcriptional trajectory with complex heterogeneous structure. Even before the split points, cells gradually co-activate opposing transcriptional programs reflecting intrinsic or extrinsic biases leading the choice of a future fate (Soldatov et al., 2019). The trajectory segment between two fate split points is a naturally discrete entity (Soldatov et al., 2019), unlike any other arbitrary definition of a progenitor cell. In line with this, the terminal definitive cell types, which form at the tips of the lineage branches, might be defined via their latest phase of developmental history representing terminal segments of the transcriptional trajectory after the last bifurcation. For instance, we can associate the entire last trajectory segment with the terminal cell fates such as mature odontoblasts, ameloblasts, or pulp cells when we assess developmental trajectories of epithelium- and mesenchyme-derived cell types in teeth.

As the evolutionary pressure on a tissue often targets embryonic development and underlying lineage trajectories, the concept of naturally discrete trajectory segments separated by fate splits acquires a new weight. When it comes to the evolution of odontodes and, more specifically, teeth, the evolutionary changes in how the epithelial or mesenchymal progenitors balance fate choices might end up driving the evolution of a shape and size of a tooth, as viewed from the developmental perspective. In line with this, the evolution of spatial patterning mechanisms (see Chapter 7) might affect cell fate choices and can be approximated as a sum of spatial biases balancing the results of the fate splits along the trajectory of a lineage.

Bringing the recent successes of dental "EvoDevo" research (Jernvall and Thesleff, 2012; Ortiz et al., 2018) into a context of a lineage trajectory at the single-cell level will require additional efforts, which as some point will aid the holistic picture of odontode development and evolution. Finally, the evolution of genomic regulatory elements responsible for odontode morphogenesis and adaptive variations of definitive cell types (ameloblasts, odontoblasts, etc.) can be studied within the framework of unbiasedly defined cell types and lineage trajectories, where the regulated genes are expressed to perform their functions (Baek and Lee, 2020).

REFERENCES

Abdo, H., L. Calvo-Enrique, J. M. Lopez, J. Song, M. D. Zhang, D. Usoskin, A. El Manira, I. Adameyko, J. Hjerling-Leffler, and P. Ernfors. 2019. Specialized cutaneous Schwann cells initiate pain sensation. *Science* 365:695–699.

Adil, A., V. Kumar, A. T. Jan, and M. Asger. 2021. Single-cell transcriptomics: Current methods and challenges in data acquisition and analysis. *Front Neurosci* 15:591122.

Ahn, J., S. Heo, J. Lee, and D. Bang. 2021. Introduction to single-cell DNA methylation profiling methods. *Biomolecules* 11.

Aldridge, S., and S. A. Teichmann. 2020. Single cell transcriptomics comes of age. *Nat Commun* 11:4307.

Andersson, A. J., F. T. Mackenzie, and N. R. Bates. 2008. Life on the margin: Implications of ocean acidification on Mg-calcite, high latitude and cold-water marine calcifiers. *Mar Ecol Prog Ser* 373:265–273.

Arendt, D., J. M. Musser, C. V. H. Baker, A. Bergman, C. Cepko, D. H. Erwin, M. Pavlicev, G. Schlosser, S. Widder, M. D. Laubichler, and G. P. Wagner. 2016. The origin and evolution of cell types. *Nat Rev Genet* 17:744–757.

Asp, M., J. Bergenstrahle, and J. Lundeberg. 2020. Spatially resolved transcriptomes-next generation tools for tissue exploration. *Bioessays* 42:e1900221.

Athanassiou-Papaefthymiou, M., D. Kim, L. Harbron, S. Papagerakis, S. Schnell, H. Harada, and P. Papagerakis. 2011. Molecular and circadian controls of ameloblasts. *Eur J Oral Sci* 119(Suppl 1):35–40.

Babajko, S., M. de La Dure-Molla, K. Jedeon, and A. Berdal. 2014. MSX2 in ameloblast cell fate and activity. *Front Physiol* 5:510.

Bae, J. M., J. C. Clarke, H. Rashid, M. D. Adhami, K. McCullough, J. S. Scott, H. Chen, K. M. Sinha, B. de Crombrugghe, and A. Javed. 2018. Specificity protein 7 is required for proliferation and differentiation of ameloblasts and odontoblasts. *J Bone Miner Res* 33:1126–1140.

Baek, S., and I. Lee. 2020. Single-cell ATAC sequencing analysis: From data preprocessing to hypothesis generation. *Comput Struct Biotechnol J* 18:1429–1439.

Balic, A. 2019. Concise review: Cellular and molecular mechanisms regulation of tooth initiation. *Stem Cells* 37:26–32.

Balic, A., and I. Thesleff. 2015. Tissue interactions regulating tooth development and renewal. *Curr Top Dev Biol* 115:157–186.

Bartlett, J. D. 2013. Dental enamel development: Proteinases and their enamel matrix substrates. *ISRN Dent* 2013:684607.

Beertsen, W., C. A. G. McCulloch, and J. Sodek. 1997. The periodontal ligament: A unique, multifunctional connective tissue. *Periodontology* 13(1):20–40. https://doi.org/10.1111/j.1600-0757.1997.tb00094.x

Bei, M. 2009. Molecular genetics of ameloblast cell lineage. *J Exp Zool B Mol Dev Evol* 312B:437–444.

Bentov, S., E. D. Aflalo, J. Tynyakov, L. Glazer, and A. Sagi. 2016. Calcium phosphate mineralization is widely applied in crustacean mandibles. *Sci Rep* 6.

Bertin, T. J. C., B. Thivichon-Prince, A. R. H. LeBlanc, M. W. Caldwell, and L. Viriot. 2018. Current perspectives on tooth implantation, attachment, and replacement in amniota. *Front Physiol* 9:1630. https://doi.org/10.3389/fphys.2018.01630

Bertrand, S., J. Fuentealba, A. Aze, C. Hudson, H. Yasuo, M. Torrejon, H. Escriva, and S. Marcellini. 2013. A dynamic history of gene duplications and losses characterizes the evolution of the SPARC family in eumetazoans. *Proc Biol Sci* 280:20122963.

Beverdam, A., A. Brouwer, M. Reijnen, J. Korving, and F. Meijlink. 2001. Severe nasal clefting and abnormal embryonic apoptosis in Alx3/Alx4 double mutant mice. *Development* 128:3975–3986.

Borrero-Lopez, O., P. J. Constantino, M. B. Bush, and B. R. Lawn. 2020. On the vital role of enamel prism interfaces and graded properties in human tooth survival. *Biol Lett* 16.

Boskey, A. L., and E. Villarreal-Ramirez. 2016. Intrinsically disordered proteins and biomineralization. *Matrix Biol* 52–54:43–59.

Brannstrom, M., and A. Astrom. 1972. The hydrodynamics of the dentine; its possible relationship to dentinal pain. *Int Dent J* 22:219–227.

Buchtova, M., J. Stembirek, K. Glocova, E. Matalova, and A. S. Tucker. 2012. Early regression of the dental lamina underlies the development of diphyodont dentitions. *J Dent Res* 91:491–498.

Chan, Y. H., Y. C. Lee, C. Y. Hung, P. J. Yang, P. C. Lai, and S. W. Feng. 2021. Three-dimensional spheroid culture enhances multipotent differentiation and stemness capacities of human dental pulp-derived mesenchymal stem cells by modulating MAPK and NF-kB signaling pathways. *Stem Cell Rev Rep* 17:1810–1826.

Chen, S., J. Gluhak-Heinrich, Y. H. Wang, Y. M. Wu, H. H. Chuang, L. Chen, G. H. Yuan, J. Dong, I. Gay, and M. MacDougall. 2009. Runx2, osx, and dspp in tooth development. *J Dent Res* 88:904–909.

Chen, S., S. Rani, Y. Wu, A. Unterbrink, T. T. Gu, J. Gluhak-Heinrich, H. H. Chuang, and M. Macdougall. 2005. Differential regulation of dentin sialophosphoprotein expression by Runx2 during odontoblast cytodifferentiation. *J Biol Chem* 280:29717–29727.

Chen, Z., Q. Zhang, H. Wang, W. Li, F. Wang, C. Wan, S. Deng, H. Chen, Y. Yin, X. Li, Z. Xie, and S. Chen. 2017. Klf5 mediates odontoblastic differentiation through regulating dentin-specific extracellular matrix gene expression during mouse tooth development. *Sci Rep* 7:46746.

Chiba, Y., K. Saito, D. Martin, E. T. Boger, C. Rhodes, K. Yoshizaki, T. Nakamura, A. Yamada, R. J. Morell, Y. Yamada, and S. Fukumoto. 2020. Single-cell RNA-sequencing from mouse incisor reveals dental epithelial cell-type specific genes. *Front Cell Dev Biol* 8:841.

Chu, Q., Y. Gao, X. Gao, Z. Dong, W. Song, Z. Xu, L. Xiang, Y. Wang, L. Zhang, M. Li, and Y. Gao. 2018. Ablation of Runx2 in ameloblasts suppresses enamel maturation in tooth development. *Sci Rep* 8:9594.

Chung, G., S. J. Jung, and S. B. Oh. 2013. Cellular and molecular mechanisms of dental nociception. *J Dent Res* 92:948–955.

Cooper, P. R., I. J. Chicca, M. J. Holder, and M. R. Milward. 2017. Inflammation and regeneration in the dentin-pulp complex: Net gain or net loss? *J Endod* 43:S87–S94.

Cope, A. L., B. C. O'Meara, and M. A. Gilchrist. 2020. Gene expression of functionally-related genes coevolves across fungal species: Detecting coevolution of gene expression using phylogenetic comparative methods. *BMC Genomics* 21:370.

Cox, C., K. Suzuki, H. Yamaguchi, J. Ruby, S. Suzuki, N. Akimoto, N. Maeda, and Y. Momoi. 2017. Sensory mechanisms in dentine: A literature review of light microscopy (LM), transmission microscopy (TEM), scanning microscopy (SEM) & electro physiological

(EP) tooth sensitivity: Is the ciliary organelle on the odontoblast the elusive primary nociceptor? *Dental, Oral and Craniofacial Research* 4:1–14.

Davit-Béal, T., F. Allizard, and J. Y. Sire. 2007. Enameloid/enamel transition through successive tooth replacements in Pleurodeles waltl (Lissamphibia, Caudata). *Cell Tissue Res* 328:167–183.

Davit-Beal, T., A. S. Tucker, and J. Y. Sire. 2009. Loss of teeth and enamel in tetrapods: Fossil record, genetic data and morphological adaptations. *J Anat* 214:477–501.

de Sousa-Romero, L., and A. M. Moreno-Fernández. 2016. Growth and transcription factors in tooth development. *Int J Oral Craniofac Sci* 2:15–29.

Dobreva, G., M. Chahrour, M. Dautzenberg, L. Chirivella, B. Kanzler, I. Farinas, G. Karsenty, and R. Grosschedl. 2006. SATB2 is a multifunctional determinant of craniofacial patterning and osteoblast differentiation. *Cell* 125:971–986.

Duverger, O., T. Ohara, P. W. Bible, A. Zah, and M. I. Morasso. 2017. DLX3-dependent regulation of ion transporters and carbonic anhydrases is crucial for enamel mineralization. *J Bone Miner Res* 32:641–653.

Eames, B. F., D. M. Medeiros, and I. Adameyko (Eds.). 2020. *Evolving Neural Crest Cells* (1st ed.). CRC Press.

Feng, J., A. Mantesso, C. De Bari, A. Nishiyama, and P. T. Sharpe. 2011. Dual origin of mesenchymal stem cells contributing to organ growth and repair. *Proc Natl Acad Sci* 108(16):6503–6508. https://doi.org/10.1073/pnas.1015449108

Ferenbach, D., and J. Hughes. 2008. Macrophages and dendritic cells: What is the difference? *Kidney Int* 74:5–7.

Filippo, A., D. C. Kalthoff, G. Billet, and H. G. Rodrigues. 2020. Evolutionary and functional implications of incisor enamel microstructure diversity in notoungulata (Placentalia, Mammalia). *J Mamm Evol* 27:211–236.

Franca, C. M., R. Riggers, J. L. Muschler, M. Widbiller, P. M. Lococo, A. Diogenes, and L. E. Bertassoni. 2019. 3D-imaging of whole neuronal and vascular networks of the human dental pulp via CLARITY and light sheet microscopy. *Sci Rep* 9:10860.

Franceschi, R. T., C. Ge, G. Xiao, H. Roca, and D. Jiang. 2009. Transcriptional regulation of osteoblasts. *Cells Tissues Organs* 189:144–152.

Fraser, G. J., R. Cerny, V. Soukup, M. Bronner-Fraser, and J. T. Streelman. 2010. The odontode explosion: The origin of tooth-like structures in vertebrates. *Bioessays* 32:808–817.

Fresia, R., P. Marangoni, T. Burstyn-Cohen, and A. Sharir. 2021. From bite to byte: Dental structures resolved at a single-cell resolution. *J Dent Res* 100:897–905.

Fried, K., C. Lillesaar, W. Sime, N. Kaukua, and M. Patarroyo. 2007. Target finding of pain nerve fibers: Neural growth mechanisms in the tooth pulp. *Physiol Behav* 92:40–45.

Fried, K., C. Nosrat, C. Lillesaar, and C. Hildebrand. 2000. Molecular signaling and pulpal nerve development. *Crit Rev Oral Biol Med* 11:318–332.

Fried, K., B. J. Sessle, and M. Devor. 2011. The paradox of pain from tooth pulp: Low-threshold 'algoneurons'? *Pain* 152:2685–2689.

Gaengler, P., and E. Metzler. 1992. The periodontal differentiation in the phylogeny of teeth-an overview. *J Periodontal Res* 27:214–225.

Gautron, J., L. Stapane, N. Le Roy, Y. Nys, A. B. Rodriguez-Navarro, and M. T. Hincke. 2021. Avian eggshell biomineralization: An update on its structure, mineralogy and protein tool kit. *BMC Mol Cell Biol* 22.

Gerli, R., I. Secciani, F. Sozio, A. Rossi, E. Weber, and G. Lorenzini. 2010. Absence of lymphatic vessels in human dental pulp: A morphological study. *Eur J Oral Sci* 118:110–117.

Gillis, J. A., E. C. Alsema, and K. E. Criswell. 2017. Trunk neural crest origin of dermal denticles in a cartilaginous fish. *Proc Natl Acad Sci U S A* 114:13200–13205.

Gollner, H., C. Dani, B. Phillips, S. Philipsen, and G. Suske. 2001. Impaired ossification in mice lacking the transcription factor Sp3. *Mech Dev* 106:77–83.

Hagemann-Jensen, M., C. Ziegenhain, P. Chen, D. Ramskold, G. J. Hendriks, A. J. M. Larsson, O. R. Faridani, and R. Sandberg. 2020. Single-cell RNA counting at allele and isoform resolution using Smart-seq3. *Nat Biotechnol* 38:708–714.

Hall, B. K., and J. A. Gillis. 2013. Incremental evolution of the neural crest, neural crest cells and neural crest-derived skeletal tissues. *J Anat* 222:19–31.

Haridy, Y., M. Osenberg, A. Hilger, I. Manke, D. Davesne, and F. Witzmann. 2021. Bone metabolism and evolutionary origin of osteocytes: Novel application of FIB-SEM tomography. *Sci Adv* 7.

Harris, B. D., M. Crow, S. Fischer, and J. Gillis. 2021. Single-cell co-expression analysis reveals that transcriptional modules are shared across cell types in the brain. *Cell Syst* 12:748–756 e3.

Haspel, G., A. Schwartz, A. Streets, D. E. Camacho, and D. Soares. 2012. By the teeth of their skin, cavefish find their way. *Curr Biol* 22:R629–R630.

Hassan, M. Q., A. Javed, M. I. Morasso, J. Karlin, M. Montecino, A. J. van Wijnen, G. S. Stein, J. L. Stein, and J. B. Lian. 2004. Dlx3 transcriptional regulation of osteoblast differentiation: Temporal recruitment of Msx2, Dlx3, and Dlx5 homeodomain proteins to chromatin of the osteocalcin gene. *Mol Cell Biol* 24:9248–9261.

Heatfield, B. M. 1970. Calcification in echinoderms—effects of temperature and diamox on incorporation of Ca-45 in-vitro by regenerating spines of strongylocentrotus-purpuratus. *Biol Bull* 139:151.

Hildebrand, C., K. Fried, F. Tuisku, and C. S. Johansson. 1995. Teeth and tooth nerves. *Prog Neurobiol* 45:165–222.

Hirasawa, T., and S. Kuratani. 2015. Evolution of the vertebrate skeleton: Morphology, embryology, and development. *Zoological Lett* 1:2.

Hodge, R. D., T. E. Bakken, J. A. Miller, K. A. Smith, E. R. Barkan, L. T. Graybuck, J. L. Close, B. Long, N. Johansen, O. Penn, Z. Yao, J. Eggermont, T. Hollt, B. P. Levi, S. I. Shehata, B. Aevermann, A. Beller, D. Bertagnolli, K. Brouner, T. Casper, C. Cobbs, R. Dalley, N. Dee, S. L. Ding, R. G. Ellenbogen, O. Fong, E. Garren, J. Goldy, R. P. Gwinn, D. Hirschstein, C. D. Keene, M. Keshk, A. L. Ko, K. Lathia, A. Mahfouz, Z. Maltzer, M. McGraw, T. N. Nguyen, J. Nyhus, J. G. Ojemann, A. Oldre, S. Parry, S. Reynolds, C. Rimorin, N. V. Shapovalova, S. Somasundaram, A. Szafer, E. R. Thomsen, M. Tieu, G. Quon, R. H. Scheuermann, R. Yuste, S. M. Sunkin, B. Lelieveldt, D. Feng, L. Ng, A. Bernard, M. Hawrylycz, J. W. Phillips, B. Tasic, H. Zeng, A. R. Jones, C. Koch, and E. S. Lein. 2019. Conserved cell types with divergent features in human versus mouse cortex. *Nature* 573:61–68.

Huang, W., S. Yang, J. Shao, and Y. P. Li. 2007. Signaling and transcriptional regulation in osteoblast commitment and differentiation. *Front Biosci* 12:3068–3092.

Ivashkin, E., and I. Adameyko. 2013. Progenitors of the protochordate ocellus as an evolutionary origin of the neural crest. *Evodevo* 4:12.

Iwasaki, Y., H. Otsuka, N. Yanagisawa, H. Hisamitsu, A. Manabe, N. Nonaka, and M. Nakamura. 2011. In situ proliferation and differentiation of macrophages in dental pulp. *Cell Tissue Res* 346:99–109.

Jenkins, K. M., and J. O. Shaw. 2020. Bite force data suggests relationship between acrodont tooth implantation and strong bite force. *PeerJ* 8:e9468.

Jernvall, J., T. Aberg, P. Kettunen, S. Keranen, and I. Thesleff. 1998. The life history of an embryonic signaling center: BMP-4 induces p21 and is associated with apoptosis in the mouse tooth enamel knot. *Development* 125:161–169.

Jernvall, J., and I. Thesleff. 2000. Reiterative signaling and patterning during mammalian tooth morphogenesis. *Mech Dev* 92:19–29.

Jernvall, J., and I. Thesleff. 2012. Tooth shape formation and tooth renewal: Evolving with the same signals. *Development* 139:3487–3497.
Jing, J., J. Feng, J. Li, H. Zhao, T. V. Ho, J. He, Y. Yuan, T. Guo, J. Du, M. Urata, P. Sharpe, and Y. Chai. 2021. Reciprocal interaction between mesenchymal stem cells and transit amplifying cells regulates tissue homeostasis. *Elife* 10.
Juuri, E., K. Saito, L. Ahtiainen, K. Seidel, M. Tummers, K. Hochedlinger, O. D. Klein, I. Thesleff, and F. Michon. 2012. Sox2+ stem cells contribute to all epithelial lineages of the tooth via Sfrp5+ progenitors. *Dev Cell* 23:317–328.
Kameneva, P., and I. Adameyko. 2019. Recent advances in our understanding of central and peripheral nervous system progenitors. *Curr Opin Cell Biol* 61:24–30.
Karatajùtè-Talimaa, V. 1995. The Mongolepidida: Scale structure and systematic position. *Geobios* 28:35–37.
Kaukua, N., M. K. Shahidi, C. Konstantinidou, V. Dyachuk, M. Kaucka, A. Furlan, Z. An, L. Wang, I. Hultman, L. Ahrlund-Richter, H. Blom, H. Brismar, N. A. Lopes, V. Pachnis, U. Suter, H. Clevers, I. Thesleff, P. Sharpe, P. Ernfors, K. Fried, and I. Adameyko. 2014. Glial origin of mesenchymal stem cells in a tooth model system. *Nature* 513:551–554.
Kawasaki, K. 2011. The SCPP gene family and the complexity of hard tissues in vertebrates. *Cells Tissues Organs* 194:108–112.
Kawasaki, K. 2018. The origin and early evolution of SCPP genes and tissue mineralization in vertebrates. Endo K., Kogure T., Nagasawa H., editors. *Biomineralization*. Springer, Singapore, pp. 157–164.
Kawasaki, K., J. N. Keating, M. Nakatomi, M. Welten, M. Mikami, I. Sasagawa, M. N. Puttick, P. C. J. Donoghue, and M. Ishiyama. 2021. Coevolution of enamel, ganoin, enameloid, and their matrix SCPP genes in osteichthyans. *iScience* 24.
Keating, J. N., C. L. Marquart, F. Marone, and P. C. J. Donoghue. 2018. The nature of aspidin and the evolutionary origin of bone. *Nat Ecol Evol* 2:1501–1506.
Kettunen, P., S. Loes, T. Furmanek, K. Fjeld, I. H. Kvinnsland, O. Behar, T. Yagi, H. Fujisawa, S. Vainio, M. Taniguchi, and K. Luukko. 2005. Coordination of trigeminal axon navigation and patterning with tooth organ formation: Epithelial-mesenchymal interactions, and epithelial Wnt4 and Tgfbeta1 regulate semaphorin 3a expression in the dental mesenchyme. *Development* 132:323–334.
Kharchenko, P. V. 2021. The triumphs and limitations of computational methods for scRNA-seq. *Nat Methods* 18:723–732.
Khatibi Shahidi, M., J. Krivanek, N. Kaukua, P. Ernfors, L. Hladik, V. Kostal, S. Masich, A. Hampl, V. Chubanov, T. Gudermann, R. A. Romanov, T. Harkany, I. Adameyko, and K. Fried. 2015. Three-dimensional imaging reveals new compartments and structural adaptations in odontoblasts. *J Dent Res* 94:945–954.
Khor, J. M., and C. A. Ettensohn. 2020. Transcription factors of the alx family: Evolutionarily conserved regulators of deuterostome skeletogenesis. *Front Genet* 11:569314.
Khor, J. M., J. Guerrero-Santoro, and C. A. Ettensohn. 2019. Genome-wide identification of binding sites and gene targets of Alx1, a pivotal regulator of echinoderm skeletogenesis. *Development* 146.
Kjaer, I. 1998. Prenatal traces of aberrant neurofacial growth. *Acta Odontol Scand* 56:326–330.
Kobayashi, T., and H. Kronenberg. 2005. Minireview: Transcriptional regulation in development of bone. *Endocrinology* 146:1012–1017.
Kollar, E. J., and G. R. Baird. 1969. The influence of the dental papilla on the development of tooth shape in embryonic mouse tooth germs. *J Embryol Exp Morphol* 21:131–148.
Komori, T. 2006. Regulation of osteoblast differentiation by transcription factors. *J Cell Biochem* 99:1233–1239.

Krivanek, J., I. Adameyko, and K. Fried. 2017. Heterogeneity and developmental connections between cell types inhabiting teeth. *Front Physiol* 8:376.

Krivanek, J., R. A. Soldatov, M. E. Kastriti, T. Chontorotzea, A. N. Herdina, J. Petersen, B. Szarowska, M. Landova, V. K. Matejova, L. I. Holla, U. Kuchler, I. V. Zdrilic, A. Vijaykumar, A. Balic, P. Marangoni, O. D. Klein, V. C. M. Neves, V. Yianni, P. T. Sharpe, T. Harkany, B. D. Metscher, M. Bajenoff, M. Mina, K. Fried, P. V. Kharchenko, and I. Adameyko. 2020. Dental cell type atlas reveals stem and differentiated cell types in mouse and human teeth. *Nat Commun* 11:4816.

Krystel-Whittemore, M., K. N. Dileepan, and J. G. Wood. 2015. Mast cell: A multi-functional master cell. *Front Immunol* 6:620.

Kuzynski, M., M. Goss, M. Bottini, M. C. Yadav, C. Mobley, T. Winters, A. Poliard, O. Kellermann, B. Lee, J. L. Millan, and D. Napierala. 2014. Dual role of the Trps1 transcription factor in dentin mineralization. *J Biol Chem* 289:27481–27493.

La Manno, G., R. Soldatov, A. Zeisel, E. Braun, H. Hochgerner, V. Petukhov, K. Lidschreiber, M. E. Kastriti, P. Lönnerberg, A. Furlan, J. Fan, L. E. Borm, Z. Liu, D. van Bruggen, J. Guo, X. He, R. Barker, E. Sundström, G. Castelo-Branco, . . . P. V. Kharchenko. 2018. RNA velocity of single cells. *Nature* 560(7719):494–498. https://doi.org/10.1038/s41586-018-0414-6

Landova Sulcova, M., O. Zahradnicek, J. Dumkova, H. Dosedelova, J. Krivanek, M. Hampl, M. Kavkova, T. Zikmund, M. Gregorovicova, D. Sedmera, J. Kaiser, A. S. Tucker, and M. Buchtova. 2020. Developmental mechanisms driving complex tooth shape in reptiles. *Dev Dyn* 249:441–464.

Larionova, D., H. Lesot, and A. Huysseune. 2021. Miniaturization: How many cells are needed to build a tooth? *Dev Dyn* 250(7):1021–1035. https://doi.org/10.1002/dvdy.300

Lavicky, J., M. Kolouskova, D. Prochazka, V. Rakultsev, M. Gonzalez-Lopez, K. Steklikova, M. Bartos, A. Vijaykumar, J. Kaiser, P. Porizka, M. Hovorakova, M. Mina, and J. Krivanek. 2021. The development of dentin microstructure is controlled by the type of adjacent epithelium. *J Bone Miner Res* 37(2):323–339.

LeBlanc, A. R., and R. R. Reisz. 2013. Periodontal ligament, cementum, and alveolar bone in the oldest herbivorous tetrapods, and their evolutionary significance. *PLoS ONE* 8:e74697.

Ledesma-Martinez, E., V. M. Mendoza-Nunez, and E. Santiago-Osorio. 2016. Mesenchymal stem cells derived from dental pulp: A review. *Stem Cells Int* 2016:4709572.

Lee, J. H., D. S. Lee, H. W. Choung, W. J. Shon, B. M. Seo, E. H. Lee, J. Y. Cho, and J. C. Park. 2011. Odontogenic differentiation of human dental pulp stem cells induced by preameloblast-derived factors. *Biomaterials* 32:9696–9706.

Lee, K., B. M. Lee, C. K. Park, Y. H. Kim, and G. Chung. 2019. Ion channels involved in tooth pain. *Int J Mol Sci* 20.

Lehnert, B. P., C. Santiago, E. L. Huey, A. J. Emanuel, S. Renauld, N. Africawala, I. Alkislar, Y. Zheng, L. Bai, C. Koutsioumpa, J. T. Hong, A. R. Magee, C. D. Harvey, and D. D. Ginty. 2021. Mechanoreceptor synapses in the brainstem shape the central representation of touch. *Cell* 184:5608–5621 e18.

Lei, H., H. Liu, Y. Ding, and L. Ge. 2014. Immunohistochemical localization of Pax6 in the developing tooth germ of mice. *J Mol Histol* 45:373–379.

Le Roy, N., L. Stapane, J. Gautron, and M. T. Hincke. 2021. Evolution of the avian eggshell biomineralization protein toolkit—New insights from multi-omics. *Front Genet* 12.

Lezot, F., B. Thomas, S. R. Greene, D. Hotton, Z. A. Yuan, B. Castaneda, A. Bolanos, M. Depew, P. Sharpe, C. W. Gibson, and A. Berdal. 2008. Physiological implications of DLX homeoproteins in enamel formation. *J Cell Physiol* 216:688–697.

Li, J., Q. Lin, Y. Lin, R. Lai, and W. Zhang. 2021. Effects of DLX3 on the osteogenic differentiation of induced pluripotent stem cellderived mesenchymal stem cells. *Mol Med Rep* 23.

Li, J., C. Parada, and Y. Chai. 2017. Cellular and molecular mechanisms of tooth root development. *Development* 144:374–384.

Li, W., Z. Zhang, and Z. M. Wang. 2020. Differential immune cell infiltrations between healthy periodontal and chronic periodontitis tissues. *BMC Oral Health* 20:293.

Liao, J., X. Lu, X. Shao, L. Zhu, and X. Fan. 2021. Uncovering an organ's molecular architecture at single-cell resolution by spatially resolved transcriptomics. *Trends Biotechnol* 39:43–58.

Lim, H. M., M. H. Nam, Y. M. Kim, and Y. K. Seo. 2021. Increasing odontoblast-like differentiation from dental pulp stem cells through increase of beta-catenin/p-GSK-3beta expression by low-frequency electromagnetic field. *Biomedicines* 9.

Liu, H., X. Yan, M. Pandya, X. Luan, and T. G. Diekwisch. 2016. Daughters of the enamel organ: Development, fate, and function of the stratum intermedium, stellate reticulum, and outer enamel epithelium. *Stem Cells Dev* 25:1580–1590.

Lohrberg, M., and J. Wilting. 2016. The lymphatic vascular system of the mouse head. *Cell Tissue Res* 366:667–677.

Longo, S. K., M. G. Guo, A. L. Ji, and P. A. Khavari. 2021. Integrating single-cell and spatial transcriptomics to elucidate intercellular tissue dynamics. *Nat Rev Genet* 22:627–644.

Lundquist, P., T. Lundgren, A. Gritli-Linde, and A. Linde. 2000. Na^+/Ca^{2+} exchanger isoforms of rat odontoblasts and osteoblasts. *Calcif Tissue Int* 67:60–67.

Luukko, K., and P. Kettunen. 2014. Coordination of tooth morphogenesis and neuronal development through tissue interactions: Lessons from mouse models. *Exp Cell Res* 325:72–77.

Luukko, K., and P. Kettunen. 2016. Integration of tooth morphogenesis and innervation by local tissue interactions, signaling networks, and semaphorin 3A. *Cell Adh Migr* 10:618–626.

Makanae, A., Y. Tajika, K. Nishimura, N. Saito, J. I. Tanaka, and A. Satoh. 2020. Neural regulation in tooth regeneration of Ambystoma mexicanum. *Sci Rep* 10:9323.

Marquez-Zacarias, P., R. M. Pineau, M. Gomez, A. Veliz-Cuba, D. Murrugarra, W. C. Ratcliff, and K. J. Niklas. 2021. Evolution of cellular differentiation: From hypotheses to models. *Trends Ecol Evol* 36:49–60.

Moxham, B. J., and B. K. Berkovitz. 1974. The effects of root transection on the unimpeded eruption rate of the rabbit mandibular incisor. *Arch Oral Biol* 19:903–909.

Murdock, D. J. E. 2020. The 'biomineralization toolkit' and the origin of animal skeletons. *Biol Rev* 95:1372–1392.

Nanci, A. 2016. *Ten Cate's Oral Histology: Development, Structure, and Function* (9th ed.). Elsevier.

Neves, V. C. M., V. Yianni, and P. T. Sharpe. 2020. Macrophage modulation of dental pulp stem cell activity during tertiary dentinogenesis. *Sci Rep* 10:20216.

Nosrat, I. V., J. Widenfalk, L. Olson, and C. A. Nosrat. 2001. Dental pulp cells produce neurotrophic factors, interact with trigeminal neurons in vitro, and rescue motoneurons after spinal cord injury. *Dev Biol* 238:120–132.

Nuti, N., C. Corallo, B. M. Chan, M. Ferrari, and B. Gerami-Naini. 2016. Multipotent differentiation of human dental pulp stem cells: A literature review. *Stem Cell Rev Rep* 12:511–523.

Nweeia, M. T., F. C. Eichmiller, P. V. Hauschka, G. A. Donahue, J. R. Orr, S. H. Ferguson, C. A. Watt, J. G. Mead, C. W. Potter, R. Dietz, A. A. Giuseppetti, S. R. Black, A. J. Trachtenberg, and W. P. Kuo. 2014. Sensory ability in the narwhal tooth organ system. *Anat Rec (Hoboken)* 297:599–617.

Ortiz, A., S. E. Bailey, G. T. Schwartz, J. J. Hublin, and M. M. Skinner. 2018. Evo-devo models of tooth development and the origin of hominoid molar diversity. *Sci Adv* 4.

Ørvig, T. 1967. Phylogeny of tooth tissues: Evolution of some calcified tissues in early vertebrates. Miles A.E.W., editor. *Structural and Chemical Organization of Teeth*. Academic Press, pp. 45–110.

Ovens, K., B. F. Eames, and I. McQuillan. 2021. Comparative analyses of gene co-expression networks: Implementations and applications in the study of evolution. *Front Genet* 12:695399.

Paes, S. M., Y. M. Pupo, B. C. Cavenago, T. Fonseca-Silva, and C. C. O. Santos. 2021. Cryopreservation of mesenchymal stem cells derived from dental pulp: A systematic review. *Restor Dent Endod* 46:e26.

Pagella, P., L. de Vargas Roditi, B. Stadlinger, A. E. Moor, and T. A. Mitsiadis. 2021. A single-cell atlas of human teeth. *iScience* 24:102405.

Qu, Q., T. Haitina, M. Zhu, and P. E. Ahlberg. 2015. New genomic and fossil data illuminate the origin of enamel. *Nature* 526:108–111.

Qu, Q., S. Sanchez, H. Blom, P. Tafforeau, and P. E. Ahlberg. 2013. Scales and tooth whorls of ancient fishes challenge distinction between external and oral 'teeth'. *PLoS ONE* 8:e71890.

Que, B. G., and G. E. Wise. 1997. Colony-stimulating factor-1 and monocyte chemotactic protein-1 chemotaxis for monocytes in the rat dental follicle. *Arch Oral Biol* 42:855–860.

Renvoisé, E., and F. Michon. 2014. An Evo-Devo perspective on ever-growing teeth in mammals and dental stem cell maintenance. *Front Physiol* 5:324.

Rincon, J. C., W. G. Young, and P. M. Bartold. 2006. The epithelial cell rests of Malassez—A role in periodontal regeneration? *J Periodontal Res* 41:245–252.

Ruben, J. A., and A. F. Bennett. 1981. Intense exercise, bone structure and blood calcium levels in vertebrates. *Nature* 291:411–413.

Sahara, N. 2014. Development of coronal cementum in hypsodont horse cheek teeth. *Anat Rec* 297(4):716–730. https://doi.org/10.1002/ar.22880

Salomon, R., D. Kaczorowski, F. Valdes-Mora, R. E. Nordon, A. Neild, N. Farbehi, N. Bartonicek, and D. Gallego-Ortega. 2019. Droplet-based single cell RNAseq tools: A practical guide. *Lab Chip* 19:1706–1727.

Sanz-Navarro, M., I. Delgado, M. Torres, T. Mustonen, F. Michon, and D. P. Rice. 2019. Dental epithelial stem cells express the developmental regulator Meis1. *Front Physiol* 10:249.

Sato, M., K. Ogura, M. Kimura, K. Nishi, M. Ando, M. Tazaki, and Y. Shibukawa. 2018. Activation of mechanosensitive transient receptor potential/piezo channels in odontoblasts generates action potentials in cocultured isolectin B4-negative medium-sized trigeminal ganglion neurons. *J Endod* 44:984–991 e2.

Seidel, K., P. Marangoni, C. Tang, B. Houshmand, W. Du, R. L. Maas, S. Murray, M. C. Oldham, and O. D. Klein. 2017. Resolving stem and progenitor cells in the adult mouse incisor through gene co-expression analysis. *ELife* 6:e24712. https://doi.org/10.7554/eLife.24712

Shadad, O., R. Chaulagain, K. Luukko, and P. Kettunen. 2019. Establishment of tooth blood supply and innervation is developmentally regulated and takes place through differential patterning processes. *J Anat* 234:465–479.

Sharir, A., P. Marangoni, R. Zilionis, M. Wan, T. Wald, J. K. Hu, K. Kawaguchi, D. Castillo-Azofeifa, L. Epstein, K. Harrington, P. Pagella, T. Mitsiadis, C. W. Siebel, A. M. Klein, and O. D. Klein. 2019. A large pool of actively cycling progenitors orchestrates self-renewal and injury repair of an ectodermal appendage. *Nat Cell Biol* 21:1102–1112.

Shekhar, K., and V. Menon. 2019. Identification of cell types from single-cell transcriptomic data. *Methods Mol Biol* 1935:45–77.

Shibukawa, Y., and T. Suzuki. 2003. Ca2+ signaling mediated by IP3-dependent Ca2+ releasing and store-operated Ca2+ channels in rat odontoblasts. *J Bone Miner Res* 18:30–38.

Sire, J. Y., P. C. J. Donoghue, and M. K. Vickaryous. 2009. Origin and evolution of the integumentary skeleton in non-tetrapod vertebrates. *J Anat* 214:409–440.
Sire, J. Y., and K. Kawasaki. 2012. Origin and evolution of bone and dentin and of their phosphorylated, acid-rich matrix proteins. *Phosphorylated Extracelular Matrix Proteins of Bone and Dentin* (vol. 2). Bentham Science Publisher, pp. 3–60.
Soldatov, R., M. Kaucka, M. E. Kastriti, J. Petersen, T. Chontorotzea, L. Englmaier, N. Akkuratova, Y. Yang, M. Haring, V. Dyachuk, C. Bock, M. Farlik, M. L. Piacentino, F. Boismoreau, M. M. Hilscher, C. Yokota, X. Qian, M. Nilsson, M. E. Bronner, L. Croci, W. Y. Hsiao, D. A. Guertin, J. F. Brunet, G. G. Consalez, P. Ernfors, K. Fried, P. V. Kharchenko, and I. Adameyko. 2019. Spatiotemporal structure of cell fate decisions in murine neural crest. *Science* 364.
Sorrentino, A., E. Malucelli, F. Rossi, C. Cappadone, G. Farruggia, C. Moscheni, A. J. Perez-Berna, J. J. Conesa, C. Colletti, N. Roveri, E. Pereiro, and S. Iotti. 2021. Calcite as a precursor of hydroxyapatite in the early biomineralization of differentiating human bone-marrow mesenchymal stem cells. *Int J Mol Sci* 22.
Stock, D. W. 2001. The genetic basis of modularity in the development and evolution of the vertebrate dentition. *Philos Trans R Soc Lond B Biol Sci* 356:1633–1653.
Sun, Z., W. Yu, M. Sanz Navarro, M. Sweat, S. Eliason, T. Sharp, H. Liu, K. Seidel, L. Zhang, M. Moreno, T. Lynch, N. E. Holton, L. Rogers, T. Neff, M. J. Goodheart, F. Michon, O. D. Klein, Y. Chai, A. Dupuy, J. F. Engelhardt, Z. Chen, and B. A. Amendt. 2016. Sox2 and Lef-1 interact with Pitx2 to regulate incisor development and stem cell renewal. *Development* 143:4115–4126.
Tabata, T., and K. Karita. 1991. Response properties of periodontal mechanosensitive neurons in the trigeminal spinal tract nucleus of the cat. *Somatosens Mot Res* 8:261–269.
Tagliabracci, V. S., J. L. Engel, J. Wen, S. E. Wiley, C. A. Worby, L. N. Kinch, J. Xiao, N. V. Grishin, and J. E. Dixon. 2012. Secreted kinase phosphorylates extracellular proteins that regulate biomineralization. *Science* 336:1150–1153.
Takahashi, S., K. Ambe, H. Kon, S. Takada, T. Ohno, and H. Watanabe. 2012. Immunohistochemical investigation of lymphatic vessel formation control in mouse tooth development: Lymphatic vessel-forming factors and receptors in tooth development in mice. *Tissue Cell* 44:170–181.
Tanay, A., and A. Sebe-Pedros. 2021. Evolutionary cell type mapping with single-cell genomics. *Trends Genet* 37:919–932.
Tao, H., H. Lin, Z. Sun, F. Pei, J. Zhang, S. Chen, H. Liu, and Z. Chen. 2019. Klf4 promotes dentinogenesis and odontoblastic differentiation via modulation of TGF-beta signaling pathway and interaction with histone acetylation. *J Bone Miner Res* 34:1502–1516.
Thesleff, I. 2003. Epithelial-mesenchymal signalling regulating tooth morphogenesis. *J Cell Sci* 116:1647–1648.
Thesleff, I. 2018. From understanding tooth development to bioengineering of teeth. *Eur J Oral Sci* 126:67–71.
Thesleff, I., S. Keranen, and J. Jernvall. 2001. Enamel knots as signaling centers linking tooth morphogenesis and odontoblast differentiation. *Adv Dent Res* 15:14–18.
Thesleff, I., and M. Tummers. 2008. Tooth organogenesis and regeneration. *StemBook*. Harvard Stem Cell Institute. http://www.ncbi.nlm.nih.gov/books/NBK27071/
Tucker, A. S., and G. J. Fraser. 2014. Evolution and developmental diversity of tooth regeneration. *Semin Cell Dev Biol* 25–26:71–80.
Tuisku, F., and C. Hildebrand. 1994. Evidence for a neural influence on tooth germ generation in a polyphyodont species. *Dev Biol* 165:1–9.
Usoskin, D., A. Furlan, S. Islam, H. Abdo, P. Lonnerberg, D. Lou, J. Hjerling-Leffler, J. Haeggstrom, O. Kharchenko, P. V. Kharchenko, S. Linnarsson, and P. Ernfors. 2015.

Unbiased classification of sensory neuron types by large-scale single-cell RNA sequencing. *Nat Neurosci* 18:145–153.

Vaskaninova, V., D. Chen, P. Tafforeau, Z. Johanson, B. Ekrt, H. Blom, and P. E. Ahlberg. 2020. Marginal dentition and multiple dermal jawbones as the ancestral condition of jawed vertebrates. *Science* 369:211–216.

Venugopalan, S. R., X. Li, M. A. Amen, S. Florez, D. Gutierrez, H. J. Cao, J. B. Wang, and B. A. Amendt. 2011. Hierarchical interactions of homeodomain and forkhead transcription factors in regulating odontogenic gene expression. *J Biol Chem* 286:21372–21383.

Vidovic, I., A. Banerjee, R. Fatahi, B. G. Matthews, N. A. Dyment, I. Kalajzic, and M. Mina. 2017. ASMA-expressing perivascular cells represent dental pulp progenitors in vivo. *J Dental Res* 96(3):323–330. https://doi.org/10.1177/0022034516678208

Virág, A. 2012. Histogenesis of the unique morphology of proboscidean ivory. *J Morphol* 273(12):1406–1423. https://doi.org/10.1002/jmor.20069

Wagner, D. E., C. Weinreb, Z. M. Collins, J. A. Briggs, S. G. Megason, and A. M. Klein. 2018. Single-cell mapping of gene expression landscapes and lineage in the zebrafish embryo. *Science* 360:981–987.

Walsh, L. J. 2003. Mast cells and oral inflammation. *Crit Rev Oral Biol Med* 14:188–198.

Wang, J., H. Sun, M. Jiang, J. Li, P. Zhang, H. Chen, Y. Mei, L. Fei, S. Lai, X. Han, X. Song, S. Xu, M. Chen, H. Ouyang, D. Zhang, G. C. Yuan, and G. Guo. 2021. Tracing cell-type evolution by cross-species comparison of cell atlases. *Cell Rep* 34:108803.

Wang, J., S. Xia, B. Arand, H. Zhu, R. Machiraju, K. Huang, H. Ji, and J. Qian. 2016. Single-cell co-expression analysis reveals distinct functional modules, co-regulation mechanisms and clinical outcomes. *PLoS Comput Biol* 12:e1004892.

Wegner, K. A., M. T. Cadena, R. Trevena, A. E. Turco, A. Gottschalk, R. B. Halberg, J. Guo, J. A. McMahon, A. P. McMahon, and C. M. Vezina. 2017. An immunohistochemical identification key for cell types in adult mouse prostatic and urethral tissue sections. *PLoS ONE* 12:e0188413.

Westergaard, B., and M. W. Ferguson. 1987. Development of the dentition in Alligator mississippiensis. Later development in the lower jaws of embryos, hatchlings and young juveniles. *J Zool* 212:191–222.

Westergaard, B., and M. W. Ferguson. 1990. Development of the dentition in Alligator mississippiensis: Upper jaw dental and craniofacial development in embryos, hatchlings, and young juveniles, with a comparison to lower jaw development. *Am J Anat* 187:393–421.

Whitlock, J. A., and J. M. Richman. 2013. Biology of tooth replacement in amniotes. *Int J Oral Sci* 5:66–70.

Whitney, M. R., K. D. Angielczyk, B. R. Peecook, and C. A. Sidor. 2021. The evolution of the synapsid tusk: Insights from dicynodont therapsid tusk histology. *Proc Royal Soc B Biol Sci* 288(1961):20211670. https://doi.org/10.1098/rspb.2021.1670

Wilmers, J., and S. Bargmann. 2020. Nature's design solutions in dental enamel: Uniting high strength and extreme damage resistance. *Acta Biomater* 107:1–24.

Xiong, J., S. Gronthos, and P. M. Bartold. 2013. Role of the epithelial cell rests of Malassez in the development, maintenance and regeneration of periodontal ligament tissues. *Periodontol* 2000 63:217–233.

Yan, F., D. R. Powell, D. J. Curtis, and N. C. Wong. 2020. From reads to insight: A hitchhiker's guide to ATAC-seq data analysis. *Genome Biol* 21:22.

Yao, S., F. Pan, V. Prpic, and G. E. Wise. 2008. Differentiation of stem cells in the dental follicle. *J Dent Res* 87:767–771.

Yianni, V., and P. T. Sharpe. 2019. Perivascular-derived mesenchymal stem cells. *J Dent Res* 98:1066–1072.

Yianni, V., and P. T. Sharpe. 2020. Transcriptomic profiling of dental pulp pericytes: An RNAseq approach. *Front Dent Med* 1:6. https://doi.org/10.3389/fdmed.2020.00006

Yoshizaki, K., S. Fukumoto, D. D. Bikle, and Y. Oda. 2020. Transcriptional regulation of dental epithelial cell fate. *Int J Mol Sci* 21.

Yu, S., J. Guo, Z. Sun, C. Lin, H. Tao, Q. Zhang, Y. Cui, H. Zuo, Y. Lin, S. Chen, H. Liu, and Z. Chen. 2021. BMP2-dependent gene regulatory network analysis reveals Klf4 as a novel transcription factor of osteoblast differentiation. *Cell Death Dis* 12:197.

Zeichner-David, M., T. Diekwisch, A. Fincham, E. Lau, M. MacDougall, J. Moradian-Oldak, J. Simmer, M. Snead, and H. C. Slavkin. 1995. Control of ameloblast differentiation. *Int J Dev Biol* 39:69–92.

Zhang, J. Y., and L. C. Li. 2015. Genetic expression and functional characterization of the RUNX2 gene in human adult bone marrow mesenchymal stem cells. *Genet Mol Res* 14:18210–18217.

Zhao, H., J. Feng, K. Seidel, S. Shi, O. Klein, P. Sharpe, and Y. Chai. 2014. Secretion of shh by a neurovascular bundle niche supports mesenchymal stem cell homeostasis in the adult mouse incisor. *Cell Stem Cell* 14:160–173.

Zhao, J., L. Faure, I. Adameyko, and P. T. Sharpe. 2021. Stem cell contributions to cementoblast differentiation in healthy periodontal ligament and periodontitis. *Stem Cells* 39:92–102.

Zhao, N., B. L. Foster, and L. F. Bonewald. 2016. The cementocyte-an osteocyte relative? *J Dent Res* 95:734–741.

Zrimec, J., C. S. Borlin, F. Buric, A. S. Muhammad, R. Chen, V. Siewers, V. Verendel, J. Nielsen, M. Topel, and A. Zelezniak. 2020. Deep learning suggests that gene expression is encoded in all parts of a co-evolving interacting gene regulatory structure. *Nat Commun* 11:6141.

2 The Conquest of the Oropharynx by Odontogenic Epithelia

Ann Huysseune, Robert Cerny and P. Eckhard Witten

2.1 INTRODUCTION: WHAT ARE TEETH AND WHERE ARE THEY FORMED?

Digging into the evolutionary origin of teeth has been at the heart of Evo-Devo research for many decades. Both teeth and skin denticles qualify as 'odontodes'. In its original concept, the term 'odontode' was used to refer to 'dermal teeth'— hard tissue units in the corium, corresponding very closely to teeth, composed principally of dentine or dentinous tissue, with or without an outer covering of enameloid, and forming ontogenetically in a single undivided mesenchymal dental papilla (Ørvig, 1967, 1977). Subsequently, the definition of Ørvig (1977) was slightly modified by Reif (1982), who defined odontodes as isolated structures that consist of a dentine cone, are covered by a hypermineralized layer, and have a bony base which serves as an attachment. Reif (1982) also expanded the definition of odontodes to encompass both 'dermal denticles' and 'teeth'. Only those organ systems that are formed by a dental lamina (for an explanation, see Section 2.4) were considered by Reif as dentitions, and their components as teeth. Reif's (1982) definition of an odontode has gained wide acceptance, but the dependence on a dental lamina for teeth is now no longer considered generally applicable. For example the dental lamina has been seriously challenged as a patterning device for the dentition in the earliest jawed vertebrates (Qu et al., 2013). Moreover, in many extant species, teeth are not developing from a dental lamina (reviewed in Huysseune et al., 2009, 2010). In an attempt to streamline the terminology, Rücklin et al. (2011) and Donoghue and Rücklin (2016) distinguish external from internal odontodes. Internal odontodes—the subject of this review—are odontodes located within the oropharyngeal cavity. Two categories of internal odontodes can be distinguished—oral and pharyngeal teeth, and oral and pharyngeal denticles—although differences between teeth and denticles are somewhat blurred as a result of their conserved structure and development. Teeth are usually large units in low numbers, adapted for food uptake, transport, and manipulation. In contrast, denticles are usually small units, in high numbers, and functionally resembling external odontodes, e.g. in providing the oropharyngeal lining with a rough surface (but see Huysseune et al. (2022), for an elaborate discussion of terminology).

DOI: 10.1201/9781003439653-2

Teeth are restricted in the vertebrate body to locations inside the mouth and pharynx (collectively called the oropharynx), and, in some instances, the esophagus. Oral teeth are those located on the mandibular arch and palatal bones, and pharyngeal teeth on any of the post-mandibular pharyngeal arches. Pharyngeal teeth have also been described in the esophagus of certain teleost species (Gilchrist, 1922; Isokawa et al., 1965). Tooth distribution in the oropharynx varies substantially across vertebrates. When a dentition is present, oral teeth are the most stable in their occurrence. Variation in which bones are toothed or non-toothed is especially prominent in the pharyngeal region (reviewed in Huysseune et al., 2022; see also Peyer, 1968; Berkovitz and Shellis, 2017). In what follows is a brief—and inevitably incomplete—summary of pharyngeal tooth/denticle distribution. For a more extensive survey, we refer to Huysseune et al. (2022).

In chondrichthyans, the pharynx can be completely covered with denticles. In extant sharks, these are frequently very small and often referred to as (mucous membrane) denticles (Peyer, 1968). They vary in their distribution from being tightly crowded to widely scattered (Nelson, 1970) or are missing altogether, for example in the now widely used model species, the catshark (*Scyliorhinus canicula*). Species close to the osteichthyan key divergence into ray-finned (actinopterygians) and lobe-finned fishes (sarcopterygians) include bichirs (Polypteridae), sturgeons (Acipenseridae), and gars (Lepisosteidae). All of these possess pharyngeal teeth, distributed over different pharyngeal arches (Clemen et al., 1998, Hilton et al., 2011, Comabella et al., 2012; note that the dentition in sturgeon is lost during ontogeny, see Pospisilova et al., 2022). In teleosts, representing the majority of actinopterygians and the most speciose group of vertebrates, pharyngeal tooth distribution is extremely variable (Nelson, 1969, 1970; Vandewalle et al., 1994). There is nevertheless a general evolutionary trend towards tooth loss on anterior pharyngeal arches (i.e. beyond the mandibular arch) and consolidation of a dentition in more posterior arches in some lineages associated to a loss of oral teeth. Like their extinct relatives, extant lobe-finned fishes, such as *Latimeria chalumnae*, have an extensive dentition including denticulated plates in the branchial region (Huysseune et al., 2022). In the transition from lobe-finned fishes to tetrapods, palatal teeth appear to have been conserved better than other oropharyngeal tooth plates (Smith and Coates, 2000). Many temnospondyl amphibians possessed branchial denticles, considered to be probably homologous to palatal and pharyngeal teeth of actinopterygians (Schoch, 2002). In extant amphibians, denticles disappeared from the pharynx and never appeared again in the ensuing lineages, the amniotes (reptiles and mammals), where teeth are restricted to the oral surface (Matsumoto and Evans, 2017). The extent of the tooth-covered area of the mouth is greatest among the most ancient groups of reptiles (Peyer, 1968). Extant reptilians have a single marginal row of teeth in the lower jaw, but a variable pattern in the upper jaw: Either a single row of marginal teeth or parallel rows of marginal and palatal teeth (Peyer, 1968; Mahler and Kearney, 2006). Marginal teeth attach to the maxillary bone, and palatal teeth attach to the palatine and/or pterygoid bone (Richman and Handrigan, 2011). In the lineage leading to the (edentulous) turtles, a palatal dentition appears to have survived much longer than the marginal dentition (Lainoff et al., 2015). Palatine teeth have been lost in mammals, leaving only a single marginal tooth row on the upper and lower jaws.

Like other odontodes, teeth are the result of a cascade of sequential and reciprocal interactions between the tooth-forming (odontogenic) epithelium and underlying mesenchyme. The source of the odontogenic mesenchyme, the neural crest, has now been well established in many vertebrates, as is discussed in Chapter 3 of this volume. Whether the odontogenic epithelium is of ectodermal or endodermal origin has been a matter of a heavy debate, connected to views that either saw teeth arising, in evolution, deep in the pharynx from endodermal epithelium ('inside-out') or from skin odontodes, and thus from ectoderm ('outside-in'). That internal odontodes in extant vertebrates develop in areas where ectoderm and endoderm meet, most notably the mouth, has undoubtedly fueled the view that internal odontodes/teeth originated from (ectoderm-derived) dermal denticles. The 'outside-in' view has now gained wide acceptance, as elaborately discussed in Huysseune et al. (2022). In particular, it has been argued more recently that, during evolution, teeth arose from external odontodes by the transfer of odontogenic competence from external to internal epithelia (Huysseune et al., 2009, 2010; Rücklin et al., 2011, 2012; Murdock et al., 2013; Donoghue and Rücklin, 2016). The question, however, remains unanswered as to how this transfer may have occurred.

The present chapter aims at reviewing new findings that may provide the start of an answer to this question. We discuss the origin, structure, and fate of the tooth-forming (odontogenic) epithelia, in particular, to uncover if and how external epithelia could have transferred odontogenic competence to internal epithelia. We do not discuss the ensuing morphogenesis of the enamel organ (i.e. the epithelial part of the tooth anlage), which has been introduced in Chapter 1, but focus on where and when teeth are initiated. External and internal odontodes, including pharyngeal teeth/denticles, are homologous structures (e.g. Sire and Huysseune, 2003; Huysseune et al., 2009, 2010; Blais et al., 2011; Debiais-Thibaud et al., 2011; Rücklin et al., 2012; Murdock et al., 2013; Donoghue and Rücklin, 2016; Haridy et al., 2019; Chen et al., 2020; Berio and Debiais-Thibaud, 2021), sharing a common evolutionary origin and thus united by more than sets of co-expressed genes (Fraser et al., 2010). We must therefore take into account not only oral but also pharyngeal teeth/denticles when attempting to explain the origin of odontogenic epithelia. Moreover, the review necessarily encompasses extant species only, as we have no information on epithelia in extinct species. Finally, the focus is on the epithelia giving rise to the primary dentition, even if abortive. The presence of a tooth anlage, as small or rudimentary as it may be, is indeed a sign that odontogenic competence is present. Once a tooth germ has been formed, signals may be propagated to generate more teeth of the same generation or to secure replacement (Sadier et al., 2020).

2.2 THE ODONTOGENIC EPITHELIA: NEW PLAYERS

New insights on the role of external epithelia in the formation of internal odontodes (teeth or denticles) stem from a number of recent observations. First, while a physical link between external epithelium and oral teeth, via the mouth, is obvious, it has become apparent that pharyngeal teeth, or internal odontodes in general, always arise in the vicinity of channels where the oropharyngeal (endodermal) epithelium meets the external (ectodermal) epithelium, be it at the pouches and ensuing gill slits (Jollie, 1968; Nelson, 1969; Oralová et al., 2020), the spiracular canal (Bjerring, 1998; Brazeau and Ahlberg, 2006), or inhalant ducts (Van der Brugghen and Janvier, 1993) (reviewed in Huysseune et al., 2022). Second, in particular taxa, dental fields have been demonstrated to be initiated close to the ectodermal/endodermal boundary (Soukup et al., 2021). Third, recent studies have discovered that the external layer of the epidermis, the periderm, partially invades the communication channels between external (skin) and oropharyngeal epithelium, at least in the actinopterygian species for which such studies were undertaken, the zebrafish (Rosa et al., 2019; Oralová et al., 2020). Obviously, periderm appears to be a new player in the story, but as explained later in the chapter, we can eventually consider zebrafish periderm as an early ectodermal layer. Because of its potential role in transfer of competence from external to internal epithelia, it is necessary to briefly review its origin (both developmentally and evolutionary) and fate. Data are, however, scarce, and limited to only a few vertebrate species.

In representatives of extant lineages that are close to the osteichthyan key divergence into ray-finned and lobe-finned fishes, such as bichir and sturgeon, the epidermis at the end of gastrulation is stratified into an upper and a basal layer. Both are considered as ectodermal layers in sturgeon (Dettlaff, 1993; Dettlaff et al., 1993), although the basal layer in bichir has been called 'subepidermis layer' (Diedhiou and Bartsch, 2009). In teleosts, the epidermis at the end of gastrulation is also bilayered, but homology of the superficial layer, or periderm, with the superficial ectodermal layer in, e.g. sturgeon, has been questioned. In zebrafish, the precursor of the periderm, the enveloping layer or EVL, has long been thought to experience an early lineage restriction (Kimmel et al., 1990). This has led to the view that the teleost periderm is an extra-embryonic layer. The belief that it is lost shortly after the embryonic period is furthermore fostered by general textbook statements that the periderm is a transient layer that is soon shed (Gilbert and Barresi, 2016). This view needs to be adjusted in the light of recent findings. Indeed, periderm cells, characterized by the expression of specific keratins such as *krt4*, persist in the zebrafish epidermis much longer than previously thought and contribute to cells of the body, be it the epidermis or breeding tubercles (Fukazawa et al., 2010; Fischer et al., 2014; Lee et al., 2014) or cells of Kupffer's vesicle (Warga and Kane, 2018). Importantly, periderm cells in zebrafish invade the oropharynx and constitute part of the oropharyngeal lining (Rosa et al., 2019). Collazo et al. (1994) suggested that the teleost EVL either constitutes a novelty, or could be homologized with the superficial ectodermal layer in other taxa, yet acquired a new fate. In amphibian (urodele) embryos, stratification of the ectoderm occurs during neurulation (Dettlaff, 1993), i.e. between stages 13 and 20 (staging according to Bordzilovskaya et al., 1989). Thus, one may regard the stratification of the ectoderm as an early (sturgeon) or late (urodele) event, together with a possible early specialization of the outer layer (in teleosts). The superficial ectodermal layer in sturgeons and urodeles would then be homologous to the teleostean periderm, a suggestion also made by Warga and Kane (2018). In contrast to early periderm differentiation in non-amniotes, the periderm in mammals develops long after gastrulation and forms by proliferation of the basal epidermal layer (M'Boneko and Merker, 1988; Byrne et al., 1994). Its ectodermal origin is therefore undisputed.

Several experiments have shown that pharyngeal tooth formation in zebrafish depends on the formation of pharyngeal pouches (Oralová et al., 2020). Using a transgenic line as well as fluorescent markers labeling the periderm, Rosa et al. (2019) discovered that the periderm in zebrafish invades the endodermal pouches (the precursors of the gill slits), with lumenization progressing concomitantly with the invading periderm (Figure 2.1, 2.2A–C). It was also observed that the periderm does not reach the midline endoderm, but instead contacts a layer of cells arising endogenously, with a similar expression profile as the invading periderm. These cells were called 'periderm-like'. Both the invading periderm and the 'periderm-like' cells form a bilayer, and, after lumenization, a single layer paves the floor and

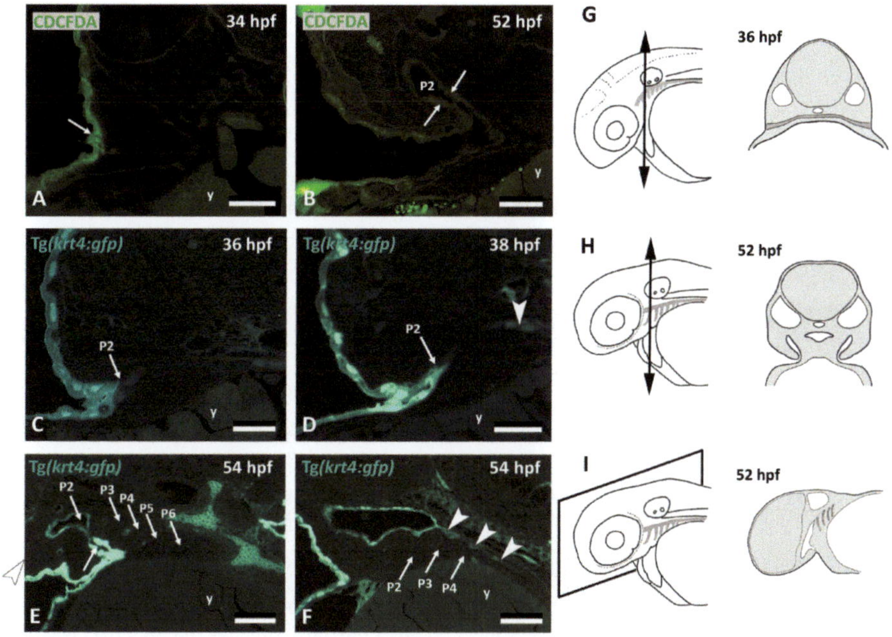

FIGURE 2.1 Periderm invasion in zebrafish starts at pharyngeal pouch 2. (A–B) Cross sections of embryonic zebrafish showing invasion of periderm via pouch 2. At 34 hpf, the fluorescent vital stain CDCFDA labels periderm on the body surface (A, arrow); at 52 hpf, periderm (arrows) has covered pouch 2 (B). (C–D) Details of the progress of the periderm using the transgenic line Tg*(krt4:gfp)*. *krt4*-positive periderm cells (arrow) start to invade pouch 2 at 36 hpf (C) and merge with 'periderm-like cells' (arrowhead) arising endogenously at 38 hpf (D). (E–F) Sagittal sections of 54 hpf zebrafish at a more lateral (E) and more medial level (F), showing how *krt4*-positive periderm cells (arrows) have invaded pouch 2 from outside and are joined by less intensely fluorescent 'periderm-like' cells (arrowheads) which have expanded posteriorly. The other pouches (indicated by P2, P3, etc.) have not yet been invaded by periderm migrating in from outside. (G–I) Schematics showing an external view and a cross section (G, H) or sagittal section (I) at the stage indicated. (G) Representative for figure parts A, C, and D; (H) Representative for figure part B and I, representative for figure parts E–F. y = yolk. Scale bars in A–D = 25 μm and in E–F = 50 μm.

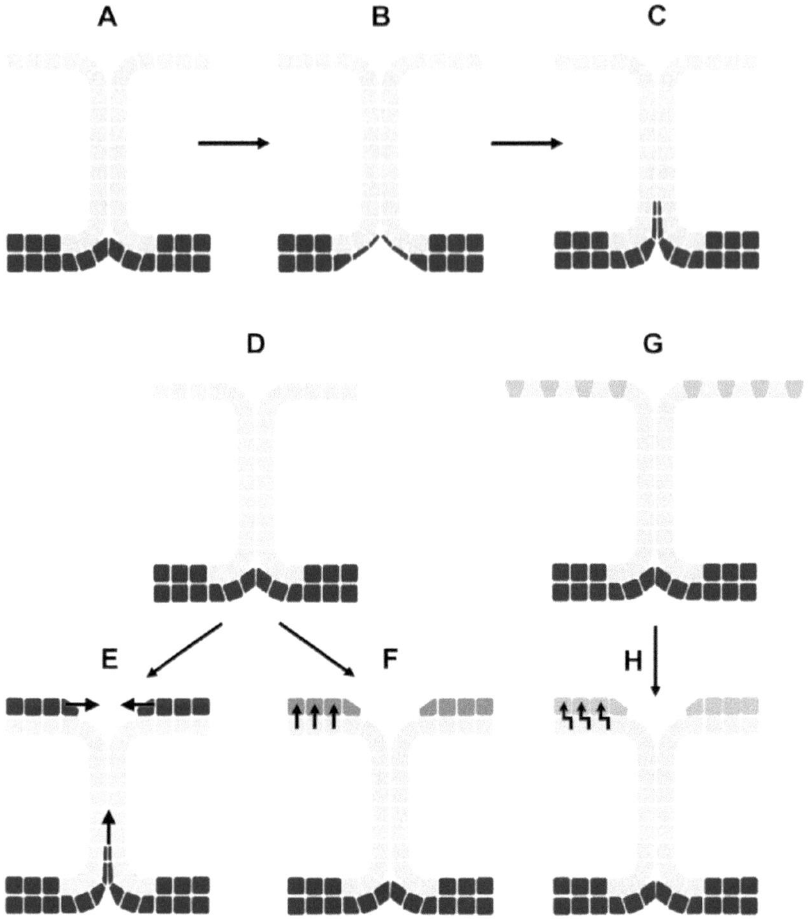

FIGURE 2.2 Ectoderm–endoderm connections and modes of stratification of the pharyngeal epithelium. (A) Pouch endoderm (light grey, bilayered) and its contact with the ectoderm (black, bilayered). (B, C) Superficial ectodermal layer partially invades the pouch. (D) Single-layered endodermal lining of the oropharynx. Stratification occurs via invasion of ectoderm (E, black, arrows) or oriented cell divisions (F, dark grey, arrows). (G) A pseudostratified endodermal epithelium can give rise to stratification via morphogenetic movements (H, median grey, arrows). Pharynx lumen to the top; exterior to the bottom in all figures.

roof, respectively, of the ensuing gill slit and pharyngeal lumen. The source of the 'periderm-like' cells could not be determined unambiguously, but they are possibly endodermal cells co-opted into a peridermal fate. These internal 'periderm-like' cells expand within the oropharyngeal cavity from anterior to posterior. In a follow-up study, Oralová et al. (2020) showed that tooth anlagen, while developing close to the pouches (zebrafish have pharyngeal teeth only), start to invaginate from midline endodermal epithelium, but not before this is covered with 'periderm-like' cells. Without these 'periderm-like' cells, tooth anlagen did not develop. Periderm was also observed to partially invade the mouth via the lips. However, because zebrafish possess no oral dentition, the potential role of this internal expansion for oral tooth development could not be assessed.

A key question is if these events in zebrafish are representative with respect to actinopterygians in general. In contrast to non-teleost actinopterygians and amphibians, teleosts have a highly derived early development, characterized by incomplete (meroblastic) cleavage and unique patterns of gastrulation and endoderm formation (Collazo et al., 1994; Cooper and Virta, 2007). To know if teleost periderm invasion represents a remnant of a once vaster invagination event or, conversely, should be considered a novelty, we need data on non-teleost actinopterygians with holoblastic cleavage (such as bichir and sturgeon). Such research is currently ongoing. Still, the example of the zebrafish shows how enamel organs, despite developing from endoderm, need the proximity of cells that share a number of characteristics with periderm (or ectoderm for that matter).

Urodele amphibians provide another interesting case that demonstrates the importance of the vicinity of ectoderm even for teeth that develop in endodermal territory. A recent study by Soukup et al. (2021), based on the use of GFP-labeled axolotls, demonstrated that the initiator-teeth of the premaxillary and vomerine tooth fields arise from ectoderm, but those of the palatine, coronoid, and dentary tooth fields (the latter specifically the medialmost teeth) stereotypically arise at the ectoderm–endoderm boundary. Interestingly, a century ago, Adams (1924) reported palatine and coronoid teeth to possess endodermal enamel organs, while the median pair of dentary teeth was considered to have enamel organs of mixed ectodermal–endodermal composition. Combining these two sets of observations, it is clear that teeth that arise from endoderm also develop in the vicinity of ectoderm. Moreover, the endodermal lining of the pharynx in the axolotl shows an interesting change of phenotype, before any evidence of tooth anlagen. While this lining is bilayered at the moment a lumen starts to appear (both in the prospective pharynx roof and floor), the surface layer loses its yolk content and starts to display apical differentiations, similar to those observed in the ectoderm (Figure 2.3). This has led Huysseune et al. (2022) to adopt

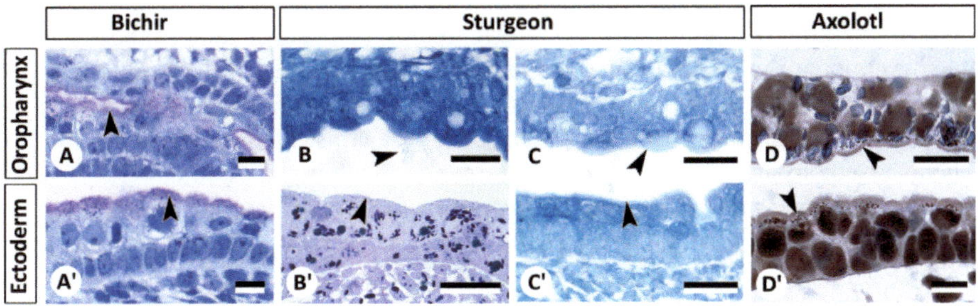

FIGURE 2.3 Stratified ectodermal and oropharyngeal epithelia. (A–D′) Apical differentiation of cells in the oropharynx (A–D) and ectoderm (A′–D′) of different taxa, as indicated. Arrowheads indicate in A and A′: Metachromatic granules; in B: Ciliated cells; in B′: Apical 'striations'; and in C, D, C′, and D′: Cuticle-like structures. Scale bars = 25 μm.

a term introduced in a short abstract of the 1970s as 'ectodermalized endoderm' (Hayashi, 1972). This concept should not be surprising in the light of recent findings claiming endoderm to be the first germ layer in evolution, while ectoderm originated as a secondary germ layer freed from ancestral feeding functions (Hashimshony et al., 2015). Thus, not only do teeth in urodeles develop either from ectoderm, but even when developing from endodermal epithelium, they do so close to the ectoderm boundary and/or have their surface layer 'ectodermalized'.

Together, these observations support the importance for tooth initiation of a persisting physical link between external and internal epithelia, if not of the presence of ectoderm itself.

2.3 ODONTOGENESIS STARTS IN STRATIFIED EPITHELIA

A recurrent feature for any vertebrate tooth anlage is that prior to the first morphological sign of tooth morphogenesis, the epithelium appears to be invariably stratified. Huysseune et al. (2022) present examples taken from all major vertebrate lineages illustrating this stratification.

In general, the stratification of epithelia can be achieved in various ways and is often a combination of several mechanisms. One way is that a migrating population covers the monolayer resting on the basal lamina (Figure 2.2D–E). Polarized (perpendicular) cell divisions constitute a perhaps more common mechanism to achieve stratification (Figure 2.2D–F). Other processes include delamination, i.e. the splitting of one cellular sheet into two more or less parallel sheets (Gilbert and Barresi, 2016). Radial intercalation is a process whereby cells detach from the basal lamina to occupy a suprabasal position (Figure 2.2G–H) (Rauzi, 2020). This has been proposed as a mechanism that builds up the *Xenopus* epidermis (a model also for other mucociliary epithelia). Here, ciliated cell precursors develop in the basal layer and next intercalate into the superficial layer (Stubbs et al., 2006; Walentek and Quigley, 2017). How stratification of odontogenic epithelia is achieved has not been considered in detail, but the expansion of the 'periderm-like' cells, described in zebrafish, provides an example of the first mechanism listed earlier.

Why should odontogenesis take place in stratified epithelia? Obviously, in a stratified epithelium, the basal layer(s) is free to undergo morphogenetic changes without compromising the integrity of the epithelium. The surface layer could, e.g. maintain a function as an osmotic regulator (i.e. in aquatic vertebrates), or protect the enamel organ-forming basal layer mechanically against damage produced by food items or other particles. An immunologic function is not excluded either. From this perspective, one may claim that, irrespective of the germ layer origin of the basal layer that produces the enamel organ, it has to be covered by another epithelial layer, whose germ layer origin does not matter any longer. The argument that we wish to make here, however, is that the stratification is part of a process by which competent external epithelia participate in tooth formation. Thus, we propose that the surface layer may carry an ectodermal/peridermal signature and play a role in initiating odontogenesis.

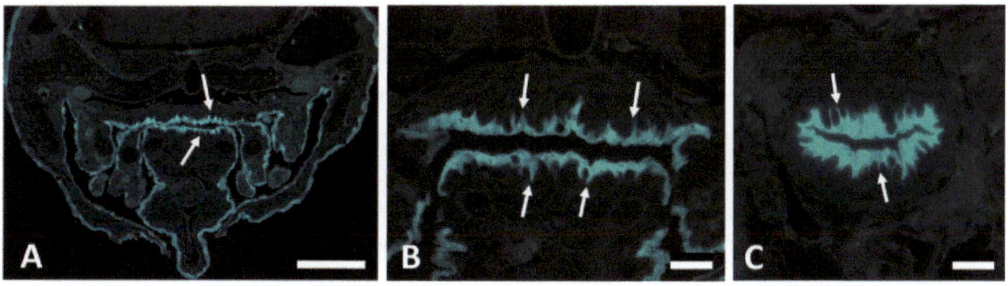

FIGURE 2.4 *krt4*-positive cells in the pharyngeal region of zebrafish. (A) Cross section at 5dpf, showing that the entire pharynx (as well as the basal epidermal layer) is lined by *krt4*-positive 'periderm-like' cells (green fluorescence). (B, C) Magnification of the pharyngeal epithelium in the posterior pharynx (B) and esophagus (C), showing cell processes extending from the superficial, 'periderm-like' cells down to the basal lamina (arrows). Scale bar in A = 100 μm; scale bar in B and C = 25 μm.

The apparent change of phenotype of the superficial layer of the endodermal epithelium in amphibian embryos has been mentioned before. The need of a 'periderm-like' layer to initiate odontogenesis from the basal layer below has been clearly shown for the pharyngeal teeth in zebrafish: Without its presence, no pharyngeal teeth form. Still, what factor(s) in the 'periderm-like' layer is responsible for this capacity remains unknown. Oralová et al. (2020) tested the potential role of Delta-Notch signalling between the two epithelial layers, both via pharmaceutical inhibition using DAPT and in mutants defective for mindbomb *(mib)*, a ubiquitin ligase essential for efficient activation of Notch signalling by Delta. However, in none of both approaches could any evidence be collected for this kind of juxtacrine signalling. In the zebrafish pharynx, long cell processes extend from the 'periderm-like' cells down to the basal lamina, squeezed between cells of the endodermal layer (Figure 2.4). While these cell processes occur in a domain of the oropharyngeal epithelium that is much more extensive than just the site of future tooth formation, their presence is intriguing and suggestive for a function that is more than mere protective.

In mice, oral periderm forms prior to the start of tooth formation and strong expression with a GFP reporter line for keratin 17 is observed in developing tooth germs at E12.5 (Richardson et al., 2014). In studies focusing on tooth development, the periderm layer is nevertheless usually not considered (e.g. Li et al., 2016). The importance of the periderm in preventing inappropriate oral epithelial adhesions (that could lead to orofacial clefts) has now become quite apparent, and its molecular control starts to be unraveled (e.g. Peyrard-Janvid et al., 2014; Sweat et al., 2020). Interestingly, a very recent study of the mouse mandible, based on single-cell transcriptomic analysis, revealed a different transcriptional profile for dental basal cells versus periderm cells (Ye et al., 2022).

In conclusion, a much broader survey is needed to uncover the origin and identity of the superficial epithelial tier(s) in areas of odontogenesis. Furthermore, to unequivocally demonstrate that the periderm is a signalling partner in odontogenesis, or perhaps participates in tooth formation as a structural component in certain taxa, functional studies will be imperative. Such studies could take advantage of tools that have been developed to selectively eliminate periderm cells (e.g. Eisenhoffer et al., 2017).

2.4 ODONTOGENESIS IN THE ORAL REGION IS RESTRICTED TO ODONTOGENIC BANDS OR DENTAL LAMINAE

Along with the question if, and how much, the periderm or 'periderm-like' cells (or other ectoderm) participates in the stratification of prospective odontogenic epithelia, a related question to be addressed is how far such an influence stretches. In other words: Is tooth-forming competence restricted to certain domains within the oropharynx, and how?

In mammals, the development of teeth is initiated from stripes (bands) of multilayered epithelium, clearly assigned to the ectoderm (Rothova et al., 2012). These bands precede the formation of the tooth placodes, the first morphological evidence of prospective tooth germs (classically to be considered at E11.5 in the mouse, Kim et al., 2017; but see Peterkova et al., 2006 for a detailed analysis of the transient dental placodes in the mouse and how to interpret them). A lot of terminological confusion surrounds the description of the epithelia involved. Although these bands have been termed 'dental lamina', they are to be distinguished from the deep downgrowth of the epithelium that generates primary (first-generation) teeth in other vertebrates; the named bands in mammals are therefore better termed odontogenic bands. Similar bands have been reported in a number of non-mammalians (sharks: Smith et al., 2009; teleosts: Fraser et al., 2004, 2008; anurans: Grieco and Hlusko, 2016; Lamoureux et al., 2018; reptiles: Vonk et al., 2008; Richman and Handrigan, 2011). In certain (but not all) taxa, these bands precede the formation of a dental lamina, a deep and mostly angled ingrowth from the epithelium into the underlying mesenchyme, from which teeth are then initiated. This is the case, e.g. in the ball python (snakes) and in the leopard gecko and the bearded dragon (two species of lizards) (Handrigan and Richman, 2010). Moreover, in all three species, the dental lamina is continuous along the maxillary and mandibular tooth-forming regions. In mammals (both in humans and mice), the distinction between prospective tooth buds and the regions between adjacent tooth buds (interdental regions) is not easy either, as emphasized by Hovorakova et al. (2018). Indeed, on sections, the epithelium of tooth buds, as well as of interdental gaps, exhibits a bud shape, which only differs by its size. A similar observation was done for a carnivore, the ferret (Jussila et al., 2014). These authors refer to the interdental regions as 'interdental lamina'. Together, these observations highlight the continuous nature not only of the dental lamina (in reptiles) but also of the odontogenic band (in mammals).

With the exception of teratomas, teeth in mammals are usually not found outside odontogenic bands or dental laminae. The anlage of the oral vestibule (epithelium) in humans, the area located externally to the dentition and lined by the gums, lips, and cheeks, has an original odontogenic potential which can be awakened under pathological conditions and gives rise to e.g. peripheral odontomas (Hovorakova et al., 2016). Moreover, in humans, teeth have been described on the palate and in the nasal cavity (Thawley and LaFerriere, 1977; Chen et al., 2002), the maxillary sinus (Erkmen et al., 1998), and in the oropharynx (Nagarajappa and Manjunatha, 2011). Some of these teeth result from miseruption and should not be considered as ectopic; others clearly represent ectopic teeth. Still, like the vestibule, these locations are, again, in areas of direct access for external epithelia.

In conclusion, while oral tooth formation in extant vertebrates is usually restricted to odontogenic bands or dental laminae, odontogenic potential can be awakened outside these areas. This suggests that tooth competence was once more widespread but that this capacity is not normally expressed anymore. The observation that a pharyngeal dentition was lost during sarcopterygian evolution likewise indicates that tooth-forming competence was originally widespread in the oropharynx. This raises two questions. First, what was the decisive factor that transferred odontode-forming competence inside the body, and did this indeed require the physical invasion of an ectodermal layer into the oropharynx? Given the established homology between external and internal odontodes, a valid assumption is that the factors responsible for odontogenic competence are also operating in external epithelia, more specifically to allow initiation and/or patterning of external odontodes or their derivatives. Such data, comparing molecular control of teeth and external odontodes, are scarce. Debiais-Thibaud et al. (2011) found very few differences in *Dlx* expression patterns between developing catshark (*Scyliorhinus canicula*) teeth and dermal denticles. Later work on this species, however, revealed a number of differences in epithelial gene expression patterns during initiation and early morphogenesis stages of teeth and scale buds (Debiais-Thibaud et al., 2015). A recent study by Mori and Nakamura (2022) identified a conserved odontode genetic regulatory network (oGRN) as well as expression of the paired like homeodomain 2 (*pitx2*) gene in developing dermal denticles of an armoured catfish (*Ancistrus* sp.), strongly suggesting the conservation of the developmental module to generate external or internal odontodes. Skin is stratified, even in an embryonic phase, and thus, external odontodes also develop from a stratified epithelium. If a role can be demonstrated of the superficial ectodermal layer of the skin (or periderm for that matter) for the development of external odontodes, then we are way closer in solving the role of stratification/periderm in the evolution of internal odontodes. Future studies could take advantage from the extensive knowledge acquired for other ectodermal appendages such as hairs and feathers. Current views revolve around activator/inhibitor mechanisms of the Turing type (e.g. Inaba et al., 2019). These can explain the formation of discrete units on the basis of initially homogeneous concentrations of molecules.

The second question is what factors restricted the tooth-forming competence to (essentially) the odontogenic bands in the oral region, and what factors limited tooth-forming capacity to certain arches in the pharyngeal region? This question is discussed in the next section.

2.5 THE DISTRIBUTION OF PHARYNGEAL TEETH: THE ROLE OF RETINOIC ACID

An analysis of pharyngeal tooth distribution in jawed vertebrates clearly shows that odontogenic competence was once widespread, involving all pharyngeal arches, but that this general competence was then regionally restricted to certain arches.

Thus, one must assume that the capacity to form teeth was lost from some arches but retained on other arches. How could this have been achieved? A few studies have been undertaken, in particular in zebrafish, that may shed light on this problem. While tinkering with several signalling molecules in zebrafish can produce supernumerary teeth (reviewed in Huysseune et al., 2022), only the manipulation of retinoic acid (RA) signalling has been shown so far to be capable of expanding the dentition to the anterior, non-tooth bearing arches. First, it was demonstrated that pharyngeal tooth formation in the cypriniform *Danio rerio* (zebrafish), but not in the characiform *Astyanax mexicanus* (Mexican tetra), nor in the beloniform *Oryzias latipes* (Japanese medaka), depends on RA (Gibert et al., 2010). Huysseune et al. (2022) have argued that the latter two species may not be different from zebrafish, as timing of the experiments demonstrating such differences was not similar for the three species. That RA plays a role in tooth initiation, at least in zebrafish, was demonstrated unequivocally in a follow-up experiment (Seritrakul et al., 2012). Zebrafish embryos exposed to exogenous RA from 24 hpf onwards showed an expansion of teeth to anterior arches. Seritrakul et al. (2012) concluded that changes in the levels of RA may be an evolutionary mechanism controlling tooth distribution. Subsequently, Gibert et al. (2015) applied exogenous RA to zebrafish later in development (56 hpf and older). This resulted in an increase in tooth number, yet limited to the tooth-bearing arch only, without anterior expansion over the other arches.

How the levels of RA in the zebrafish pharynx eventually translate into the presence of teeth on the last arch only is still an open question. One possibility is through the interaction with sonic hedgehog (SHH), a signalling molecule from the hedgehog family of paracrine factors (reviewed in Seppala et al., 2017; Qi and Li, 2020). In the mouse tongue, antero-posterior patterning of the lingual epithelium is controlled by antagonistic SHH and RA activities (El Shahawy et al., 2017). These authors showed that SHH antagonizes RA signalling by maintaining the expression of the RA-degrading enzymes *Cyp26a1* and *Cyp26c1*. However, *Cyp26a1/c1* may not be a direct transcriptional target of *Shh* signalling. Rather, SHH appears to be responsible for the maintenance and/or reinforcement of *Cyp26* expression, but not its induction (El Shahawy et al., 2017). In the zebrafish, the expression of the RA-degrading enzyme *cyp26b1* shows an extreme gradient in the pharynx at 24 and 30hpf, with highest levels around the posterior pharyngeal arch (Zhao et al., 2005) suggesting lowest levels of RA in the posterior pharynx (note that the first morphological evidence of the first tooth is seen at around 48 hpf). The signal diminishes in the posterior pharynx by 48 hpf (Zhao et al., 2005). Interestingly, at around 48 hpf, the expression of the RA-synthesizing enzyme, *aldh1a2*, is prominent in the posterior pharynx, albeit positioned rather laterally (Grandel et al., 2002; Gibert et al., 2010). Possibly, only at the level of the posterior pharyngeal arch, enough ALDH1a2 is formed to counteract CYP26b1 and allow tooth formation. Alternatively, *aldh1a2* and *cyp26b1* may have more subtle spatio-temporal patterns that have not been detected with the whole mounts used. Whether SHH has a role in this is presently not known.

Summarizing, a detailed analysis at a fine spatial and temporal scale, along the A–P axis of the pharynx, of the expression domains of genes implicated in

RA metabolism will be required to test whether a linkage exists between RA and SHH. Such a test will advance our knowledge on how the pattern of pharyngeal teeth in zebrafish, or in other gnathostomes possessing pharyngeal teeth, can be established.

2.6 CONCLUSIONS AND DIRECTIONS FOR FUTURE RESEARCH

In this review, we have focused on the role of the epithelium in odontogenesis, both from an evolutionary and developmental perspective. Following heavy debates regarding the germ layer origin of odontogenic epithelia and concomitant hypotheses on where teeth originated, opinions have converged on the proposal that odontogenic competence was transferred from external epithelia into the oropharynx, the only tooth-bearing region of the alimentary canal. A recent key observation, in zebrafish, is that periderm invades the zebrafish oropharynx to take part in the covering epithelium and that 'peridem-like' cells, possibly coopted from the endoderm, are necessary for tooth induction. A crucial point is whether the process that is observed in zebrafish is common to teleosts, and if the same process—or variations—are rooted deep in actinopterygian, and possibly even gnathostome, phylogeny. This question is currently being addressed by studies on non-teleost actinopterygians. Irrespective of whether or not periderm or 'periderm-like' cells cover the prospective enamel organ-forming epithelium, odontogenic epithelia are stratified. We propose that at least one of these strata bears an ectodermal signature and that this is part of the strategy by which external and internal epithelia are linked. Periderm, as the outer stratified layer, may well have been the epithelium that originally transferred competence into the oropharynx through its multiple openings (mouth and gill slits), but its role may have been coopted by endoderm. We hypothesize that the presence of periderm, or 'periderm-like' cells, or 'ectodermalized endoderm' is required to initiate tooth formation.

Once tooth competence is established in the oropharynx, the question is: What defines odontogenic versus non-odontogenic regions? And what caused tooth loss on specific arches, starting from a plesiomorphic condition where all pharyngeal arches are toothed? Whether or not teeth actually develop on certain arches may be determined by an interplay of RA and SHH, with a prominent role for the pouches and their contact with the ectoderm. To test such an association, it will be necessary to investigate, on a fine spatial and temporal scale, the presence of RA-metabolizing enzymes and of *Shh* signalling. Forthcoming single-cell resolution studies might allow to characterize the molecular control of oral epithelia stratification and to test our assumptions on the specific role of the stratified odontogenic epithelia in the conquest of the oropharynx.

ACKNOWLEDGMENTS

The authors wish to thank the reviewer for the insightful comments on an earlier draft of the manuscript. Ann Huysseune and Robert Cerny gratefully acknowledge a grant from The Czech Science Foundation GACR (project no 22–25061S).

P. Eckhard Witten acknowledges funding from the European Union's Horizon 2020 Research and Innovation Programme, Marie Sklodowska-Curie grant agreement no 766347.

REFERENCES

Adams, A.E. 1924. An experimental study of the development of the mouth in the amphibian embryo. *Journal of Experimental Zoology* 40: 311–379.

Berio, F. and Debiais-Thibaud, M. 2021. Evolutionary developmental genetics of teeth and odontodes in jawed vertebrates: A perspective from the study of elasmobranchs. *Journal of Fish Biology* 98: 906–918.

Berkovitz, B. and Shellis, P. 2017. *The Teeth of Non-Mammalian Vertebrates*. Academic Press, Elsevier, London, San Diego, Cambridge, Oxford.

Bjerring, H.C. 1998. The fates of spiracular allostoses in mammals. *Acta Zoologica (Stockholm)* 79: 51–67.

Blais, S.A., MacKenzie, L.A. and Wilson, M.V.H. 2011. Tooth-like scales in early Devonian eugnathostomes and the 'outside-in' hypothesis for the origins of teeth in vertebrates. *Journal of Vertebrate Paleontology* 31: 1189–1199.

Bordzilovskaya, N.P., Dettlaff, T.A., Duhon, S. and Malacinski, G. 1989. Developmental stage series of axolotl embryos. In J.B. Armstrong and G.M. Malacinski (eds.), *Developmental Biology of the Axolotl*. Oxford University Press, New York.

Brazeau, M. and Ahlberg, P. 2006. Tetrapod-like middle ear architecture in a Devonian fish. *Nature* 439: 318–321.

Byrne, C., Tainsky, M. and Fuchs, E. 1994. Programming gene expression in developing epidermis. *Development* 120: 2369–2383.

Chen, A., Huang, J.-K., Cheng, S.-J. and Sheu, C.-Y. 2002. Nasal teeth: Report of three cases. *American Journal of Neuroradiology* 23: 671–673.

Chen, D., Blom, H., Sanchez, S., Tafforeau, P., Märss, T. and Ahlberg, P.E. 2020. The developmental relationship between teeth and dermal odontodes in the most primitive bony fish *Lophosteus*. *eLife* 9: e60985.

Clemen, G., Bartsch, P. and Wacker, K. 1998. Dentition and dentigerous bones in juveniles and adults of *Polypterus senegalus* (Cladistia, Actinopterygii). *Annals of Anatomy* 180: 211–221.

Collazo, A., Bolker, J.A. and Keller, R. 1994. A phylogenetic perspective on teleost gastrulation. *American Naturalist* 144: 133–152.

Comabella, Y., Franyutti, A.H., Hurtado, A., Canabal, J. and Garcia-Galano, T. 2012. Ontogenetic development of the digestive tract in Cuban gar (*Atractosteus tristoechus*) larvae. *Reviews in Fish Biology and Fisheries* 23: 245–260.

Cooper, M.S. and Virta, V.C. 2007. Evolution of gastrulation in the ray-finned (actinopterygian) fishes. *Journal of Experimental Zoology (Mol. Dev. Evol.)* 308B: 591–608.

Debiais-Thibaud, M., Chiori, R., Enault, S., Oulion, S., Germon, I., Martinand-Mari, C., Casane, D. and Borday-Birraux, V. 2015. Tooth and scale morphogenesis in shark: An alternative process to the mammalian enamel knot system. *BMC Evolutionary Biology* 15: 292.

Debiais-Thibaud, M., Oulion, S., Bourrat, F., Laurenti, P., Casane, D. and Borday-Birraux, V. 2011. The homology of odontodes in gnathostomes: Insights from *Dlx* gene expression in the dogfish, *Scyliorhinus canicula*. *BMC Evolutionary Biology* 11: 307.

Dettlaff, T.A. 1993. Evolution of the histological and functional structure of ectoderm, chordamesoderm and their derivatives in Anamnia. *Roux's Archives in Developmental Biology* 203: 3–9.

Dettlaff, T.A., Ginsburg, A.S. and Schmalhausen, O.I. 1993. *Sturgeon Fishes: Developmental Biology and Aquaculture*. Springer-Verlag, Berlin, Heidelberg, pp. 49–154.

Diedhiou, S. and Bartsch, P. 2009. Staging of the early development of *Polypterus* (Cladistia: Actinopterygii). In Y.W. Kunz, C.A. Luer and B.G. Kapoor (eds.), *Development of Non-Teleost Fishes*. Science Publishers Enfield, Jersey and Plymouth.

Donoghue, P.C.J. and Rücklin, M. 2016. The ins and outs of the evolutionary origin of teeth. *Evolution and Development* 18: 19–30.

Eisenhoffer, G.T., Slattum, G., Ruiz, O.E., Otsuna, H., Bryan, C.D., Lopez, J., Wagner, D.S., Bonkowsky, J.L., Chien, C.-B., Dorsky, R.I. and Rosenblatt, J. 2017. A toolbox to study epidermal cell types in zebrafish. *Journal of Cell Science* 130: 269–277.

El Shahawy, M., Reibring, C.-G., Neben, C.L., Hallberg, K., Marangoni, P., Harfe, B.D., Klein, O.D., Linde, A. and Gritli-Linde, A. 2017. Cell fate specification in the lingual epithelium is controlled by antagonistic activities of Sonic hedgehog and retinoic acid. *PLoS Genetics* 13: e1006914.

Erkmen, N., Ölmez, S. and Önerci, M. 1998. Supernumerary tooth in the maxillary sinus: Case report. *Australian Dental Journal* 43: 385–386.

Fischer, B., Metzger, M., Richardson, R., Knyphausen, P., Ramezani, T., Franzen, R., Schmelzer, E., Bloch, W., Carney, T.J. and Hammerschmidt, M. 2014. p53 and TAp63 promote keratinocyte proliferation and differentiation in breeding tubercles of the zebrafish. *PLoS Genetics* 10: e1004048.

Fraser, G.J., Bloomquist, R.F. and Streelman, J.T. 2008. A periodic pattern generator for dental diversity. *BMC Biology* 6: 32.

Fraser, G.J., Cerny, R., Soukup, V., Bronner-Fraser, M. and Streelman, J.T. 2010. The odontode explosion: The origin of tooth-like structures in vertebrates. *BioEssays* 32: 808–817.

Fraser, G.J., Graham, A. and Smith, M.M. 2004. Conserved deployment of genes during odontogenesis across osteichthyans. *Proceedings of the Royal Society of London B* 271: 2311–2317.

Fukazawa, C., Santiago, C., Park, K.M., Deery, W.J., de la Torre Canny, S.G., Holterhoff, C.K. and Wagner, D.S. 2010. *poky/chuk/ikk1* is required for differentiation of the zebrafish embryonic epidermis. *Developmental Biology* 346: 272–283.

Gibert, Y., Bernard, L., Debiais-Thibaud, M., Bourrat, F., Joly, J.-S., Pottin, K., Meyer, A., Retaux, S., Stock, D.W., Jackman, W.R., Seritrakul, P., Begemann, G. and Laudet, V. 2010. Formation of oral and pharyngeal dentition in teleosts depends on differential recruitment of retinoic acid signalling. *FASEB Journal* 24: 3298–3309.

Gibert, Y., Samarut, E., Pasco-Viel, E., Bernard, L., Borday-Birraux, V., Sadier, A., Labbé, C., Viriot, L. and Laudet, V. 2015. Altered retinoic acid signalling underpins dentition evolution. *Proceedings of the Royal Society B* 282: 20142764.

Gilbert, S.F. and Barresi, M.J.F. 2016. *Developmental Biology*. 11th Edition. Sinauer Associates, Sunderland, MA.

Gilchrist, J.D.F. 1922. XXVII.—Note on the oesophageal teeth of the Stromateidæ. *Annals and Magazine of Natural History* 9(51): 249–255.

Grandel, H., Lun, K., Rauch, G.J., Rhinn, M., Piotrowski, T., Houart, C., Sordino, P., Kuchler, A.M., Schulte-Merker, S., Geisler, R., Holder, N., Wilson, S.W. and Brand, M. 2002. Retinoic acid signalling in the zebrafish embryo is necessary during pre-segmentation stages to pattern the anterior-posterior axis of the CNS and to induce a pectoral fin bud. *Development* 129: 2851–2865.

Grieco, T.M. and Hlusko, L.J. 2016. Insight from frogs: Sonic hedgehog gene expression and a re-evaluation of the vertebrate odontogenic band. *Anatomical Record* 299: 1099–1109.

Handrigan, G.R. and Richman, J.M. 2010. Autocrine and paracrine Shh signalling are necessary for tooth morphogenesis, but not tooth replacement in snakes and lizards (Squamata). *Developmental Biology* 337: 171–186.

Haridy, Y., Gee, B.M., Witzmann, F., Bevitt, J.J. and Reisz, R.R. 2019. Retention of fish-like odontode overgrowth in Permian tetrapod dentition supports outside-in theory of tooth origins. *Biology Letters* 15: 20190514.

Hashimshony, T., Feder, M., Levin, M., Hall, B.K. and Yanai, I. 2015. Spatiotemporal transcriptomics reveals the evolutionary history of the endoderm germ layer. *Nature* 519: 219–222.

Hayashi, H. 1972. Opening mechanism of stomodeum cavity and transformation of yolk platelet proteins to various subcellular structures in developing newt embryos, *Triturus pyrrhogaster. Journal of Electron Microscopy* 21: 246 (abstract).

Hilton, E.J., Grande, L. and Bemis, W.E. 2011. Skeletal anatomy of the shortnose sturgeon, *Acipenser brevirostrum* Lesueur, 1818, and the systematics of sturgeons (Acipenseriformes, Acipenseridae). *Fieldiana Life Earth Sciences* 3: 1–168.

Hovorakova, M., Lesot, H., Peterka, M. and Peterkova, R. 2018. Early development of the human dentition revisited. *Journal of Anatomy* 233: 135–145.

Hovorakova, M., Lochovska, K., Zahradnicek, O., Domonkosova Tibenska, K., Dornhoferova, M., Horakova-Smrckova, L. and Bodorikova, S. 2016. One odontogenic cell-population contributes to the development of the mouse incisors and of the oral vestibule. *PLoS ONE* 11(9): e0162523.

Huysseune, A., Cerny, R. and Witten, P.E. 2022. The conundrum of pharyngeal teeth origin: The role of germ layers, pouches, and gill slits. *Biological Reviews* 97: 414–447.

Huysseune, A., Sire, J.-Y. and Witten, P.E. 2009. Evolutionary and developmental origins of the vertebrate dentition. *Journal of Anatomy* 214: 465–476.

Huysseune, A., Sire, J.-Y. and Witten, P.E. 2010. A revised hypothesis on the evolutionary origin of the vertebrate dentition. *Journal of Applied Ichthyology* 26: 152–155.

Inaba, M., Harn, H.I.-C. and Chuong, C.-M. 2019. Turing patterning with and without a global wave. *PLoS Biology* 17(3): e3000195.

Isokawa, S., Kubota, K., Kosakai, T., Satomura, I., Tsubouchi, M. and Sera, A. 1965. Some contributions to study of esophageal sacs and teeth of fishes. *Journal of the Nihon University School of Dentistry* 7: 103–111.

Jollie, M. 1968. Some implications of the acceptance of a delamination principle. In T. Ørvig (ed.), *Current Problems of Lower Vertebrate Phylogeny. Proceedings of the Fourth Nobel Symposium, Stockholm, 1967.* Almqvist and Wiksell, Stockholm and New York.

Jussila, M., Yanez, X.C. and Thesleff, I. 2014. Initiation of teeth from the dental lamina in the ferret. *Differentiation* 87: 32–43.

Kim, R., Green, J.B.A. and Klein, O.D. 2017. From snapshots to movies: Understanding early tooth development in four dimensions. *Developmental Dynamics* 246: 442–450.

Kimmel, C.B., Warga, R.M. and Schilling, T.F. 1990. Origin and organization of the zebrafish fate map. *Development* 108: 581–594.

Lainoff, A.J., Moustakas-Verho, J.E., Hu, D., Kallonen, A., Marcucio, R.S. and Hlusko, L.J. 2015. A comparative examination of odontogenic gene expression in both toothed and toothless amniotes. *Journal of Experimental Zoology (Mol. Dev. Evol.)* 324B: 255–269.

Lamoureux, D.O., LeBlanc, A.R.H. and Caldwell, M.W. 2018. Tooth germ initiation patterns in a developing dentition: An in vivo study of *Xenopus laevis* tadpoles. *Journal of Morphology* 279: 616–625.

Lee, R.T.H., Asharani, P.V. and Carney, T.J. 2014. Basal keratinocytes contribute to all strata of the adult zebrafish dermis. *PLoS ONE* 9: e84858.

Li, J., Chatzeli, L., Panousopoulou, E., Tucker, A.S. and Green, J.B.A. 2016. Epithelial stratification and placode invagination are separable functions in early morphogenesis of the molar tooth. *Development* 143: 670–681.

Mahler, D.L. and Kearney, M. 2006. The palatal dentition in squamate reptiles: Morphology, development, attachment, and replacement. *Fieldiana, Zoology (New Series)* 108: 1540.

Matsumoto, R. and Evans, S.E. 2017. The palatal dentition of tetrapods and its functional significance. *Journal of Anatomy* 230: 47–65.

M'Boneko, V. and Merker, H.-J. 1988. Development and morphology of the periderm of mouse embryos. *Acta Anatomica* 133: 325–336.

Mori, S. and Nakamura, T. 2022. Redeployment of odontode gene regulatory network underlies dermal denticle formation and evolution in suckermouth armored catfish. *Scientific Reports* 12: 6172.

Murdock, D.J.E., Dong, X.-P., Repetski, J.E., Marone, F., Stampanoni, M. and Donoghue, P.C.J. 2013. The origin of conodonts and of vertebrate mineralized skeletons. *Nature* 502: 546–549.

Nagarajappa, D. and Manjunatha, B.S. 2011. Tooth in oropharynx. *Journal of Oral and Maxillofacial Pathology* 15: 346–347.

Nelson, G.J. 1969. Gill arches and the phylogeny of fishes, with notes on the classification of vertebrates. *Bulletin of the American Museum of Natural History* 141: 475–552.

Nelson, G.J. 1970. Pharyngeal denticles (placoid scales) of sharks, with notes on the dermal skeleton of vertebrates. *American Museum Novitates* 2413: 1–26.

Oralová, V., Rosa, J.T., Larionova, D., Witten, P.E. and Huysseune, A. 2020. Multiple epithelia are required to form teeth deep in the pharynx. *Proceedings of the National Academy of Science of the United States of America* 117: 11503–11512.

Ørvig, T. 1967. Phylogeny of tooth tissues: Evolution of some calcified tissues in early vertebrates. In A.E.W. Miles (ed.), *Structural and Chemical Organization of Teeth*. Vol. 1. Academic Press, London.

Ørvig, T. 1977. A survey of odontodes ('dermal teeth') from developmental, structural, functional, and phyletic points of view. In S.M. Andrews, R.S. Miles and A.D. Walker (eds.), *Problems in Vertebrate Evolution*. Academic Press, London.

Peterkova, R., Lesot, H. and Peterka, M. 2006. Phylogenetic memory of developing mammalian dentition. *Journal of Experimental Zoology (Mol. Dev. Evol.)* 306B: 234–250.

Peyer, B. 1968. *Comparative Odontology*. The University of Chicago, Chicago and London.

Peyrard-Janvid, M., Leslie, E.J., Kousa, Y.A., Smith, T.L., Dunnwald, M., Magnusson, M., Lentz, B.A., Unneberg, P., Fransson, I., Koillinen, H.K., Rautio, J., Pegelow, M., Karsten, A., Basel-Vanagaite, L., Gordon, W., Andersen, B., Svensson, T., Murray, J.C., Cornell, R.A., Kere, J. and Schutte, B.C. 2014. Dominant mutations in *GRHL3* cause Van der Woude syndrome and disrupt oral periderm development. *American Journal of Human Genetics* 94: 23–32.

Pospisilova, A., Stundl, J., Brejcha, J., Metscher, B.D., Psenicka, M., Cerny, R. and Soukup, V. 2022. The remarkable dynamics in the establishment, rearrangement, and loss of dentition during the ontogeny of the sterlet sturgeon. *Developmental Dynamics* 251: 826–845.

Qi, X. and Li, X. 2020. Mechanistic insights into the generation and transduction of hedgehog signalling. *Trends in Biochemical Sciences* 45: 397–410.

Qu, Q., Sanchez, S., Blom, H., Tafforeau, P. and Ahlberg, P.E. 2013. Scales and tooth whorls of ancient fishes challenge distinction between external and oral 'teeth'. *PLoS ONE* 8: e71890.

Rauzi, M. 2020. Cell intercalation in a simple epithelium. *Philosophical Transactions of the Royal Society B* 375: 20190552.

Reif, W.E. 1982. Evolution of dermal skeleton and dentition in vertebrates. The odontode regulation theory. *Evolutionary Biology* 15: 287–368.

Richardson, R.J., Hammond, N.L., Coulombe, P.A., Saloranta, C., Nousiainen, H.O., Salonen, R., Berry, A., Hanley, N., Headon, D., Karikoski, R. and Dixon, M.J. 2014. Periderm prevents pathological epithelial adhesions during embryogenesis. *The Journal of Clinical Investigation* 124: 3891–3900.

Richman, J.M. and Handrigan, G.R. 2011. Reptilian tooth development. *Genesis* 49: 247–260.

Rosa, J.T., Oralová, V., Larionova, D., Eisenhoffer, G.T., Witten, P.E. and Huysseune, A. 2019. Periderm invasion contributes to epithelial formation in the teleost pharynx. *Scientific Reports* 9: 10082.

Rothova, M., Thompson, H., Lickert, H. and Tucker, A.S. 2012. Lineage tracing of the endoderm during oral development. *Developmental Dynamics* 241: 1183–1191.

Rücklin, M., Donoghue, P.C.J., Johanson, Z., Trinajstic, K., Marone, F. and Stampanoni, M. 2012. Development of teeth and jaws in the earliest jawed vertebrates. *Nature* 491: 748–751.

Rücklin, M., Giles, S., Janvier, P. and Donoghue, P.C.J. 2011. Teeth before jaws? Comparative analysis of the structure and development of the external and internal scales in the extinct jawless vertebrate *Loganellia scotica*. *Evolution and Development* 13: 523–532.

Sadier, A., Jackman, W.R., Laudet, V. and Gibert, Y. 2020. The vertebrate tooth row: Is it initiated by a single organizing tooth? *BioEssays* 42: 1900229.

Schoch, R.R. 2002. The evolution of metamorphosis in temnospondyls. *Lethaia* 35: 309–327.

Seppala, M., Fraser, G.J., Birjandi, A.A., Xavier, G.M. and Cobourne, M.T. 2017. Sonic hedgehog signalling and development of the dentition. *Journal of Developmental Biology* 5: 6.

Seritrakul, P., Samarut, E., Lama, T.T.S., Gibert, Y., Laudet, V. and Jackman, W.R. 2012. Retinoic acid expands the evolutionarily reduced dentition of zebrafish. *FASEB Journal* 26: 5014–5024.

Sire, J.-Y. and Huysseune, A. 2003. Formation of dermal skeletal and dental tissues in fish: A comparative and evolutionary approach. *Biological Reviews* 78: 219–249.

Smith, M.M. and Coates, M.I. 2000. Evolutionary origins of teeth and jaws: Developmental models and phylogenetic patterns. In M. Teaford, M. Smith and M. Ferguson (eds.), *Development, Function and Evolution of Teeth*. Cambridge University Press, Cambridge.

Smith, M.M., Fraser, G.J., Chaplin, N., Hobbs, C. and Graham, A. 2009. Reiterative pattern of sonic hedgehog expression in the catshark dentition reveals a phylogenetic template for jawed vertebrates. *Proceedings of the Royal Society B-Biological Sciences* 276: 1225–1233.

Soukup, V., Tazaki, A., Yamazaki, Y., Pospisilova, A., Epperlein H.-H., Tanaka, E. and Cerny, R. 2021. Oral and palatal dentition of axolotl arises from a common tooth-competent zone along the ecto-endodermal boundary. *Frontiers in Cell and Developmental Biology* 8: 622308.

Stubbs, J.L., Davidson, L., Keller, R. and Kintner, C. 2006. Radial intercalation of ciliated cells during *Xenopus* skin development. *Development* 133: 2507–2515.

Sweat, Y.Y., Sweat, M., Yu, W., Sanz-Navarro, M., Zhang, L., Sun, Z., Eliason, S., Klein, O.D., Michon, F., Chen, Z. and Amendt, B.A. 2020. *Sox2* controls periderm and rugae development to inhibit oral adhesions. *Journal of Dental Research* 99: 1397–1405.

Thawley, S.E. and LaFerriere, K.A. 1977. Supernumerary nasal tooth. *The Laryngoscope* 87: 1770–1773.

Van der Brugghen, W. and Janvier, P. 1993. Denticles in thelodonts. *Nature* 364: 107.

Vandewalle, P., Huysseune, A., Aerts, P. and Verraes, W. 1994. The pharyngeal apparatus in teleost feeding. In V. Bels, M. Chardon and P. Vandewalle (eds.), *Biomechanics of Feeding in Vertebrates. Advances in Comparative and Environmental Physiology*. Vol. 18, pp. 59–92. Springer Verlag, Berlin and Heidelberg.

Vonk, F.J., Admiraal, J.F., Jackson, K., Reshef, R., de Bakker, M.A.G., Vanderschoot, K., van den Berge, I., van Atten, M., Burgerhout, E., Beck, A., Mirtschin, P.J., Kochva, E., Witte, F., Fry, B.G., Woods, A.E. and Richardson, M.K. 2008. Evolutionary origin and development of snake fangs. *Nature* 454: 630–633.

Walentek, P. and Quigley, I.K. 2017. What we can learn from a tadpole about ciliopathies and airway diseases: Using systems biology in *Xenopus* to study cilia and mucociliary epithelia. *Genesis* 55: e23001.

Warga, R.M. and Kane, D.A. 2018. Wilson cell origin for Kupffer's vesicle in the zebrafish. *Developmental Dynamics* 247: 1057–1069.

Ye, Q., Bhojwani, A. and Hu, J.K. 2022. Understanding the development of oral epithelial organs through single cell transcriptomic analysis. *Development* 149: dev200539.

Zhao, Q., Dobbs-McAuliffe, B. and Linney, E. 2005. Expression of *cyp26b1* during zebrafish early development. *Gene Expression Patterns* 5: 363–369.

3 The Neural Crest and the Development of Odontoskeletogenic Potential along the Body Axis

Jan Stundl and Marianne E. Bronner

3.1 INTRODUCTION

The evolutionary success of vertebrates is often associated with the acquisition of two defining cell populations: The neural crest (NC) and ectodermal placodes. Both are thought to be critical for the innovation of the highly specialized 'new head' of vertebrates that is not present in non-vertebrate chordates (Gans and Northcutt, 1983; Northcutt and Gans, 1983; Northcutt, 2005; Bronner-Fraser, 2008). Both NC and cranial placodes cells arise within the dorsal ectoderm at the 'neural plate border' between the neural plate (future CNS) and non-neural ectoderm (future epidermis). NC cells then undergo an epithelial-to-mesenchymal transition, enabling them to migrate and differentiate into diverse cell types (Le Douarin and Kalcheim, 1999). As the embryo develops, the developmental potential of embryonic cells becomes progressively restricted in each germ layer from a pluripotent state to multipotent to differentiated state. Multipotency reflects the ability to give rise to numerous cell types within a specific lineage. Surprisingly, NC cells challenge this paradigm because they transiently reactivate a pluripotency program that allows them to possess a developmental potential that is broader than that of their germ layer of origin (Zalc et al., 2021). Thus, the NC can be considered as a 're-activated' ectoderm, whose derivatives enhanced the vertebrate body plan with key evolutionary innovations, such as jaws, advanced dentition, or protective dermal armor.

Indeed, the NC is a unique multipotent/pluripotent population whose derivatives encompass ectodermal-, mesodermal-, and endodermal-like cell types. Thus, the NC possesses exceptionally broad developmental potential that transcends those normally ascribed to a germ layer. Moreover, similar to mesoderm, the NC gives rise to mesenchymal cells that extensively migrate throughout the body to form various cell types. Indeed, of the 411 different human cell types, the NC gives rise to 47 (cf. endoderm gives rise to 57 cell types) (Vickaryous and Hall, 2006) and, unlike other germ layers, the NC contributes to every basic tissue type. For all these reasons, the

 DOI: 10.1201/9781003439653-3

NC is referred to as the fourth germ layer (in addition to the canonical germ layers: Ectoderm, endoderm, and mesoderm). Thus, vertebrates are considered as quadroblastic organisms (Hall, 1998, 2000, 2008).

Initiating during gastrulation, the NC is induced within the neural plate border, a broad stripe of ectoderm located between the prospective neural plate and epidermis (Figure 3.1A). Although the neural plate border represents one of the most developmentally important regions of the vertebrate embryos, it is not distinguishable by a specific set of genes but rather is defined by unique gene interactions within the neural plate border and surrounding tissues (Williams et al., 2022). Three main transcription factors have been considered to be 'markers/specifiers' of the neural plate border (Figure 3.1A): *Pax3/7*, *Zic*, and *Msx1/2*. The expression of these 'specifiers' is largely conserved across vertebrates (Meulemans and Bronner-Fraser, 2004; Basch et al., 2006; Betters et al., 2018), suggesting that the specification of the neural plate border region likely evolved prior to the appearance of NC and cranial placodes in non-vertebrate chordates. Indeed, it has been shown that the orthologues of these neural plate border 'specifiers' define the lateral neural border identity in invertebrates, including some protostomes (Wada et al., 1998; Yu et al., 2008; Stolfi et al., 2015; Li et al., 2017; Arendt, 2018). This suggests that the molecular machinery responsible for neural plate border specification is deeply conserved among chordates and might be shared with protostomes, consistent with a possible origin in stem bilaterians (Li et al., 2017; Stundl et al., 2021). In addition to the similarity of neural plate border patterning, this ectodermal domain can generate migratory cell derivatives such as pigment cells and peripheral neuronal subtypes in several non-vertebrate bilaterians that appear to be specified by conserved terminal selector genes (Li et al., 2017; Zhao et al., 2019; Stundl et al., 2021). Thus, the neural plate border and lateral neural border, respectively, have significant developmental potential to produce migratory progeny; thus, it is likely that the NC module was assembled within this ectodermal domain at the onset of vertebrate evolution.

In vertebrate embryos, the neural plate border elevates to form neural folds at the dorsal aspect of the neural tube as the embryo undergoes neurulation, culminating with a closed and cavitated neural tube which forms the CNS of all vertebrates (Figure 3.1A–D). The premigratory NC cells reside within the neural folds, and these cells undergo an epithelial-to-mesenchymal transition (EMT) and delaminate after the fusion (Figure 3.1C) or elevation of the neural folds depending upon the vertebrate species (Theveneau and Mayor, 2012; Stundl et al., 2020). EMT is the cellular process whereby the prospective NC cells lose their epithelial arrangement and gain mesenchymal traits to move away from the CNS and migrate to their final destinations (Nieto, 2011). The process of EMT requires several steps, including changes in the extracellular matrix, cell junctions, adhesion properties, and interplay of several signaling pathways (Alkobtawi and Monsoro-Burq, 2020). Interestingly, several NC specifier genes (e.g., *Snail1/2, FoxD3*, or *Twist1*) also comprise a significant part of the gene regulatory network (GRN) module that drives EMT (Figure 3.1C), indicating that the identity and behavior of NC may be interlinked to the process of EMT. However, NC EMT differs at the cellular and molecular level depending upon location along the neural axis; for example, there are differences in expression of EMT factors, signaling pathways, cadherin regulation, and cell-cycle control between the cranial versus trunk axial levels (Strobl-Mazzulla and Bronner, 2012; Simões-Costa and Bronner, 2015).

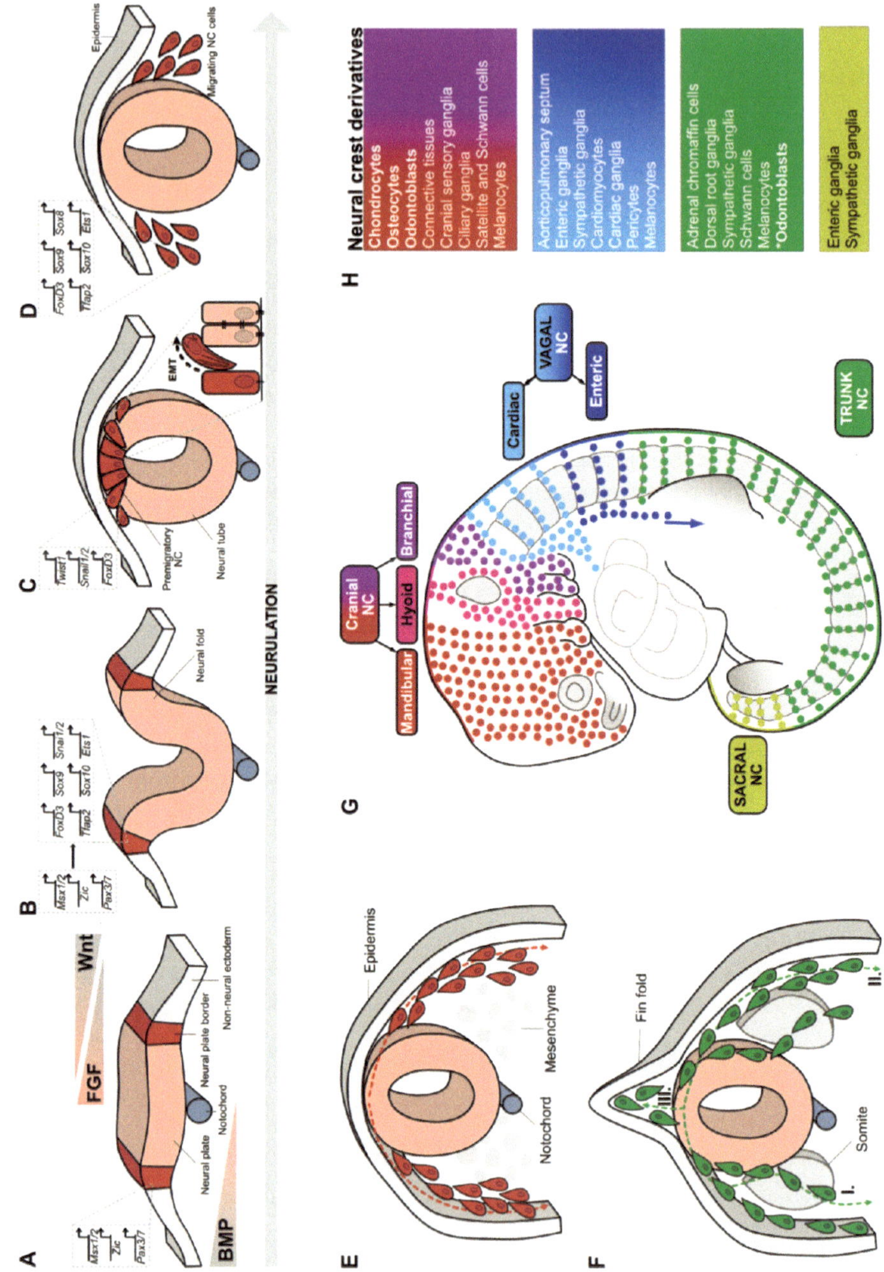
A
B
C
D
E
F
G
H
BMP
FGF
Wnt
Msx1/2
Zic
Pax3/7
FoxD3
Sox9
Snail1/2
Tfap2
Sox10
Ets1
Twist1
Sox8
Neural plate
Notochord
Neural plate border
Non-neural ectoderm
Neural fold
Premigratory NC
Neural tube
EMT
Epidermis
Migrating NC cells
NEURULATION
Mesenchyme
Fin fold
Somite
I.
II.
III.
Cranial NC
Mandibular
Hyoid
Branchial
Cardiac
VAGAL NC
Enteric
TRUNK NC
SACRAL NC
Neural crest derivatives
Chondrocytes
Osteocytes
Odontoblasts
Connective tissues
Cranial sensory ganglia
Ciliary ganglia
Satellite and Schwann cells
Melanocytes
Aorticopulmonary septum
Enteric ganglia
Sympathetic ganglia
Cardiomyocytes
Cardiac ganglia
Pericytes
Melanocytes
Adrenal chromaffin cells
Dorsal root ganglia
Sympathetic ganglia
Schwann cells
Melanocytes
*Odontoblasts
Enteric ganglia
Sympathetic ganglia

FIGURE 3.1 Generalized model of vertebrate neurulation and axial regionalization of the NC and its derivatives along the anteroposterior axis. (A) The neural crest (NC) is induced by signals (mainly BMPs, Wnt, and/or FGFs) in the neural plate border (red region), which upregulate transcription factors called 'neural plate border specifiers' (dashed rectangle) in the ectodermal domain located between neural ectoderm (salmon pink) and non-neural ectoderm (grey). (B) Subsequently, the neural plate border elevates to form neural folds, and the prospective NC expresses a battery of the 'neural crest specifiers' (dashed rectangle). (C) The premigratory NC cells reside within the neural folds, and these cells undergo an epithelial-to-mesenchymal transition (EMT) and delaminate; some NC specifiers also drive EMT (enlarged part of the scheme). (D) NC cells express *bona fide* NC specifiers (dashed rectangle) as they migrate throughout the periphery to distant locations. (E) Cranial NC cells (red) migrate dorsolaterally in the space between the epidermis and mesodermal–endodermal layers. (F) The trunk NC cells (green) migrate in a segmental chainlike fashion, following three major migratory pathways (I. to III.; the third pathway is characteristic of vertebrates such as lamprey or *Xenopus* that have fin folds). (G) The NC can be subdivided into several unique subpopulations along the body axis, designated on the basis of their axial level of origin as cranial, vagal, trunk, and sacral NC, reflecting their axial level of origin. Cranial NC cells migrate throughout the head region in three broad distinct migratory streams termed mandibular (red), hyoid (pink), and branchial (purple). The vagal NC (blue) is a transitional NC subpopulation between the head and trunk regions; the vagal NC can be further subdivided into the cardiac (light blue) and enteric (dark blue) NC subpopulations. The largest NC subpopulation in the postcranial region is trunk NC (green), and the last well-defined subpopulation is the sacral NC (mustard yellow). The borders of individual NC subpopulations vary in different vertebrates. Lateral view on the mammalian embryo. (H) Major derivatives of NC subpopulations and the skeletogenic derivatives are in bold. Asterisk marks possible contribution of trunk NC to the odontoblasts of little skate dermal denticles. A to D are modified from Simões-Costa and Bronner (2015) and E, H are modified from Rothstein et al. (2018).

Following delamination, NC cells migrate throughout the space between the epidermis, neural tube, and somites (Figure 3.1D–F) to distant locations, exhibiting individual or collective cell migration and populating tissues derived from all germ layers. Although the divergence of migration mode is related to their axial level of origin, several studies suggest that NC cells in anamniotes prefer a more collective migration whereas those in amniotes adapt individual cell migration (reviewed in Piacentino et al., 2020). During migration, NC cells are guided by highly coordinated and cooperative mechanisms that can be divided into two categories: Intrinsic mechanisms (repulsion or co-attraction between cells and cell motility) and extrinsic mechanisms (positive and negative extracellular factors including secreted factors and matrix components, tissue fluidity, interactions with molecules at the surface of other cell populations) (reviewed in Gouignard et al., 2018; Szabó and Mayor, 2018; Shellard and Mayor, 2019). Cranial NC cells migrate through the head region in three broad and distinct migratory streams toward their final destinations. These streams are termed mandibular, hyoid, and branchial according to the anteroposterior axis (Figure 3.1G) (Le Douarin and Kalcheim, 1999). Although cranial NC migratory patterns appear to be generally conserved across vertebrates, there are several examples of heterochronies in the timing of NC cells emergence, which strongly influence migratory patterns as well as derivatives (Stundl et al., 2019, 2020). In contrast, the trunk NC cells in amniotes migrate in a segmental chainlike fashion, following two major migratory pathways: A ventromedial route between the neural tube and around or through the somites (Figure 3.1F) and a dorsolateral route below the dorsal ectoderm (cf. cranial NC (Figure 3.1E)) (Serbedzija et al., 1989; Kulesa and Gammil, 2010; Szabó and Mayor, 2018). Despite the general presence of two distinct migratory pathways, there are differences in the number of cells or migratory patterns of individual trunk NC subpopulations among the vertebrates. For example, some cells following the ventromedial route migrate between the somites then turn laterally toward the ectoderm to give rise to pigment cells which primarily arise from the dorsolateral subpopulation (Collazo et al., 1993). In addition, in some vertebrates, such as lamprey, amphibians, or zebrafish, trunk NC cells undergo dorsal migration to contribute to the fin folds (Figure 3.1F) (Collazo et al., 1993; Smith et al., 1994; Häming et al., 2011). However, other studies suggest that the trunk NC cells do not generate mesenchymal cells of the fin fold but rather that these cells are derived from mesoderm (Lee et al., 2013a, 2013b; Taniguchi et al., 2015).

Migratory NC cells express NC specifier genes, including *Sox10*, *Tfap2*, *FoxD3*, and *Ets1* (Figure 3.1D; Khudyakov and Bronner-Fraser, 2009; Simões-Costa and Bronner, 2015; Martik and Bronner, 2017). The NC specification module utilizes a conserved set of genetic interactions in all vertebrates (Sauka-Spengler et al., 2007; Nikitina et al., 2008; Martik and Bronner, 2021); however, at different axial levels, some NC specifiers are controlled by distinct enhancers which serve as 'switches' within individual modules of GRNs coordinating expression of downstream targets (Betancur et al., 2010; Simões-Costa et al., 2012; Murko and Bronner, 2017). One example is the cis-regulation of *FoxD3*, which has two enhancers (NC1 and NC2) mediating expression in the head versus the trunk region of chicken embryos. Whereas NC1 is active in the head and requires *Ets1* inputs, NC2 requires input from *Zic1*. NC1 is activated in the NC cells before their delamination. Subsequently, its activity decreases in migrating cells, in which an activated NC2 enhancer maintains

FoxD3 expression (Simões-Costa et al., 2012). These results demonstrate that cis-regulatory elements may influence the spatiotemporal expression of NC specifiers and also developmental fates of distinct NC subpopulations. Based on these data, it is likely that the molecular toolkit forming the NC GRN may differ between individual NC subpopulations that correspond to the regionalization of the body axis.

3.2 DISTINCT NEURAL CREST SUBPOPULATIONS WITH DIFFERENT DEVELOPMENTAL POTENTIAL ALONG THE ANTEROPOSTERIOR AXIS

Along the body axis of vertebrates, the NC is not a single homogenous cell population. The first pioneering studies recognized two major NC subpopulations—cranial and trunk. Although this simple classification is still often used in recent literature, the NC can be further subdivided into several subpopulations along the body axis, designated on the basis of their axial level of origin as cranial, vagal including cardiac, trunk, and sacral NC (Figure 3.1G; Le Douarin and Kalcheim, 1999). Whereas individual NC subpopulations give rise to specific derivatives, melanocytes and peripheral glia are produced by all axial levels in the embryo (Figure 3.1H; Le Douarin and Kalcheim, 1999), perhaps because these cell types are present throughout the whole body. In contrast, other NC derivatives are specific to particular axial levels: That is, cranial NC produces cartilage, cardiac NC cells contribute to ectomesenchymal components of the heart, and vagal and sacral NC cells uniquely contribute to the enteric nervous system. This indicates that positional cues along the neural axis may influence the developmental fate of NC cells (Rothstein et al., 2018; Rocha et al., 2020). Although all NC subpopulations share common GRN modules, there also exist regional differences between various NC subpopulations originating at different axial levels.

Notably, the cranial NC subpopulation has a markedly different repertoire of developmental fates from the trunk NC. The most apparent is the ability to produce ectomesenchymal derivatives, such as chondroblasts, osteoblasts, or odontoblasts (Le Douarin and Kalcheim, 1999). This raises the intriguing possibility that this unique developmental fate may be the consequence of the deployment of axial-specific circuits. Indeed, based on transcriptional profiling, the chick cranial NC is enriched in transcriptional factors (*Dmbx1*, *Brn3c*, and *Lhx5* expressed throughout NC specification and *Tfap2b*, *Sox8*, and *Ets1* expressed in migrating NC cells; cranial crest-specific subcircuit) that are missing in early trunk NC (Simões-Costa et al., 2014; Simoes-Costa and Bronner, 2016). Importantly, ectopic expression of three transcription factors of this cranial NC-specific subcircuit in the trunk NC region was able to alter the fate of trunk NC cells into 'cranial-like' cells. In addition, this reprogramed trunk NC was able to upregulate genes associated with chondrogenic potential, such as *Alx1* or *Runx2*. After transplantation into the cranial region, the reprogrammed cells gave rise to cartilage condensations, something that chick trunk NC does not normally do *in vivo* (Simoes-Costa and Bronner, 2016). Similarly, a cardiac crest-specific subcircuit comprising *Sox8*, *Ets1*, and *Tgif1* transcription factors is expressed specifically in delaminating cardiac NC cells and is sufficient to reprogram the trunk NC fate such that these cells can rescue persistent *truncus arteriosus* when transplanted in place of ablated cardiac NC (Gandhi et al., 2020). Thus, both these studies demonstrate that the developmental fate of NC subpopulations can be defined by a specific subcircuit that has the ability to alter the fate of other NC subpopulations. We can therefore speculate that the entire proto-neural crest had a more or less similar set of transcription factors constituting simple NC GRN, and differences among the NC subpopulations originated by involving additional transcription factors conferring specific lineage identity.

Indeed, transcriptional profiling at the bulk and single-cell levels can reveal how developmental processes may have changed during vertebrate evolution. Recently, comparative analysis of cranial and trunk NC GRNs of several vertebrates has shown that the cranial NC of lamprey (jawless vertebrate) is similar to trunk NC GRN of jawed vertebrates. Based on this evidence, the authors proposed that the trunk NC might better represent the ancestral form of NC of early vertebrates. Thus, the cranial NC subpopulation is likely to have evolved by progressive addition of new genes into an ancient trunk-like NC, assembling a novel axial-specific regulatory circuit (Figure 3.2) (Martik et al., 2019).

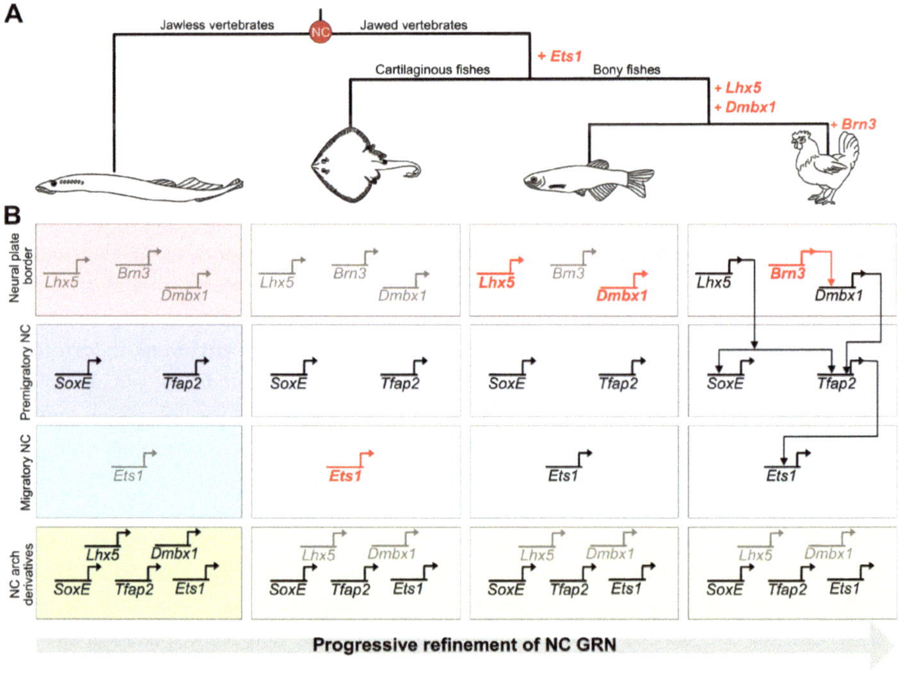

FIGURE 3.2 Model of the evolution of region-specific NC GRN subcircuits during vertebrate evolution. (A) Simplified cladogram of extant vertebrates (lamprey, little skate, zebrafish, chicken) highlighting progressive addition of certain transcription factors (red) to the GRN during the course of vertebrate evolution. (B) The cranial NC subpopulation is likely to have evolved by progressive addition of new genes (red) into an 'ancient' (represented by lamprey) more uniform and trunk-like NC, culminating in the addition of an axial level-specific regulatory circuit. Black—expressed transcription factors; grey—not expressed transcription factors. Based on the data in Martik et al. (2019).

However, as Rocha et al. (2020) remarked, although lampreys lack the cranial crest-specific circuit, NC cells still give rise to pharyngeal cartilages. Moreover, the cranial NC cells of lamprey migrate much like those of amphibians like *Xenopus* and are distinctly different from the migratory routes in the lamprey trunk. Thus, lamprey's simplified NC GRN still displays some distinction between cranial and trunk NC, consistent with the fact that lampreys have continued to evolve over the millennia. Furthermore, comparing extant lampreys and jawed vertebrates may be biased because lampreys do not form bone or dentin. Some studies suggest that anaspids (a group of fossil jawless vertebrates) with bony dermal shields belong among stem jawless vertebrates (Miyashita et al., 2019, 2021). Thus, modern lampreys may not represent the ancestral NC GRN of the common ancestor of jawless and jawed vertebrates, and the absence of abilities to form ectomesenchymal derivatives might be a result of secondary reductions. Nevertheless, it is tempting to speculate that the vertebrate ancestor had a single uniform and trunk-like NC along the entire rostrocaudal axis and that different NC subpopulations and their developmental repertoires evolved by a progressive refinement of NC GRN subcircuits during vertebrate evolution.

3.3 SKELETAL BIOMINERALIZATION IS TIGHTLY ASSOCIATED WITH THE ACQUISITION OF NEURAL CREST

Typically, skeletogenic ability has been primarily associated with the cranial NC subpopulation, which forms most of the craniofacial skeleton. Due to interactions with surrounding tissues, the NC cells generate remarkable craniofacial diversity from filter-feeding sieves to powerful toothy jaws. Although the formation of cartilage was essential for vertebrate evolution (pharyngeal skeleton and its derivatives), cartilaginous tissues similar to vertebrate cartilage have been found in several invertebrates such as cephalopods, hemichordates, polychaetes, and many others (Cole and Hall, 2004; Eames et al., 2020). Recently, studies in cephalopods and horseshoe crab have suggested that vertebrate cartilage did not elaborate *de novo* but instead evolved from the chondrogenic GRN and effector programs in a common bilaterian ancestor (Tarazona et al., 2016; Adameyko, 2020). Key support for this claim is found in a seminal work on amphioxus and chordates that lack NC cells but still possess cellular vertebrate-like cartilage in the oral cirri (Jandzik et al., 2015). The authors suggest that amphioxus oral cartilages might have developed from the coelomic mesothelium implying that the chondrogenetic program was deployed in mesendoderm and later in evolution coopted by ancestral NC. Thus, this might be a part of a general cooption of a 'mesodermal' competence node comprising multipotent progenitor fates and states that in turn led to the ability to give rise to the majority of mesodermal derivatives, including skeletogenic cell types (Adameyko, 2020). However, the evolutionary scenario underlying acquisition of the vertebrate mineralization program is unclear, and its *de novo* elaboration cannot be ruled out. Many genes associated with mineralization of vertebrate hard tissues are members of the secretory calcium-binding phosphoprotein family (SCPP) that evolved from SPARCL1 and SPARC genes associated with fibrillar collagens and also present in non-vertebrate chordates (Kawasaki et al., 2004). Thus, the first mineralization may have occurred in the dermis as the reinforcement by calcium deposition into fibrillar collagens (see Chapter 4); this partially stiff dermis could represent the first dermal armor protecting brain and other inner organs (Kawasaki et al., 2004). Kawasaki and Weiss (2008) proposed that SPARCL1 diverged from SPARC roughly contemporaneous with the origin of skeletal mineralization, and, following tandem duplication, gave rise to many SCPP genes associated with the formation of various vertebrate mineralized tissues (Ryll et al., 2014; Kawasaki, 2018). Although it is unclear how the skeletogenic program was incorporated into the NC GRN, it is tempting to speculate that the diversification of SCPP genes in stem jawed vertebrates might correlate with the acquisition of dermal armor consisting of either dentin, bone, or both tissues.

While bone and dentin represent evolutionary novelties of vertebrates, the evolutionary origin of dentin-secreting cells (odontoblasts) and bone-secreting cells (osteoblasts) remains enigmatic (Eames et al., 2020). These cell types are similar in several aspects, exhibiting cell processes and secreting extracellular matrix proteins. However, odontoblasts can be distinguished from osteoblasts on the basis of expressions of specific matrix genes (e.g., *Dspp* or *Bgp*). While odontoblasts are located in the pulp cavity and project long oriented processes into the mineralized matrix, the osteoblasts are surrounded by the secreted matrix (Kawasaki et al., 2004; Franz-Odendaal et al., 2006; Kawasaki, 2009; Hall and Gillis, 2013).

Interestingly, it has been postulated that odontoblasts might represent an evolutionary modification of neuroglial fate. Accordingly, they may have evolved from sensory receptors transmitting chemical and temperature changes from the water, which were protected by a collagenous and proteinaceous matrix that eventually mineralized at a later time. Indeed, odontoblasts are sensory cells innervated by pain fibers and expressing thermo-, mechanosensory, and voltage-gated ion channels (Okumura et al., 2005; Baker, 2008; Magloire et al., 2009; Ivashkin and Adameyko, 2013). Interestingly, osteoblasts can also respond to mechanical stimuli using the same members of the gene family as odontoblasts, such as *Piezo* (Sun et al., 2019); however, the underlying mechanisms remain unknown.

Because there is no evidence for dentin or dentin-like tissues outside of vertebrates, dentin is unambiguously the first hard tissue that evolved in vertebrates and is an exclusive invention of the NC. The first dentinous tissue was identified in small fragments of an enigmatic vertebrate *Anatolepis* (Late Cambrian to the Early Ordovician period ~490 mya to 450 mya) (Repetski, 1978; Smith et al., 1996). These fragments were ornamented with elongated tubercles, which might be odontodes. Indeed, the evolution of dentin is tightly connected with the formation of the odontodes, which represent the first hard tissue structure and the basic unit of the dermal armor of early vertebrates. Odontodes have a long evolutionary history and comprise dentin surrounding a pulp cavity sometimes covered by hypermineralized tissues (enamel or enameloid) and/or attached by a vascularized bone-like tissue. These hard structures are thought to have evolved in the skin of the first vertebrates and later formed the ornamentation of postcranial dermal plates (Ørvig, 1977; Reif, 1982; Donoghue, 2002; Chen et al., 2020). The primary function of dermal odontodes in the exoskeleton of early vertebrates is a subject of ongoing debate. Some authors suggested that it was sensory function served by odontoblasts (Baker and Bronner-Fraser, 1997; Fraser et al., 2010). Indeed, besides dentin production, maturing odontoblasts mediate the first steps of mechanical, thermal, and chemical sensitivity (Brannstrom, 1963; Shibukawa et al., 2015; Nishiyama et al., 2016; Solé-Magdalena et al., 2018), which is well known to everyone as tooth pain. However, the mechanism responsible for dentin sensitivity is still not fully understood despite several proposed hypotheses. The recent well-described hypothesis is the odontoblast hydrodynamic receptor theory (Shibukawa et al., 2015). Dentin is permeated by many tubules, the pulpal edges of which are filled with odontoblasts that extend their long perfectly oriented processes into the dentinal tubules. The detection of various stimuli on the dentin surface results from fluid flow into the tubules, which induces membrane deformation of odontoblast cell processes and activates releasing of neurotransmitters ATP and glutamate (Nishiyama et al., 2016). Thus, odontoblasts and their cellular processes in particular, which were initially protected by the secretion of the collagen matrix and subsequently mineralized (Baker and Bronner-Fraser, 1997), are the key components of this sensory transduction sequence. Thus, the dermal odontodes might represent one of the first sensory organ systems of early vertebrates, which helped them to detect temperature and chemical signals in surrounding water. Indeed, it has been shown that the dermal odontodes of cave-dwelling catfish have mechano- and/or chemo-sensory ability which enhances their sensory adaptability to extreme environments (Haspel et al., 2012).

It has been hypothesized that the evolutionary development of osteoblasts may be a result of further transformation of the odontogenic program within the NC. Thus, the developmental capacity to mineralize may have shifted deeper into the dermis such that bones would form at the odontode base, an ability that was subsequently coopted into the mesoderm (Ivashkin and Adameyko, 2013; Adameyko, 2020). However, this represents only one possible explanation because the bones might have evolved from mature cartilage or from deuterostome mineralization GRN; thus, the evolutionary origin of bone remains unresolved (Gómez-Picos and Eames, 2015; Eames et al., 2020). Although dentin-producing odontoblasts and bone-producing osteoblasts might share the same evolutionary history, it is also possible that both cell types represent an evolutionary modification of NC neuroglial fate. Because neurosensory organs were distributed throughout the skin surface of first soft-bodied vertebrates, a possible neuroglial origin of odonto-skeletal cell types may help explain why the dermal skeleton evolved in early vertebrates.

Regarding the view that dentin possibly represents the first mineralized tissue of vertebrates that arose within the dermal exoskeleton as odontodes, it is tempting to speculate that biomineralization was originally specific to NC before the cooption into the mesoderm. The development of odontodes is characterized by mutual interactions of the NC-derived mesenchyme and the epithelium. Tooth development represents one of the best models for understanding the evolutionary role of the NC in the formation of hard tissues and odontodes. The tooth morphogenesis begins as a placode (epithelial thickening) which proliferates and invades the mesenchyme whose cells subsequently start to condense and form the initial tooth germ. Subsequent development is characterized by a differentiation of epithelium-derived ameloblasts and NC-derived mesenchyme not only into the odontoblasts but also into the dental pulp, cementum, alveolar bone, and periodontal ligament (Koussoulakou et al., 2009; Krivanek et al., 2020; Jing et al., 2022). Odontode development is controlled by the so-called odontogenic GRN (oGRN) that involves the interplay of NC-derived mesenchyme and an epithelium derived from both ectoderm and endoderm (Figure 3.3; Fraser et al., 2010; Berio and Debiais-Thibaud, 2021).

Although the oGRN has been defined on the basis of the analysis of teeth, the best-studied odontodes, the individual genes of the oGRN are involved in formation of dermal odontodes, including shark and armored catfish denticles (Debiais-Thibaud et al., 2011; Martin et al., 2016; Cooper et al., 2017; Berio and Debiais-Thibaud, 2021; Rivera-Rivera et al., 2021; Mori and Nakamura, 2022; Cooper et al., 2023). Because different forms of odontodes apparently share the expression of most components of oGRN, it is likely to have a deep gnathostome origin. This is consistent with the outside-in theory (scale-to-teeth) suggesting the dermal odontodes arose first as component of dermal exoskeleton prior to tooth formation in oropharyngeal cavity (Reif, 1982; Huysseune et al., 2010; Donoghue and Rücklin, 2016; Haridy et al., 2019). Instead of a simple migration of dermal odontodes into the oropharyngeal cavity to form teeth, it is possible that they evolved by coopting the oGRN into the Sox2-positive progenitors within the oropharyngeal cavity, which were previously associated with the formation of a different epithelial organ, the taste buds. Furthermore, it seems that the cooption of *Sox2* enabled increased regeneration capacity for endless tooth regeneration (polyphyodonty) which by contrast is reduced in denticles (Martin et al., 2016). We can therefore speculate that different types of odontodes are likely to have evolved by progressive addition of new genes into an ancestral oGRN. Rather than the epithelium which arises from various germ layers (Soukup et al., 2008), it is likely that NC-derived mesenchyme has the odontode-competent potential. This possibility is supported by experimental grafting of NC cells from mice to chick embryos which demonstrated that the NC-derived mesenchyme is capable of interpreting epithelial signals and initiating a dental developmental program in toothless avian embryos (Mitsiadis et al., 2003). Overall, this suggests that the odontocompetent NC was critical for the formation of the first dermal odontodes which served diverse functions, including protection, prey processing, and/or possible sensory modalities.

3.4 SCHWANN CELL PRECURSORS, CELLS WITH NEURAL CREST-LIKE DEVELOPMENTAL POTENTIAL

The NC has an enormous developmental potential to produce a vast array of cell types, representing a major driver of vertebrate evolution. Because of its broad developmental potential to form diverse cell types including ectomesenchyme, the NC is often referred to as the fourth germ layer, and vertebrates are considered as quadroblastic organisms. NC-derived cell types can be divided into two major categories: Ectomesenchymal (including chondroblasts, osteoblasts, odontoblasts, pulp cells, adipocytes, pericytes, cardiomyocytes, and many others) and non-ectomesenchymal cell types (different types of neurons, glia, or various pigmented cells) (Figure 3.4). Schwann cells and their precursors (SCPs) are fascinating NC derivatives, which can dedifferentiate or remain in a stem-like state. While other NC-derived cells progressively differentiate during development, SCPs retain expression of a number of core neural crest transcription factors (e.g., *Sox10* and *FoxD3*) and are transient and multipotent progenitor cells distributed along nerve processes of the peripheral nervous

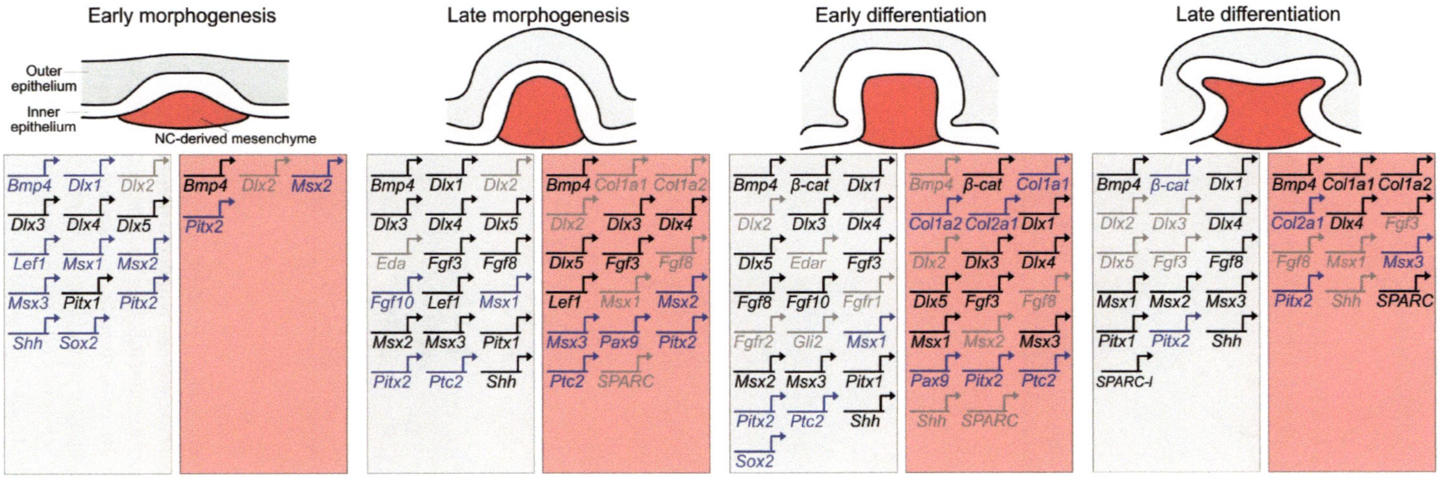

FIGURE 3.3 The odontode development (dermal denticle of catshark) with underlying odontogenic gene regulatory network. Simplified schematic representations of four developmental stages of catshark dermal denticles with highlighted NC-derived mesenchyme (red) and competent epithelium (grey). Early morphogenesis corresponds to the bud stage, late morphogenesis to the cap stage, early differentiation to the early bell stage, and late differentiation to the late-bell stage. The rectangles denote transcription factors expressed at each of these developmental stages in the epithelium or NC-derived mesenchyme, respectively. Black—expression in both dermal (denticle) and oral (tooth) odontodes; blue—expression only in tooth; grey—expression only in denticle. Modified from Berio and Debiais-Thibaud (2021).

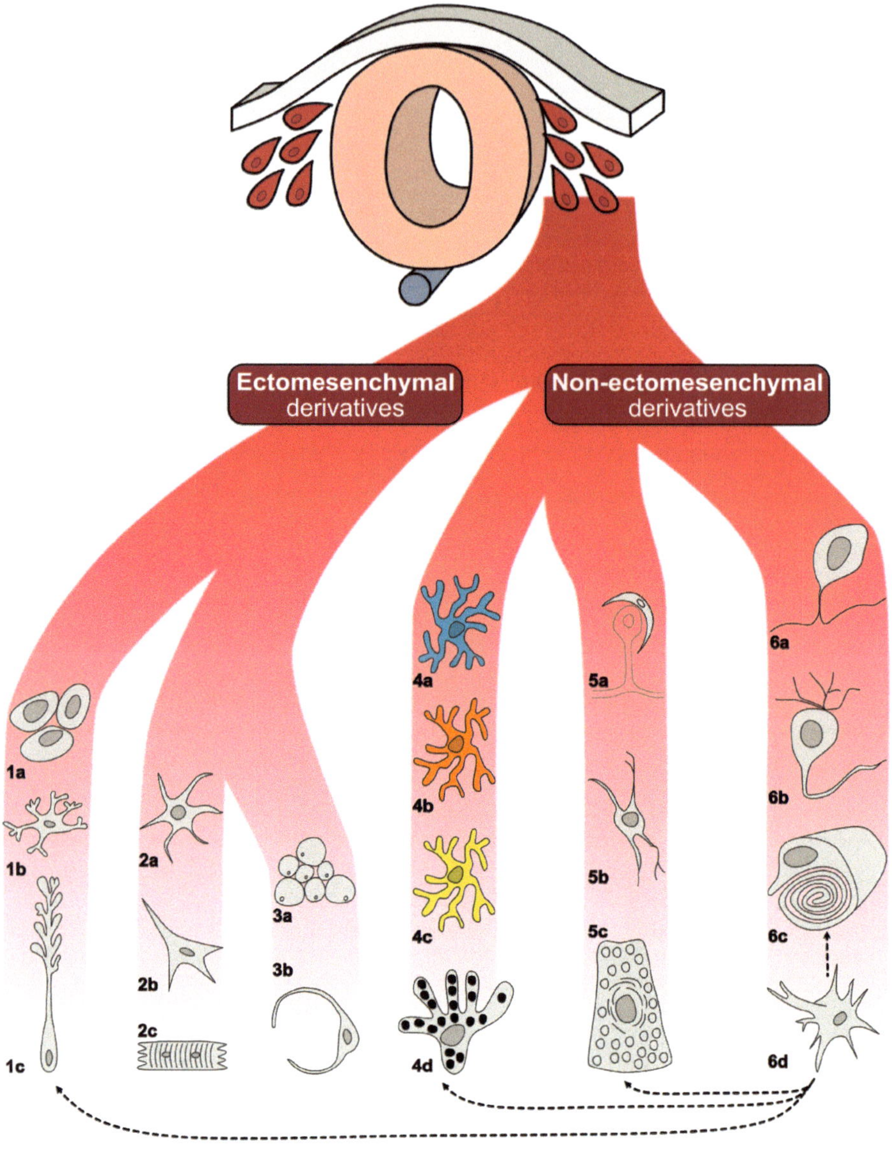

FIGURE 3.4 Neural crest gives rise to a broad range of cell types. Individual NC subpopulations produce both common and axial level-specific cell types. NC derivatives can be divided into two major categories: Ectomesenchymal and non-ectomesenchymal derivatives. 1a, chondroblasts; 1b, osteocytes; 1c, odontoblasts; 2a, fibroblasts; 2b, smooth muscles; 2c, cardiomyocytes; 3a, adipocytes; 3b, pericytes; 4a, iridiophores; 4b, erythrophores; 4c, xanthophores; 4d, melanocytes; 5a, satellite cells; 5b, glia; 5c, chromaffin cells; 6a, sensory neurons; 6b, autonomic neurons; 6c, Schwann cells; 6d, Schwann cell precursors (SCPs). Dashed lines highlight the ability of SCPs to differentiate to many NC cell types.

system. During development, SCPs migrate along the sprouting nerves and differentiate into mature non-myelinating and myelinating Schwann cells, wrapping axons of peripheral nerves and enabling saltatory conduction (Jessen et al., 2015; Solovieva and Bronner, 2021). However, some SCPs detach from nerve fibers and retain the ability to differentiate into numerous derivatives pre-natally (and, in some cases, postnatally) such as pigment cells (Adameyko et al., 2009), enteric neurons (Uesaka et al., 2015), chromaffin cells (Furlan et al., 2017; Kastriti et al., 2019), and many other cell types (reviewed in Solovieva and Bronner, 2021) (Figure 3.4).

Interestingly, SCPs can also generate skeletogenic derivatives, such as odontoblasts (Kaukua et al., 2014) or chondrocytes and osteocytes (Xie et al., 2019). Recent work from Xie et al. (2019) has demonstrated that SCPs leaving the peripheral nerves are capable of giving rise to a minor proportion of chondrocytes and osteocytes of intramembranous and endochondral bones in the craniofacial region but occasionally also contribute to trunk skeletal elements, such as ribs and the scapula. Such ability of NC derivatives to contribute to skeletal elements in the trunk region goes against the classical concept of assuming a solely mesodermal origin of bony trunk elements, except for the reported dual origin of the scapula in mice (Matsuoka et al., 2005). On this basis, it is intriguing to speculate that the NC may have the potential to participate in bone development outside of the head region, at least at the level of SCPs-derived cells. Taken together, SCPs are notably similar to the multipotent NC cells from which they are derived (Furlan and Adameyko, 2018; Kastriti et al., 2022). But unlike the NC, they produce multiple derivatives at a later phase of the development than that normally associated with NC differentiation (Figure 3.4). Moreover, they may provide a targeted source of mesenchymal progenitor cells during late and postnatal development, which might be significant for longer developing and larger animals (Xie et al., 2019). Several recent studies have also demonstrated that the SCPs are a crucial cell population participating in the regeneration of several tissues, such as skeletal muscles (Pessina et al., 2014) and the heart (Mahmoud et al., 2015). Moreover, it seems that SCPs also play a crucial role in skin repair and digit tip regeneration (Johnston et al., 2016). The authors have shown that the ablation of SCPs inhibits mesenchymal precursor proliferation in the blastema, leading to impaired bone and nail regeneration (Johnston et al., 2016). In addition to bones, SCPs participate in tooth formation and give rise to odontoblasts and pulp cells similar to NC cells. Furthermore, the injury of adult incisors revealed that SCPs generate multipotent mesenchymal stem cells that are involved in tooth regeneration, at least in mice (Kaukua et al., 2014). Based on all these studies, it seems that the SCPs may represent *bona fide* NC-like stem cells participating in adult tissue regeneration.

3.5 NEURAL CREST AS A GENERATOR OF ODONTOSKELETOGENIC POTENTIAL ALONG THE BODY AXIS

The jaws, bones, or odontodes represent key traits that contributed to the evolutionary success of early vertebrates by enabling the transition of vertebrate ancestors from soft-bodied filter feeders to active predators with a biting feeding apparatus

(Gans and Northcutt, 1983; Northcutt and Gans, 1983), all features connected with the NC. Indeed, several NC-derived cell types give rise to novel tissues that emerged in the vertebrate lineage during NC evolution, such as dentin, likely representing the first example of vertebrates hard tissue that emerged in the external dermal armor of stem gnathostomes (Smith and Hall, 1990). Thus, in contrast to the generally accepted idea that only the cranial NC gives rise to skeletogenic derivatives, it is probable that the NC had skeletogenic potential along the entire anteroposterior axis of ancestral vertebrates. The concept of cranial specific NC skeletogenic potential is a generalization based on evidence primarily from model organisms such as mice or chicken, which lack extensive postcranial skeleton; this raises the question of when and which part of the NC acquired skeletogenic potential during vertebrate evolution.

The first description of the skeletogenic ability of the NC was provided by Julia Platt in her seminal studies where she used the presence of yolk droplets in ectodermal but not mesodermal cells of a salamander (*Necturus maculosus*) to follow their fate through the developing embryos. She found that mesenchymal cells arise from ectodermal cells containing yolk droplets, and these cells contribute not only to cranial ganglia and pharyngeal cartilages but also to odontoblasts and pulp cells (Platt, 1893, 1897). Platt described these cells as arising from the placodal ectoderm lying lateral to the neural tube, the population now known as the neural crest. Unfortunately, Platt's work was not well-accepted because her results did not fit with the germ layer theory that bones, cartilage, and dentin were associated with mesoderm. However, Platt's results were later confirmed by tissue transplantation experiments in amphibians in studies by several other authors (Stone, 1929; Raven, 1931; Hörstadius and Sellman, 1946, etc.). Essential for understanding NC contributions during embryonic development was the establishment of quail-chick chimera by Nicole Le Douarin. These chimerae revealed numerous NC derivatives, the presence of NC subpopulations along the body axis, and their developmental potential; they also demonstrate that each rhombomere of the hindbrain contributes to distinct segments of the skull (Couly et al., 1993; Kontges and Lumsden, 1996; Le Douarin and Kalcheim, 1999). Another type of avian chimerae, quail-duck, revealed that the premigratory NC cells of the donor contain inherent information about the final craniofacial phenotype (e.g., size of the skeletal elements such as beak) in host species, despite environmental and tissue interactions (Schneider and Helms, 2003; Eames and Schneider, 2008; Lwigale and Schneider, 2008).

As mentioned before, one of the striking differences among NC subpopulations is the ability to produce skeletogenic derivatives. In birds and mammals, this feature is associated with the cranial NC. Based on the quail-chick grafting approaches, it has been suggested that the skeletogenic ability of the NC extends only to the level of somite 5 at least in chicken embryos (Le Lièvre and Le Douarin, 1975). The developmental potential of trunk NC was also analyzed by using quail-chick chimera when the quail donor 'trunk' neural tube was transplanted into the place of the hindbrain in the host chicken embryo, and the experimental embryos produced migrating NC cells giving rise to ectomesenchymal derivatives but not to any skeletogenic cell types such as chondroblasts or osteoblasts. By contrast, the transplantation of quail cranial neural tube into the trunk region of chick embryos not only resulted in the formation of ordinary derivatives of trunk NC but also formed ectopic cartilage nodules (Le Douarin and Teillet, 1974; Le Lievre et al., 1980; Nakamura and Ayer-le Lievre, 1982). Moreover, when aggregates of loose donor cells from the trunk region were cultured and subsequently transplanted to the host maxillary and mandibular prominences in the developing chick head, the transplanted cells were able to participate in the Meckel's and scleral cartilages formation (McGonnell and Graham, 2002). However, one cannot rule out the possibility that culturing of cells before implantation might have changed their developmental abilities. Consistent with this, avian trunk NC is also able to produce the skeletogenic derivatives under certain culture conditions *in vitro* (McGonnell and Graham, 2002; Abzhanov et al., 2003; Ido and Ito, 2006; Coelho-Aguiar et al., 2013) or when it is reprogrammed using cranial NC transcription factors (Simoes-Costa and Bronner, 2016; see Section 3.2). Thus, skeletogenic potential is an intrinsic and distinguishing feature of cranial NC, although trunk NC may have a residual ability to produce ectomesenchyme under appropriate conditions.

While the evolutionary significance of the cranial NC (e.g., jaws bearing teeth or skull protecting sensory organs and brain) is clear, the trunk NC also may have contributed in important ways to vertebrate evolution as many early vertebrates had bodies covered by extensive dermal armor consisting of bone (osteogenic unit), dentin (odontogenic unit), or both tissues (Figure 3.5). The first evidence of the dermal armor in the fossil record may be the Cambrian-Ordovician fossil fragments of *Anatolepis* (Repetski, 1978; Smith et al., 1996; Ahlberg and Haitina, 2020). The *Anatolepis* 'armor' resembles odonto-osteogenic composition because its exoskeleton was formed by individual tubercles capped by dentinous tissue and joined by a lamellar tissue that might represent a precursor of bone (Figure 3.5) (Smith et al., 1996). Although this dermal armor has been lost, reduced, or modified in the majority of extant vertebrates, its residual elements are retained in the form of various types of scales, dermal denticles, spines, or other forms of exoskeleton (reviewed in Smith and Hall, 1993; Sire et al., 2009; Vickaryous and Sire, 2009; Rocha et al., 2020).

The first sign of the potential trunk NC's ability to form postcranial skeletal elements was reported in the hard shell turtle (*Trachemys scripta*) (Clark et al., 2001; Gilbert et al., 2007; Cebra-Thomas et al., 2007, 2013). These studies suggested that the plastron bones located on the ventral surface of the turtle shell are derived from late-migrating trunk NC cells which also expressed some markers characteristic of chick NC cells (e.g., HNK-1 epitope or PDGFRα). Similarly, the HNK-1 and PDGFRα antibodies also positively stained cells of the nuchal bone, the most anterior bone of turtle carapace (Gilbert et al., 2007). Furthermore, HNK-1 expression was detected in cells of alligator gastralia which are thought to be evolutionary homologs of the plastron bones (Gilbert et al., 2007). Recently, similar findings have been reported of cells expressing 'NC' markers participating in the formation of armadillo osteoderms (exoskeletal bones) (Krmpotic et al., 2021). Finally, Soldatov and coauthors (2019) reported the existence of a cryptic bifurcation between neuro-glial and mesenchymal fates in the murine posterior trunk neural crest. In that bifurcation, the clearly emerging mesenchymal bias is weak and does not take over a cell population, which results in an abrogation of the mesenchymal branch and causes the formation of a single neuro-glial branch instead of the fate split toward mesenchymal (skeletogenic) and non-mesenchymal progenies. Based on that, hypothetically, if a strong signaling or autonomous pro-mesenchymal bias would prevail, the trunk NC cells could form a skeletogenic population. Thus, it is possible that the trunk NC has the potential to participate in the postcranial bones formation in certain amniotes under specific conditions (Figure 3.5); however, no single gene can be used to identify a cell population or its fate, suggesting that further validation is necessary.

The first report of trunk NC potential to form skeletal elements was suggested by the observations that DiI-labelled cells originating from the trunk neural tube populate the caudal fin in zebrafish (Smith et al., 1994). A subsequent study using Sox10:cre transgenic lines that enabled long-term fate-mapping of trunk NC cells and demonstrated previously suggested a contribution of trunk NC to the bony fin rays (lepidotrichia) of zebrafish (Kague et al., 2012). Further suggestion that teleost scales are derived from NC cells was based on the observation that NC-derived tumor pigment cells of goldfish can differentiate and form scales *in vitro* (Matsumoto et al., 1983). However, subsequent studies using NC and mesoderm specific transgenic lines suggested that the bone-forming cells in fin rays and scales are derived from mesoderm in both zebrafish and medaka, rather than the NC (Lee et al., 2013a, 2013b; Mongera and Nüsslein-Volhard, 2013; Shimada et al., 2013). Based on these results, it has been concluded that the fin rays and scales of teleosts are mesoderm-derived, and the skeletogenic ability of trunk NC might be associated only with the cranial region (Figure 3.5). That said, it is important to note that most teleosts (including zebrafish or medaka) have highly derived scales consisting of elasmodin, which is not an ordinary bone but a different tissue of enigmatic origin (e.g., some authors suggested that it is evolved from the dentinous layer (Sire and Akimenko, 2004)). The elasmoid scales may have completely lost the basal bone of the ancestral type of scale and evolved a novel tissue, elasmodin. Thus, the developmental evidence about embryonic origin of elasmoid scales may not be relevant for other ray-finned fishes that have different types of scales or different forms of dermal exoskeleton such as scutes or odontodes.

Other teleosts have a wide array of exoskeletal elements such as the armor of poachers or lumpsuckers (Kruppert et al., 2020; Woodruff et al., 2022), spines of puffer fishes (Shono et al., 2019), ctenial spines of cichlids, dermal denticles of gasterosteiforms (including sticklebacks or seahorses), or odontodes of armored catfishes (Sire and Huysseune, 2003). The latter are particularly interesting because their composition and development are similar to teeth (Bhatti, 1936; Rivera-Rivera and Montoya-Burgos, 2017; Mori and Nakamura, 2022). It has also been suggested that the catfish odontodes grow in close association not only with dermal bony plates but also with other bony structures, all of which may trigger odontogenesis. Moreover, based on the evidence that dental structures are derived from NC, it has been proposed that the dermal odontodes might be derived from trunk NC cells (Rivera-Rivera and Montoya-Burgos, 2017). The dermal odontodes are retained in the form of dermal denticles (placoid scales) in cartilaginous fishes (including sharks, skates, and rays). Unlike catfish odontodes, the denticles of crown cartilaginous fishes are composed of typical dentin coated with enameloid (odontogenic unit), and the basal osteogenic unit (bony part) is missing (Miyake et al., 1999; Welten

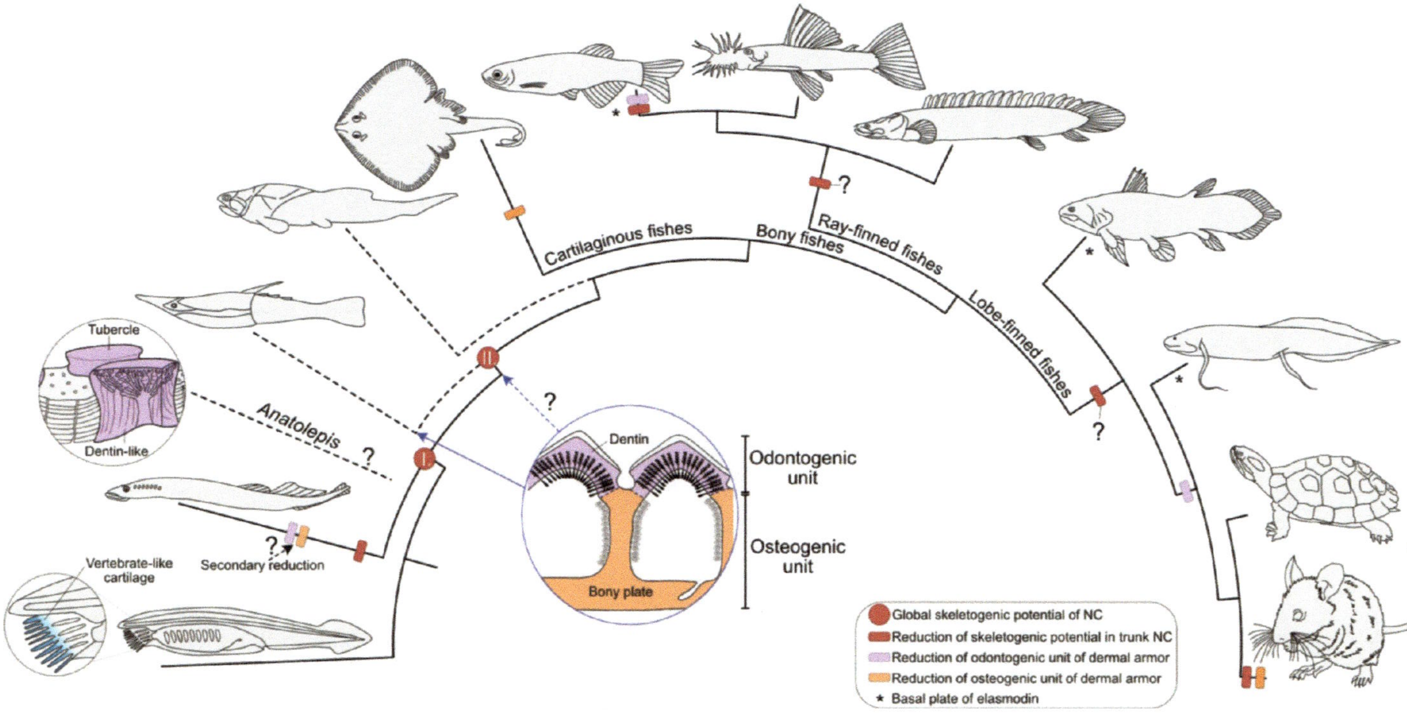

FIGURE 3.5 Hypothesis of the evolution of NC odontoskeletogenic potential. Simplified evolutionary tree of vertebrates (from left to the right: Outgroup—amphioxus, lamprey, *Anatolepis*, heterostracans, placoderms, little skate, zebrafish, armored catfish, bichir, latimeria, lungfish, hardshell turtle, mouse) showing the evolution of NC skeletogenic ability. The skeletogenic potential of NC along the entire body axis was present in the ancestor of vertebrates (I.) or in the ancestor of jawed vertebrates (II.), but this potential was reduced in trunk NC subpopulation (red rounded rectangle) in several vertebrates (e.g., zebrafish). Amphioxus has vertebrate-like cartilage in the oral cirri (blue cartilage in the circle), and the chondrogenic program was coopted later in evolution by ancestral NC, perhaps leading to the ability to give rise to the majority of mesodermal derivatives, including skeletogenic cell types. The dermal armor of early vertebrates, such as seen in heterostracans (blue circle), was composed of bone (osteogenic unit) and dentin (odontogenic unit). The odontogenic unit of the dermal armor of early vertebrates has been suggested to be NC-derived, whereas the osteogenic unit is mesoderm-derived. During the vertebrate evolution, the dermal armor was built by differential loss and elaboration of osteogenic (orange rounded rectangle) and odontogenic units (purple rounded rectangle). The fossil fragments of *Anatolepis* represent the first evidence of dentin-like tissue (purple) in the fossil record: Its dermal armor was formed by individual tubercles capped by dentinous tissue (purple) and joined by a lamellar tissue.

et al., 2015). Recently, lineage tracing in the little skate (*Leucoraja erinacea*) has shown that trunk NC has the ability to give rise to dentin-secreting odontoblasts of dermal denticles which cover the entire body (Gillis et al., 2017). Thus, this suggests the retention or redeployment of skeletogenic developmental programs, at least in cartilaginous fishes (Figure 3.5). Based on the developmental data from the little skate and zebrafish, it has been suggested that the odontogenic layer of the dermal armor of early vertebrates might be derived from trunk NC and the osteogenic layer from paraxial mesoderm. The dermal armor was built by differential loss and elaboration of osteogenic and odontogenic units during vertebrate evolution; this has led to a wide variety of dermal armor found in living and extinct vertebrates (Figure 3.5) (Smith et al., 1993; Shimada et al., 2013; Gillis et al., 2017; Rocha et al., 2020). Thus, it is tempting to speculate that the trunk NC might have had skeletogenic ability along the anteroposterior axis in ancestral vertebrates, but this potential was later gradually suppressed during the course of vertebrate evolution (Figure 3.5).

A caveat is this concept has been formulated on the basis of the developmental data from highly derived teleost scales and one type of dentin of cartilaginous fishes. Although the elasmodin is roughly regarded as bone, its evolutionary origin is still enigmatic (Sire and Akimenko, 2004). It is intriguing to speculate that elasmodin is produced by the mesoderm-derived cells, and the teleost elasmoid scales may represent a specialized condition. Thus, important insight may come from non-teleost fishes or armored catfishes to test the ability of trunk NC to give rise to osteoblasts in the postcranial region. Similarly, the claim about NC-derived dentin is based on the observation from the acellular dentin that is similar to the orthodentin in the teeth of most extant vertebrates. However, the origin of cellular dentin that is commonly found in early vertebrates or osteodentin of teeth of some extant sharks (lamniforms) and rays (batoids) or teleosts (pike) (Jambura et al., 2018; Thangadurai et al., 2022) remains unknown.

Ray-finned fishes are the most numerous and remarkably diversified vertebrates with a highly variable postcranial exoskeleton that has been retained in the form of various scales, ray fins, scutes, or odontodes. Thus, their extensive exoskeletons may represent good model systems for testing the possible skeletogenic abilities of trunk NC cells. Given that non-teleost fishes have an important phylogenetic position within ray-finned fishes as well as different types of scales derived from the ancestral rhomboid scales of early osteichthyans—bichirs with ganoid scales (exoskeletal bone coated with dentin and ganoin layers), sturgeons with bony scutes, gars with ganoid scales (missing the dentin layer) (Sire and Huysseune, 2003; Sire et al., 2009), these may represent better model systems for addressing the skeletogenic potential of trunk NC and the origin of dermal armor (see our recent work Stundl et al., 2023).

3.6 CONCLUSION AND PERSPECTIVES

This chapter attempts to summarize our current knowledge about a fascinating embryonic population, the neural crest (NC), and its odontoskeletogenic potential in vertebrates. A conundrum in the field revolves around whether the NC's ability to form hard tissues, such as bone or dentin, is specific to the cranial NC or an ancestral characteristic of the entire NC but is later restricted to the head in some vertebrate groups. Dentin is unambiguously the first hard tissue that appeared in the vertebrate fossil record and represents an exclusive invention of the neural crest. Based on developmental evidence, it has been suggested that odontoblasts are NC-derived in the cranial (teeth/oral odontodes) and trunk (dermal denticles/odontodes; Gillis et al., 2017) region. Evidence of NC-derived dermal denticles is a bit inconclusive as these data were based on faint CM-DiI staining; however, it is important to note that this type of long-term lineage tracing is very challenging and remains the best lineage tracing method for non-model organisms. Future research using single-cell RNA-sequencing together with spatial transcriptomics may clarify the odontogenic ability of trunk NC by providing developmental trajectories of the denticles cell populations, as has been done for tooth morphogenesis (e.g., Jing et al., 2022). Furthermore, it would be interesting to analyze the embryonic origin of different types of dentin tissues such as osteodentin (see Chapter 5 for more details on the microstructures of these tissues). While unlikely, it cannot be ruled out that different forms of dentin and the odontogenic process may not be unique to NC. However, it is tempting to speculate that the odontocompetent NC was a crucial developmental generator for the formation of the first odontodes and the dermal armor of early vertebrates.

All types of odontodes (odontode-like structures) develop from an epithelial placode through well-defined developmental stages. It has been suggested that they share a common odontogenic gene regulatory network (oGRN) (see Section 3.3) responsible for development of various dental structures, such as teeth and denticles. However, it is important to note that this claim is based primarily on the topographic expressions of a few core genes whereas interactions within individual developmental units and analyses of kernels that distinguish individual 'odontode fates' within the oGRN remain unresolved. In addition, it is noteworthy that feathers and other epithelial appendages developed by epithelial–mesenchymal interactions express a significant number of the genes in common with the oGRN (Lai et al., 2018; Cooper et al., 2019). Although the feathers are recognized as an epithelial structure, chimeric experiments revealed that NC is the developmental source of species-specific patterning information for feathers in the head (Eames and Schneider, 2005) and perhaps in trunk region as well since dermis/mesenchyme influences species-specific patterning of feathers (Dhouailly, 1967; Schneider, 2018). A possible additional perspective might be gained via an analysis of keratinized teeth in lampreys in which the mesenchyme seems to participate (Lethbridge and Potter, 1981). It is intriguing to speculate that vertebrate integumentary appendages develop via epithelial–mesenchymal interactions that might share a basic GRN of deep vertebrate origin. This could have been later modified by the cooption of additional factors into already existing networks which enabled the development of hard tissue odontodes. Overall, this highlights the importance of further research on the oGRN in order to provide valuable insights into the evolution of epithelial–mesenchymal units *sensu lato* and the specific role of NC-derived mesenchymal cells.

While there is evidence for some odontogenic activity of trunk NC in cartilaginous fishes, the potential to form bones remains enigmatic and controversial (e.g., Kague et al. (2012), but see Lee et al. (2013a, 2013b); Mongera and Nüsslein-Volhard (2013); Shimada et al. (2013)). Similarly, several studies on hard shell turtles suggest that later emigrating trunk NC cells may have the ability to give rise to osteoblasts of nuchal and plastron bones (e.g., Cebra-Thomas et al., 2013) but remain inconclusive. It has been suggested that the odontogenic unit of the dermal armor of early vertebrates was NC-derived, whereas the osteogenic unit was mesoderm-derived. However, the origin of the osteogenic unit is based on data from zebrafish elasmoid scales which are highly derived. Going forward, critical insights may be provided from phylogenetically important non-teleost fishes with ganoid scales and scutes and other ray-finned fishes such as armored catfishes (Mori and Nakamura, 2022; Rivera-Rivera et al., 2021). Dermal odontodes of catfishes may represent an interesting model system as they have been suggested to have mechanosensory function in one cave-dwelling catfish (Haspel et al., 2012), concordant with the findings that NC-derived and SCPs-derived odontoblasts have mechanosensory properties. Thus, catfish may represent an excellent model organism (easy to breed in the lab and laying tens of eggs) for testing important Evo-Devo questions regarding 'the sensory odontodes hypothesis.' Going forward, the elucidation of the role of NC in the formation of odontodes and the evolutionary origin of ancient dermal armor will require comparative, interdisciplinary approaches combining disparate fields such as developmental biology, evolutionary genomics, paleontology, or morphology. The application of ever improving technologies (CRISPR-Cas technologies, single-cell RNA-seq, or ATAC-seq, etc.) to modern developmental biology and paleontology (e.g., synchrotron microtomography) holds the promise of producing exciting answers to the role of the neural crest in the formation of the dermal exoskeleton.

ACKNOWLEDGMENTS

We would like to thank Donglei Chen and Per Erik Ahlberg for organizing this book volume; we are greatly in debt to them for their immense patience. Jan Stundl is supported by funding from the European Union's Horizon 2020 research and innovation program under the Marie Skłodowska-Curie grant agreement No. 897949 and NIH grant R35NS111564 to Marianne E. Bronner.

REFERENCES

Abzhanov, A., Tzahor, E., Lassar, A. B., & Tabin, C. J. (2003). Dissimilar regulation of cell differentiation in mesencephalic (cranial) and sacral (trunk) neural crest cells in vitro. *Development, 130*, 4567–4579. https://doi.org/10.1242/dev.00673

Adameyko, I. (2020). Elaboration of fates in neural crest lineage during evolution. In B. F. Eames, D. M. Medeiros, & I. Adameyko (Eds.), *Evolving Neural Crest Cells* (pp. 5–66). CRC Press, Boca Raton.

Adameyko, I., Lallemend, F., Aquino, J. B., Pereira, J. A., Topilko, P., Müller, T., Fritz, N., Beljajeva, A., Mochii, M., Liste, I., Usoskin, D., Suter, U., Birchmeier, C., & Ernfors, P. (2009). Schwann cell precursors from nerve innervation are a cellular origin of melanocytes in skin. *Cell, 139*, 366–379. https://doi.org/10.1016/j.cell.2009.07.049

Ahlberg, P. E., & Haitina, T. (2020). Neural crest in fossil vertebrates: What, if anything, can we know? In B. F. Eames, D. M. Medeiros, & I. Adameyko (Eds.), *Evolving Neural Crest Cells* (pp. 243–264). CRC Press, Boca Raton.

Alkobtawi, M., & Monsoro-Burq, A. H. (2020). The neural crest, a vertebrate invention. In B. F. Eames, D. M. Medeiros, & I. Adameyko (Eds.), *Evolving Neural Crest Cells* (pp. 5–66). CRC Press, Boca Raton.

Arendt, D. (2018). Animal evolution: Convergent nerve cords? *Current Biology*, *28*, R225–R227. https://doi.org/10.1016/j.cub.2018.01.056

Baker, C. V. H. (2008). The evolution and elaboration of vertebrate neural crest cells. *Current Opinion in Genetics & Development*, *18*, 536–543. https://doi.org/10.1016/j.gde.2008.11.006

Baker, C. V. H., & Bronner-Fraser, M. (1997). The origins of the neural crest. Part II: An evolutionary perspective. *Mechanisms of Development*, *69*, 13–29. https://doi.org/10.1016/S0925-4773(97)00129-9

Basch, M. L., Bronner-Fraser, M., & García-Castro, M. I. (2006). Specification of the neural crest occurs during gastrulation and requires Pax7. *Nature*, *441*, 218–222. https://doi.org/10.1038/nature04684

Berio, F., & Debiais-Thibaud, M. (2021). Evolutionary developmental genetics of teeth and odontodes in jawed vertebrates: A perspective from the study of elasmobranchs. *Journal of Fish Biology*, *98*, 906–918. https://doi.org/10.1111/jfb.14225

Betancur, P., Bronner-Fraser, M., & Sauka-Spengler, T. (2010). Genomic code for *Sox10* activation reveals a key regulatory enhancer for cranial neural crest. *Proceedings of the National Academy of Sciences*, *107*, 3570–3575. https://doi.org/10.1073/pnas.0906596107

Betters, E., Charney, R. M., & Garcia-Castro, M. I. (2018). Early specification and development of rabbit neural crest cells. *Developmental Biology*, *444*, S181–S192. https://doi.org/10.1016/j.ydbio.2018.06.012

Bhatti, H. K. (1936). *The Integument and Dermal Skeleton of Siluroidea* [Doctoral dissertation, University of Cambridge]. University of Cambridge Repository. https://doi.org/10.17863/CAM.34268

Brannstrom, M. (1963). Dentin sensitivity and aspiration of odontoblasts. *Journal of American Dental Association*, *66*, 366–370. https://doi.org/10.14219/jada.archive.1963.0104

Bronner-Fraser, M. (2008). On the trail of the 'new head' in Les Treilles. *Development*, *135*, 2995–2999. https://doi.org/10.1242/dev.019901

Cebra-Thomas, J. A., Betters, E., Yin, M., Plafkin, C., McDow, K., & Gilbert, S. F. (2007). Evidence that a late-emerging population of trunk neural crest cells forms the plastron bones in the turtle *Trachemys scripta*. *Evolution & Development*, *9*, 267–277. https://doi.org/10.1111/j.1525-142X.2007.00159.x

Cebra-Thomas, J. A., Terrell, A., Branyan, K., Shah, S., Rice, R., Gyi, L., Yin, M., Hu, Y., Mangat, G., Simonet, J., Betters, E., & Gilbert, S. F. (2013). Late-emigrating trunk neural crest cells in turtle embryos generate an osteogenic ectomesenchyme in the plastron. *Developmental Dynamics*, *242*, 1223–1235. https://doi.org/10.1002/dvdy.24018

Chen, D., Blom, H., Sanchez, S., Tafforeau, P., Märss, T., & Ahlberg, P. E. (2020). The developmental relationship between teeth and dermal odontodes in the most primitive bony fish *Lophosteus*. *Elife*, *9*, e60985. https://doi.org/10.7554/eLife.60985

Clark, K., Bender, G., Murray, B. P., Panfilio, K., Cook, S., Davis, R., Murnen, K., Tuan, R. S., & Gilbert, S. F. (2001). Evidence for the neural crest origin of turtle plastron bones. *Genesis*, *31*, 111–117. https://doi.org/10.1002/gene.10012

Coelho-Aguiar, J. M., Le Douarin, N. M., & Dupin, E. (2013). Environmental factors unveil dormant developmental capacities in multipotent progenitors of the trunk neural crest. *Developmental Biology*, *384*, 13–25. https://doi.org/10.1016/j.ydbio.2013.09.030

Cole, A. G., & Hall, B. K. (2004). The nature and significance of invertebrate cartilages revisited: Distribution and histology of cartilage and cartilage-like tissues within the Metazoa. *Zoology*, *107*, 261–273. https://doi.org/10.1016/j.zool.2004.05.001

Collazo, A., Bronner-Fraser, M., & Fraser, S. E. (1993). Vital dye labelling of *Xenopus laevis* trunk neural crest reveals multipotency and novel pathways of migration. *Development*, *118*, 363–376. https://doi.org/10.1242/dev.118.2.363

Cooper, R. L., Lloyd, V. J., Di-Poï, N., Fletcher, A. G., Barrett, P. M., & Fraser, G. J. (2019). Conserved gene signalling and a derived patterning mechanism underlie the development of avian footpad scales. *Evodevo*, *10*, 19. https://doi.org/10.1186/s13227-019-0130-9

Cooper, R. L., Martin, K. J., Rasch, L. J., & Fraser, G. J. (2017). Developing an ancient epithelial appendage: FGF signalling regulates early tail denticle formation in sharks. *EvoDevo*, *8*, 8. https://doi.org/10.1186/s13227-017-0071-0

Cooper, R. L., Nicklin, E. F., Rasch, L. J., & Fraser, G. J. (2023). Teeth outside the mouth: The evolution and development of shark denticles. *Evolution & Development*, *25*, 54–72. https://doi.org/10.1111/ede.12427

Couly, G. F., Coltey, P. M., & Le Douarin, N. M. (1993). The triple origin of skull in higher vertebrates: A study in quail-chick chimeras. *Development*, *117*, 409–429. https://doi.org/10.1242/dev.117.2.409

Debiais-Thibaud, M., Oulion, S., Bourrat, F., Laurenti, P., Casane, D., & Borday-Birraux, V. (2011). The homology of odontodes in gnathostomes: Insights from *Dlx* gene expression in the dogfish, *Scyliorhinus canicula*. *BMC Evolutionary Biology*, *11*, 307. https://doi.org/10.1186/1471-2148-11-307

Dhouailly, D. (1967). Analysis of the factors in the specific differenciation of the neoptile feathers in the duck and chicken. *Journal of Embryology and Experimental Morphology*, *18*, 389–400.

Donoghue, P. C. J. (2002). Evolution of development of the vertebrate dermal and oral skeletons: Unraveling concepts, regulatory theories, and homologies. *Paleobiology*, *28*, 474–507. https://doi.org/10.1666/0094-8373(2002)028<0474:EODOTV>2.0.CO;2

Donoghue, P. C. J., & Rücklin, M. (2016). The ins and outs of the evolutionary origin of teeth. *Evolution & Development*, *18*, 19–30. https://doi.org/10.1111/ede.12099

Eames, B. F., Gomez-Picos, P., & Jandzik, D. (2020). On the evolution of skeletal cells before and after neural crest. In B. F. Eames, D. M. Medeiros, & I. Adameyko (Eds.), *Evolving Neural Crest Cells* (pp. 185–218). CRC Press, Boca Raton.

Eames, B. F., & Schneider, R. A. (2005). Quail-duck chimeras reveal spatiotemporal plasticity in molecular and histogenic programs of cranial feather development. *Development*, *132*, 1499–1509. https://doi.org/10.1242/dev.01719

Eames, B. F., & Schneider, R. A. (2008). The genesis of cartilage size and shape during development and evolution. *Development*, *135*, 3947–3958. https://doi.org/10.1242/dev.023309

Franz-Odendaal, T. A., Hall, B. K., & Witten, P. E. (2006). Buried alive: How osteoblasts become osteocytes. *Developmental Dynamics*, *235*, 176–190. https://doi.org/10.1002/dvdy.20603

Fraser, G. J., Cerny, R., Soukup, V., Bronner-Fraser, M., & Streelman, J. T. (2010). The odontode explosion: The origin of tooth-like structures in vertebrates. *BioEssays*, *32*, 808–817. https://doi.org/10.1002/bies.200900151

Furlan, A., & Adameyko, I. (2018). Schwann cell precursor: A neural crest cell in disguise? *Developmental Biology*, *444*, S25–S35. https://doi.org/10.1016/j.ydbio.2018.02.008

Furlan, A., Dyachuk, V., Kastriti, M. E., Calvo-Enrique, L., Abdo, H., Hadjab, S., Chontorotzea, T., Akkuratova, N., Usoskin, D., Kamenev, D., Petersen, J., Sunadome, K., Memic, F., Marklund, U., Fried, K., Topilko, P., Lallemend, F., Kharchenko, P. V., Ernfors, P., & Adameyko, I. (2017). Multipotent peripheral glial cells generate neuroendocrine cells of the adrenal medulla. *Science*, *357*, eaal3753. https://doi.org/10.1126/science.aal3753

Gandhi, S., Ezin, M., & Bronner, M. E. (2020). Reprogramming axial level identity to rescue neural-crest-related congenital heart defects. *Developmental Cell*, *53*, 300–315.e4. https://doi.org/10.1016/j.devcel.2020.04.005

Gans, C., & Northcutt, R. G. (1983). Neural crest and the origin of vertebrates: A new head. *Science*, *220*, 268–274. https://doi.org/10.1126/science.220.4594.268

Gilbert, S. F., Bender, G., Betters, E., Yin, M., & Cebra-Thomas, J. A. (2007). The contribution of neural crest cells to the nuchal bone and plastron of the turtle shell. *Integrative & Comparative Biology*, *47*, 401–408. https://doi.org/10.1093/icb/icm020

Gillis, J. A., Alsema, E. C., & Criswell, K. E. (2017). Trunk neural crest origin of dermal denticles in a cartilaginous fish. *Proceedings of the National Academy of Sciences*, *114*, 13200–13205. https://doi.org/10.1073/pnas.1713827114

Gómez-Picos, P., & Eames, B. F. (2015). On the evolutionary relationship between chondrocytes and osteoblasts. *Frontiers in Genetics*, *6*, 297. https://doi.org/10.3389/fgene.2015.00297

Gouignard, N., Andrieu, C., & Theveneau, E. (2018). Neural crest delamination and migration: Looking forward to the next 150 years. *Genesis*, *56*, e23107. https://doi.org/10.1002/dvg.23107

Hall, B. K. (1998). Germ layers and the germ-layer theory revisited: Primary and secondary germ layers, neural crest as a fourth germ layer, homology, and demise of the germ-layer theory. *Evolutionary Biology*, *30*, 121–186. https://doi.org/10.1007/978-1-4899-1751-5_5

Hall, B. K. (2000). The neural crest as a fourth germ layer and vertebrates as quadroblastic not triploblastic. *Evolution & Development*, *2*, 3–5. https://doi.org/10.1046/j.1525-142x.2000.00032.x

Hall, B. K. (2008). The neural crest and neural crest cells: Discovery and significance for theories of embryonic organization. *Journal of Biosciences*, *33*, 781–793. https://doi.org/10.1007/s12038-008-0098-4

Hall, B. K., & Gillis, J. A. (2013). Incremental evolution of the neural crest, neural crest cells and neural crest-derived skeletal tissues. *Journal of Anatomy*, *222*, 19–31. https://doi.org/10.1111/j.1469-7580.2012.01495.x

Häming, D., Simoes-Costa, M., Uy, B., Valencia, J., Sauka-Spengler, T., & Bronner-Fraser, M. (2011). Expression of sympathetic nervous system genes in lamprey suggests their recruitment for specification of a new vertebrate feature. *PLoS ONE*, *6*, e26543. https://doi.org/10.1371/journal.pone.0026543

Haridy, Y., Gee, B. M., Witzmann, F., Bevitt, J. J., & Reisz, R. R. (2019). Retention of fish-like odontode overgrowth in Permian tetrapod dentition supports outside-in theory of tooth origins. *Biology Letters*, *15*, 20190514. https://doi.org/10.1098/rsbl.2019.0514

Haspel, G., Schwartz, A., Streets, A., Camacho, D. E., & Soares, D. (2012). By the teeth of their skin, cavefish find their way. *Current Biology*, *22*, R629–R630. https://doi.org/10.1016/j.cub.2012.06.035

Hörstadius, S., & Sellman, S. (1946). Experimentelle Untersuchungen über die Determination des knorpeligen Kopfskelettes bei Urodelen. *Nova Acta Regiae Societatis Scientiarum Upsaliensis*, *13*, 1–170.

Huysseune, A., Sire, J. Y., & Witten, P. E. (2010). A revised hypothesis on the evolutionary origin of the vertebrate dentition. *Journal of Applied Ichthyology*, *26*, 152–155. https://doi.org/10.1111/j.1439-0426.2010.01395.x

Ido, A., & Ito, K. (2006). Expression of chondrogenic potential of mouse trunk neural crest cells by FGF2 treatment. *Developmental Dynamics*, *235*, 361–367. https://doi.org/10.1002/dvdy.20635

Ivashkin, E., & Adameyko, I. (2013). Progenitors of the protochordate ocellus as an evolutionary origin of the neural crest. *EvoDevo*, *4*, 12. https://doi.org/10.1186/2041-9139-4-12

Jambura, P. L., Pfaff, C., Underwood, C. J., Ward, D. J., & Kriwet, J. (2018). Tooth mineralization and histology patterns in extinct and extant snaggletooth sharks, *Hemipristis* (Carcharhiniformes, Hemigaleidae)—Evolutionary significance or ecological adaptation? *PLoS ONE*, *13*, e0200951. https://doi.org/10.1371/journal.pone.0200951

Jandzik, D., Garnett, A. T., Square, T. A., Cattell, M. V., Yu, J.-K., & Medeiros, D. M. (2015). Evolution of the new vertebrate head by co-option of an ancient chordate skeletal tissue. *Nature, 518*, 534–537. https://doi.org/10.1038/nature14000

Jessen, K. R., Mirsky, R., & Lloyd, A. C. (2015). Schwann cells: Development and role in nerve repair. *Cold Spring Harbor Perspectives in Biology, 7*, a020487. https://doi.org/10.1101/cshperspect.a020487

Jing, J., Feng, J., Yuan, Y., Guo, T., Lei, J., Pei, F., Ho, T. V., & Chai, Y. (2022). Spatiotemporal single-cell regulatory atlas reveals neural crest lineage diversification and cellular function during tooth morphogenesis. *Nature Communications, 13*, 4803. https://doi.org/10.1038/s41467-022-32490-y

Johnston, A. P. W., Yuzwa, S. A., Carr, M. J., Mahmud, N., Storer, M. A., Krause, M. P., Jones, K., Paul, S., Kaplan, D. R., & Miller, F. D. (2016). Dedifferentiated Schwann cell precursors secreting paracrine factors are required for regeneration of the mammalian digit tip. *Cell Stem Cell, 19*, 433–448. https://doi.org/10.1016/j.stem.2016.06.002

Kague, E., Gallagher, M., Burke, S., Parsons, M., Franz-Odendaal, T., & Fisher, S. (2012). Skeletogenic fate of zebrafish cranial and trunk neural crest. *PLoS ONE, 7*, e47394. https://doi.org/10.1371/journal.pone.0047394

Kastriti, M. E., Faure, L., Von Ahsen, D., Bouderlique, T. G., Boström, J., Solovieva, T., Jackson, C., Bronner, M., Meijer, D., Hadjab, S., Lallemend, F., Erickson, A., Kaucka, M., Dyachuk, V., Perlmann, T., Lahti, L., Krivanek, J., Brunet, J. F., Fried, K., & Adameyko, I. (2022). Schwann cell precursors represent a neural crest-like state with biased multipotency. *The EMBO Journal, 41*, e108780. https://doi.org/10.15252/embj.2021108780

Kastriti, M. E., Kameneva, P., Kamenev, D., Dyachuk, V., Furlan, A., Hampl, M., Memic, F., Marklund, U., Lallemend, F., Hadjab, S., Calvo-Enrique, L., Ernfors, P., Fried, K., & Adameyko, I. (2019). Schwann cell precursors generate the majority of chromaffin cells in zuckerkandl organ and some sympathetic neurons in paraganglia. *Frontiers in Molecular Neuroscience, 12*, 6. https://doi.org/10.3389/fnmol.2019.00006

Kaukua, N., Shahidi, M. K., Konstantinidou, C., Dyachuk, V., Kaucka, M., Furlan, A., An, Z., Wang, L., Hultman, I., Ahrlund-Richter, L., Blom, H., Brismar, H., Lopes, N. A., Pachnis, V., Suter, U., Clevers, H., Thesleff, I., Sharpe, P., Ernfors, P., Fried, K., & Adameyko, I. (2014). Glial origin of mesenchymal stem cells in a tooth model system. *Nature*, 513, 551–554. https://doi.org/10.1038/nature13536

Kawasaki, K. (2009). The SCPP gene repertoire in bony vertebrates and graded differences in mineralized tissues. *Development Genes and Evolution, 219*, 147–157. https://doi.org/10.1007/s00427-009-0276-x

Kawasaki, K. (2018). The origin and early evolution of SCPP genes and tissue mineralization in vertebrates. In K. Endo, T. Kogure, & H. Nagasawa (Eds.), *Biomineralization*. Springer, Singapore. https://doi.org/10.1007/978-981-13-1002-7_17

Kawasaki, K., Suzuki, T., & Weiss, K. M. (2004). Genetic basis for the evolution of vertebrate mineralized tissue. *Proceedings of the National Academy of Sciences, 101*, 11356–11361. https://doi.org/10.1073/pnas.0404279101

Kawasaki, K., & Weiss, K. M. (2008). SCPP gene evolution and the dental mineralization continuum. *Journal of Dental Research, 87*, 520–531. https://doi.org/10.1177/154405910808700608

Khudyakov, J., & Bronner-Fraser, M. (2009). Comprehensive spatiotemporal analysis of early chick neural crest network genes. *Developmental Dynamics, 238*, 716–723. https://doi.org/10.1002/dvdy.21881

Kontges, G., & Lumsden, A. (1996). Rhombencephalic neural crest segmentation is preserved throughout craniofacial ontogeny. *Development, 122*, 3229–3242. https://doi.org/10.1242/dev.122.10.3229

Koussoulakou, D. S., Margaritis, L. H., & Koussoulakos, S. L. (2009). A curriculum vitae of teeth: Evolution, generation, regeneration. *International Journal of Biological Sciences*, *5*, 226–243. https://doi.org/10.7150/ijbs.5.226

Krivanek, J., Soldatov, R. A., Kastriti, M. E., Chontorotzea, T., Herdina, A. N., Petersen, J., Szarowska, B., Landova, M., Matejova, V. K., Holla, L. I., Kuchler, U., Zdrilic, I. V., Vijaykumar, A., Balic, A., Marangoni, P., Klein, O. D., Neves, V. C. M., Yianni, V., Sharpe, P. T., Harkany, T., Metscher, B. D., Bajénoff, M., Mina, M., Fried, K., Kharchenko, P. V., & Adameyko, I. (2020). Dental cell type atlas reveals stem and differentiated cell types in mouse and human teeth. *Nature Communications*, *11*, 4816. https://doi.org/10.1038/s41467-020-18512-7

Krmpotic, C. M., Nishida, F., Galliari, F. C., et al. (2021). The dorsal integument of the southern long-nosed armadillo dasypus hybridus (Cingulata, Xenarthra), and a possible neural crest origin of the osteoderms. Discussing evolutive consequences for amniota. *Journal of Mammalian Evolution*, *28*, 635–645. https://doi.org/10.1007/s10914-021-09538-9

Kruppert, S., Chu, F., Stewart, M. C., Schmitz, L., & Summers, A. P. (2020). Ontogeny and potential function of poacher armor (Actinopterygii: Agonidae). *Journal of Morphology*, *281*, 1018–1028. https://doi.org/10.1002/jmor.21223

Kulesa, P. M., & Gammill, L. S. (2010). Neural crest migration: Patterns, phases and signals. *Developmental Biology*, *344*, 566–568. https://doi.org/10.1016/j.ydbio.2010.05.005

Lai, Y. C., Liang, Y. C., Jiang, T. X., Widelitz, R. B., Wu, P., & Chuong, C. M. (2018). Transcriptome analyses of reprogrammed feather/scale chimeric explants revealed co-expressed epithelial gene networks during organ specification. *BMC Genomics*, *19*, 780. https://doi.org/10.1186/s12864-018-5184-x

Le Douarin, N. M., & Kalcheim, C. (1999). *The Neural Crest* (2nd ed.). Cambridge University Press, Cambridge. https://doi.org/10.1017/CBO9780511897948

Le Douarin, N. M., & Teillet, M. A. (1974). Experimental analysis of the migration and differentiation of neuroblasts of the autonomic nervous system and of neurectodermal mesenchymal derivatives, using a biological cell marking technique. *Developmental Biology*, *41*, 162–184. https://doi.org/10.1016/0012-1606(74)90291-7

Lee, R. T. H., Knapik, E. W., Thiery, J. P., & Carney, T. J. (2013a). An exclusively mesodermal origin of fin mesenchyme demonstrates that zebrafish trunk neural crest does not generate ectomesenchyme. *Development*, *140*, 2923–2932. https://doi.org/10.1242/dev.093534

Lee, R. T. H., Thiery, J. P., & Carney, T. J. (2013b). Dermal fin rays and scales derive from mesoderm, not neural crest. *Current Biology*, *23*, R336–R337. https://doi.org/10.1016/j.cub.2013.02.055

Le Lièvre, C. S., & Le Douarin, N. M. (1975). Mesenchymal derivatives of the neural crest: Analysis of chimaeric quail and chick embryos. *Development*, *34*, 125–154. https://doi.org/10.1242/dev.34.1.125

Le Lievre, C. S., Schweizer, G. G., Ziller, C. M., & Le Douarin, N. M. (1980). Restrictions of developmental capabilities in neural crest cell derivatives as tested by *in vivo* transplantation experiments. *Developmental Biology*, *77*, 362–378. https://doi.org/10.1016/0012-1606(80)90481-9

Lethbridge, R. C., & Potter, I. C. (1981). The development of teeth and associated feeding structures during the metamorphosis of the lamprey, Geotria australis. *Acta Zoologica*, *62*, 201–214. https://doi.org/10.1111/j.1463-6395.1981.tb00629.x

Li, Y., Zhao, D., Horie, T., Chen, G., Bao, H., Chen, S., Liu, W., Horie, R., Liang, T., Dong, B., Feng, Q., Tao, Q., & Liu, X. (2017). Conserved gene regulatory module specifies lateral neural borders across bilaterians. *Proceedings of the National Academy of Sciences*, *114*, E6352–E6360. https://doi.org/10.1073/pnas.1704194114

Lwigale, P. Y., & Schneider, R. A. (2008). Other chimeras: Quail—duck and mouse—chick. *Methods in Cell Biology*, *87*, 59–74. https://doi.org/10.1016/S0091-679X(08)00203-3

Magloire, H., Couble, M.-L., Thivichon-Prince, B., Maurin, J.-C., & Bleicher, F. (2009). Odontoblast: A mechano-sensory cell. *Journal of Experimental Zoology Part B: Molecular and Developmental Evolution*, *312B*, 416–424. https://doi.org/10.1002/jez.b.21264

Mahmoud, A. I., O'Meara, C. C., Gemberling, M., Zhao, L., Bryant, D. M., Zheng, R., Gannon, J. B., Cai, L., Choi, W.-Y., Egnaczyk, G. F., Burns, C. E., Burns, C. G., MacRae, C. A., Poss, K. D., & Lee, R. T. (2015). Nerves regulate cardiomyocyte proliferation and heart regeneration. *Developmental Cell*, *34*, 387–399. https://doi.org/10.1016/j.devcel.2015.06.017

Martik, M. L., & Bronner, M. E. (2017). Regulatory logic underlying diversification of the neural crest. *Trends in Genetics*, *33*, 715–727. https://doi.org/10.1016/j.tig.2017.07.015

Martik, M. L., & Bronner, M. E. (2021). Riding the crest to get a head: Neural crest evolution in vertebrates. *Nature Reviews Neuroscience*, *22*, 616–626. https://doi.org/10.1038/s41583-021-00503-2

Martik, M. L., Gandhi, S., Uy, B. R., Gillis, J. A., Green, S. A., Simoes-Costa, M., & Bronner, M. E. (2019). Evolution of the new head by gradual acquisition of neural crest regulatory circuits. *Nature*, *574*, 675–678. https://doi.org/10.1038/s41586-019-1691-4

Martin, K. J., Rasch, L. J., Cooper, R. L., Metscher, B. D., Johanson, Z., & Fraser, G. J. (2016). Sox2+ progenitors in sharks link taste development with the evolution of regenerative teeth from denticles. *Proceedings of the National Academy of Sciences*, *113*, 14769–14774. https://doi.org/10.1073/pnas.1612354113

Matsumoto, J., Lynch, T. J., Grabowski, S., Richards, C. M., Lo, S. L., Clark, C., Kern, D., Taylor, J. D., Tchen, T. T., Ishikawa, T., Masahito, P., & Takayama, S. (1983). Fish tumor pigment cells: Differentiation and comparison to their normal counterparts. *American Zoologist*, *23*, 569–580. https://doi.org/10.1093/icb/23.3.569

Matsuoka, T., Ahlberg, P. E., Kessaris, N., Iannarelli, P., Dennehy, U., Richardson, W. D., McMahon, A. P., & Koentges, G. (2005). Neural crest origins of the neck and shoulder. *Nature*, *436*, 347–355. https://doi.org/10.1038/nature03837

McGonnell, I. M., & Graham, A. (2002). Trunk neural crest has skeletogenic potential. *Current Biology*, *12*, 767–771. https://doi.org/10.1016/S0960-9822(02)00818-7

Meulemans, D., & Bronner-Fraser, M. (2004). Gene-regulatory interactions in neural crest evolution and development. *Developmental Cell*, *7*, 291–299. https://doi.org/10.1016/j.devcel.2004.08.007

Mitsiadis, T. A., Chéraud, Y., Sharpe, P., & Fontaine-Pérus, J. (2003). Development of teeth in chick embryos after mouse neural crest transplantations. *Proceedings of the National Academy of Sciences*, *100*, 6541–6545. https://doi.org/10.1073/pnas.1137104100

Miyake, T., Vaglia, J. L., Taylor, L. H., & Hall, B. K. (1999). Development of dermal denticles in skates (Chondrichthyes, Batoidea): Patterning and cellular differentiation. *Journal of Morphology*, *241*, 61–81. https://doi.org/10.1002/(SICI)1097-4687(199907)241:1<61::AID-JMOR4>3.0.CO;2-S

Miyashita, T., Coates, M. I., Farrar, R., Larson, P., Manning, P. L., Wogelius, R. A., Edwards, N. P., Anné, J., Bergmann, U., Palmer, A. R., & Currie, P. J. (2019). Hagfish from the Cretaceous Tethys Sea and a reconciliation of the morphological—molecular conflict in early vertebrate phylogeny. *Proceedings of the National Academy of Sciences*, *116*, 2146–2151. https://doi.org/10.1073/pnas.1814794116

Miyashita, T., Gess, R. W., Tietjen, K., & Coates, M. I. (2021). Non-ammocoete larvae of Palaeozoic stem lampreys. *Nature*, *591*, 408–412. https://doi.org/10.1038/s41586-021-03305-9

Mongera, A., & Nüsslein-Volhard, C. (2013). Scales of fish arise from mesoderm. *Current Biology, 23*, R338–R339. https://doi.org/10.1016/j.cub.2013.02.056

Mori, S., & Nakamura, T. (2022). Redeployment of odontode gene regulatory network underlies dermal denticle formation and evolution in suckermouth armored catfish. *Scientific Reports, 12*, 6172. https://doi.org/10.1038/s41598-022-10222-y

Murko, C., & Bronner, M. E. (2017). Tissue specific regulation of the chick Sox10E1 enhancer by different Sox family members. *Developmental Biology, 422*, 47–57. https://doi.org/10.1016/j.ydbio.2016.12.004

Nakamura, H., & Ayer-Le Lievre, C. S. (1982). Mesectodermal capabilities of the trunk neural crest of birds. *Journal of Embryology and Experimental Morphology, 70*, 1–18.

Nieto, M. A. (2011). The ins and outs of the epithelial to mesenchymal transition in health and disease. *Annual Reviews of Cell and Developmental Biology, 27*, 347–376. https://doi.org/10.1146/annurev-cellbio-092910-154036

Nikitina, N., Sauka-Spengler, T., & Bronner-Fraser, M. (2008). Dissecting early regulatory relationships in the lamprey neural crest gene network. *Proceedings of the National Academy of Sciences, 105*, 20083–20088. https://doi.org/10.1073/pnas.0806009105

Nishiyama, A., Sato, M., Kimura, M., Katakura, A., Tazaki, M., & Shibukawa, Y. (2016). Intercellular signal communication among odontoblasts and trigeminal ganglion neurons via glutamate. *Cell Calcium, 60*, 341–355. https://doi.org/10.1016/j.ceca.2016.07.003

Northcutt, R. G. (2005). The new head hypothesis revisited. *Journal of Experimental Zoology Part B: Molecular and Developmental Evolution, 304B*, 274–297. https://doi.org/10.1002/jez.b.21063

Northcutt, R. G., & Gans, C. (1983). The genesis of neural crest and epidermal placodes: A reinterpretation of vertebrate origins. *The Quarterly Review of Biology, 58*, 1–28. https://doi.org/10.1086/413055

Okumura, R., Shima, K., Muramatsu, T., Nakagawa, K., Shimono, M., Suzuki, T., Magloire, H., & Shibukawa, Y. (2005). The odontoblast as a sensory receptor cell? The expression of TRPV1 (VR-1) channels. *Archives of Histology and Cytology, 68*, 251–257. https://doi.org/10.1679/aohc.68.251

Ørvig, T. (1977). A survey of odontodes ('dermal teeth') from developmental, structural, functional, and phyletic points of view. In S. M. Andrews, R. S. Miles, & A. D. Walker (Eds.), *Problems in Vertebrate Evolution* (pp. 5–66). Academic Press, London.

Pessina, P., Cabrera, D., Morales, M. G., Riquelme, C. A., Gutiérrez, J., Serrano, A. L., Brandan, E., & Muñoz-Cánoves, P. (2014). Novel and optimized strategies for inducing fibrosis *in vivo*: Focus on Duchenne Muscular Dystrophy. *Skeletal Muscle, 4*, 7. https://doi.org/10.1186/2044-5040-4-7

Piacentino, M. L., Li, Y., & Bronner, M. E. (2020). Epithelial-to-mesenchymal transition and different migration strategies as viewed from the neural crest. *Current Opinion in Cell Biology, 66*, 43–50. https://doi.org/10.1016/j.ceb.2020.05.001

Platt, J. B. (1893). Ectodermic origin of the cartilages of the head. *Anatomischer Anzeiger, 8*, 506–509.

Platt, J. B. (1897). The development of the cartilaginous skull and of the branchial and hypoglossal musculature in *Necturus*. *Morphologisches Jahrbuch, 25*, 377–464.

Raven, C. P. (1931). Zur entwicklung der Ganglienleiste. I. Die Kinematik der Ganglienleistenentwicklung bei den Urodelen. *Wilhelm Roux' Archiv für Entwicklungsmechanik der Organismen, 125*, 210–292. https://doi.org/10.1007/BF00576356

Reif, W.E. (1982). Evolution of dermal skeleton and dentition in vertebrates. In M. K. Hecht, B. Wallace, & G. T. Prance (Eds.), *Evolutionary Biology Volume 15* (pp. 287–368). Springer, New York. https://doi.org/10.1007/978-1-4615-6968-8_7

Repetski, J. E. (1978). A fish from the Upper Cambrian of North America. *Science*, *200*, 529–531. https://doi.org/10.1126/science.200.4341.529

Rivera-Rivera, C. J., Guevara-Delgadillo, N. I., Bahechar, I. A., Shea, C. A., & Montoya-Burgo, J. I. (2021). Loricarioid catfish evolved skin denticles that recapitulate teeth at the structural, developmental, and genetic levels. *bioRxiv*. https://doi.org/10.1101/2021.05.17.444419

Rivera-Rivera, C. J., & Montoya-Burgos, J. I. (2017). Trunk dental tissue evolved independently from underlying dermal bony plates but is associated with surface bones in living odontode-bearing catfish. *Proceedings of the Royal Society B: Biological Sciences*, *284*, 20171831. https://doi.org/10.1098/rspb.2017.1831

Rocha, M., Beiriger, A., Kushkowski, E. E., Miyashita, T., Singh, N., Venkataraman, V., & Prince, V. E. (2020). From head to tail: Regionalization of the neural crest. *Development*, *147*, dev193888. https://doi.org/10.1242/dev.193888

Rothstein, M., Bhattacharya, D., & Simoes-Costa, M. (2018). The molecular basis of neural crest axial identity. *Developmental Biology*, *444*, S170–S180. https://doi.org/10.1016/j.ydbio.2018.07.026

Ryll, B., Sanchez, S., Haitina, T., Tafforeau, P., & Ahlberg, P. E. (2014). The genome of Callorhinchus and the fossil record: A new perspective on SCPP gene evolution in gnathostomes. *Evolution & Development*, *16*, 123–124. https://doi.org/10.1111/ede.12071

Sauka-Spengler, T., Meulemans, D., Jones, M., & Bronner-Fraser, M. (2007). Ancient evolutionary origin of the neural crest gene regulatory network. *Developmental Cell*, *13*, 405–420. https://doi.org/10.1016/j.devcel.2007.08.005

Schneider, R. A. (2018). Neural crest and the origin of species-specific pattern. *Genesis*, *56*, e23219. https://doi.org/10.1002/dvg.23219

Schneider, R. A., & Helms, J. A. (2003). The cellular and molecular origins of beak morphology. *Science*, *299*, 565–568. https://doi.org/10.1126/science.1077827

Serbedzija, G. N., Bronner-Fraser, M., & Fraser, S. E. (1989). A vital dye analysis of the timing and pathways of avian trunk neural crest cell migration. *Development*, *106*, 809–816. https://doi.org/10.1242/dev.106.4.809

Shellard, A., & Mayor, R. (2019). Integrating chemical and mechanical signals in neural crest cell migration. *Current Opinion in Genetics & Development*, *57*, 16–24. https://doi.org/10.1016/j.gde.2019.06.004

Shibukawa, Y., Sato, M., Kimura, M., Sobhan, U., Shimada, M., Nishiyama, A., Kawaguchi, A., Soya, M., Kuroda, H., Katakura, A., Ichinohe, T., & Tazaki, M. (2015). Odontoblasts as sensory receptors: Transient receptor potential channels, pannexin-1, and ionotropic ATP receptors mediate intercellular odontoblast-neuron signal transduction. *Pflügers Archiv—European Journal of Physiology*, *467*, 843–863. https://doi.org/10.1007/s00424-014-1551-x

Shimada, A., Kawanishi, T., Kaneko, T., Yoshihara, H., Yano, T., Inohaya, K., Kinoshita, M., Kamei, Y., Tamura, K., & Takeda, H. (2013). Trunk exoskeleton in teleosts is mesodermal in origin. *Nature Communications*, *4*, 1639. https://doi.org/10.1038/ncomms2643

Shono, T., Thiery, A. P., Cooper, R. L., Kurokawa, D., Britz, R., Okabe, M., & Fraser, G. J. (2019). Evolution and developmental diversity of skin spines in pufferfishes. *iScience*, *19*, 1248–1259. https://doi.org/10.1016/j.isci.2019.06.003

Simões-Costa, M. S., & Bronner, M. E. (2015). Establishing neural crest identity: A gene regulatory recipe. *Development*, *142*, 242–257. https://doi.org/10.1242/dev.105445

Simoes-Costa, M. S., & Bronner, M. E. (2016). Reprogramming of avian neural crest axial identity and cell fate. *Science*, *352*, 1570–1573. https://doi.org/10.1126/science.aaf2729

Simões-Costa, M. S., McKeown, S. J., Tan-Cabugao, J., Sauka-Spengler, T., & Bronner, M. E. (2012). Dynamic and differential regulation of stem cell factor FoxD3 in the neural crest is encrypted in the genome. *PLoS Genetics*, *8*, e1003142. https://doi.org/10.1371/journal.pgen.1003142

Simões-Costa, M. S., Tan-Cabugao, J., Antoshechkin, I., Sauka-Spengler, T., & Bronner, M. E. (2014). Transcriptome analysis reveals novel players in the cranial neural crest gene regulatory network. *Genome Research*, *24*, 281–290. https://doi.org/10.1101/gr.161182.113

Sire, J.-Y., & Akimenko, M.-A. (2004). Scale development in fish: A review, with description of sonic hedgehog (shh) expression in the zebrafish (*Danio rerio*). *The International Journal of Developmental Biology*, *48*, 233–247. https://doi.org/10.1387/ijdb.15272389

Sire, J.-Y., Donoghue, P. C. J., & Vickaryous, M. K. (2009). Origin and evolution of the integumentary skeleton in non-tetrapod vertebrates. *Journal of Anatomy*, *214*, 409–440. https://doi.org/10.1111/j.1469-7580.2009.01046.x

Sire, J.-Y., & Huysseune, A. (2003). Formation of dermal skeletal and dental tissues in fish: A comparative and evolutionary approach. *Biological Reviews*, *78*, 219–249. https://doi.org/10.1017/S1464793102006073

Smith, M. M., & Hall, B. K. (1990). Development and evolutionary origins of vertebrate skeletogenic and odontogenic tissues. *Biological Reviews*, *65*, 277–373. https://doi.org/10.1111/j.1469-185X.1990.tb01427.x

Smith, M. M., & Hall, B. K. (1993). A developmental model for evolution of the vertebrate exoskeleton and teeth. *Evolutionary Biology*, *27*, 387–448. https://doi.org/10.1007/978-1-4615-2878-4_10

Smith, M. P., Hickman, A., Amanze, D., Lumsden, A., & Thorogood, P. (1994). Trunk neural crest origin of caudal fin mesenchyme in the zebrafish *Brachydanio rerio*. *Proceedings of the Royal Society B: Biological Sciences*, *256*, 137–145. https://doi.org/10.1098/rspb.1994.0061

Smith, M. P., Sansom, I. J., & Repetski, J. E. (1996). Histology of the first fish. *Nature*, *380*, 702–704. https://doi.org/10.1038/380702a0

Soldatov, R., Kaucka, M., Kastriti, M. E., Petersen, J., Chontorotzea, T., Englmaier, L., Akkuratova, N., Yang, Y., Häring, M., Dyachuk, V., Bock, C., Farlik, M., Piacentino, M. L., Boismoreau, F., Hilscher, M. M., Yokota, C., Qian, X., Nilsson, M., Bronner, M. E., . . . Adameyko, I. (2019). Spatiotemporal structure of cell fate decisions in murine neural crest. *Science*, *364*, eaas9536. https://doi.org/10.1126/science.aas9536

Solé-Magdalena, A., Martínez-Alonso, M., Coronado, C. A., Junquera, L. M., Cobo, J., & Vega, J. A. (2018). Molecular basis of dental sensitivity: The odontoblasts are multisensory cells and express multifunctional ion channels. *Annals of Anatomy—Anatomischer Anzeiger*, *215*, 20–29. https://doi.org/10.1016/j.aanat.2017.09.006

Solovieva, T., & Bronner, M. (2021). Schwann cell precursors: Where they come from and where they go. *Cells & Development*, *166*, 203686. https://doi.org/10.1016/j.cdev.2021.203686

Soukup, V., Epperlein, H.-H., Horácek, I., & Cerny, R. (2008). Dual epithelial origin of vertebrate oral teeth. *Nature*, *455*, 795–798. https://doi.org/10.1038/nature07304

Stolfi, A., Ryan, K., Meinertzhagen, I. A., & Christiaen, L. (2015). Migratory neuronal progenitors arise from the neural plate borders in tunicates. *Nature*, *527*, 371–374. https://doi.org/10.1038/nature15758

Stone, L. S. (1929). Experiments showing the role of migrating neural crest (mesectoderm) in the formation of head skeleton and loose connective tissue in *Rana palustris*. *Wilhelm Roux' Archiv für Entwicklungsmechanik der Organismen*, *118*, 40–77. https://doi.org/10.1007/BF02108871

Strobl-Mazzulla, P. H., & Bronner, M. E. (2012). A PHD12—Snail2 repressive complex epigenetically mediates neural crest epithelial-to-mesenchymal transition. *Journal of Cell Biology*, *198*, 999–1010. https://doi.org/10.1083/jcb.201203098

Stundl, J., Bertucci, P. Y., Lauri, A., Arendt, D., & Bronner, M. E. (2021). Evolution of new cell types at the lateral neural border. *Current Topics in Developmental Biology*, *141*, 173–205. https://doi.org/10.1016/bs.ctdb.2020.11.005

Stundl, J., Martik, M. L., Chen, D., Raja, D. A, Franěk, R., Pospisilova, A., Pšenička, M., Metscher, B. D., Braasch, I., Haitina, T., Cerny, R., Ahlberg, P. E., Bronner, M. E. (2023). Ancient vertebrate dermal armor evolved from trunk neural crest. *Proceedings of the National Academy of Sciences*, 120, e2221120120, https://doi.org/10.1073/pnas.2221120120

Stundl, J., Pospisilova, A., Jandzik, D., Fabian, P., Dobiasova, B., Minarik, M., Metscher, B. D., Soukup, V., & Cerny, R. (2019). Bichir external gills arise via heterochronic shift that accelerates hyoid arch development. *eLife*, *8*, e43531. https://doi.org/10.7554/eLife.43531

Stundl, J., Pospisilova, A., Matějková, T., Psenicka, M., Bronner, M. E., & Cerny, R. (2020). Migratory patterns and evolutionary plasticity of cranial neural crest cells in ray-finned fishes. *Developmental Biology*, *467*, 14–29. https://doi.org/10.1016/j.ydbio.2020.08.007

Sun, W., Chi, S., Li, Y., Ling, S., Tan, Y., Xu, Y., Jiang, F., Li, J., Liu, C., Zhong, G., Cao, D., Jin, X., Zhao, D., Gao, X., Liu, Z., Xiao, B., & Li, Y. (2019). The mechanosensitive Piezo1 channel is required for bone formation. *Elife*, *8*, e47454. https://doi.org/10.7554/eLife.47454

Szabó, A., & Mayor, R. (2018). Mechanisms of neural crest migration. *Annual Reviews of Genetics*, *52*, 43–63. https://doi.org/10.1146/annurev-genet-120417-031559

Taniguchi, Y., Kurth, T., Medeiros, D. M., Tazaki, A., Ramm, R., & Epperlein, H.-H. (2015). Mesodermal origin of median fin mesenchyme and tail muscle in amphibian larvae. *Scientific Reports*, *5*, 11428. https://doi.org/10.1038/srep11428

Tarazona, O. A., Slota, L. A., Lopez, D. H., Zhang, G., & Cohn, M. J. (2016). The genetic program for cartilage development has deep homology within Bilateria. *Nature*, *533*, 86–89. https://doi.org/10.1038/nature17398

Thangadurai, S., Brumfeld, V., Milgram, J., Li, L., & Shahar, R. (2022). Osteodentin in the Atlantic wolffish (*Anarhichas lupus*): Dentin or bone? *Journal of Morphology*, *283*, 219–235. https://doi.org/10.1002/jmor.21438

Theveneau, E., & Mayor, R. (2012). Neural crest delamination and migration: From epithelium-to-mesenchyme transition to collective cell migration. *Developmental Biology*, *366*, 34–54. https://doi.org/10.1016/j.ydbio.2011.12.041

Uesaka, T., Nagashimada, M., & Enomoto, H. (2015). Neuronal differentiation in Schwann cell lineage underlies postnatal neurogenesis in the enteric nervous system. *Journal of Neuroscience*, *35*, 9879–9888. https://doi.org/10.1523/JNEUROSCI.1239-15.2015

Vickaryous, M. K., & Hall, B. K. (2006). Human cell type diversity, evolution, development, and classification with special reference to cells derived from the neural crest. *Biological Reviews*, *81*, 425–455. https://doi.org/10.1017/S1464793106007068

Vickaryous, M. K., & Sire, J.-Y. (2009). The integumentary skeleton of tetrapods: Origin, evolution, and development. *Journal of Anatomy*, *214*, 441–464. https://doi.org/10.1111/j.1469-7580.2008.01043.x

Wada, H., Saiga, H., Satoh, N., & Holland, P. W. (1998). Tripartite organization of the ancestral chordate brain and the antiquity of placodes: Insights from ascidian *Pax-2/5/8*, *Hox* and *Otx* genes. *Development*, *125*, 1113–1122. https://doi.org/10.1242/dev.125.6.1113

Welten, M., Smith, M. M., Underwood, C., & Johanson, Z. (2015). Evolutionary origins and development of saw-teeth on the sawfish and sawshark rostrum (Elasmobranchii; Chondrichthyes). *Royal Society Open Science*, *2*, 150189. https://doi.org/10.1098/rsos.150189

Williams, R. M., Lukoseviciute, M., Sauka-Spengler, T., & Bronner, M. E. (2022). Single-cell atlas of early chick development reveals gradual segregation of neural crest lineage from the neural plate border during neurulation. *eLife*, *11*, e74464. https://doi.org/10.7554/eLife.74464

Woodruff, E. C., Huie, J. M., Summers, A. P., & Cohen, K. E. (2022). Pacific Spiny Lumpsucker armor—Development, damage, and defense in the intertidal. *Journal of Morphology*, *283*, 164–173. https://doi.org/10.1002/jmor.21435

Xie, M., Kamenev, D., Kaucka, M., Kastriti, M. E., Zhou, B., Artemov, A. V., Storer, M., Fried, K., Adameyko, I., Dyachuk, V., & Chagin, A. S. (2019). Schwann cell precursors contribute to skeletal formation during embryonic development in mice and zebrafish. *Proceedings of the National Academy of Sciences*, *116*, 15068–15073. https://doi.org/10.1073/pnas.1900038116

Yu, J.-K., Meulemans, D., McKeown, S. J., & Bronner-Fraser, M. (2008). Insights from the amphioxus genome on the origin of vertebrate neural crest. *Genome Research*, *18*, 1127–1132. https://doi.org/10.1101/gr.076208.108

Zalc, A., Sinha, R., Gulati, G. S., Wesche, D. J., Daszczuk, P., Swigut, T., Weissman, I. L., & Wysocka, J. (2021). Reactivation of the pluripotency program precedes formation of the cranial neural crest. *Science*, *371*, eabb4776. https://doi.org/10.1126/science.abb4776

Zhao, D., Chen, S., & Liu, X. (2019). Lateral neural borders as precursors of peripheral nervous systems: A comparative view across bilaterians. *Development, Growth & Differentiation*, *61*, 58–72. https://doi.org/10.1111/dgd.12585

4 Evolutionary Genomics of Odontode Tissues

Tatjana Haitina and Mélanie Debiais-Thibaud

4.1 INTRODUCTION

One of the shared derived characters of vertebrates is their dermoskeleton that evolved into a variety of forms (Janvier, 1996; Donoghue, 2002). An important component of a dermoskeleton is the hypermineralized enamel and/or dentin-built units that may be fused into armour plates, or individualized in smaller scales or fin spines (Donoghue, 2002). These units may or may not be associated with dermal bone in the building of a dermoskeleton (Donoghue, 2002). Teeth and tooth-like structures were first grouped together for their shared histological characteristics of being made internally of dentin and externally of enamel or enameloid (Stensiö, 1961). The initial definition of the 'odontode' (Ørvig, 1967) included all types of dentin-built units to the exclusion of teeth, because of the very specific function of the dentition. However, several lines of evidence have shown shared developmental processes between teeth and odontodes, which were interpreted in the context of their evolution from a single ancestral developmental pathway (Reif, 1982; Huysseune and Sire, 1998; Smith and Coates, 1998; Sire and Huysseune, 2003; Huysseune et al., 2009; Fraser et al., 2010; Donoghue and Rücklin, 2016; Chen et al., 2020). Over the past ten years, Evo-Devo researchers have gathered genetic data that have widely supported the hypothesis of a common developmental pathway for oral teeth and a variety of other odontodes (Debiais-Thibaud et al., 2007, 2011; Cooper et al., 2017; Oralová et al., 2020; Berio and Debiais-Thibaud, 2021), although this pathway may display some divergence between lineages and/or between structures within a given species (Debiais-Thibaud et al., 2015; Berio and Debiais-Thibaud, 2021; Square et al., 2021). These shared developmental processes were described in a variety of teeth and other odontodes. They rely heavily on epithelial–mesenchymal interactions that will: (1) Lead to the definition of an epithelial placode associated with the condensation of mesenchyme (reviewed in Balic, 2019; Huysseune et al., 2022) and (2) generate synchronized cell proliferation (morphogenesis) and differentiation in both epithelial and mesenchymal compartments (reviewed in Jernvall and Thesleff, 2012). These successive steps of morphogenesis will achieve the building of an organ where the epithelial–mesenchymal junction foreruns the shape of the definitive mineralized organ. The differentiating cells of the epithelial and mesodermal compartments will then synthesize their specific extracellular matrix (ECM) proteins that will work as a spatial template that will then be mostly degraded to leave space for calcium phosphate crystals during the mineralization process.

 DOI: 10.1201/9781003439653-4

In jawed vertebrates, the mineralized odontodes are under strong selection for their function not only in feeding (teeth: Reisz, 2006; Hulsey et al., 2008; Melstrom, 2017), but also in protection (scales, fin spines: Song et al., 2011; Yang et al., 2013) and reproduction (Gorman & Hulsey, 2020; Graham et al., 2020; teeth: Randau et al., 2013). As a result, a number of independent events of tooth, scale, or odontode tissue loss have taken place during the evolution of jawed vertebrates (Davit-Béal et al., 2009; Liu et al., 2016; Qu et al., 2021). Dentin in gar scales was lost, odontodes were lost in the scales of both teleosts and lungfishes (Sire et al., 2009), and teeth became the single type of odontode in tetrapods (Witzmann, 2009). Tooth loss might involve the induction of secondary feeding apparatus, such as the tube-like long snout in sea horses and pipefishes (Lin et al., 2016), the elongated sticky tongues in toads, and the keratinized elements growing from the epithelium, like baleen plates in baleen whales, or modified jawbones covered by keratin, like beaks in birds, turtles, and platypuses (Davit-Béal et al., 2009). There are also described examples of enamel layer loss in aardvark and sloths and enamel layer thinning in armadillos, which is subsequently worn down (Davit-Béal et al., 2009).

The traditional way to analyse these mineralized structures is by tissue histology, being applied to both fossil and extant vertebrates. In extant vertebrates, one can use gene expression by *in situ* hybridization or transcriptome sequencing, and protein detection by immunohistochemistry or quantitative proteomics indicating that gene transcripts and proteins, respectively, are present during the formation of odontodes. Furthermore, the rapidly growing number of chromosome-level vertebrate genome assemblies adds a new layer to this comparative analysis of mineralized odontode tissues, allowing comparative chromosomal synteny analysis and the identification of homologous genes. The genes that are the focus of the genomic analysis encode proteins which can be divided into: (i) Proteins secreted within the matrix and present in the mineralized tissues besides the bioapatite crystals, collagens; (ii) proteins first secreted within the matrix and then to a large degree removed from the matrix during the mineralization process, such as the secretory calcium-binding phosphoproteins; (iii) proteins acting as enzymes during the formation of odontode layers, for instance matrix metallopeptidases. In this chapter, we will provide examples of how the analysis of these genes in the sequenced vertebrate genomes (being combined with other methods of odontode tissue analysis in fossil and extant vertebrates) enables unravelling the evolution of tissues in the mineralized odontodes.

4.2 TOOTH AND ODONTODE CELLS: CONSERVED FEATURES IN EXTANT JAWED VERTEBRATES

Ameloblasts and odontoblasts are the two major differentiating cell types that are involved in the secretion and mineralization of extracellular matrix components of odontode tissues (Figure 4.1). In the epithelial compartment, ameloblasts differentiate from the epithelial lineage and face the basal lamina that separates them from the mesenchymal compartment. While the odontogenic epithelium may have an ectodermal, endodermal, or mixed embryonic origin depending on the taxon, the presence of ectodermally derived cells seems to be consistently associated with the ability to

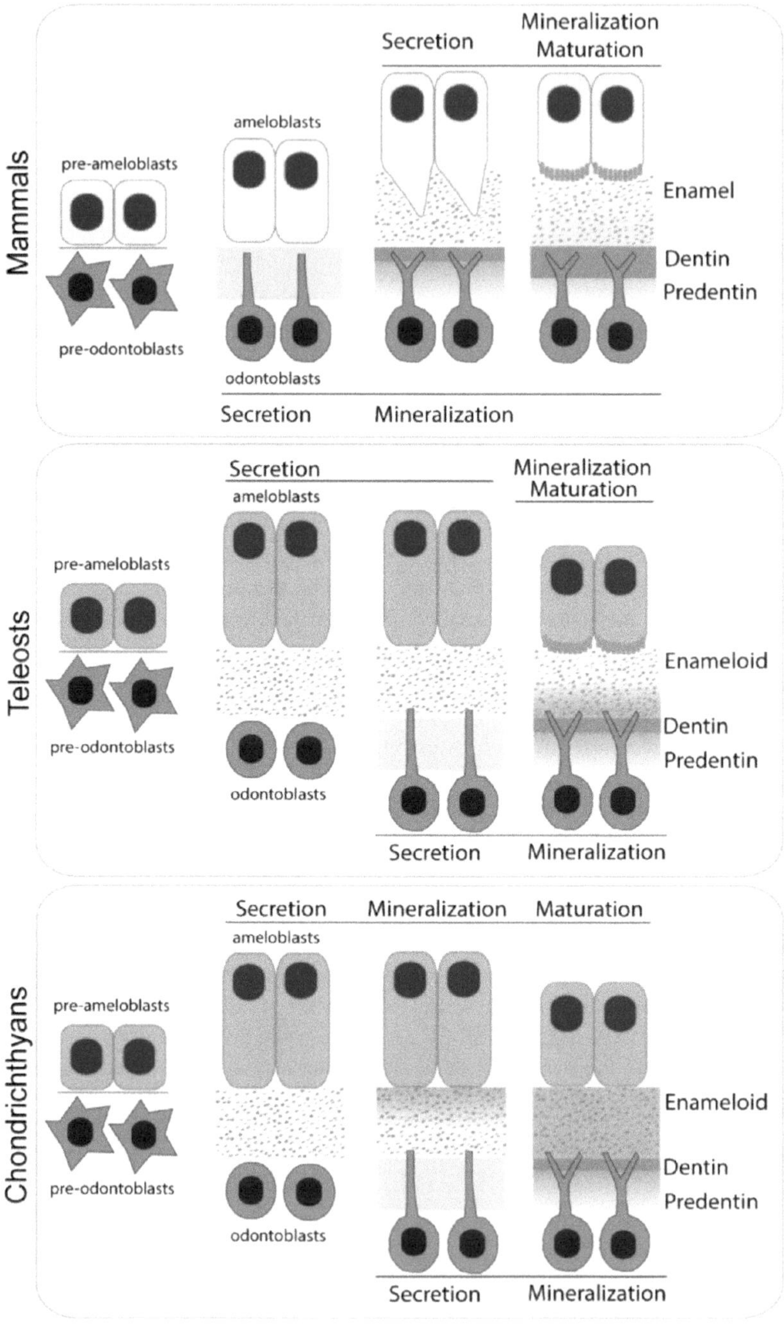

FIGURE 4.1 Enamel–dentin formation in mammals and enameloid–dentin formation in actinopterygian and chondrichthyan fishes. A line between the pre-ameloblasts and pre-odontoblasts indicates the basal lamina. Gradient shows progressive mineralization in the secreted matrices of the dentin (light grey) or enamel or enameloid (dotted background).

initiate tooth/odontode development (Huysseune et al., 2022). Differentiating ameloblasts were described in mammals as going first through a secretory stage where cells are columnar (Warshawsky, 1968) then grow cell expansions (Tomes' processes) on their basal pole (Sander, 2000). The subsequent maturation stage involves apoptosis of some ameloblasts, while the remnants take a cuboidal shape, synthesize a new basal lamina, and display a ruffled basal pole (Smith and Nanci, 1995; Smith, 1998). The secretory stage involves a partial degradation of the initial basal lamina for the ameloblasts to contact mesenchymal cells (Slavkin and Bringas, 1976) and the secretion and initial mineralization of the enamel matrix (Sasaki et al., 1997), while the maturation phase involves a degradation of enamel protein components to finalize the hypermineralization of the enamel (Smith, 1998). In the vast majority of cases, ameloblasts are lost with tooth/odontode eruption. Therefore, their function is restricted in time, and enamel/enameloid regeneration is not possible once the organ is erupted and functional.

Although many of these processes are described in the greatest detail in mammals (Simmer et al., 2021), only some aspects of mammalian ameloblasts in the teeth appear fully conserved in other vertebrates (Sire et al., 1987; Sasagawa, 1995, 2002a, 2002b; Sasagawa and Ishiyama, 2005; Davit-Béal et al., 2007; Hermyt et al., 2021). In particular, a phase of maturation with high activity of ameloblasts in secreting the degradation enzymes, targeting the enameloid matrix, appears conserved in ray-finned (Actinopterygii) and cartilaginous (Chondrichthyes) fishes (Sasagawa, 1997, 2002a, 2002b; Sasagawa and Ishiyama, 2005). However, several cellular characteristics seem to be specific to mammals, such as the existence of Tomes' processes leading to the specific prism organization of enamel (Sander, 2000). In contrast, matrix vesicles and collagen fibres were observed in the enameloid matrix of actinopterygian and chondrichthyan fishes, which is not a characteristic of mammalian enamel (Sasagawa, 1997, 2002a; Sasagawa and Ishiyama, 2005).

In the mesenchymal compartment, cells differentiate to generate the dentin-secreting cells: Odontoblasts (Arana-Chavez and Massa, 2004). These odontoblasts are neural crest-derived mesenchymal cells that migrate and condense to form the early tooth and other odontodes (Lumsden, 1988; Gillis et al., 2017). Their differentiation involves cell polarization and then a secretory phase. Odontoblasts function not only during tooth development but also in a longer-term mode with some capacities to maintain and produce dentin over the organ's lifetime (Arana-Chavez and Massa, 2004).

Odontoblasts share many characteristics with osteoblasts (Ørvig, 1967), including differentiation via expression of the Runx2 transcription factor (Komori, 2010), the extension of the cell membrane into projections (a single process in odontoblasts (Arana-Chavez and Massa, 2004), many dendritic projections by osteoblasts (Bonewald, 2011)), and the association of genes involved in matrix synthesis and mineralization (Chen et al., 2009). In mammals, the odontoblast cell processes are initiated close to the dentin–enamel junction and elongate while the dentin matrix is secreted and then mineralized, pushing the cell body away as it sits in the dental pulp cavity (Goldberg et al., 2011). Growth generates parallel tubules embedded in the dentin matrix, giving the name *orthodentin* to this mineralized tissue in mammals and making dentin a live tissue (Goldberg et al., 2011). Outside of mammals, several types of dentin were described from variations of several characteristics (Ørvig, 1967): Ramifications of the odontoblast process in the dentin matrix (in some chondrichthyan fishes); the presence of cell bodies in the dentin matrix (in some actinopterygian fishes, stem gnathostomes, and stem chondrichthyans); and the presence of blood vessels through the dentin matrix (*vasodentin* in some teleost fishes and in some mammals (Herold, 1970; Kalthoff, 2011). Evolution towards focal hypermineralization of dentin is also recorded in some vertebrates by the specialization of some odontoblasts (e.g. in lungfishes (Oka et al., 2017) and holocephalans (Smith et al., 2019)), new data of which will be presented in Chapter 5.

Another important variation in vertebrates is the relative timing of secretion and mineralization of dentin versus enamel or enameloid (Figure 4.1). In mammalian teeth, the secretory activity of odontoblasts producing predentin starts earlier than that of ameloblasts producing enamel, and dentin mineralizes before enamel (Arana-Chavez and Massa, 2004). However, this observation is not common in taxa possessing enameloid, where enameloid is a first layer deposited by odontoblasts (in some amphibians Davit-Béal et al., 2007), where enameloid starts mineralizing first and centripetally (in chondrichthyans, Sasagawa, 2002a), or where mineralization starts at the junction between enameloid and dentin and further propagates in both directions (in teleosts Sasagawa and Ishiyama, 2005).

Additional matrix-secreting and mineralizing cells, cementoblasts, differentiate from the mesenchymal compartment (reviewed in Zhao et al., 2016). Cementum is a mineralized tissue added to some surfaces of the tooth root, and cementoblasts behave globally as osteoblasts/osteocytes. Although cementum has been much less studied due to its presence mainly limited to amniotes, more details will be provided in Chapter 6.

4.3 TOOTH AND DERMAL ODONTODE TISSUES: VARIABLE HISTOLOGICAL FEATURES IN JAWED VERTEBRATES

Mammalian tooth crowns are covered with enamel, a hypermineralized tissue where bioapatite crystals generate a mineral composition estimated at 96% (Nancy and Ten Cate, 2003), and are supramolecularly organized as rod and interrod depending on ameloblast Tomes' processes activity and cell movements (Nanci and Warshawsky, 1984; Beniash et al., 2019). The analysis of the presence of enamel tissue in fossils is possible thanks to the frequently good preservation of histology and the birefringence of enamel in polarized light microscopy due to its ordered supramolecular structure (Schmidt et al., 1971). As a result, enamel evolution could be traced in the closest relatives of tetrapods, the sarcopterygian fish, where, in contrast to tetrapods, both oral and dermal odontodes can be present. Enamel covers the teeth of extant coelacanth *Latimeria chalumnae* (Smith, 1978) and Australian lungfish *Neoceratodus forsteri* (Satchell et al., 2000; Barry and Kemp, 2007), as well as the teeth of fossil sarcopterygian onychodontiform fish (Mann et al., 2017). Enamel is also found in association with dentine and complex pore-canal network, together forming cosmine (Mondéjar-Fernández, 2018). Cosmine is described on the dermal skeleton of many osteichthyans, including on teeth and dermal odontodes of *Youngolepis* (Chang and Smith, 1992) and the scales of *Megalichtys hibberti* (Schultze, 2016), two extinct sarcopterygian fish.

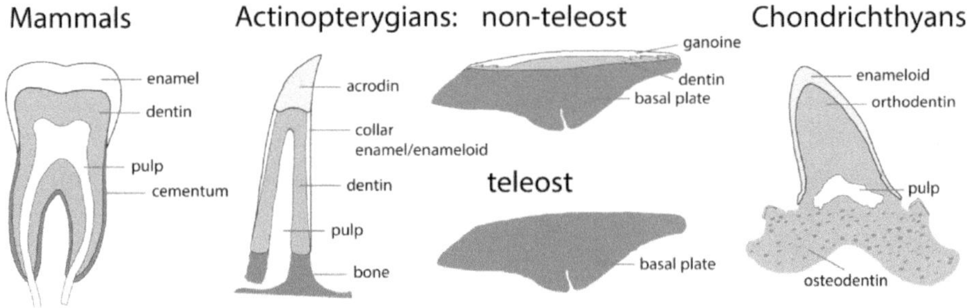

FIGURE 4.2 Tooth and scale tissues in mammals, non-teleost actinopterygians, teleosts, and chondrichthyans. Dentin layer is lost in gar *Lepisosteus oculatus* scales (Sire, 1995).

In actinopterygians, several hypermineralized tissues can be found including collar enamel, collar enameloid, acrodin or cap enameloid, and ganoine or multilayered enamel (Figure 4.2). In bichirs and gars, groups of non-teleost actinopterygians, the hypermineralized tissues homologous to enamel are present on the top of the scales, named ganoine, and around the shaft of the tooth, named collar enamel (Sire et al., 2009; Sasagawa et al., 2013). Under polarized light, ganoine, like enamel, is strongly birefringent (Gallo, 2005). Ganoine or multilayered enamel on scales was also present in the extinct fishes of classes Neopterygii: *Halecomorphi*, *Siemensichthys*, *Lophionotus* and Teleostei: *Gondwanapleuropholis longimaxillaris* (Goodrich, 1907; Schaeffer et al., 1950; Ebert et al., 2020), whereas extant teleosts have lost this hypermineralized layer on their scales. The tip of the teeth in actinopterygians is covered with acrodin, namely an enameloid cap (Schultze, 2016). Enamel evolution can be further traced to the stem group of osteichthyans. Enamel is present on the scales of stem osteichthyan *Andreolepis* and both the scales and dermal bones of stem sarcopterygian *Psarolepis*, but not on the teeth of both taxa (Qu et al., 2015).

Enameloid is a mineralized tissue that differs from enamel in several ways. Enameloid matrix is thought to be secreted by both the inner dental epithelium cells and the odontoblasts (Poole, 1967; Shellis and Miles, 1974), while enamel matrix is considered being of exclusive ameloblastic origin. Mature enameloid is characterized by embedded odontoblast processes and can also contain collagen fibres and giant fibres (Prostak and Skobe, 1988), while enamel matrix is devoid of odontoblast processes and collagen fibres. Enameloid matrix is deposited first, and its mineralization process starts during the predentin deposition, while enamel is usually mineralized after dentin.

Enameloid has been identified in tetrapods, only in the pre-metamorphic teeth of the tetrapod newt *Pleurodeles watl* (Davit-Béal et al., 2007). Their enameloid is the first layer deposited by odontoblasts, before ameloblasts deposit true enamel. The actinopterygian acrodin is also considered an enameloid tissue. Enameloid covers the scales, fin spines, and teeth of all cartilaginous fishes or chondrichthyans, where enamel tissue has never been described (Figure 4.2). Enameloid was already present in the earliest vertebrate dermoskeleton from the Ordovician: Arandaspids (Sansom et al., 2005) and *Astraspis* (Lemierre and Germain, 2019); in the most basal vertebrates anaspids, which is controversial being stem gnathostomes or stem lampreys (Keating and Donoghue, 2016) and in many jawed stem gnathostomes (placoderms), including the most primitive vertebrate teeth of *Radotina* (Vaškaninová et al., 2020). However, for the late Devonian acanthodians *Halimacanthodes ahlbergi* and *Triazeugacanthus affinis*, a well-mineralized birefringent crown layer has been described on scales. This layer was identified as enameloid (Burrow, 2011) or as ganoine (Chevrinais et al., 2017), which makes the phylogenetic position of the acanthodian group as stem chondrichthyans very important to the interpretation of the enamel/enameloid evolution. Currently, the lack of phylogenetic continuum of enameloid presence among heterostracans, chondrichthyans, and actinopterygians suggests convergent evolution of enameloids (Bendix-Almgreen and Bang, 1996; Sasagawa, 2002a).

Dentin is a mineralized tissue beneath enamel or enameloid with 70% mineral content and 20% organic content. Dentin is first secreted in the form of the predentin matrix by differentiating odontoblasts (Aiello and Dean, 1990). The odontoblasts have long cell processes that remain in tubules after the cells retreat from enamel–dentin junction. As the straight dentin tubules are a histological key character of the orthodentin, the composition of the tissue is not homogenous, containing both collagenous intertubular dentin and non-collagenous peritubular dentin, the latter is laid down around the odontoblast processes inside the tubules. While the organic matrix of intertubular dentin mainly consists of collagen type I, additional components are represented by proteoglycans, including small integrin-binding ligand, N-linked glycoproteins (SIBLINGs) and others (Goldberg et al., 2011).

The dentin layer formed after eruption is called secondary dentin, which can protect the pulp from exposure. As dentin is deposited towards the pulp center, in the non-shedding teeth of early vertebrates, the pulp is often limited in canals rather than in a wide chamber, and the pulp canals in older teeth can be completely infilled by dentin. Thus, the level of dentin mineralization and infill, as well as the dentin increments, can provide the information on life history (Vaškaninová et al., 2020). Dentin is an exceptionally well-described mineralized tissue in both living and fossil vertebrates, starting from *Anatolepis* from the early Ordovician (Smith et al., 1996), and with different types of dentin found in vertebrate odontodes (orthodentin, vasodentin, osteodentin, and others).

4.4 TOOTH/ODONTODE BIOMINERALIZATION, SHARED ANCESTRAL PROCESSES?

Despite taxonomical variation in the histology of dentines and of enamel/enameloids, several aspects of the biomineralization process are expected to be conserved both within and between these tissues. However, detailed data are currently available mostly for mammals (He et al., 2019; Simmer et al., 2021). Three main categories of extracellular matrix components are secreted and regulated by the biomineralizing cells, odontoblasts and ameloblasts, to generate the proper conditions for regulated apatite crystal growth (reviewed in Moradian-Oldak and George, 2021). First, structural proteins are secreted, and their orientation becomes a template for the organization of the future mineral. These proteins are mainly collagenous in dentin and non-collagenous in mammalian enamel and are secreted with collagen-associated proteoglycans (Robey, 1996; Houari et al., 2014). Second, calcium-interacting proteins are secreted, so that the locally high concentration of calcium allows local nucleation of calcium phosphate, initiating the subsequent crystal growth. These non-collagenous proteins interact with calcium and between themselves so that calcium and phosphate precipitation is globally inhibited except in specific sites (Moradian-Oldak and George, 2021). Third, ions are transported allowing the regulation of matrix pH and the necessary high quantity of calcium and phosphate (Linde, 1995; Lacruz et al., 2010; Bronckers, 2017; Kimura et al., 2021). This aspect of the biomineralizing cell activity implies their ability to store or massively transport these ions during the time of secretion (Nurbaeva et al., 2017; Chen et al., 2021). One of the last steps in the acquisition of the final hypermineralized extracellular matrices is the wave of protein degradation that will allow the extensive growth of mineral crystals. This phase of maturation is major in enamel (in association with ameloblast shape change; Smith, 1998) where degradation and endocytosis of the protein content in the matrix leave space for crystal growth (Maruyama et al., 2016; Pham et al., 2017). This process is less important in dentin where the mineral forms in the gaps within and between the collagen fibers without complete degradation of the collagen fibers (Satoyoshi et al., 2001; Tjäderhane and Haapasalo, 2009).

As a first attempt to understand how these biomineralization processes evolved in vertebrates, we will here focus on exploring the evolution of selected protein families involved in the secretory and maturation processes of dentin and enamel in mammals. We will take advantage of recent advances in the sequencing of many jawed vertebrate genomes, including cartilaginous and actinopterygian fishes, to describe their genomic conservation and evolution. Functional and expression data will also be compared, where available. These data are presented in the following four sections: General ECM structural components (fibrillar collagens and proteoglycans) produced during the secretory stage (Section 4.5); Matrix mineralization components (Section 4.6); The secretory calcium-binding phosphoprotein family (Section 4.7); and Matrix degradation components (Section 4.8).

4.5 GENERAL ECM STRUCTURAL COMPONENTS

4.5.1 Fibrillar and Minor Collagens

From a total of 28 collagen types in humans, fibrillar collagens are of types I, II, III, V, XI, XXIV, and XXVII (Bella and Hulmes, 2017). Mutations in genes coding for type I collagen (*COL1A1* and *COL1A2*) are responsible for dentinogenesis imperfecta in humans (Rauch et al., 2010). Type I collagen has been shown to make up the major protein content of predentin and dentin, similar to bone matrix, and is said to reach up to 95% of the collagen content of functional dentin (Butler, 1995; Robey, 1996), ensuring elasticity in this mineralized tissue (Ibrahim et al., 2019). More recent proteomics data identified a greater diversity of collagen types, notably the presence of type VI collagen (Feridouni Khamaneh et al., 2021). The major function of type I collagen is to structure the extracellular matrix of predentin, specifically in the intertubular space, and generate fibres on which the mineral will first nucleate and then crystallize (He et al., 2019). The basal lamina that builds the dentin–enamel junction (DEJ) integrates type IV and type VI collagens (Webb et al., 1998). In sharp contrast, fibrillar collagens are considered not to be expressed by mammalian ameloblasts, although type I collagen fibres go through the DEJ in definitive teeth (Lin et al., 1993), making type I collagen a minor protein found in proteomic studies of enamel (Farah et al., 2010).

COL1A1 and COL1A2 genes involved in type I collagen fibrillogenesis are highly conserved over vertebrate lineages in terms of structure, sequence, and synteny (Morvan-Dubois et al., 2003; Enault et al., 2015). Because the presence of type I collagen fibres in enameloid is a major argument for differentiating it from enamel, several studies have tested the expression of type I collagen genes by ameloblasts, or the presence of type I collagen fibres in enameloid matrix. Type I collagen genes are expressed in odontoblasts but not ameloblasts of teeth and scales in a shark species (Enault et al., 2018). In amphibians, *col1a1* is expressed by ameloblasts and odontoblasts during the early synthesis of enameloid matrix (Assaraf-Weill et al., 2014), and type I collagen fibres are detected in the enameloid matrix (Kogaya, 1999). In teleosts, both secretory ameloblasts and odontoblasts express *col1a1* (Huysseune et al., 2008; Kawasaki, 2009).

In all jawed vertebrates, a strong expression of type I collagen genes by odontoblasts supports the conservation of the major protein component of dentin. The type II collagen gene *col2a1* was shown to be expressed at low levels by odontoblasts in a shark and a ray species (Enault et al., 2018) possibly associated with the synthesis of the DEJ matrix, which has been very poorly studied outside of mammals.

4.5.2 Proteoglycans

Most of the identified proteoglycans in dentinogenesis belong to the family of Small Leucine-Rich Proteoglycans (SLRPs): Decorin (DCN), biglycan (BGN), and fibromodulin (FMOD) (Goldberg et al., 2011). These proteoglycans function in type I collagen fibrillogenesis and are associated with mineral deposition (Milan et al., 2004, 2005; Goldberg et al., 2006). These proteoglycans are secreted by odontoblasts (Embery et al., 2001; Goldberg et al., 2011); they are co-expressed with type I collagen in predentin and subsequently partly degraded; the same proteoglycans are additionally present at the mineralization front (at the border between predentin and the mineralized dentin) and appear stable in the mineralized dentin matrix. More recent proteomic analyses and gene expression studies have uncovered a wider diversity of SLRPs in dentin matrix, including lumican (LUM), asporin (ASPN), osteoglycin (OGN), proline/arginine-rich end leucine-rich repeat protein (PRELP), and osteomodulin (OMD) (Randilini et al., 2020; Feridouni Khamaneh et al., 2021). Asporin was shown to function in the mineralization of dentin (Lee et al., 2011), but functional data on other SLRPs is still missing (reviewed in Listik et al., 2019). Additional (non-SLRP) dentin proteoglycans are aggrecan (ACAN) and versican (VCAN), large aggregating proteoglycans that are detected in the dentin matrix (Green et al., 2019; Feridouni Khamaneh et al., 2021), although versican was only detected as fragments (Waddington et al., 2003; Ruggeri et al., 2009). The function of these proteoglycans in dentin is not known, and the versican protein core can vary depending on transcript variants (Xu et al., 2019). The function of chondroitin sulfate glycosaminoglycan is however known in regulating mineralization along collagen fibres (He et al., 2020).

In ameloblasts, only *Omd* and *Lum* expression was detected, the former in secretory ameloblasts, the latter with higher levels at the maturation stage (Buchaille et al., 2000; Simmer et al., 2014), also suggesting a function in the mineralization process. Fine proteomic data suggest a strong association between the dentin–enamel junction and the presence of BGN, LUM, OMD ASPN, and OGN (Green et al., 2019).

These SLRP genes are found in all jawed vertebrates (Park et al., 2008; Costa et al., 2018), but very little is known about their sites of expression and function outside of mammals. Some data have linked the expression of some SLRPs to the skeletal development in actinopterygian fishes, often associated with mineralization (Pedersen et al., 2013; Hannesson et al., 2015), suggesting overall conservation of their function in relation with type I collagen. Further research is needed in this regard. Similarly, the molecular evolution of lecticans (that include aggrecan and versican) has been well described in vertebrates and displays very good conservation within jawed vertebrates (Root et al., 2021) where limited data supports their conserved expression in the skeletal extracellular matrices (Kang et al., 2004; Marconi et al., 2020), making them strong candidates in the study of tooth and odontode development.

4.6 MATRIX MINERALIZATION COMPONENTS

The mineralization process involves the sequestration of calcium and phosphate at high quantity in the enamel/enameloid and dentin matrices, which is performed by proteins often well-known from endoskeletal mineralization (bone or cartilage), revealing ancestral aspects of the biomineralization process.

4.6.1 The Matrix- and Bone-Gla Proteins

Matrix-Gla protein (MGP) and bone-Gla protein (BGP, also known as Osteocalcin) are both vitamin K-dependent (VKD) proteins, where the Gla-domain can be carboxylated and then bind calcium (Yáñez et al., 2012). In mammals, BGP is best known for its function in bone development, while MGP is found in a wide range of organs. BGP is both a positive and negative regulator of mineralization in extracellular matrices, while MGP inhibits this process (Kaipatur et al., 2008). They are both expressed during tooth development (Howe and Webster, 1994) and are detected in dentin in proteomic studies (Green et al., 2019). BGP functions in linking special sites of the type I collagen fibres, so that calcium phosphate nucleates at the initiation of mineralization (Chen et al., 2015; Ustriyana et al., 2021). BGP was not detected in the enamel matrix but was in the enamel–dentin junction (Green et al., 2019). MGP proteins were detected both in maturation enamel and DEJ (Green et al., 2019) despite the *Mgp* gene expression being higher in secretory ameloblasts than maturation ameloblasts (Lacruz et al., 2011). Previous studies suggested that the presence of BGP in the enamel matrix was from the transport from odontoblast processes (Papagerakis et al., 2002). Odontoblasts would therefore be involved in the expression of both genes, while ameloblasts express only *Mgp*.

Several studies have described the evolutionary dynamics of the *Mgp*/*Bgp* genes outside of mammals (Cancela et al., 2014; Leurs et al., 2021). The single *Bgp* gene found in mammals (although in three recent tandem duplicates in the mouse) is actually one (*Bgp1*) out of two duplicates, which was shown to have translocated from its initial locus in some tetrapod ancestor (Figure 4.3; Leurs et al., 2021). All vertebrate *Mgp* and *Bgp* genes are thought to originate from a single locus in the last common ancestor of jawed vertebrates (Leurs et al., 2021). A *Bgp2* gene was identified in osteichthyans, but no functional data is available yet, and this duplicate was convergently lost in mammals and actinopterygians (Leurs et al., 2021). *Bgp1* in teleosts was kept as two duplicates after the teleost whole-genome duplication: *bgp1a* and *bgp1b* (also

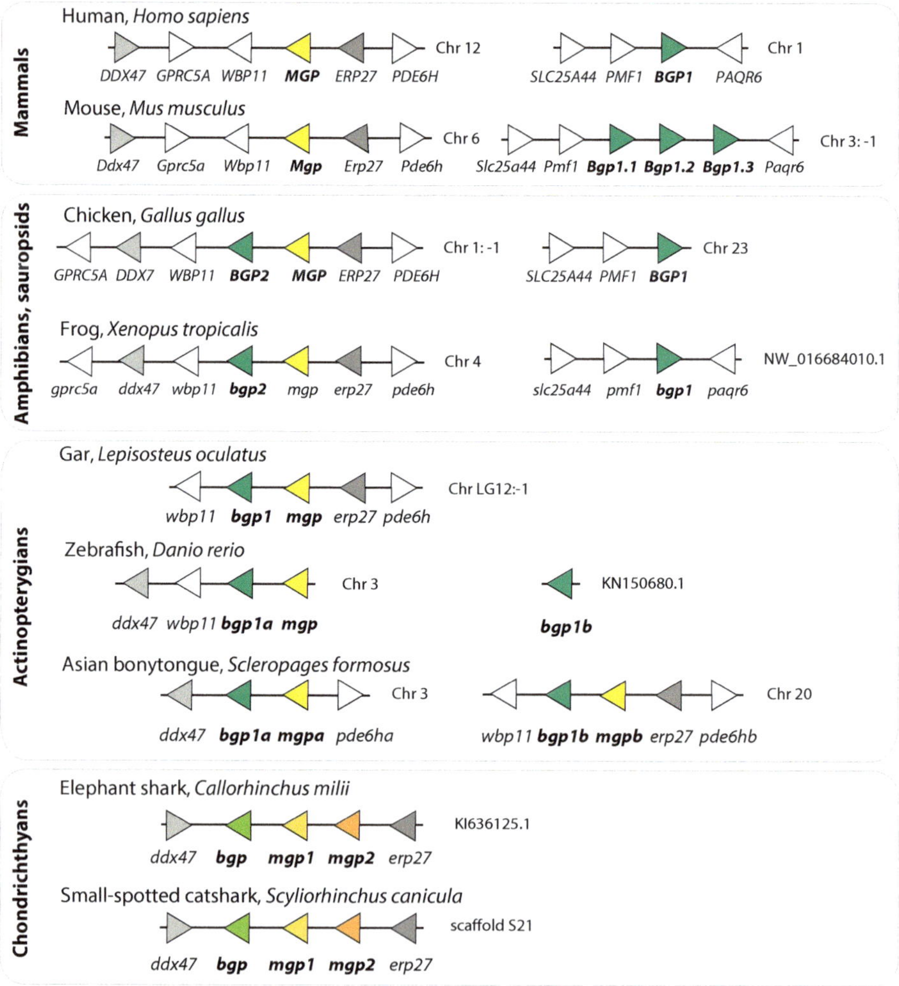

FIGURE 4.3 Conserved chromosomal synteny of *Mgp/Bgp* family and neighbouring genes in jawed vertebrates (modified after Leurs et al., 2021).

named osteocalcin2, Cancela et al., 2014). The teleost genes *bgp1a* and *mgp* are expressed in odontoblasts over dentin secretion and mineralization (Ortiz-Delgado et al., 2005). Chondrichthyans have only one *bgp* gene which displays expression in dental or scale odontoblasts during the secretion/maturation stage (Leurs et al., 2021), making Bgp an ancestral protein in jawed vertebrate dentin. Of the two chondrichthyan *mgp* genes, only *mgp2* is weakly expressed in tooth tissues (Leurs et al., 2021).

4.6.2 The Specific Case of Type X Collagen

Type X collagen is best known in mammals for its function in hypertrophic chondrocytes and endochondral ossification (Mäkitie et al., 2010; Tsang et al., 2015). Several lines of evidence in mammals suggest that the function of type X collagen is the anchoring of matrix vesicles in the cartilaginous matrix before mineralization (Kwan et al., 1997). In mammals, type X collagen is not involved in the matrix of either dentin or enamel (Feridouni Khamaneh et al., 2021), as inferred by the absence of identification in proteomic data (see Green et al., 2019).

Recent data available for species belonging to cartilaginous fishes, actinopterygian fishes, and non-mammalian tetrapods have shown that ameloblasts and odontoblasts do express the type X collagen gene (*Col10a1*) at the matrix secretion step (Debiais-Thibaud et al., 2019). This wide distribution implies that mammalian species secondarily lost this type X collagen secretion in dental matrices and that type X collagen is an ancestral component of dental matrices for all jawed vertebrates (Debiais-Thibaud et al., 2019). The genetic locus for the single *Col10a1* gene is conserved in osteichthyans, and both duplicates generated by the teleost whole-genome duplication were described (Figure 4.4; Debiais-Thibaud et al., 2019). This locus has however evolved through a series of local tandem duplications, leading to up to six *col10a1* genes in the genomes of cartilaginous fishes, with the expression of all duplicates in either ameloblasts and/or odontoblasts, suggesting a major function of type X collagen in the composition of the dental matrices (Debiais-Thibaud et al., 2019). It is noteworthy that type X collagen is expressed by both ameloblasts and odontoblasts, apparently in great quantity, so it is expected to be a major protein component of both enamel/enameloid and dentin, where its specific function is completely unknown.

4.6.3 SPARC (Osteonectin) and SPARC-like Proteins

The secreted protein acidic and rich in cysteine (SPARC) and SPARC-like protein 1 (SPARCL1, also named Hevin) are extracellular proteins with functions in interactions between the cells and their extracellular environment, in particular with collagen fibres (Brekken and Sage, 2001; Sullivan et al., 2008). Other members of the same protein family include SMOC and SPOCK, with similar protein domains and functions (Viloria et al., 2016). They are expressed in a wide range of tissue types (Klingler et al., 2020); however, *Sparc* expression, in particular, is well known as a marker of osteoblast differentiation and the activation of the mineralization process through bioapatite and collagen binding (Zhu et al., 2020). It was also identified

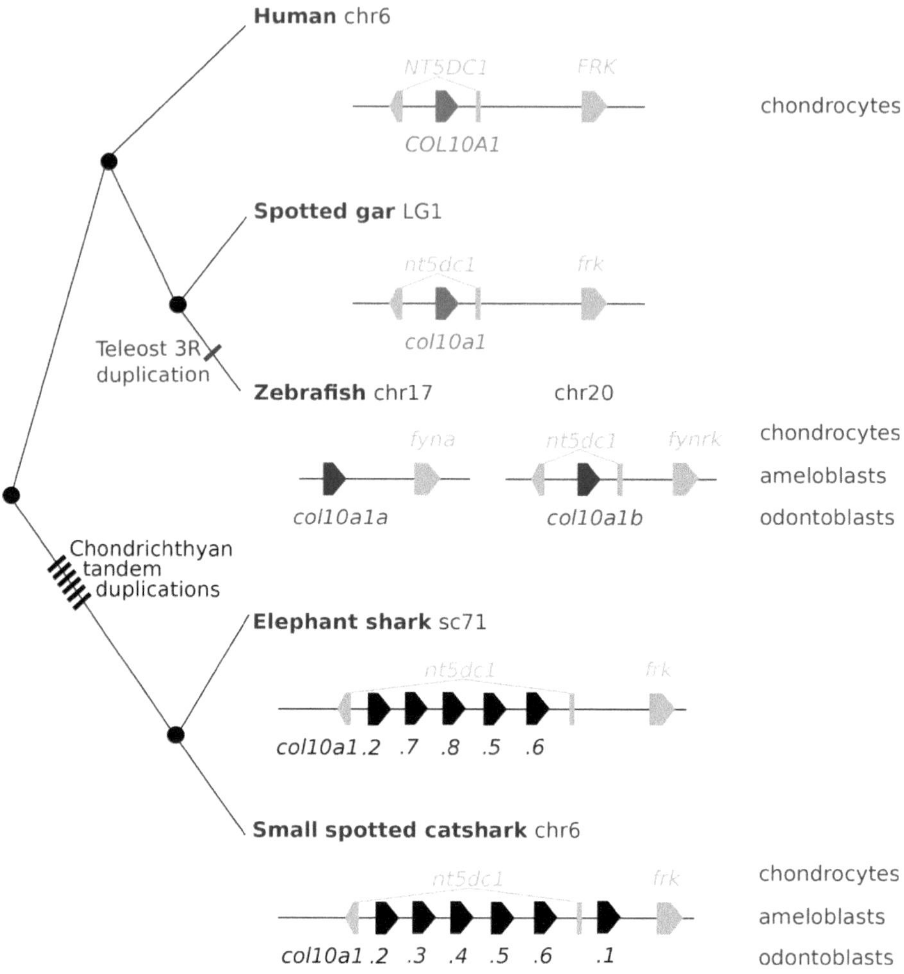

FIGURE 4.4 Conserved chromosomal synteny of *Col10a1* and neighbouring genes in jawed vertebrates (modified after Debiais-Thibaud et al., 2019). The synteny data are modified by the addition of genomic data sScyCan1.1 (GCF_902713615.1) for *Scyliorhinus canicula*.

as a marker of odontoblast secretion and dentin composition (Sommer et al., 1996; Papagerakis et al., 2002). SPARCL1 is less studied in mineralized tissues, but both these proteins were detected in the dentin by proteomic studies (Green et al., 2019). Although the SPARCL1 protein is not detected in early phases of enamel secretion (Green et al., 2019), the *SparcL1* gene was found upregulated at the maturation stage (Lacruz et al., 2011), but there has been no illustration of this expression in ameloblasts. However, a recent single-cell RNA-sequencing analysis revealed the expression of *SparcL1* in ventral epithelium cells of mouse incisor (Sharir et al., 2019). The function of SPARCL1 in tooth tissues is unknown, and, as a consequence, the range of research on this topic is scarce. Likewise, very poorly studied in the biology of tooth development, *Smoc2* was transcriptionally upregulated in ameloblasts at the maturation stage (Lacruz et al., 2011) and appears to be an activator of the ameloblast differentiation (Wang et al., 2018).

The *Sparc* and *SparcL* genes are considered to have evolved at the time of the two rounds of genomic duplications that preceded jawed vertebrate evolution (Enault et al., 2018). Two additional Sparc-related genes, *SparcR1* and *SparcR2*, also seem to have originated during the two rounds of vertebrate whole-genome duplications (Kawasaki et al., 2017), but they will not be further discussed as their function is virtually unknown.

The genomic locus of the *Sparc* gene is highly conserved among jawed vertebrates, but the *SparcL* locus went through major phases of tandem gene duplications/losses (see the next section for the evolutionary history of SCPP genes; Enault et al., 2018). The mammalian *SparcL1* is actually one out of two copies found in bony fishes, although the *SparcL2* (*SparcL1L1*) gene, which arose through tandem duplication, was only found in non-tetrapod sarcopterygian and non-teleost actinopterygian fishes; a single gene (*sparcl*) is found in chondrichthyans (Qu et al., 2015; Enault et al., 2018; Kawasaki, 2018). Quite similar to the situation in mammals, the expression of *sparc* by odontoblasts has been shown in chondrichthyans and an amphibian (Enault et al., 2018), suggesting that expression pattern may be broadly conserved between osteichthyans and chondrichthyans. Expression of *sparcl* in ameloblasts was shown in chondrichthyans (Enault et al., 2018), but no comparison can be further proposed since there is no available expression data of *SparcL1* or *SparcL2* in non-mammalian bony vertebrates.

4.6.4 Extracellular Phosphatases

The tissue non-specific alkaline phosphatase (ALPL) is the phosphatase associated with biomineralization in the skeleton, including teeth (reviewed in Villa-Suárez et al., 2021). Mutations in the *Alpl* gene lead to several defects of mineralization including in the dentin and cementum (Kramer et al., 2021). ALPL is a membrane-bound protein that interacts with phosphate in the ECM (Villa-Suárez et al., 2021). The ALPL protein is detected in the dentin matrix (Green et al., 2019). In bone and dentin, ALPL is responsible for the degradation of pyrophosphates (that inhibit mineralization) into inorganic phosphate (Villa-Suárez et al., 2021). As such, it is one of the tools in the general biomineralization toolkit. However, the expression of the *Alpl* gene was not detected in the ameloblasts (Lacruz et al., 2011), even though ALPL is active in the surrounding epithelial strata for the transport of phosphate to the enamel organ (Wöltgens et al., 1995).

Another phosphatase, the testicular acid phosphatase (ACPT or ACP4), was detected in the enamel matrix at the time of mineralization, supposedly to function in the dephosphorylation of specific enamel proteins that allow the maturation of enamel (Green et al., 2019). ACPT is also active in odontoblasts at an early stage of their differentiation and secretion (Choi et al., 2016). *ACPT* was shown to be missing from the genomes of mammals that lost the enamel layer or the teeth (Mu et al., 2021).

4.7 THE SECRETORY CALCIUM-BINDING PHOSPHOPROTEIN FAMILY

The secretory calcium-binding phosphoprotein (SCPP) gene family in mammals encodes enamel, dentin, and bone matrix proteins, as well as milk caseins and saliva proteins (Kawasaki and Weiss, 2003). All human SCPP genes, except amelogenins (AMEL) located on X- and Y-chromosomes, are located on chromosome 4q13-q22 in two clusters, one of P/Q-rich and another of acidic SCPPs, separated by SPARC-like 1 (SPARCL1) (Figure 4.5). The SCPP genes are considered to be tandem copies

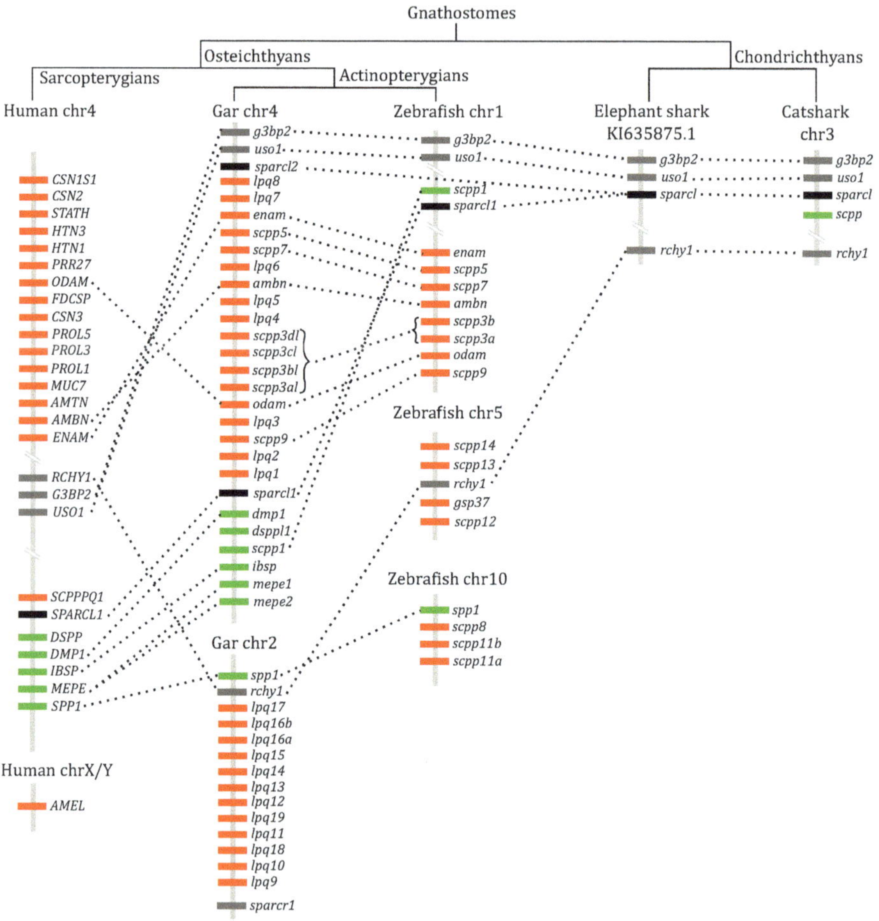

FIGURE 4.5 The arrangement of *SCPP* genes, *SPARCL* family genes, *USO1*, *G3BP2*, and *RCHY1* in the human, gar, zebrafish, catshark and elephant shark genomes (modified after Kawasaki, 2018; Leurs et al., 2022). Parallel duplication events led to a single *scpp* in elasmobranchs but many tandem *scpp* genes in osteichthyans. Acidic *SCPP*, P/Q-rich *SCPP*, and *SPARCL/L1*/L2 genes are displayed as green, orange, and black boxes, respectively; genes conserved in synteny are shown as grey boxes; dashed lines represent orthology relationships between genes. Catshark genome data from NCBI assembly sScyCan1.1 (GCF_902713615.1).

originating from *SparcL1* (Kawasaki and Weiss, 2003) with the first copy of the *SparcL1* gene most probably already present in stem gnathostomes (Ryll et al., 2014). The number of *SCPP* genes varies greatly in the analysed sequenced genomes of bony fishes with a larger number of genes reported in gars, bichir, and reedfish (Kawasaki et al., 2017; Mikami et al., 2022). On the other hand, within chondrichthyans, holocephalans have a single *sparcl* gene in this locus, and SCPP genes are lacking (Ryll et al., 2014; Venkatesh et al., 2014). However, in elasmobranch fishes, a convergent duplication of *sparcl* generated a convergent *scpp* gene with protein characteristics similar to the bony fish *SCPP* genes and expression in secretory ameloblasts (Leurs et al., 2022). The main current hypothesis is therefore that an ancestral *scpp* gene occurred in the gnathostome stem group that duplicated further to produce the SCPP gene family in the osteichthyans (Ryll et al., 2014). That ancestral *scpp* was lost in early chondrichthyans, but a parallel duplication of *sparcl* led to the single *scpp* in elasmobranchs (Leurs et al., 2022). The great number of duplications in osteichthyans led to many SCPPs, typically classified as P/Q-rich or acidic SCPPs.

4.7.1 P/Q-Rich SCPPs

P/Q-rich SCPPs include enamel matrix proteins: Amelogenin (AMEL), Ameloblastin (AMBN), and Enamelin (ENAM) that are secreted by ameloblasts in the secretory stage and are crucial for enamel formation. Mutations in human enamel matrix protein genes lead to amelogenesis imperfecta (Collier et al., 1997; Rajpar, 2001; Poulter et al., 2014), and gene inactivation in mice lead to similar phenotypes displaying disorganized hypoplastic enamel (Gibson et al., 2001; Rajpar, 2001; Fukumoto et al., 2004).

Amelogenin constitutes 90% of the enamel protein matrix (Fincham et al., 1999). In addition to secretory stage ameloblasts, the expression of amelogenin in mammals was also reported in bone, cartilage, and bone marrow cells (Haze et al., 2007), as well as cementum (Nuñez et al., 2010). *Amel* is present in tetrapods and lobe-finned fishes (Kawasaki and Amemiya, 2014), and its expression has been characterized in amphibians and reptiles where, similar to mammals, it was restricted to differentiating ameloblasts during matrix deposition including the larval stage when enameloid matrix is present (Assaraf-Weill et al., 2013). N- and C-terminal regions of Amel are conserved, but there can be sequence variation in the central region that does not seem to affect the spatiotemporal expression profile of *Amel*. Amel does not contribute to ganoine or collar enamel in actinopterygians, as *Amel* is absent in their genomes. Interestingly, in *Xenopus tropicalis*, *Amel* is expressed in the skin at the peak of the metamorphosis and the adult dermis prior to the deposition of calcium in the dermal layer (Okada et al., 2013). Pseudogenization of *Amel* and in many cases also *Ambn* and *Enam* is associated with the loss of teeth in several extant tetrapod clades including birds, turtles, true toads, and others (Meredith et al., 2013; Shaheen et al., 2021).

AMBN is a cell adhesion molecule expressed by secretory ameloblasts, but, in contrast to AMEL, only small amounts of AMBN are detected in the enamel matrix (Krebsbach et al., 1996). A broad distribution of *Ambn* in mesenchymal tissues has also been described, including pulpal mesenchymal cells during early odontogenesis and differentiating odontoblasts prior to mantle dentin mineralization (Fong et al., 1998). It is also detected in bone overlying connective tissues, osteoblasts, chondrocytes, and perichondrium (Spahr et al., 2006) as well as cementoblasts (Nuñez et al., 2010). Several studies have shown that AMBN plays a role in early bone formation, long bone growth, and mineralization (Tamburstuen et al., 2011; Lu et al., 2016). Overexpression in mice also shows that it plays a role in the regulation of cranial bone growth and suture closure (Atsawasuwan et al., 2013). *Ambn* is present in tetrapods, sarcopterygian fishes, and non-teleost actinopterygians (Kawasaki and Amemiya, 2014; Qu et al., 2015; Thompson et al., 2021). Following the SCPP family expansion in osteichthyans, actinopterygians display a large number of tandemly duplicated SCPP genes, not found in tetrapods. Initially annotated as actinopterygian-specific, the teleost *scpp6* gene was proposed as *AMBN* orthologue, following *Ambn* identification in the spotted gar genome (Qu et al., 2015).

ENAM is the largest of enamel matrix proteins, though constituting only 5% of the forming enamel matrix (Uchida et al., 1991). Extensive analysis of *Enam* homozygous mutant mice that completely lack enamel revealed that ENAM plays an additional role in bone metabolism (Fuchs et al., 2012). *Enam* expression has also been detected in odontoblasts but at low levels (Nagano et al., 2003). Similar to other enamel matrix protein-coding genes, *Enam* is conserved in tetrapods and sarcopterygians (Kawasaki and Amemiya, 2014). Its orthologue was identified in the spotted gar genome where it also encodes a large protein, but the sequence is divergent from the lobe-finned fishes (Qu et al., 2015).

Initially, there was no *Enam* orthologue identified in teleosts, but additional sequence comparison of teleost *fa93e10* to gar and tetrapod *Enam* revealed some sequence similarity in the N-terminal region of the encoded protein in addition to a similar chromosomal location in actinopterygians (Kawasaki et al., 2017). Recent studies have annotated teleost *fa93e10* as an *ENAM* orthologue; however, some divergence in the protein function is expected compared to tetrapod ENAM and also to non-teleost actinopterygian Enam since there is no enamel nor ganoine present in extant teleost odontodes.

Comparative expression analysis has been done for a non-teleost actinopterygian gar and the teleost zebrafish for *ambn*, *enam*, and the actinopterygian-specific *scpp5* gene. In this gar species, the expression of all three genes was associated exclusively with ameloblasts during matrix formation and detected in scale ganoine, teeth collar enamel, and acrodin (Kawasaki et al., 2021). It is therefore possible that as an intermediate tissue, acrodin would form with contributions from both ameloblasts and odontoblasts (Shellis and Miles, 1974). Zebrafish *ambn* expression showed a similar exclusive association with ameloblasts while *enam* and *scpp5* were expressed by both ameloblasts and odontoblasts associated with acrodin and collar enameloid formation (Kawasaki et al., 2021; Rosa et al., 2021).

Scpp5 is conserved in a majority of actinopterygian genomes; it is however missing or pseudogenized in seahorse and sea dragon (syngnathid) genomes together with other P/Q-rich SCPP genes (Lin et al., 2016; Qu et al., 2021). The loss of *scpp5* in syngnathids is thought to be associated with the loss of teeth, as zebrafish homozygous mutants for *scpp5* were found to have a reduced number of teeth related to the proposed failure of tooth replacement (Qu et al., 2021). Additional SCPP genes are lost in syngnathids and could also be linked to tooth loss. In sturgeon and paddlefish (Acipenseriformes) genomes, *scpp5* was lost together with *enam*, but *scpp6* is still present (Cheng et al., 2021). SCPP gene losses in Acipenseriformes are thought to be associated with the reduction of hypermineralized tissue in scales and/or teeth in these species (Mikami et al., 2022), together with sturgeon teeth being present only in larval stages (Bemis et al., 2002). On the other hand, *scpp5* is present whereas *enam* and *scpp6* are lost in the genome of the teleost channel catfish, which is thought to be associated with the loss of scales (Liu et al., 2016). The loss of *scpp6* linked to the loss of scales can be further proposed in cavefish, fugu, and icefish (Kawasaki, 2009; Zhang et al., 2020; Thompson et al., 2021). However, in all analysed cases of tooth and/or scale loss, additional SCPPs are lost in parallel while others remain, making it difficult to associate any specific SCPP gene to a specific odontode type or a specific mineralized layer. Many more studies on SCPP gene expression and proteome analysis in different odontode tissues are needed, ideally also distinguishing expression in either ameloblasts and/or odontoblasts in order to make firmer conclusions about SCPP function in ganoine versus collar enamel versus acrodin versus collar enameloid matrix formation.

Mammalian P/Q-rich SCPPs also include amelotin (AMTN) and odontogenic ameloblast-associated protein (ODAM) that are produced by maturation stage ameloblasts and junctional epithelium (Moffatt et al., 2006, 2008). Detailed analysis revealed the presence of ODAM in mouse alveolar bone osteoblasts; secretory- and maturation-stage ameloblasts; the enamel matrix; the interface between ameloblasts and enamel layer; predentin; and odontoblast processes, with variation in its cellular/extracellular localization (Lee et al., 2012). Deletion of exons 3–6 in human *AMTN* is linked to amelogenesis imperfecta (Smith et al., 2016). *Amtn* inactivation in mice results in hypomineralized enamel with structural changes (Nakayama et al., 2015). Inactivation of *Odam* did not result in any visible effect on enamel but did affect the junctional epithelium (Wazen et al., 2015). In several mammals lacking teeth or enamel, the inactivating mutations for *AMTN* and *ODAM* have been reported (Meredith et al., 2014; Gasse et al., 2015). In addition, *ODAM* is inactivated in toothed whales, and *AMTN* is pseudogenized in toothless turtles and birds (Gasse et al., 2015; Springer et al., 2019). *Amtn* and *Odam* are present in tetrapods and coelacanth (Kawasaki and Amemiya, 2014). Gene expression analyses indicate that *Amtn* is expressed throughout amelogenesis in non-mammalian tetrapods (Gasse et al., 2015). Both *amtn* and *odam* are present in non-teleost actinopterygians (Qu et al., 2015), while only *odam* has been identified in teleosts. Expression of *odam* has been described in ameloblasts of inner dental epithelium during the maturation stage of enamel in frog tadpoles and of enameloid (acrodin) in fugu and zebrafish (Kawasaki et al., 2005; Kawasaki, 2009). AMTN and ODAM, together with another P/Q-rich SCPP member, the secretory calcium-binding phosphoprotein proline-glutamine rich 1 gene (SCPPPQ1), are implicated in the function of the basal lamina through the epithelial cell to mineral attachment (Fouillen et al., 2017).

There is an additional gene located between two SCPP clusters on human chromosome 4, *C4orf26*, that encodes Odontogenesis Associated Phosphoprotein (ODAPH), linked to the hypomaturation form of amelogenesis imperfecta (Parry et al., 2012). A recent analysis of homozygous mouse mutants revealed that ODAPH is essential for the transition from secretory to maturation stage ameloblasts and potentially for the adhesion of maturing ameloblasts to the enamel surface (Liang et al., 2021). Gene-inactivating mutations have been identified in this gene in toothless placental mammals, but the gene remained functional in other mammals with partial enamel loss, including aardvark, sloth, and armadillo (Springer et al., 2016). *Odaph* is also identified in several tetrapods and gar; however, its function in non-mammalian vertebrates remains to be described (Liang et al., 2021).

4.7.2 The SIBLING Family of Acidic SCPPs

The small integrin-binding ligand, N-linked glycoprotein (SIBLING) family of acidic SCPPs in human includes dentin sialophosphoprotein (DSPP), dentin matrix acidic phosphoprotein 1 (DMP1), integrin-binding sialoprotein (IBSP), matrix extracellular phosphoglycoprotein (MEPE), and secreted phosphoprotein 1 (SPP1), also called Osteopontin. The SIBLING family is implicated in dentin and bone formation, and its members are rich in aspartic acid, glutamic acid, and serine residues and have a characteristic RGD motif promoting cell adhesion (Kawasaki and Weiss, 2003).

DSPP is a precursor protein for dentin sialoprotein (DSP) and dentin phosphoprotein (DPP) with specific roles in dentin formation (Suzuki et al., 2009). Mutations in human DSPP lead to dentinogenesis imperfecta II and dentin dysplasia (Zhang et al., 2001; McKnight et al., 2008). *Dspp* is expressed by odontoblasts, also transiently by pre-ameloblasts, and at a much lower level in bone (Ritchie et al., 1997; Qin et al., 2002). In agreement with the expression profile, *Dspp* knockout mice show various tooth phenotypes including decreased mineralization of the dentin; changes in the predentin, pulp, and alveolar bone; as well as periodontal defects and accelerated secretion and maturation of enamel (Verdelis et al., 2016; Shi et al., 2020). *Dspp* has been independently inactivated in three lineages of edentulous amniotes: Birds, turtles, and the Chinese pangolin and is linked to the loss of dentin layer (Meredith et al., 2014). *Dspp* has not been identified outside of amniotes. However, a *Dspp-like1* gene was described in the coelacanth (Kawasaki and Amemiya, 2014) and non-teleost actinopterygians (Mikami et al., 2022). Gars have lost dentin layer on scales but retained tooth dentin: *Dspp-like1* is present in their genome; however, RNA-seq data show very low expression in both jaw and skin tissues (Mikami et al., 2022), not supporting functional conservation of Dspp-like1 as compared to amniote Dspp. The fact that a *Dspp* ortholog is absent outside the amniotes is striking considering the conservation of dentin in osteichthyan, chondrichthyan, and early vertebrate dermal skeleton (Smith and Sansom, 2000).

Dmp1 is associated with bone and, to a lesser degree, with dentin. It is expressed not only in young odontoblasts, but also in maturation ameloblasts, cementoblasts, and osteoblasts or osteocytes (D'souza et al., 1997; Macdougall et al., 1998; Toyosawa et al., 2001). Mutations in *Dmp1* are associated with rickets in human and sheep (Farrow et al., 2009; Zhao et al., 2011), and gene inactivation in mouse leads to osteomalacia and rickets phenotypes (Feng et al., 2006). Other members of the SIBLING family (Ibsp, SPP1, and Mepe) play prominent roles in bone. *Ibsp* and *Spp1* are regulated by Runx2 in terminal hypertrophic chondrocytes and immature osteoblasts (Komori, 2010), and the mutagenesis of zebrafish *spp1* results in reduced bone formation (Venkatesh et al., 2014). Matrix extracellular phosphoglycoprotein (*MEPE*) is highly expressed in osteocytes in human alveolar bone. Both DMP1 and MEPE are responsive to mechanical loading in the tooth movement model (Gluhak-Heinrich et al., 2003, 2007). *Dmp1* is lost in paddlefish and sturgeon genomes possibly associated with the retention of the cartilaginous skeleton. Members of the teleost family Syngnathidae have fused jaws and dermal armour plates and lack teeth and scales, and only *scpp1* and *spp1* remain in the SCPP cluster. Functional analysis of acidic SCPPs in non-teleost actinopterygians will help to clarify the evolution of dentin and bone.

4.8 MATRIX DEGRADATION COMPONENTS

After the secretion of the matrix-building proteins in enamel/enameloids and dentin, a phase of protein degradation is necessary for the proper phase of mineralization. This phase of degradation is major in enamel and correlates with specific cell shapes of ameloblasts, leading to a specific tooth stage of development named enamel maturation. In mammals, the enamel matrix made of AMEL, AMBN, and ENAM proteins will mineralize through a complex process involving C-terminal cleavage by the enamelysin protease, N-terminal self-assembly of amelogenin, and enamel crystal nucleation and growth (reveiwed in Pandya and Diekwisch, 2021). A continuous matrix degradation by proteases (e.g. KLK4) and replacement with bioapatite crystals then allow a final 96% mineral composition of enamel (Nancy and Ten Cate, 2003). Although less pervasive and not linked to known cell shape changes, matrix protein degradation is also found in dentin. The members of several protein families are active over this degradation process in mammals, which we will describe here in more detail. Most of them belong to the large group of zinc-domain metalloproteases (metzincins, Huxley-Jones et al., 2007): The matrixins, also known as Matrix MetalloProteinases (MMPs); the ADAMTS (A Disintegrin And Metalloproteinase with ThromboSpondin motif, a family that belongs to the adamalysin metzincins); and astacins (that include BMP1/TLL proteins and meprins). Other proteases will also be presented here: The cathepsins, kallikreins, and another poorly described proteinase—PHosphate regulating Endopeptidase homolog X-linked (PHEX, a membrane metalloendopeptidase). In this section, we will not give prospective evolutionary considerations on the role of these various gene families in the evolution of jawed vertebrate enamel/enameloid or dentin, as there is extremely limited knowledge about these gene families outside of mammals.

4.8.1 MATRIXINS: THE MATRIX METALLOPROTEINASE FAMILY

MMPs are zinc-dependent endopeptidases with a wide range of protein targets (although they were initially identified either as collagenases or with other specific targets), and they are encoded by a family of 24 members in the human genome (Puente et al., 2003). They are secreted as inactive proproteins that undergo cleavage to be functional as proteases in a very large array of tissues in which they regulate the ECM dynamics (Hannas et al., 2007). In the ECM, they can be inhibited by

Tissue Inhibitors of MetalloProteinases (TIMPs) with four members of this family in vertebrates (Brew, 2019).

During mammalian enamel secretion, ameloblasts produce Matrix Metallopeptidase 20 (MMP20, enamelysin) (Lu et al., 2008; Simmer et al., 2014) that targets several enamel matrix proteins for cleavage: Amelogenins (Ryu et al., 1999) and ameloblastin (Iwata et al., 2007). MMP20 is then sequestered in the enamel matrix until the end of the mineralization process (Green et al., 2019). The mutation of *Mmp20* leads to hypomineralization of the enamel matrix due to incorrect growth of amorphous calcium phosphate ribbons as a result of improper enamel matrix processing (Lu et al., 2008; Bartlett et al., 2021). The function of MMP20 in dentin secretion and mineralization is much less studied, as it was thought that the secretion of MMP20 by odontoblasts was targeting the enamel proteins at the enamel–dentin junction (Fukae et al., 2002). However, a recent study showed that changes in the amino acid sequence of MMP20 are also associated with thin and hypomineralized dentin (Wang et al., 2020).

MMP2 is found both in the enamel and dentin matrix (Green et al., 2019), while MMP9 is found in dentin only. In the dentin, MMPs were described to function in collagen degradation (Carrilho et al., 2020) and proteoglycan degradation (Fukae et al., 2002). MMP3 was also proposed to be acting in proteoglycan degradation of both the enamel and dentin matrices (Hall et al., 1999). The expression of several other MMPs is enhanced in ameloblasts at the maturation stage: *Mmp13*, *Mmp15*, and *Mmp23* (Simmer et al., 2014), with *Mmp8* and *Mmp13* also shown to be expressed in periodontium (Tsubota et al., 2002). However, there is limited functional data for these MMPs in each compartment. MMP2, MMP8, and MMP9 may have peptidase activity against the dentin matrix component DSPP (Tsuchiya et al., 2011).

As functional inhibitors of the MMP proteins, TIMPs are important for better understanding the MMP function. From four members, only TIMP1 was identified in dentin (Gobbi et al., 2021) while only *Timp3* was shown to be more highly expressed in maturing versus secretory ameloblasts (Simmer et al., 2014).

From an evolutionary perspective, the differential regulation of *MMP20* gene expression is proposed to be linked to the thickening of enamel in the human lineage (Horvath et al., 2014) and, similar to the effect of the tooth loss on enamel matrix protein genes, *MMP20* is a pseudogene in baleen whales (Meredith et al., 2011). As previously stated, further evolutionary considerations, outside of *Mmp20* or outside of mammals, cannot be currently drawn because the phylogenetic history of this family has been poorly described in jawed vertebrates (Huxley-Jones et al., 2007; Kawasaki and Suzuki, 2011), and the functional understanding of MMP proteins has until now been only centred on a few members.

4.8.2 Adamalysins: The ADAM and ADAMTS Families

Among adamalysins (that include the families of ADAM, ADAMTS, and class III snake venoms, Huxley-Jones et al., 2007), ADAMTS proteins, with 19 members in human, have been identified in the matrices of enamel or dentin in mammals. *Adamts1*, *Adamts2*, and *Adamts17* show higher levels of expression in maturation ameloblasts, while *Adamts9* showed a higher level of expression in secretory ameloblasts (Simmer et al., 2014). The expression of *Adamts1*, *4*, and *5* was also described in the odontoblasts and cementoblasts (Sone et al., 2005). All these genes encode widely expressed metalloproteinases that cleave hyalectans (aggrecan, versican, brevican, or neurocan) (Kelwick et al., 2015; Dancevic et al., 2016), except for ADAMTS2, which targets procollagens, and ADAMTS17, which targets mucin-proteoglycans (Dancevic et al., 2016). Particularly in dentin, ADAMTS proteins were suspected to cleave versican (Sone et al., 2005). The expression of *Adam1a* (higher in secretory ameloblasts) and of *Adam9* (higher in maturation ameloblasts) was also reported (Simmer et al., 2014), but there is, to our best knowledge, no analysis of the function of these membrane-bound metalloproteases in enamel or dentin matrices.

4.8.3 Astacins: The Bmp1/Tll and Meprins

Three out of six members of this gene family are expressed in odontoblasts: *Bmp1* (that generates various isoforms including Bmp1 and Tolloid), *Mep1a*, and *Mep1b*, and all of their protein products are able to cleave the DSPP into the dentin-specific fragments (Tsuchiya et al., 2011). These three astacins are also able to cleave several other components of the dentin matrix, including not only DMP1 (Steiglitz et al., 2004; von Marschall and Fisher, 2010) but also procollagen (Scott et al., 1999). The inactivation of *Bmp1* in odontoblasts led to thinner dentin and disorganized dentinal tubules, showing its function in the secretion and processing of the dentin matrix, prior to the mineralization phase (Zhang et al., 2017).

4.8.4 Non-metzincin Proteases: Peptide Release and Ground Matrix Degradation

4.8.4.1 The Cathepsin Family

Eleven members of the lysosomal cysteine cathepsins exist in the human genome (Turk et al., 2000; Rossi et al., 2004). Although most of these proteases have a wide range of expression and are well-known for their activity in immune response, Cathepsin K functions in the degradation of the mineralized extracellular matrix by osteoclasts (Rossi et al., 2004). Two of them are expressed in both ameloblasts and odontoblasts over the course of mineralization of the enamel and dentin matrices: Cathepsin K and Cathepsin B (Jiang et al., 2017; Carrilho et al., 2020). Two other records from proteomic data report Cathepsin L in the enamel and dentin matrix (Green et al., 2019), and a higher expression of Cathepsin C in ameloblasts at the maturation stage versus secretory stage was identified using transcriptomics (Simmer et al., 2014). Cathepsins K and B cleave enamel proteins (Smid et al., 2001; Jiang et al., 2017) while Cathepsin B cleaves denatured collagen (Carrilho et al., 2020).

4.8.4.2 Kallikreins

The degradation of protein content in the enamel is linked to the activity of KLK4, one member of the Kallikrein serine proteases whose expression is widely enhanced in maturation ameloblasts (Simmer and Hu, 2002; Lu et al., 2008; Simmer et al.,

2014). *Klk4* is even more strongly expressed by odontoblasts (Nagano et al., 2003), and the protease is found deposited at the enamel–dentin junction and in the dentin matrix (Green et al., 2019), probably as a trigger for enamel maturation (Fukae et al., 2002). Mutations in *Klk4* were associated with a soft enamel phenotype, where enamel proteins are not degraded during the maturation phase, leading to impaired mineralization (Hart, 2004). *KLK4* is found in placental mammal genomes and originated from a gene duplication that also gave rise to *KLK5* (Kawasaki et al., 2014). This *KLK4* duplicate underwent rapid evolution, most probably in parallel with the divergence of very specific mammalian enamel (Kawasaki et al., 2014). The Kallikrein family itself might be a recently evolved family in amniotes, potentially related to the trypsin gene family (Kawasaki et al., 2014). More recent transcriptome sequencing data showed *Klk8* was more highly expressed in maturation ameloblasts than during the secretory stage, but no further data is available on this member of the family (Simmer et al., 2014), suggesting the involvement of other Kallikreins in tooth matrix maturation in mammals.

4.8.4.3 PHEX

The Phosphate-regulating Endopeptidase homolog X-linked (*PHEX*) gene was first identified as a locus of mutation linked to hypophosphatemia and hypomineralization of several tissues including dentin (reviewed in Foster et al., 2014). The encoded protein is an endopeptidase, which cleaves peptides generated by the previous cleavage of the MEPE, DMP1, and OPN proteins in the dentin matrix (Addison et al., 2008; Coyac et al., 2018). As these fragments are inhibitors for mineral deposition, PHEX acts as a peptidase that activates mineralization (Reznikov et al., 2020). The *PHEX* gene seems to be conserved in bony vertebrates (Bianchetti et al., 2002), and similar endopeptidases were identified in other mammalian/amniote genomes (Rowe, 2012), but no comparative analyses have been made outside of mammals on the function and expression of PHEX.

4.9 CONCLUDING REMARKS

The recent enormous increase in chromosome-level genome assemblies makes it possible to follow the evolution of gene families across the whole diversity of jawed vertebrates. Genome-wide comparative studies are now expected to give a wider view of odontode evolution following excellent descriptions of evolutionary patterns of odontode cells, tissues, and developmental processes in jawed vertebrates. Here, we showed that the orchestrated rapid development of complex odontodes requires a specialized network of interacting genes from a diversity of gene families during matrix secretion and mineralization (Figure 4.6).

Wide comparative analyses are expected to uncover unidentified actors of tooth/odontode evolution, because previous studies have always focused on candidate genes that used mammals as a reference point. However, mammalian dentition is highly derived and is probably poorly representative of the ancestral odontode development genetic toolkit. With the loss of all odontodes but teeth, the leveraging of some selective pressures initially existing on scale developmental processes probably allowed rapid dental evolution, with an increase in the complexity of the tooth

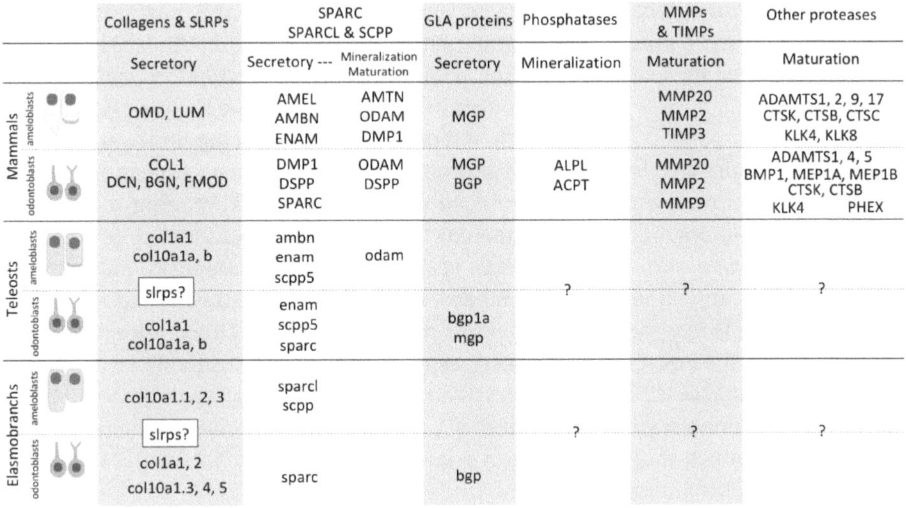

FIGURE 4.6 The summary of ameloblast and odontoblast molecular markers in mammals, teleosts, and elasmobranchs during secretion, mineralization, and maturation stages.

occlusal surface. Lastly, jawed vertebrates outside of mammals often display continuous odontode replacement during their lifetime: They require a stable production of enough matrix material and a balanced mineralization process but no such morphological specialization or complexity as mammalian teeth. Some odontode-related gene families such as the SCPP gene family were proposed to evolve in conjunction with odontode evolution, and the diversity of enamel-related tissues in osteichthyans or the loss of enamel and dentin in teleost scales. Several gene families may have been lost during mammalian evolution, which will be identified only if comparative analyses do not use a candidate-gene approach.

To do so, the molecular evolution of genes involved in tooth and odontode matrix secretion, mineralization, and maturation should be studied in more detail with a focus on non-mammalian vertebrates, since there are now many more chromosome-level vertebrate genome sequences available. In addition, the description of gene families without detailed data on the expression pattern of these genes is still preventing further inference of evolutionary scenarios. As a consequence, functional studies using non-mammalian animal models, genome editing to produce gene knock-outs, RNA-sequencing, and proteome analysis will be crucial to understand the function in non-mammal and non-tetrapod species. To escape the candidate-gene approach, transcriptomic and proteomic analyses are expected to generate non-biased results. Single-cell RNA-sequencing data, revealing cellular hierarchies of the continuously growing mouse incisor, have been recently made available (Sharir et al., 2019; Chiba et al., 2020; Krivanek et al., 2020), which should be extended to non-model organisms. Further development in spatial transcriptomics, *in situ* sequencing, and 3D RNA profiling will help to measure, map, and compare the gene activity between odontode tissue layers and identify additional molecular markers (Larsson et al., 2021; Mayeur et al., 2021).

Finally, a more global view of jawed vertebrate diversity, as is now possible with genomic data and a renewed interest in non-model organisms in the lab, is essential to gain a better understanding of the genomic basis of tooth/odontode evolution. From current data, chondrichthyans appear to lack most of the known odontode tissue matrix components, despite teeth and dermal denticles being under major selective pressures in these organisms. In addition, major evolutionary transitions in teeth and scales occurred with the diversification of teleost fishes, making the study of non-teleost actinopterygians fundamental for comparing evolutionary processes at the level of jawed vertebrates. Jawless vertebrates such as lampreys and hagfishes do not possess any mineralized odontodes, making chondrichthyans and non-teleost actinopterygians key models in the description of the mineralized odontode evolution in extant species. Genome-editing experiments are not available in chondrichthyans due to inaccessible early embryos and only are just starting to be developed in non-teleost actinopterygians (Stundl et al., 2022): In the near future, single-cell and spatial multi-omics analyses will be crucial for identifying the molecular markers of early odontode evolution.

REFERENCES

Addison, W. N., Y. Nakano, T. Loisel, P. Crine, and M. D. McKee. 2008. MEPE-ASARM peptides control extracellular matrix mineralization by binding to hydroxyapatite: An inhibition regulated by PHEX cleavage of ASARM. *Journal of Bone and Mineral Research* 23:1638–1649.

Aiello, L., and C. Dean. 1990. The microanatomy and development of teeth; pp. 106–132 in *An Introduction to Human Evolutionary Anatomy*. Elsevier, San Diego.

Arana-Chavez, V. E., and L. F. Massa. 2004. Odontoblasts: The cells forming and maintaining dentine. *The International Journal of Biochemistry & Cell Biology* 36:1367–1373.

Assaraf-Weill, N., B. Gasse, N. Al-Hashimi, S. Delgado, J.-Y. Sire, and T. Davit-Béal. 2013. Conservation of amelogenin gene expression during tetrapod evolution. *Journal of Experimental Zoology Part B: Molecular and Developmental Evolution* 320B:200–209.

Assaraf-Weill, N., B. Gasse, J. Silvent, C. Bardet, J.-Y. Sire, and T. Davit-Béal. 2014. Ameloblasts express type I collagen during amelogenesis. *Journal of Dental Research* 93:502–507.

Atsawasuwan, P., X. Lu, Y. Ito, Y. Zhang, C. A. Evans, and X. Luan. 2013. Ameloblastin inhibits cranial suture closure by modulating Msx2 expression and proliferation. *PloS ONE* 8:e52800.

Balic, A. 2019. Concise review: Cellular and molecular mechanisms regulation of tooth initiation. *Stem Cells* 37:26–32.

Barry, J. C., and A. Kemp. 2007. High resolution transmission electron microscopy of developing enamel in the Australian lungfish, *Neoceratodus forsteri* (Osteichthyes: Dipnoi). *Tissue and Cell* 39:387–398.

Bartlett, J. D., C. E. Smith, Y. Hu, A. Ikeda, M. Strauss, T. Liang, Y.-H. Hsu, A. H. Trout, D. W. McComb, R. C. Freeman, J. P. Simmer, and J. C.-C. Hu. 2021. MMP20-generated amelogenin cleavage products prevent formation of fan-shaped enamel malformations. *Scientific Reports* 11:10570.

Bella, J., and D. J. S. Hulmes. 2017. Fibrillar collagens; pp. 457–490 in D. Parry and J. Squire (eds.), *Fibrous Proteins: Structures and Mechanisms*, 1st ed. Springer, Cham, Switzerland.

Bemis, W. E., E. K. Findeis, and L. Grande. 2002. An overview of Acipenseriformes; pp. 25–71 in *Sturgeon Biodiversity and Conservation*. Kluwer Academic Publishers, Dordrecht.

Bendix-Almgreen, S. E., and B. S. Bang. 1996. *Aspects of Enameloid Ontogeny in Fossil and Recent Selachians with Comments on Enameloids and Their Occurrence During Early Lower Vertebrate Phylogeny*. DFG Online Series. Geological Society of Denmark, Copenhagen.

Beniash, E., C. A. Stifler, C.-Y. Sun, G. S. Jung, Z. Qin, M. J. Buehler, and P. U. P. A. Gilbert. 2019. The hidden structure of human enamel. *Nature Communications* 10:4383.

Berio, F., and M. Debiais-Thibaud. 2021. Evolutionary developmental genetics of teeth and odontodes in jawed vertebrates: A perspective from the study of elasmobranchs. *Journal of Fish Biology* 98:906–918.

Bianchetti, L., C. Oudet, and O. Poch. 2002. M13 endopeptidases: New conserved motifs correlated with structure, and simultaneous phylogenetic occurrence of PHEX and the bony fish. *Proteins: Structure, Function, and Genetics* 47:481–488.

Bonewald, L. F. 2011. The amazing osteocyte. *Journal of Bone and Mineral Research* 26:229–238.

Brekken, R. A., and E. H. Sage. 2001. SPARC, a matricellular protein: At the crossroads of cell—matrix communication. *Matrix Biology* 19:815–827.

Brew, K. 2019. Reflections on the evolution of the vertebrate tissue inhibitors of metalloproteinases. *The FASEB Journal* 33:71–87.

Bronckers, A. L. J. J. 2017. Ion transport by ameloblasts during amelogenesis. *Journal of Dental Research* 96:243–253.

Buchaille, R., M. Couble, H. Magloire, and F. Bleicher. 2000. Expression of the small leucine-rich proteoglycan osteoadherin/osteomodulin in human dental pulp and developing rat teeth. *Bone* 27:265–270.

Burrow, C. J. 2011. A partial articulated acanthodian from the Silurian of New Brunswick, Canada. *Canadian Journal of Earth Sciences* 48:1329–1341.

Butler, W. T. 1995. Dentin matrix proteins and dentinogenesis. *Connective Tissue Research* 33:59–65.

Cancela, M. L., V. Laizé, and N. Conceição. 2014. Matrix Gla protein and osteocalcin: From gene duplication to neofunctionalization. *Archives of Biochemistry and Biophysics* 561:56–63.

Carrilho, M. R., P. Scaffa, V. Oliveira, L. Tjäderhane, I. L. Tersariol, D. H. Pashley, F. Tay, and F. D. Nascimento. 2020. Insights into cathepsin-B activity in mature dentin matrix. *Archives of Oral Biology* 117:104830.

Chang, M., and M. Smith. 1992. Is Youngolepis a porolepiiform? *Journal of Vertebrate Paleontology*:294–312.

Chen, D., H. Blom, S. Sanchez, P. Tafforeau, T. Märss, and P. E. Ahlberg. 2020. The developmental relationship between teeth and dermal odontodes in the most primitive bony fish Lophosteus. *Elife* 9.

Chen, L., R. Jacquet, E. Lowder, and W. J. Landis. 2015. Refinement of collagen—mineral interaction: A possible role for osteocalcin in apatite crystal nucleation, growth and development. *Bone* 71:7–16.

Chen, S., J. Gluhak-Heinrich, Y. H. Wang, Y. M. Wu, H. H. Chuang, L. Chen, G. Yuan, J. Dong, I. Gay, and M. MacDougall. 2009. Runx2, osx, and dspp in tooth development. *Journal of Dental Research* 88:904–909.

Chen, Y., R. Koshy, E. Guirado, and A. George. 2021. STIM1 a calcium sensor promotes the assembly of an ECM that contains Extracellular vesicles and factors that modulate mineralization. *Acta Biomaterialia* 120:224–239.

Cheng, P., Y. Huang, Y. Lv, H. Du, Z. Ruan, C. Li, H. Ye, H. Zhang, J. Wu, C. Wang, R. Ruan, Y. Li, C. Bian, X. You, C. Shi, K. Han, J. Xu, Q. Shi, and Q. Wei. 2021. The American paddlefish genome provides novel insights into chromosomal evolution and bone mineralization in early vertebrates. *Molecular Biology and Evolution* 38:1595–1607.

Chevrinais, M., J.-Y. Sire, and R. Cloutier. 2017. From body scale ontogeny to species ontogeny: Histological and morphological assessment of the Late Devonian acanthodian *Triazeugacanthus affinis* from Miguasha, Canada. *PloS ONE* 12:e0174655.

Chiba, Y., K. Saito, D. Martin, E. T. Boger, C. Rhodes, K. Yoshizaki, T. Nakamura, A. Yamada, R. J. Morell, Y. Yamada, and S. Fukumoto. 2020. Single-cell RNA-sequencing from mouse incisor reveals dental epithelial cell-type specific genes. *Frontiers in Cell and Developmental Biology* 8.

Choi, H., T.-H. Kim, C.-Y. Yun, J.-W. Kim, and E.-S. Cho. 2016. Testicular acid phosphatase induces odontoblast differentiation and mineralization. *Cell and Tissue Research* 364:95–103.

Collier, P. M., J. J. Sauk, J. Rosenbloom, Z. A. Yuan, and C. W. Gibson. 1997. An amelogenin gene defect associated with human X-linked amelogenesis imperfecta. *Archives of Oral Biology* 42:235–242.

Cooper, R. L., K. J. Martin, L. J. Rasch, and G. J. Fraser. 2017. Developing an ancient epithelial appendage: FGF signalling regulates early tail denticle formation in sharks. *EvoDevo* 8:8.

Costa, R. A., R. S. T. Martins, E. Capilla, L. Anjos, and D. M. Power. 2018. Vertebrate SLRP family evolution and the subfunctionalization of osteoglycin gene duplicates in teleost fish. *BMC Evolutionary Biology* 18:191.

Coyac, B. R., B. Hoac, P. Chafey, G. Falgayrac, L. Slimani, P. S. Rowe, G. Penel, A. Linglart, M. D. McKee, C. Chaussain, and C. Bardet. 2018. Defective mineralization in x-linked hypophosphatemia dental pulp cell cultures. *Journal of Dental Research* 97:184–191.

Dancevic, C. M., D. R. McCulloch, and A. C. Ward. 2016. The ADAMTS hyalectanase family: Biological insights from diverse species. *Biochemical Journal* 473:2011–2022.

Davit-Béal, T., F. Allizard, and J.-Y. Sire. 2007. Enameloid/enamel transition through successive tooth replacements in *Pleurodeles waltl* (Lissamphibia, Caudata). *Cell and Tissue Research* 328:167–183.

Davit-Béal, T., A. S. Tucker, and J.-Y. Sire. 2009. Loss of teeth and enamel in tetrapods: Fossil record, genetic data and morphological adaptations. *Journal of Anatomy* 214:477–501.

Debiais-Thibaud, M., V. Borday-Birraux, I. Germon, F. Bourrat, C. J. Metcalfe, D. Casane, and P. Laurenti. 2007. Development of oral and pharyngeal teeth in the medaka (*Oryzias latipes*): Comparison of morphology and expression of *eve1* gene. *Journal of Experimental Zoology. Part B, Molecular and Developmental Evolution* 708:693–708.

Debiais-Thibaud, M., R. Chiori, S. Enault, S. Oulion, I. Germon, C. Martinand-Mari, D. Casane, and V. Borday-Birraux. 2015. Tooth and scale morphogenesis in shark: An alternative process to the mammalian enamel knot system. *BMC Evolutionary Biology* 15:292.

Debiais-Thibaud, M., S. Oulion, F. Bourrat, P. Laurenti, D. Casane, and V. Borday-Birraux. 2011. The homology of odontodes in gnathostomes: Insights from Dlx gene expression in the dogfish, *Scyliorhinus canicula*. *BMC Evolutionary Biology* 11:307.

Debiais-Thibaud, M., P. Simion, S. Ventéo, D. Muñoz, S. Marcellini, S. Mazan, and T. Haitina. 2019. Skeletal mineralization in association with type X collagen expression is an ancestral feature for jawed vertebrates. *Molecular Biology and Evolution* 36:2265–2276.

Donoghue, P. C. J. 2002. Evolution of development of the vertebrate dermal and oral skeletons: Unraveling concepts, regulatory theories, and homologies. *Paleobiology* 28:474507.

Donoghue, P. C. J., and M. Rücklin. 2016. The ins and outs of the evolutionary origin of teeth. *Evolution & Development* 18:19–30.

D'souza, R. N., A. Cavender, G. Sunavala, J. Alvarez, T. Ohshima, A. B. Kulkarni, and M. MacDougall. 1997. Gene expression patterns of murine dentin matrix protein 1 (Dmp1) and dentin sialophosphoprotein (DSPP) suggest distinct developmental functions in vivo. *Journal of Bone and Mineral Research* 12:2040–2049.

Ebert, M., D. Thies, and R. B. Hauff. 2020. First evidence of ganoin-scaled Halecomorphi (Neopterygii) in the Lower Jurassic of Holzmaden and Ohmden, Germany. *Neues Jahrbuch Für Geologie Und Paläontologie—Abhandlungen* 295:307–326.

Embery, G., R. Hall, R. Waddington, D. Septier, and M. Goldberg. 2001. Proteoglycans in dentinogenesis. *Critical Reviews in Oral Biology & Medicine* 12:331–349.

Enault, S., D. N. Muñoz, W. T. A. F. Silva, V. Borday-Birraux, M. Bonade, S. Oulion, S. Ventéo, S. Marcellini, and M. Debiais-Thibaud. 2015. Molecular footprinting of skeletal tissues in the catshark *Scyliorhinus canicula* and the clawed frog *Xenopus tropicalis* identifies conserved and derived features of vertebrate calcification. *Frontiers in Genetics* 6:1–14.

Enault, S., D. N. Muñoz, P. Simion, S. Ventéo, J. Sire, and S. Marcellini. 2018. Evolution of dental tissue mineralization: An analysis of the jawed vertebrate SPARC and SPARC-L families. *BMC Evolutionary Biology* 18:1–12.

Farah, R. A., B. C. Monk, M. V. Swain, and B. K. Drummond. 2010. Protein content of molar—incisor hypomineralisation enamel. *Journal of Dentistry* 38:591–596.

Farrow, E. G., S. I. Davis, L. M. Ward, L. J. Summers, J. S. Bubbear, R. Keen, T. C. B. Stamp, L. R. I. Baker, L. F. Bonewald, and K. E. White. 2009. Molecular analysis of DMP1 mutants causing autosomal recessive hypophosphatemic rickets. *Bone* 44:287–294.

Feng, J. Q., L. M. Ward, S. Liu, Y. Lu, Y. Xie, B. Yuan, X. Yu, F. Rauch, S. I. Davis, S. Zhang, H. Rios, M. K. Drezner, L. D. Quarles, L. F. Bonewald, and K. E. White. 2006. Loss of DMP1 causes rickets and osteomalacia and identifies a role for osteocytes in mineral metabolism. *Nature Genetics* 38:1310–1315.

Feridouni Khamaneh, Y., P. Kiani, R. J. D. Miller, H. Schlüter, and R. E. Friedrich. 2021. Complementing the pulp proteome via sampling with a picosecond infrared laser (PIRL). *Clinical Oral Investigations* 25:6757–6768.

Fincham, A. G., J. Moradian-Oldak, and J. P. Simmer. 1999. The structural biology of the developing dental enamel matrix. *Journal of Structural Biology* 126:270–299.

Fong, C. D., R. Černý, L. Hammarström, and I. Slaby. 1998. Sequential expression of an amelin gene in mesenchymal and epithelial cells during odontogenesis in rats. *European Journal of Oral Sciences* 106:324–330.

Foster, B. L., F. H. Nociti, and M. J. Somerman. 2014. The rachitic tooth. *Endocrine Reviews* 35:1–34.

Fouillen, A., J. Dos Santos Neves, C. Mary, J.-D. Castonguay, P. Moffatt, C. Baron, and A. Nanci. 2017. Interactions of AMTN, ODAM and SCPPPQ1 proteins of a specialized basal lamina that attaches epithelial cells to tooth mineral. *Scientific Reports* 7:46683.

Fraser, G. J., R. Cerny, V. Soukup, M. Bronner-Fraser, and J. T. Streelman. 2010. The odontode explosion: The origin of tooth-like structures in vertebrates. *BioEssays* 32:808–817.

Fuchs, H., S. Sabrautzki, H. Seedorf, B. Rathkolb, J. Rozman, W. Hans, R. Schneider, M. Klaften, S. M. Hölter, L. Becker, M. Klempt, R. Elvert, W. Wurst, T. Klopstock, M. Klingenspor, E. Wolf, V. Gailus-Durner, and M. H. de Angelis. 2012. Does enamelin have pleiotropic effects on organs other than the teeth? Lessons from a phenotyping screen of two enamelin-mutant mouse lines. *European Journal of Oral Sciences* 120:269–277.

Fukae, M., T. Tanabe, T. Nagano, H. Ando, Y. Yamakoshi, M. Yamada, J. P. Simmer, and S. Oida. 2002. Odontoblasts enhance the maturation of enamel crystals by secreting EMSP1 at the enamel-dentin junction. *Journal of Dental Research* 81:668–672.

Fukumoto, S., T. Kiba, B. Hall, N. Iehara, T. Nakamura, G. Longenecker, P. H. Krebsbach, A. Nanci, A. B. Kulkarni, and Y. Yamada. 2004. Ameloblastin is a cell adhesion molecule required for maintaining the differentiation state of ameloblasts. *Journal of Cell Biology* 167:973–983.

Gallo, V. 2005. Redescription of *Lepidotes piauhyensis* Roxo and Löfgren, 1936 (Neopterygii, Semionotiformes, Semionotidae) from the Late Jurassic—Early Cretaceous of Brazil. *Journal of Vertebrate Paleontology* 25:757–769.

Gasse, B., Y. Chiari, J. Silvent, T. Davit-Béal, and J.-Y. Sire. 2015. Amelotin: An enamel matrix protein that experienced distinct evolutionary histories in amphibians, sauropsids and mammals. *BMC Evolutionary Biology* 15:47.

Gibson, C. W., Z.-A. Yuan, B. Hall, G. Longenecker, E. Chen, T. Thyagarajan, T. Sreenath, J. T. Wright, S. Decker, R. Piddington, G. Harrison, and A. B. Kulkarni. 2001. Amelogenin-deficient mice display an amelogenesis imperfecta phenotype. *Journal of Biological Chemistry* 276:31871–31875.

Gillis, J. A., E. C. Alsema, and K. E. Criswell. 2017. Trunk neural crest origin of dermal denticles in a cartilaginous fish. *Proceedings of the National Academy of Sciences* 114:13200–13205.

Gluhak-Heinrich, J., D. Pavlin, W. Yang, M. MacDougall, and S. E. Harris. 2007. MEPE expression in osteocytes during orthodontic tooth movement. *Archives of Oral Biology* 52:684–690.

Gluhak-Heinrich, J., L. Ye, L. F. Bonewald, J. Q. Feng, M. MacDougall, S. E. Harris, and D. Pavlin. 2003. Mechanical loading stimulates dentin matrix protein 1 (DMP1) expression in osteocytes in vivo. *Journal of Bone and Mineral Research* 18:807–817.

Gobbi, P., T. Maravic, A. Comba, C. Mazzitelli, E. Mancuso, M. Falconi, L. Breschi, and A. Mazzoni. 2021. Biochemical and immunohistochemical analysis of tissue inhibitor of metalloproteinases-1 in human sound dentin. *Clinical Oral Investigations* 25:5067–5075.

Goldberg, M., A. B. Kulkarni, M. Young, and A. Boskey. 2011. Dentin: Structure, composition and mineralization: The role of dentin ECM in dentin formation and mineralization. *Front Biosci (Elite Ed)* 3:711–735.

Goldberg, M., D. Septier, Å. Oldberg, M. F. Young, and L. G. Ameye. 2006. Fibromodulin-deficient mice display impaired collagen fibrillogenesis in predentin as well as altered dentin mineralization and enamel formation. *Journal of Histochemistry & Cytochemistry* 54:525–537.

Goodrich, E. S. 1907. On the scales of fish, living and extinct, and their importance in classification. *Proceedings of the Zoological Society of London* 77:751–773.

Gorman, C. E., and C. D. Hulsey. 2020. Non-trophic functional ecology of vertebrate teeth: A review. *Integrative and Comparative Biology* 60:665–675.

Graham, Z. A., E. Garde, M. P. Heide-Jørgensen, and A. V. Palaoro. 2020. The longer the better: Evidence that narwhal tusks are sexually selected. *Biology Letters* 16:20190950.

Green, D. R., F. Schulte, K.-H. Lee, M. K. Pugach, M. Hardt, and F. B. Bidlack. 2019. Mapping the tooth enamel proteome and amelogenin phosphorylation onto mineralizing porcine tooth crowns. *Frontiers in Physiology* 10.

Hall, R., D. Septier, G. Embery, and M. Goldberg. 1999. Stromelysin-1 (MMP-3) in forming enamel and predentine in rat incisor-coordinated distribution with proteoglycans suggests a functional role. *Histochemical Journal* 31:761–770.

Hannas, A. R., J. C. Pereira, J. M. Granjeiro, and L. Tjäderhane. 2007. The role of matrix metalloproteinases in the oral environment. *Acta Odontologica Scandinavica* 65:1–13.

Hannesson, K. O., E. Ytteborg, H. Takle, G. Enersen, G. Bæverfjord, and M. E. Pedersen. 2015. Sulphated glycosaminoglycans and proteoglycans in the developing vertebral column of juvenile Atlantic salmon (*Salmo salar*). *Fish Physiology and Biochemistry* 41:1029–1051.

Hart, P. S. 2004. Mutation in kallikrein 4 causes autosomal recessive hypomaturation amelogenesis imperfecta. *Journal of Medical Genetics* 41:545–549.

Haze, A., A. L. Taylor, A. Blumenfeld, E. Rosenfeld, Y. Leiser, L. Dafni, B. Shay, Y. Gruenbaum-Cohen, E. Fermon, S. Haegewald, J.-P. Bernimoulin, and D. Deutsch. 2007. Amelogenin expression in long bone and cartilage cells and in bone marrow progenitor cells. *The Anatomical Record: Advances in Integrative Anatomy and Evolutionary Biology* 290:455–460.

He, H., C. Shao, Z. Mu, C. Mao, J. Sun, C. Chen, R. Tang, and X. Gu. 2020. Promotion effect of immobilized chondroitin sulfate on intrafibrillar mineralization of collagen. *Carbohydrate Polymers* 229:115547.

He, L., Y. Hao, L. Zhen, H. Liu, M. Shao, X. Xu, K. Liang, Y. Gao, H. Yuan, J. Li, J. Li, L. Cheng, and C. van Loveren. 2019. Biomineralization of dentin. *Journal of Structural Biology* 207:115–122.

Hermyt, M., B. Metscher, and W. Rupik. 2021. Ultrastructural studies of developing egg tooth in grass snake *Natrix natrix* (Squamata, Serpentes) embryos, supported by X-ray microtomography analysis. *Zoology* 146:125913.

Herold, R. C. 1970. Vasodentine and mantle dentine in teleost fish teeth. A comparative microradiographic analysis. *Archives of Oral Biology* 15:71–85.

Horvath, J. E., G. L. Ramachandran, O. Fedrigo, W. J. Nielsen, C. C. Babbitt, E. M. St. Clair, L. W. Pfefferle, J. Jernvall, G. A. Wray, and C. E. Wall. 2014. Genetic comparisons yield insight into the evolution of enamel thickness during human evolution. *Journal of Human Evolution* 73:75–87.

Houari, S., T. Wurtz, D. Ferbus, D. Chateau, A. Dessombz, A. Berdal, and S. Babajko. 2014. Asporin and the mineralization process in fluoride-treated rats. *Journal of Bone and Mineral Research* 29:1446–1455.

Howe, A. M., and W. S. Webster. 1994. Vitamin K—its essential role in craniofacial development. *Australian Dental Journal* 39:88–92.

Hulsey, C. D., R. J. Roberts, A. S. P. Lin, R. Guldberg, and J. T. Streelman. 2008. Convergence in a mechanically complex phenotype: Detecting structural adaptations for crushing in cichlid fish. *Evolution* 62:1587–1599.

Huxley-Jones, J., T.-K. Clarke, C. Beck, G. Toubaris, D. L. Robertson, and R. P. Boot-Handford. 2007. The evolution of the vertebrate metzincins; insights from *Ciona intestinalis* and *Danio rerio*. *BMC Evolutionary Biology* 7:63.

Huysseune, A., R. Cerny, and P. E. Witten. 2022. The conundrum of pharyngeal teeth origin: The role of germ layers, pouches, and gill slits. *Biological Reviews* 97:414–447.

Huysseune, A., and J.-Y. Sire. 1998. Evolution of patterns and processes in teeth and tooth-related tissues in non-mammalian vertebrates. *European Journal of Oral Sciences* 106:437–481.

Huysseune, A., J.-Y. Sire, and P. E. Witten. 2009. Evolutionary and developmental origins of the vertebrate dentition. *Journal of Anatomy* 214:465–476.

Huysseune, A., H. Takle, M. Soenens, K. Taerwe, and P. E. Witten. 2008. Unique and shared gene expression patterns in Atlantic salmon (*Salmo salar*) tooth development. *Development Genes and Evolution* 218:427–437.

Ibrahim, S., A. P. Strange, S. Aguayo, A. Shinawi, N. Harith, N. Mohamed-Ibrahim, S. Siddiqui, S. Parekh, and L. Bozec. 2019. Phenotypic properties of collagen in dentinogenesis imperfecta associated with osteogenesis imperfecta. *International Journal of Nanomedicine* 14:9423–9435.

Iwata, T., Y. Yamakoshi, J. C.-C. Hu, I. Ishikawa, J. D. Bartlett, P. H. Krebsbach, and J. P. Simmer. 2007. Processing of ameloblastin by MMP-20. *Journal of Dental Research* 86:153–157.

Janvier, P. 1996. *Early Vertebrates.* Oxford University Press Inc., New York, 393 pp.

Jernvall, J., and I. Thesleff. 2012. Tooth shape formation and tooth renewal: Evolving with the same signals. *Development (Cambridge, England)* 139:3487–3497.

Jiang, T., F. Liu, W.-G. Wang, X. Jiang, X. Wen, K.-J. Hu, and Y. Xue. 2017. Distribution of cathepsin K in late stage of tooth germ development and its function in degrading enamel matrix proteins in mouse. *PloS ONE* 12:e0169857.

Kaipatur, N. R., M. Murshed, and M. D. McKee. 2008. Matrix Gla protein inhibition of tooth mineralization. *Journal of Dental Research* 87:839–844.

Kalthoff, D. C. 2011. Microstructure of dental hard tissues in fossil and recent xenarthrans (Mammalia: Folivora and Cingulata). *Journal of Morphology* 272:641–661.

Kang, J. S., T. Oohashi, Y. Kawakami, Y. Bekku, J. C. Izpisúa Belmonte, and Y. Ninomiya. 2004. Characterization of dermacan, a novel zebrafish lectican gene, expressed in dermal bones. *Mechanisms of Development* 121:301–312.

Kawasaki, K. 2009. The SCPP gene repertoire in bony vertebrates and graded differences in mineralized tissues. *Development Genes and Evolution* 219:147–157.

Kawasaki, K. 2018. The origin and early evolution of SCPP genes and tissue mineralization in vertebrates; pp. 157–164 in *Biomineralization.* Springer Singapore, Singapore.

Kawasaki, K., and C. T. Amemiya. 2014. SCPP genes in the coelacanth: Tissue mineralization genes shared by sarcopterygians. *Journal of Experimental Zoology Part B: Molecular and Developmental Evolution* 322:390–402.

Kawasaki, K., J. C.-C. Hu, and J. P. Simmer. 2014. Evolution of *Klk4* and enamel maturation in eutherians. *Biological Chemistry* 395:1003–1013.

Kawasaki, K., J. N. Keating, M. Nakatomi, M. Welten, M. Mikami, I. Sasagawa, M. N. Puttick, P. C. J. J. Donoghue, and M. Ishiyama. 2021. Coevolution of enamel, ganoin, enameloid, and their matrix SCPP genes in osteichthyans. *Iscience* 24:102023.

Kawasaki, K., M. Mikami, M. Nakatomi, I. Braasch, P. Batzel, J. H. Postlethwait, A. Sato, I. Sasagawa, and M. Ishiyama. 2017. SCPP genes and their relatives in gar: Rapid expansion of mineralization genes in osteichthyans. *Journal of Experimental Zoology Part B: Molecular and Developmental Evolution* 328:645–665.

Kawasaki, K., and T. Suzuki. 2011. Molecular evolution of matrix metalloproteinase 20. *European Journal of Oral Sciences* 119:247–253.

Kawasaki, K., T. Suzuki, and K. M. Weiss. 2005. Phenogenetic drift in evolution: The changing genetic basis of vertebrate teeth. *Proceedings of the National Academy of Sciences of the United States of America* 102:18063–18068.

Kawasaki, K., and K. M. Weiss. 2003. Mineralized tissue and vertebrate evolution: The secretory calcium-binding phosphoprotein gene cluster. *Proceedings of the National Academy of Sciences* 100:4060–4065.

Keating, J. N., and P. C. J. Donoghue. 2016. Histology and affinity of anaspids, and the early evolution of the vertebrate dermal skeleton. *Proceedings of the Royal Society of London B: Biological Sciences* 283.

Kelwick, R., I. Desanlis, G. N. Wheeler, and D. R. Edwards. 2015. The ADAMTS (A Disintegrin and Metalloproteinase with Thrombospondin motifs) family. *Genome Biology* 16:113.

Kimura, M., H. Mochizuki, R. Satou, M. Iwasaki, E. Kokubu, K. Kono, S. Nomura, T. Sakurai, H. Kuroda, and Y. Shibukawa. 2021. Plasma membrane Ca2+—ATPase in rat and human odontoblasts mediates dentin mineralization. *Biomolecules* 11:1010.

Klingler, A., D. Regensburger, C. Tenkerian, N. Britzen-Laurent, A. Hartmann, M. Stürzl, and E. Naschberger. 2020. Species-, organ- and cell-type-dependent expression of SPARCL1 in human and mouse tissues. *PloS ONE* 15:e0233422.

Kogaya, Y. 1999. Immunohistochemical localisation of amelogenin-like proteins and type I collagen and histochemical demonstration of sulphated glycoconjugates in developing enameloid and enamel matrices of the larval urodele (*Triturus pyrrhogaster*) teeth. *Journal of Anatomy* 195:455–464.

Komori, T. 2010. Regulation of bone development and extracellular matrix protein genes by RUNX2. *Cell and Tissue Research* 339:189–195.

Kramer, K., M. B. Chavez, A. T. Tran, F. Farah, M. H. Tan, T. N. Kolli, E. J. L. dos Santos, H. F. Wimer, J. L. Millán, L. J. Suva, D. Gaddy, and B. L. Foster. 2021. Dental defects in the primary dentition associated with hypophosphatasia from biallelic ALPL mutations. *Bone* 143:115732.

Krebsbach, P. H., S. K. Lee, Y. Matsuki, C. A. Kozak, K. M. Yamada, and Y. Yamada. 1996. Full-length sequence, localization, and chromosomal mapping of ameloblastin. *Journal of Biological Chemistry* 271:4431–4435.

Krivanek, J., R. A. Soldatov, M. E. Kastriti, T. Chontorotzea, A. N. Herdina, J. Petersen, B. Szarowska, M. Landova, V. K. Matejova, L. I. Holla, U. Kuchler, I. V. Zdrilic, A. Vijaykumar, A. Balic, P. Marangoni, O. D. Klein, V. C. M. Neves, V. Yianni, P. T. Sharpe, T. Harkany, B. D. Metscher, M. Bajénoff, M. Mina, K. Fried, P. V. Kharchenko, and I. Adameyko. 2020. Dental cell type atlas reveals stem and differentiated cell types in mouse and human teeth. *Nature Communications* 11:4816.

Kwan, K. M., M. K. M. Pang, S. Zhou, S. K. Cowan, et al. 1997. Abnormal compartimentalization of cartilage matrix components in mice lacking collagen X: Implications for function. *Journal of Cell Biology* 136:459–471.

Lacruz, R. S., A. Nanci, I. Kurtz, J. T. Wright, and M. L. Paine. 2010. Regulation of pH during Amelogenesis. *Calcified Tissue International* 86:91–103.

Lacruz, R. S., C. E. Smith, Y.-B. Chen, M. J. Hubbard, J. G. Hacia, and M. L. Paine. 2011. Gene-expression analysis of early- and late-maturation-stage rat enamel organ. *European Journal of Oral Sciences* 119:149–157.

Larsson, L., J. Frisén, and J. Lundeberg. 2021. Spatially resolved transcriptomics adds a new dimension to genomics. *Nature Methods* 18:15–18.

Lee, E.-H., H.-J. Park, J.-H. Jeong, Y.-J. Kim, D.-W. Cha, D.-K. Kwon, S.-H. Lee, and J.-Y. Cho. 2011. The role of asporin in mineralization of human dental pulp stem cells. *Journal of Cellular Physiology* 226:1676–1682.

Lee, H.-K., S.-J. Park, H.-J. Oh, J.-W. Kim, H.-S. Bae, and J.-C. Park. 2012. Expression pattern, subcellular localization, and functional implications of ODAM in ameloblasts, odontoblasts, osteoblasts, and various cancer cells. *Gene Expression Patterns* 12:102–108.

Lemierre, A., and D. Germain. 2019. A new mineralized tissue in the early vertebrate *Astraspis*. *Journal of Anatomy* 235:1105–1113.

Leurs, N., C. Martinand-Mari, S. Marcellini, and M. Debiais-Thibaud. 2022. Parallel evolution of ameloblastic *scpp* genes in bony and cartilaginous vertebrates. *Molecular Biology and Evolution* 39.

Leurs, N., C. Martinand-Mari, S. Ventéo, T. Haitina, and M. Debiais-Thibaud. 2021. Evolution of matrix gla and bone gla protein genes in jawed vertebrates. *Frontiers in Genetics* 12.

Liang, T., Y. Hu, K. Kawasaki, H. Zhang, C. Zhang, T. L. Saunders, J. P. Simmer, and J. C.-C. Hu. 2021. Odontogenesis-associated phosphoprotein truncation blocks ameloblast transition into maturation in OdaphC41*/C41* mice. *Scientific Reports* 11:1132.

Lin, C. P., W. H. Douglas, and S. L. Erlandsen. 1993. Scanning electron microscopy of type I collagen at the dentin-enamel junction of human teeth. *Journal of Histochemistry & Cytochemistry* 41:381–388.

Lin, Q., S. Fan, Y. Zhang, M. Xu, H. Zhang, Y. Yang, A. P. Lee, J. M. Woltering, V. Ravi, H. M. Gunter, W. Luo, Z. Gao, Z. W. Lim, G. Qin, R. F. Schneider, X. Wang, P. Xiong, G. Li, K. Wang, J. Min, C. Zhang, Y. Qiu, J. Bai, W. He, C. Bian, X. Zhang, D. Shan, H. Qu, Y. Sun, Q. Gao, L. Huang, Q. Shi, A. Meyer, and B. Venkatesh. 2016. The seahorse genome and the evolution of its specialized morphology. *Nature* 540:395–399.

Linde, A. 1995. Dentin mineralization and the role of odontoblasts in calcium transport. *Connective Tissue Research* 33:163–170.

Listik, E., J. Azevedo Marques Gaschler, M. Matias, M. F. Neuppmann Feres, L. Toma, and A. C. Raphaelli Nahás-Scocate. 2019. Proteoglycans and dental biology: The first review. *Carbohydrate Polymers* 225:115199.

Liu, Z., S. Liu, J. Yao, L. Bao, J. Zhang, Y. Li, C. Jiang, L. Sun, R. Wang, Y. Zhang, T. Zhou, Q. Zeng, Q. Fu, S. Gao, N. Li, S. Koren, Y. Jiang, A. Zimin, P. Xu, A. M. Phillippy, X. Geng, L. Song, F. Sun, C. Li, X. Wang, A. Chen, Y. Jin, Z. Yuan, Y. Yang, S. Tan, E. Peatman, J. Lu, Z. Qin, R. Dunham, Z. Li, T. Sonstegard, J. Feng, R. G. Danzmann, S. Schroeder, B. Scheffler, M. V. Duke, L. Ballard, H. Kucuktas, L. Kaltenboeck, H. Liu, J. Armbruster, Y. Xie, M. L. Kirby, Y. Tian, M. E. Flanagan, W. Mu, and G. C. Waldbieser. 2016. The channel catfish genome sequence provides insights into the evolution of scale formation in teleosts. *Nature Communications* 7:11757.

Lu, X., S. Fukumoto, Y. Yamada, C. A. Evans, T. G. Diekwisch, and X. Luan. 2016. Ameloblastin, an extracellular matrix protein, affects long bone growth and mineralization. *Journal of Bone and Mineral Research* 31:1235–1246.

Lu, Y., P. Papagerakis, Y. Yamakoshi, J. C.-C. Hu, J. D. Bartlett, and J. P. Simmer. 2008. Functions of KLK4 and MMP-20 in dental enamel formation. *Bchm* 389:695–700.

Lumsden, A. G. 1988. Spatial organization of the epithelium and the role of neural crest cells in the initiation of the mammalian tooth germ. *Development (Cambridge, England)* 103(Suppl):155–169.

Macdougall, M., T. T. Gu, X. Luan, D. Simmons, and J. Chen. 1998. Identification of a novel isoform of mouse dentin matrix protein 1: Spatial expression in mineralized tissues. *Journal of Bone and Mineral Research* 13:422–431.

Mäkitie, O., M. Susic, and W. G. Cole. 2010. Early-onset metaphyseal chondrodysplasia type Schmid associated with a COL10A1 frame-shift mutation and impaired trimerization of wild-type α1(X) protein chains. *Journal of Orthopaedic Research* 28:1497–1501.

Mann, A., D. Rudkin, D. C. Evans, and M. Laflamme. 2017. A large onychodontiform (Osteichthyes: Sarcopterygii) apex predator from the Eifelian-aged Dundee Formation of Ontario, Canada. *Canadian Journal of Earth Sciences* 54:233–241.

Marconi, A., A. Hancock-Ronemus, and J. A. Gillis. 2020. Adult chondrogenesis and spontaneous cartilage repair in the skate, *Leucoraja erinacea*. *Elife* 9:1–26.

Maruyama, K., A. Henmi, H. Okata, and Y. Sasano. 2016. Analysis of calcium, phosphorus, and carbon concentrations during developmental calcification of dentin and enamel in rat incisors using scanning electron microscopy with energy dispersive X-ray spectroscopy (SEM-EDX). *Journal of Oral Biosciences* 58:173–179.

Mayeur, H., M. Lanoizelet, A. Quillien, A. Menuet, L. Michel, K. J. Martin, S. Dejean, P. Blader, S. Mazan, and R. Lagadec. 2021. When bigger is better: 3D RNA profiling of the developing head in the Catshark *Scyliorhinus canicula*. *Frontiers in Cell and Developmental Biology* 9.

McKnight, D. A., J. P. Simmer, P. S. Hart, T. C. Hart, and L. W. Fisher. 2008. Overlapping *DSPP* mutations cause dentin dysplasia and dentinogenesis imperfecta. *Journal of Dental Research* 87:1108–1111.

Melstrom, K. M. 2017. The relationship between diet and tooth complexity in living dentigerous saurians. *Journal of Morphology* 278:500–522.

Meredith, R. W., J. Gatesy, J. Cheng, and M. S. Springer. 2011. Pseudogenization of the tooth gene enamelysin (*MMP20*) in the common ancestor of extant baleen whales. *Proceedings of the Royal Society B: Biological Sciences* 278:993–1002.

Meredith, R. W., J. Gatesy, and M. S. Springer. 2013. Molecular decay of enamel matrix protein genes in turtles and other edentulous amniotes. *BMC Evolutionary Biology* 13:20.

Meredith, R. W., G. Zhang, M. T. P. Gilbert, E. D. Jarvis, and M. S. Springer. 2014. Evidence for a single loss of mineralized teeth in the common avian ancestor. *Science* 346.

Mikami, M., T. Ineno, A. W. Thompson, I. Braasch, M. Ishiyama, and K. Kawasaki. 2022. Convergent losses of SCPP genes and ganoid scales among non-teleost actinopterygians. *Gene* 811:146091.

Milan, A. M., R. V. Sugars, G. Embery, and R. J. Waddington. 2004. Dentinal proteoglycans demonstrate an increasing order of affinity for hydroxyapatite crystals during the transition of predentine to dentine. *Calcified Tissue International* 75:197–204.

Milan, A. M., R. V. Sugars, G. Embery, and R. J. Waddington. 2005. Modulation of collagen fibrillogenesis by dentinal proteoglycans. *Calcified Tissue International* 76:127–135.

Moffatt, P., C. E. Smith, R. St-Arnaud, and A. Nanci. 2008. Characterization of Apin, a secreted protein highly expressed in tooth-associated epithelia. *Journal of Cellular Biochemistry* 103:941–956.

Moffatt, P., C. E. Smith, R. St-Arnaud, D. Simmons, J. T. Wright, and A. Nanci. 2006. Cloning of rat amelotin and localization of the protein to the basal lamina of maturation stage ameloblasts and junctional epithelium. *Biochemical Journal* 399:37–46.

Mondéjar-Fernández, J. 2018. On cosmine: Its origins, biology and implications for sarcopterygian interrelationships. *Cybium* 42:41–65.

Moradian-Oldak, J., and A. George. 2021. Biomineralization of enamel and dentin mediated by matrix proteins. *Journal of Dental Research* 100:1020–1029.

Morvan-Dubois, G., D. Le Guellec, R. Garrone, L. Zylberberg, and L. Bonnaud. 2003. Phylogenetic analysis of vertebrate fibrillar collagen locates the position of zebrafish alpha3(I) and suggests an evolutionary link between collagen alpha chains and hox clusters. *Journal of Molecular Evolution* 57:501–514.

Mu, Y., X. Huang, R. Liu, Y. Gai, N. Liang, D. Yin, L. Shan, S. Xu, and G. Yang. 2021. *ACPT* gene is inactivated in mammalian lineages that lack enamel or teeth. *PeerJ* 9:e10219.

Nagano, T., S. Oida, H. Ando, K. Gomi, T. Arai, and M. Fukae. 2003. Relative levels of mRNA encoding enamel proteins in enamel organ epithelia and odontoblasts. *Journal of Dental Research* 82:982–986.

Nakayama, Y., J. Holcroft, and B. Ganss. 2015. Enamel hypomineralization and structural defects in amelotin-deficient mice. *Journal of Dental Research* 94:697–705.

Nanci, A., and H. Warshawsky. 1984. Characterization of putative secretory sites on ameloblasts of the rat incisor. *American Journal of Anatomy* 171:163–189.

Nancy, A., and A. R. Ten Cate. 2003. *Ten Cate's Oral Histology: Development, Structure, and Function*. Mosby Elsevier, St Louis.

Nuñez, J., M. Sanz, L. Hoz-Rodríguez, M. Zeichner-David, and H. Arzate. 2010. Human cementoblasts express enamel-associated molecules in vitro and in vivo. *Journal of Periodontal Research* 45:809–814.

Nurbaeva, M. K., M. Eckstein, S. Feske, and R. S. Lacruz. 2017. Ca^{2+} transport and signalling in enamel cells. *The Journal of Physiology* 595:3015–3039.

Oka, S., I. Sasagawa, and M. Ishiyama. 2017. Histochemical and immunohistochemical examination of odontoblasts (petroblasts) in petrodentine formation of lungfish. *Archives of Oral Biology* 83:222–229.

Okada, M., I. Tazawa, K. Nakajima, and Y. Yaoita. 2013. Expression of the amelogenin gene in the skin of *Xenopus tropicalis*. *Zoological Science* 30:154–159.

Oralová, V., J. T. Rosa, D. Larionova, P. E. Witten, and A. Huysseune. 2020. Multiple epithelia are required to develop teeth deep inside the pharynx. *Proceedings of the National Academy of Sciences* 117:11503–11512.

Ortiz-Delgado, J. B., D. C. Simes, P. J. Gavaia, C. Sarasquete, and M. L. Cancela. 2005. Osteocalcin and matrix GLA protein in developing teleost teeth: Identification of sites of mRNA and protein accumulation at single cell resolution. *Histochemistry and Cell Biology* 124:123–130.

Ørvig, T. 1967. Phylogeny of tooth tissue: Evolution of some calcified tissues in early vertebrates; pp. 45–110 in A. E. W. Miles (ed.), *Structural and Chemical Organisation of Teeth*. Academic Press, New York and London.

Pandya, M., and T. G. H. Diekwisch. 2021. Amelogenesis: Transformation of a protein-mineral matrix into tooth enamel. *Journal of Structural Biology* 213:107809.

Papagerakis, P., A. Berdal, M. Mesbah, M. Peuchmaur, L. Malaval, J. Nydegger, J. Simmer, and M. Macdougall. 2002. Investigation of osteocalcin, osteonectin, and dentin sialophosphoprotein in developing human teeth. *Bone* 30:377–385.

Park, H., J. Huxley-Jones, R. P. Boot-Handford, P. N. Bishop, T. K. Attwood, and J. Bella. 2008. LRRCE: A leucine-rich repeat cysteine capping motif unique to the chordate lineage. *BMC Genomics* 9:599.

Parry, D. A., S. J. Brookes, C. V. Logan, J. A. Poulter, W. El-Sayed, S. Al-Bahlani, S. Al Harasi, J. Sayed, E. M. Raïf, R. C. Shore, M. Dashash, M. Barron, J. E. Morgan, I. M. Carr, G. R. Taylor, C. A. Johnson, M. J. Aldred, M. J. Dixon, J. T. Wright, J. Kirkham, C. F. Inglehearn, and A. J. Mighell. 2012. Mutations in C4orf26, encoding a peptide with in vitro hydroxyapatite crystal nucleation and growth activity, cause amelogenesis imperfecta. *The American Journal of Human Genetics* 91:565–571.

Pedersen, M., E. Ytteborg, A. Kohler, G. Baeverfjord, G. Enersen, B. Ruyter, H. Takle, and K. Hannesson. 2013. Small leucine-rich proteoglycans in the vertebrae of Atlantic salmon Salmo salar. *Diseases of Aquatic Organisms* 106:57–68.

Pham, C.-D., C. E. Smith, Y. Hu, J. C.-C. Hu, J. P. Simmer, and Y.-H. P. Chun. 2017. Endocytosis and enamel formation. *Frontiers in Physiology* 8.

Poole, D. F. G. 1967. Phylogeny of tooth tissues: Enameloid and enamel in recent vertebrates, with a note on the history of cementum; pp. 111–149 in E. W. Miles (ed.), *Structural and Chemical Organisation of Teeth*. Academic Press, New York and London.

Poulter, J. A., G. Murillo, S. J. Brookes, C. E. L. Smith, D. A. Parry, S. Silva, J. Kirkham, C. F. Inglehearn, and A. J. Mighell. 2014. Deletion of ameloblastin exon 6 is associated with amelogenesis imperfecta. *Human Molecular Genetics* 23:5317–5324.

Prostak, K. S., and Z. Skobe. 1988. Ultrastructure of odontogenic cells during enameloid matrix synthesis in tooth buds from an elasmobranch, *Raja erinacae*. *The American Journal of Anatomy* 182:59–72.

Puente, X. S., L. M. Sánchez, C. M. Overall, and C. López-Otín. 2003. Human and mouse proteases: A comparative genomic approach. *Nature Reviews Genetics* 4:544–558.

Qin, C., J. C. Brunn, E. Cadena, A. Ridall, H. Tsujigiwa, H. Nagatsuka, N. Nagai, and W. T. Butler. 2002. The expression of dentin sialophosphoprotein gene in bone. *Journal of Dental Research* 81:392–394.

Qu, M., Y. Liu, Y. Zhang, S. Wan, V. Ravi, G. Qin, H. Jiang, X. Wang, H. Zhang, B. Zhang, Z. Gao, A. Huysseune, Z. Zhang, H. Zhang, Z. Chen, H. Yu, Y. Wu, L. Tang, C. Li, J. Zhong, L. Ma, F. Wang, H. Zheng, J. Yin, P. E. Witten, A. Meyer, B. Venkatesh, and Q. Lin. 2021. Seadragon genome analysis provides insights into its phenotype and sex determination locus. *Science Advances* 7.

Qu, Q., T. Haitina, M. Zhu, and P. E. Ahlberg. 2015. New genomic and fossil data illuminate the origin of enamel. *Nature* 526:108–111.

Rajpar, M. H. 2001. Mutation of the gene encoding the enamel-specific protein, enamelin, causes autosomal-dominant amelogenesis imperfecta. *Human Molecular Genetics* 10:1673–1677.

Randau, M., C. Carbone, and S. T. Turvey. 2013. Canine evolution in sabretoothed carnivores: Natural selection or sexual selection? *PloS ONE* 8:e72868.

Randilini, A., K. Fujikawa, and S. Shibata. 2020. Expression, localization and synthesis of small leucine-rich proteoglycans in developing mouse molar tooth germ. *European Journal of Histochemistry* 64.

Rauch, F., L. Lalic, P. Roughley, and F. H. Glorieux. 2010. Genotype—phenotype correlations in nonlethal osteogenesis imperfecta caused by mutations in the helical domain of collagen type I. *European Journal of Human Genetics* 18:642–647.

Reif, W.-E. 1982. Evolution of dermal skeleton and dentitions in vertebrates: The odontode-regulation theory. *Evolutionary Biology* 15:287–368.

Reisz, R. R. 2006. Origin of dental occlusion in tetrapods: Signal for terrestrial vertebrate evolution? *Journal of Experimental Zoology Part B: Molecular and Developmental Evolution* 306B:261–277.

Reznikov, N., B. Hoac, D. J. Buss, W. N. Addison, N. M. T. Barros, and M. D. McKee. 2020. Biological stenciling of mineralization in the skeleton: Local enzymatic removal of inhibitors in the extracellular matrix. *Bone* 138:115447.

Ritchie, H. H., J. E. Berry, M. J. Somerman, C. T. Hanks, A. L. J. J. Bronckers, D. Hotton, P. Papagerakis, A. Berdal, and W. T. Butler. 1997. Dentin sialoprotein (DSP) transcripts: Developmentally-sustained expression in odontoblasts and transient expression in pre-ameloblasts. *European Journal of Oral Sciences* 105:405–413.

Robey, P. G. 1996. Vertebrate mineralized matrix proteins: Structure and function. *Connective Tissue Research* 35:131–136.

Root, Z. D., D. Jandzik, C. Allen, M. Brewer, M. Romášek, T. Square, and D. M. Medeiros. 2021. Lamprey lecticans link new vertebrate genes to the origin and elaboration of vertebrate tissues. *Developmental Biology* 476:282–293.

Rosa, J. T., P. E. Witten, and A. Huysseune. 2021. Cells at the edge: The dentin—bone interface in zebrafish teeth. *Frontiers in Physiology* 12.

Rossi, A., Q. Deveraux, B. Turk, and A. Sali. 2004. Comprehensive search for cysteine cathepsins in the human genome. *Biological Chemistry* 385.

Rowe, P. S. N. 2012. The chicken or the egg: PHEX, FGF23 and SIBLINGs unscrambled. *Cell Biochemistry and Function* 30:355–375.

Ruggeri, A., G. Orsini, A. Mazzoni, F. Nato, V. Papa, M. Piccirilli, A. Putignano, G. Mazzotti, E. De Stefano Dorigo, and L. Breschi. 2009. Immunohistochemical and biochemical assay of versican in human sound predentine/dentine matrix. *European Journal of Histochemistry* 53:15.

Ryll, B., S. Sanchez, T. Haitina, P. Tafforeau, and P. E. Ahlberg. 2014. The genome of Callorhinchus and the fossil record: A new perspective on SCPP gene evolution in gnathostomes. *Evolution & Development* 16:123–124.

Ryu, O. H., A. G. Fincham, C.-C. Hu, C. Zhang, Q. Qian, J. D. Bartlett, and J. P. Simmer. 1999. Characterization of recombinant pig enamelysin activity and cleavage of recombinant pig and mouse amelogenins. *Journal of Dental Research* 78:743–750.

Sander, P. M. 2000. Prismless enamel in amniotes: Terminology, function, and evolution; pp. 92–106 in *Development, Function and Evolution of Teeth.* Cambridge University Press, Cambridge.

Sansom, I. J., P. C. Donoghue, and G. Albanesi. 2005. Histology and affinity of the earliest armoured vertebrate. *Biology Letters* 1:446–449.

Sasagawa, I. 1995. Fine structure of tooth germs during the formation of enameloid matrix in *Tilapia nilotica*, a teleost fish. *Archives of Oral Biology* 40:801–814.

Sasagawa, I. 1997. Fine structure of the cap enameloid and of the dental epithelial cells during enameloid mineralisation and early maturation stages in the tilapia, a teleost. *Journal of Anatomy* 190:589–600.

Sasagawa, I. 2002a. Mineralization patterns in elasmobranch fish. *Microscopy Research and Technique* 59:396–407.

Sasagawa, I. 2002b. Fine structural and cytochemical observations of dental epithelial cells during the enameloid formation stages in red stingrays *Dasyatis akajei. Journal of Morphology* 252:170–182.

Sasagawa, I., and M. Ishiyama. 2005. Fine structural and cytochemical observations on the dental epithelial cells during cap enameloid formation stages in *Polypterus senegalus*, a bony fish (Actinopterygii). *Connective Tissue Research* 46:33–52.

Sasagawa, I., M. Ishiyama, H. Yokosuka, M. Mikami, S. Ichiro, I. Mikio, Y. Hiroyuki, and M. Masato. 2013. Teeth and ganoid scales in Polypterus and Lepisosteus, the basic actinopterygian fish: An approach to understand the origin of the tooth enamel. *Journal of Oral Biosciences* 55:76–84.

Sasaki, T., M. Takagi, and T. Yanagisawa. 1997. Structure and function of secretory ameloblasts in enamel formation. *Ciba Foundation Symposium* 205:32–50.

Satchell, P. G., C. F. Shuler, and T. G. H. Diekwisch. 2000. True enamel covering in teeth of the Australian lungfish *Neoceratodus forsteri. Cell and Tissue Research* 299:27–37.

Satoyoshi, M., A. Kawata, T. Koizumi, K. Inoue, S. Itohara, T. Teranaka, and Y. Mikunitakagaki. 2001. Matrix metalloproteinase-2 in dentin matrix mineralization. *Journal of Endodontics* 27:462–466.

Schaeffer, B., D. H. Dunkle, and C. D. Walcott. 1950. *A Semionotid Fish from the Chinle Formation, with Consideration of Its Relationships* (American Museum Novitates, ed.). American Museum of Natural History, New York.

Schmidt, W. J., A. Keil, D. F. G. Poole, and A. I. Darling. 1971. *Polarizing Microscopy of Dental Tissues Theory, Methods, and Results from the Structural Analysis of Normal and Diseased Hard, Dental Tissues and Tissues Associated with Them in Man and Other Vertebrates.* Pergamon, Oxford.

Schultze, H.-P. 2016. Scales, enamel, cosmine, ganoine, and early osteichthyans. *Comptes Rendus Palevol* 15:83–102.

Scott, I. C., I. L. Blitz, W. N. Pappano, Y. Imamura, T. G. Clark, B. M. Steiglitz, C. L. Thomas, S. A. Maas, K. Takahara, K. W. Y. Cho, and D. S. Greenspan. 1999. Mammalian BMP-1/Tolloid-related metalloproteinases, including novel family member mammalian Tolloid-like 2, have differential enzymatic activities and distributions of expression relevant to patterning and skeletogenesis. *Developmental Biology* 213:283–300.

Shaheen, J., A. B. Mudd, T. G. H. Diekwisch, and J. Abramyan. 2021. Pseudogenized amelogenin reveals early tooth loss in true toads (Anura: Bufonidae). *Integrative and Comparative Biology* 61:1933–1945.

Sharir, A., P. Marangoni, R. Zilionis, M. Wan, T. Wald, J. K. Hu, K. Kawaguchi, D. Castillo-Azofeifa, L. Epstein, K. Harrington, P. Pagella, T. Mitsiadis, C. W. Siebel, A. M. Klein, and O. D. Klein. 2019. A large pool of actively cycling progenitors orchestrates self-renewal and injury repair of an ectodermal appendage. *Nature Cell Biology* 21:1102–1112.

Shellis, R. P., and A. E. W. Miles. 1974. Autoradiographic study of the formation of enameloid and dentine matrices in teleost fishes using tritiated amino acids. *Proceedings of the Royal Society of London. Series B. Biological Sciences* 185:51–72.

Shi, C., N. Ma, W. Zhang, J. Ye, H. Shi, D. Xiang, C. Wu, L. Song, N. Zhang, and Q. Liu. 2020. Haploinsufficiency of Dspp gene causes dentin dysplasia type II in mice. *Frontiers in Physiology* 11.

Simmer, J. P., and J. C.-C. Hu. 2002. Expression, structure, and function of enamel proteinases. *Connective Tissue Research* 43:441–449.

Simmer, J. P., J. C.-C. Hu, Y. Hu, S. Zhang, T. Liang, S.-K. Wang, J.-W. Kim, Y. Yamakoshi, Y.-H. Chun, J. D. Bartlett, and C. E. Smith. 2021. A genetic model for the secretory stage of dental enamel formation. *Journal of Structural Biology* 213:107805.

Simmer, J. P., A. S. Richardson, S.-K. Wang, B. M. Reid, Y. Bai, Y. Hu, and J. C.-C. Hu. 2014. Ameloblast transcriptome changes from secretory to maturation stages. *Connective Tissue Research* 55:29–32.

Sire, J.-Y. 1995. Ganoine formation in the scales of primitive actinopterygian fishes, lepisosteids and polypterids. *Connective Tissue Research* 33:213–222.

Sire, J.-Y., P. C. J. Donoghue, and M. K. Vickaryous. 2009. Origin and evolution of the integumentary skeleton in non-tetrapod vertebrates. *Journal of Anatomy* 214:409–440.

Sire, J.-Y., J. Géraudie, and L. Zylberberg. 1987. On the origin of ganoine: Histological and ultrastructural data on the experimental regeneration of the scales of *Calamoichthys calabaricus* (Osteichthyes, Brachyopterygii, Polypteridae). *The American Journal of Anatomy* 180:391–402.

Sire, J.-Y., and A. Huysseune. 2003. Formation of dermal skeletal and dental tissues in fish: A comparative and evolutionary approach. *Biological Reviews of the Cambridge Philosophical Society* 78:S1464793102006073.

Slavkin, H. C., and P. Bringas. 1976. Epithelial-mesenchyme interactions during odontogenesis. *Developmental Biology* 50:428–442.

Smid, J. R., W. G. Young, and P. A. Monsour. 2001. Dipeptidyl-peptidase II and cathepsin B activities in amelogenesis of the rat incisor. *European Journal of Oral Sciences* 109:260–266.

Smith, C. E. L. 1998. Cellular and chemical events during enamel maturation. *Critical Reviews in Oral Biology & Medicine* 9:128–161.

Smith, C. E. L., G. Murillo, S. J. Brookes, J. A. Poulter, S. Silva, J. Kirkham, C. F. Inglehearn, and A. J. Mighell. 2016. Deletion of amelotin exons 3–6 is associated with amelogenesis imperfecta. *Human Molecular Genetics* 25:3578–3587.

Smith, C. E. L., and A. Nanci. 1995. Overview of morphological changes in enamel organ cells associated with major events in amelogenesis. *The International Journal of Developmental Biology* 39:153–161.

Smith, M. M. 1978. Enamel in the oral teeth of *Latimeria chalumnae* (Pisces: Actinistia): A scanning electron microscope study. *Journal of Zoology* 185:355–369.

Smith, M. M., and M. I. Coates. 1998. Evolutionary origins of the vertebrate dentition: Phylogenetic patterns and developmental evolution. *European Journal of Oral Sciences* 106:482–500.

Smith, M. M., and I. J. Sansom. 2000. Evolutionary origins of dentine in the fossil record of early vertebrates: Diversity, development and function; pp. 65–81 in *Development, Function and Evolution of Teeth*. Cambridge University Press, Cambridge.

Smith, M. M., C. Underwood, T. Goral, C. Healy, and Z. Johanson. 2019. Growth and mineralogy in dental plates of the holocephalan *Harriotta raleighana* (Chondrichthyes): Novel dentine and conserved patterning combine to create a unique chondrichthyan dentition. *Zoological Letters* 5:1–30.

Smith, M. P., I. J. Sansom, and J. E. Repetski. 1996. Histology of the first fish. *Nature* 380:702–704.

Sommer, B., M. Bickel, W. Hofstetter, and A. Wetterwald. 1996. Expression of matrix proteins during the development mineralized tissues. *Bone* 19:371–380.

Sone, S., M. Nakamura, Y. Maruya, I. Takahashi, I. Mizoguchi, H. Mayanagi, and Y. Sasano. 2005. Expression of versican and ADAMTS during rat tooth eruption. *Journal of Molecular Histology* 36:281–288.

Song, J., C. Ortiz, and M. C. Boyce. 2011. Threat-protection mechanics of an armored fish. *Journal of the Mechanical Behavior of Biomedical Materials* 4:699–712.

Spahr, A., S. P. Lyngstadaas, I. Slaby, and G. Pezeshki. 2006. Ameloblastin expression during craniofacial bone formation in rats. *European Journal of Oral Sciences* 114:504–511.

Springer, M. S., C. A. Emerling, J. Gatesy, J. Randall, M. A. Collin, N. Hecker, M. Hiller, and F. Delsuc. 2019. Odontogenic ameloblast-associated (ODAM) is inactivated in toothless/enamelless placental mammals and toothed whales. *BMC Evolutionary Biology* 19:31.

Springer, M. S., J. Starrett, P. A. Morin, A. Lanzetti, C. Hayashi, and J. Gatesy. 2016. Inactivation of C4orf26 in toothless placental mammals. *Molecular Phylogenetics and Evolution* 95:34–45.

Square, T. A., S. Sundaram, E. J. Mackey, and C. T. Miller. 2021. Distinct tooth regeneration systems deploy a conserved battery of genes. *EvoDevo* 12:4.

Steiglitz, B. M., M. Ayala, K. Narayanan, A. George, and D. S. Greenspan. 2004. Bone morphogenetic protein-1/Tolloid-like proteinases process dentin matrix protein-1. *Journal of Biological Chemistry* 279:980–986.

Stensiö, E. A. 1961. Permian vertebrates; pp. 231–247 in G. O. Raash (ed.), *Geology of the Arctic*. University of Toronto, Toronto.

Stundl, J., V. Soukup, R. Franěk, A. Pospisilova, V. Psutkova, M. Pšenička, R. Cerny, M. E. Bronner, D. M. Medeiros, and D. Jandzik. 2022. Efficient CRISPR mutagenesis in sturgeon demonstrates its utility in large, slow-maturing vertebrates. *Frontiers in Cell and Developmental Biology* 10.

Sullivan, M. M., P. A. Puolakkainen, T. H. Barker, S. E. Funk, and E. H. Sage. 2008. Altered tissue repair in hevin-null mice: Inhibition of fibroblast migration by a matricellular SPARC homolog. *Wound Repair and Regeneration* 16:310–319.

Suzuki, S., T. Sreenath, N. Haruyama, C. Honeycutt, A. Terse, A. Cho, T. Kohler, R. Müller, M. Goldberg, and A. B. Kulkarni. 2009. Dentin sialoprotein and dentin phosphoprotein have distinct roles in dentin mineralization. *Matrix Biology* 28:221–229.

Tamburstuen, M. V., J. E. Reseland, A. Spahr, S. J. Brookes, G. Kvalheim, I. Slaby, M. L. Snead, and S. P. Lyngstadaas. 2011. Ameloblastin expression and putative autoregulation in mesenchymal cells suggest a role in early bone formation and repair. *Bone* 48:406–413.

Thompson, A. W., M. B. Hawkins, E. Parey, D. J. Wcisel, T. Ota, K. Kawasaki, E. Funk, M. Losilla, O. E. Fitch, Q. Pan, R. Feron, A. Louis, J. Montfort, M. Milhes, B. L. Racicot, K. L. Childs, Q. Fontenot, A. Ferrara, S. R. David, A. R. McCune, A. Dornburg, J. A. Yoder, Y. Guiguen, H. Roest Crollius, C. Berthelot, M. P. Harris, and I. Braasch. 2021. The bowfin genome illuminates the developmental evolution of ray-finned fishes. *Nature Genetics* 53:1373–1384.

Tjäderhane, L., and M. Haapasalo. 2009. The dentin-pulp border: A dynamic interface between hard and soft tissues. *Endodontic Topics* 20:52–84.

Toyosawa, S., S. Shintani, T. Fujiwara, T. Ooshima, A. Sato, N. Ijuhin, and T. Komori. 2001. Dentin matrix protein 1 is predominantly expressed in chicken and rat osteocytes but not in osteoblasts. *Journal of Bone and Mineral Research* 16:2017–2026.

Tsang, K. Y., D. Chan, and K. S. E. Cheah. 2015. Fate of growth plate hypertrophic chondrocytes: Death or lineage extension? *Development, Growth & Differentiation* 57:179–192.

Tsubota, M., Y. Sasano, I. Takahashi, M. Kagayama, and H. Shimauchi. 2002. Expression of MMP-8 and MMP-13 mRNAs in rat periodontium during tooth eruption. *Journal of Dental Research* 81:673–678.

Tsuchiya, S., J. P. Simmer, J. C.-C. Hu, A. S. Richardson, F. Yamakoshi, and Y. Yamakoshi. 2011. Astacin proteases cleave dentin sialophosphoprotein (Dspp) to generate dentin phosphoprotein (Dpp). *Journal of Bone and Mineral Research* 26:220–228.

Turk, B., D. Turk, and V. Turk. 2000. Lysosomal cysteine proteases: More than scavengers. *Biochimica et Biophysica Acta (BBA)—Protein Structure and Molecular Enzymology* 1477:98–111.

Uchida, T., T. Tanabe, M. Fukae, and M. Shimizuuh. 1991. Immunocytochemical and immunochemical detection of a 32kDa nonamelogenin and related proteins in porcine tooth germs. *Archives of Histology and Cytology* 54:527–538.

Ustriyana, P., F. Schulte, F. Gombedza, A. Gil-Bona, S. Paruchuri, F. B. Bidlack, M. Hardt, W. J. Landis, and N. Sahai. 2021. Spatial survey of non-collagenous proteins in mineralizing and non-mineralizing vertebrate tissues ex vivo. *Bone Reports* 14:100754.

Vaškaninová, V., D. Chen, P. Tafforeau, Z. Johanson, B. Ekrt, H. Blom, and P. E. Ahlberg. 2020. Marginal dentition and multiple dermal jawbones as the ancestral condition of jawed vertebrates. *Science* 369:211–216.

Venkatesh, B., A. P. Lee, V. Ravi, A. K. Maurya, M. M. Lian, J. B. Swann, Y. Ohta, M. F. Flajnik, Y. Sutoh, M. Kasahara, S. Hoon, V. Gangu, S. W. Roy, M. Irimia, V. Korzh, I. Kondrychyn, Z. W. Lim, B.-H. Tay, S. Tohari, K. W. Kong, S. Ho, B. Lorente-Galdos, J. Quilez, T. Marques-Bonet, B. J. Raney, P. W. Ingham, A. Tay, L. W. Hillier, P. Minx, T. Boehm, R. K. Wilson, S. Brenner, and W. C. Warren. 2014. Elephant shark genome provides unique insights into gnathostome evolution. *Nature* 505:174–179.

Verdelis, K., H. L. Szabo-Rogers, Y. Xu, R. Chong, R. Kang, B. J. Cusack, P. Jani, A. L. Boskey, C. Qin, and E. Beniash. 2016. Accelerated enamel mineralization in Dspp mutant mice. *Matrix Biology* 52–54:246–259.

Villa-Suárez, J. M., C. García-Fontana, F. Andújar-Vera, S. González-Salvatierra, T. de Haro-Muñoz, V. Contreras-Bolívar, B. García-Fontana, and M. Muñoz-Torres. 2021. Hypophosphatasia: A unique disorder of bone mineralization. *International Journal of Molecular Sciences* 22:4303.

Viloria, K., A. Munasinghe, S. Asher, R. Bogyere, L. Jones, and N. J. Hill. 2016. A holistic approach to dissecting SPARC family protein complexity reveals FSTL-1 as an inhibitor of pancreatic cancer cell growth. *Scientific Reports* 6:37839.

Von Marschall, Z., and L. W. Fisher. 2010. Dentin sialophosphoprotein (DSPP) is cleaved into its two natural dentin matrix products by three isoforms of bone morphogenetic protein-1 (BMP1). *Matrix Biology* 29:295–303.

Waddington, R. J., R. C. Hall, G. Embery, and D. M. Lloyd. 2003. Changing profiles of proteoglycans in the transition of predentine to dentine. *Matrix Biology* 22:153–161.

Wang, S., H. Zhang, M. B. Chavez, Y. Hu, F. Seymen, M. Koruyucu, Y. Kasimoglu, C. D. Colvin, T. N. Kolli, M. H. Tan, Y. Wang, P. Lu, J. Kim, B. L. Foster, J. D. Bartlett, J. P. Simmer, and J. C.-C. Hu. 2020. Dental malformations associated with biallelic *MMP20* mutations. *Molecular Genetics & Genomic Medicine* 8.

Wang, Y., H. Chang, H. Liu, Y. Liu, D. Han, J. Xing, H. Zhao, and H. Feng. 2018. Mmu-miR-1963 negatively regulates the ameloblast differentiation of LS8 cell line by directly targeting Smoc2 3'UTR. *Experimental Cell Research* 362:444–449.

Warshawsky, H. 1968. The fine structure of secretory ameloblasts in rat incisors. *The Anatomical Record* 161:211–229.

Wazen, R., P. Moffatt, K. Ponce, S. Kuroda, C. Nishio, and A. Nanci. 2015. Inactivation of the odontogenic ameloblast-associated gene affects the integrity of the junctional epithelium and gingival healing. *European Cells and Materials* 30:187–199.

Webb, P. P., B. J. Moxham, J. R. Ralphs, and M. Benjamin. 1998. Immunolocalisation of collagens in the developing rat molar tooth. *European Journal of Oral Sciences* 106:147–155.

Witzmann, F. 2009. Comparative histology of sculptured dermal bones in basal tetrapods, and the implications for the soft tissue dermis. *Palaeodiversity* 2:233–270.

Wöltgens, J. H., D. M. Lyaruu, A. L. Bronckers, T. J. Bervoets, and M. Van Duin. 1995. Biomineralization during early stages of the developing tooth in vitro with special reference to secretory stage of amelogenesis. *The International Journal of Developmental Biology* 39:203–212.

Xu, F., W. Cai, W. Chen, L. Li, X. Li, and B. Jiang. 2019. Expression of different isoforms of versican during the development of mouse mandibular first molars. *Journal of Histochemistry & Cytochemistry* 67:471–480.

Yáñez, M., J. Gil-Longo, and M. Campos-Toimil. 2012. Chapter 19: Calcium binding proteins. *Advances in Experimental Medicine and Biology* 740:461–482.

Yang, W., B. Gludovatz, E. A. Zimmermann, H. A. Bale, R. O. Ritchie, and M. A. Meyers. 2013. Structure and fracture resistance of alligator gar (Atractosteus spatula) armored fish scales. *Acta Biomaterialia* 9:5876–5889.

Zhang, H., P. Jani, T. Liang, Y. Lu, and C. Qin. 2017. Inactivation of bone morphogenetic protein 1 (Bmp1) and tolloid-like 1 (Tll1) in cells expressing type I collagen leads to dental and periodontal defects in mice. *Journal of Molecular Histology* 48:83–98.

Zhang, J., J. Qi, F. Shi, H. Pan, M. Liu, R. Tian, Y. Geng, H. Li, Y. Qu, J. Chen, I. Seim, and M. Li. 2020. Insights into the evolution of neoteny from the genome of the Asian icefish protosalanx chinensis. *Iscience* 23:101267.

Zhang, X., J. Zhao, C. Li, S. Gao, C. Qiu, P. Liu, G. Wu, B. Qiang, W. H. Y. Lo, and Y. Shen. 2001. DSPP mutation in dentinogenesis imperfecta Shields type II. *Nature Genetics* 27:151–152.

Zhao, N., B. L. Foster, and L. F. Bonewald. 2016. The cementocyte—An osteocyte relative? *Journal of Dental Research* 95:734–741.

Zhao, X., K. E. Dittmer, H. T. Blair, K. G. Thompson, M. F. Rothschild, and D. J. Garrick. 2011. A novel nonsense mutation in the DMP1 gene identified by a genome-wide association study is responsible for inherited rickets in Corriedale sheep. *PloS ONE* 6:e21739.

Zhu, Y., Y. Gu, C. Jiang, and L. Chen. 2020. Osteonectin regulates the extracellular matrix mineralization of osteoblasts through P38 signaling pathway. *Journal of Cellular Physiology* 235:2220–2231.

5 Odontoblast Repertoire Delivers Significantly Different Dental Tissues from Pluripotent Neural Crest-Derived Cells

Moya Meredith Smith, Aaron R. H. LeBlanc, Charlie Underwood and Zerina Johanson

5.1 INTRODUCTION

This chapter concerns the cells active in the process of biomineralisation of dental hard tissues, specifically as part of their repertoire forming diverse types of dentine, from early in the fossil record to living taxa. This is the account of the cells that make and maintain the dentine tissue in both extinct and extant teeth. These tissues, with their great functional diversity, are made entirely by pluripotent odontoblast cells. Because odontoblasts can be arranged in a tightly joined layer, they can operate systems that control the level of mineralisation, a role that is well-characterised as "a process by which a microenvironment is created by cells that serve to regulate the initiation, propagation, maturation and termination of inorganic crystallites of a precise chemical composition and shape" (Fincham et al. 2000, p. 55). The organic matrix made by odontoblasts provides this microenvironment for mineralisation, either within collagen fibrils (inotropic), the other parts of the organic matrix, phosphoproteins, and soluble calcium phosphates.

Odontoblasts can fulfil their pluripotential, neural crest-derived mesenchymal inheritance within all the types of dentines and enameloids, as outlined in this chapter. In essence, we investigate odontoblast phenotypes by interpreting the pattern of their cell processes in the mineralised tissue, as suggested by Smith and Sansom (2000). One can also interpret developmental aspects in mature tissue from semi-thin sections from both fossil and extant dental tissues, for example in early-developing teeth compared to older functional ones in chondrichthyan tooth whorls representing all mineralising developmental families.

We focus on the variation among vertebrates, both jawless and jawed forms. Broad-scale comparisons such as these rely on the basic assumption proposed by Moss (1968, pp. 49–50) that "the induction of ectomesenchymal cells into odontoblasts

DOI: 10.1201/9781003439653-5

may have a range of expressivity" due to variations in the inductive signal or the competency to respond (Smith and Sansom 2000, p. 70). Competency to make such different dental tissues reflects the essential property of cranial neural crest (CNC, ectomesenchyme) even after the cells have migrated to each site for dentine production, with molecular signals known for various subpopulations of cells, at least in the mouse (see Chapter 1). We will describe the variation in the role of odontoblasts, while they remain within the tubules as cytoplasmic processes, as changes in their secretory activity through the observed products they make, recorded as 'fixed' (i.e. cell processes within the tubules) in the mineralised dentine and enameloid types produced by the odontoblasts. One extreme odontoblast product is that found in dental plates of holocephalans (Chondrichthyes) as a derived type of dentine with a mineral density not far below that of enamel (see later in the chapter, and Smith et al. 2020). These data, known from observations in extant forms, could provide a new model for answering questions of molecular signalling and competence outside of mammals.

Thus, we will outline changes through evolutionary time of the dentine types that odontoblasts produced and their arrangement within denticles (odontodes), teeth, and dental plates. Even in some of the earliest vertebrates (Ordovician microvertebrates often referred to as 'chondrichthyan-like'), dentine tissues have enormous variety (Smith and Sansom 2000; Sansom et al. 2012; Andreev et al. 2015, 2016; Sansom and Andreev 2018). Some of these dentine types are comparable to those in odontodes in the dermal armour of the major jawless vertebrate groups Heterostraci (early Silurian to Late Devonian) and some Thelodonti (Ordovician to Late Devonian), almost an orthodentine (dentine tubules parallel and normal to the outer surface (Figure 5.1A, tub). In other thelodonts, the dentine is different, with wide pulpal tubes instead of an extensive pulp cavity, perhaps allowing clusters of cells to be located together as well as lining the pulp tubes (Figure 5.1C, pc; Žigaitė et al. 2013). The latter type of dentine has effectively a divided pulp cavity as a form of tubate dentine (Figure 5.1C, E), which is also recorded in the Ordovician microvertebrates (Andreev et al. 2020), and intriguingly, in the early developing denticles in the tails of certain living sharks (Johanson et al. 2008), and in the holocephalan dentition, discussed further later in the chapter.

Some living vertebrates, including the holocephalan chondrichthyans and some pufferfish, are unique in having lost or highly modified their teeth but retain odontoblasts that instead deposit highly mineralised dentine in a variety of forms, which form the functional (oral) feeding surface. Other holocephalan and pufferfish taxa retain teeth, thus revealing an evolutionary modification of odontoblast function within these two groups.

In these examples, dentine functions to build feeding structures. But, in another heterostracan, *Psammolepis* (Devonian), odontoblasts were able to migrate away from the formative front of the dentine and then be co-opted to repair breaks in acellular bone (Johanson et al. 2013) or simply to strengthen the dermal armour against wear. All these examples reflect the extensive diversity of dentine and the plasticity of the committed form of odontoblast to modify and perform different functions, resulting in different tissues, with their resultant morphologies.

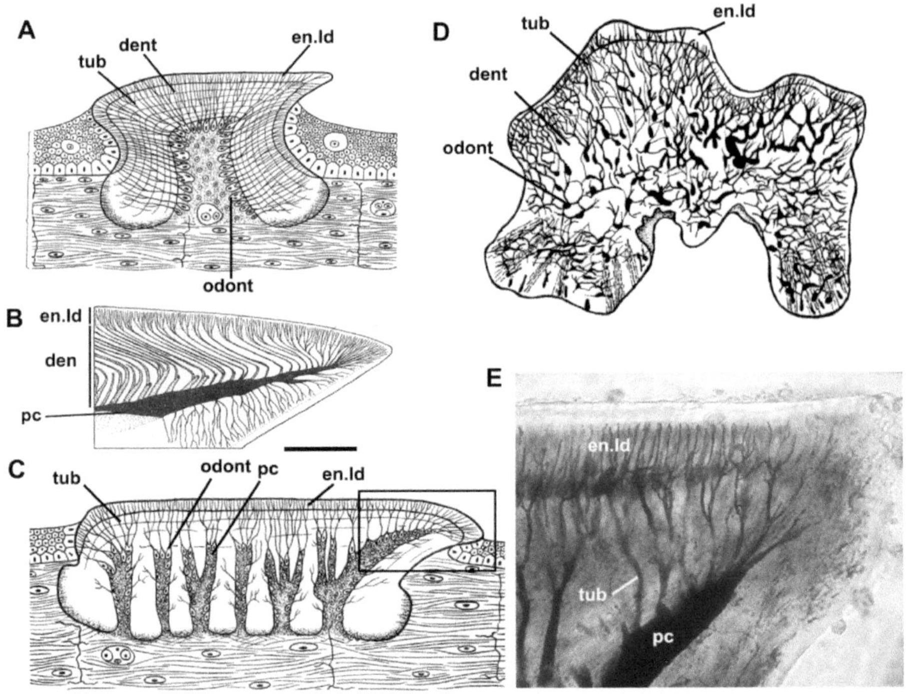

FIGURE 5.1 Types of dentine in dermal odontodes of early jawless microvertebrates. (A) *Thelodus* sp. (Thelodonti) showing normal arrangement of tubules and an undivided pulp cavity (orthodentine). (B) *Thelodus laevis*, cusp point, detail of tubule branching, bending, and extending into outer layer. (C) *Trimerolepis* sp. (Thelodonti), tubular dentine from multiple pulp cavities. (D) *Skiichthys halsteadi* with odontocyte cell spaces trapped in dentine (mesodentine). (E) *Trimerolepis lituanica* (Thelodonti) and ZNM, with tubules extending into enameloid, field equivalent to box in C; LGI-LT-nr-304. *Abbreviations*: den, dentine; en.ld, enameloid; odont, odontodes; pc, pulp cavity; tub, dentine tubules; Zeiss Nomarski optics. Scale bar B = 0.1 mm. Images taken from: (B) Gross, W. 1967. Uber Thelodontier-Schuppen. Palaeontographica Abteilung A 127, 1–6. Used with permission from Schweizerbart Publishing (www.schweizerbart.de/journals/pala); (D) Smith, M. M. and Sansom, I. 1997. Exoskeletal micro-remains of an Ordovician fish from the Harding Sandstone of Colorado. Palaeontology 40, 645–658. Used with permission from the Palaeontological Association; Žigaitė et al. (2013) (A, C, E); and Karatajūte-Talimaa (1978) (A, B, C).

5.2 MODEL OF MORPHOGENETIC UNITS FORMED FROM CRANIAL NEURAL CREST (CNC)

Regarding the embryonic origins of the odontoblast cells, the axiom is that one set of CNC-derived mesenchyme cells is ring-fenced during development to form all types of dentine in dermal odontodes and oral teeth both as hard tissues of vertebrates (Lumsden 1988; Smith and Hall 1990, 1993; Kundrát et al. 2008). Originating from cranial neural crest, they are migratory from their origins and pluripotent cells that can maximise their production of dentine through migration by moving away from the formative dentine (or enameloid) front, remaining at its surface, but with the unique potential to change cell function.

As they migrate, odontoblasts are responsible for the secretion of organic matrices and the maintenance of high levels of calcium and phosphate from an extensive system of ramifying tubules. These tubules contain cytoplasm from the migrating cell body as it moves away from the formative front. Initially, at the basal lamina at the junction with dental epithelium, the odontoblast cell body is retained for life within the pulpal and vascular spaces and is able to be reactivated for repair, or joined by perivascular stem cells, for example, in the dermal bony armour as pleromic repair tissue in the heterostracan *Psammolepis* (Smith and Sansom 2000, figure 5; Johanson et al. 2013). This can also be seen in human dental repair tissue (Nanci 2017), where the extended mineralisation of dentine by infection-affected odontoblasts can make this region more irregular in structure, or even translucent and sclerotic, as mineral deposition occludes the odontoblast process within the tubule. A similar process is also utilised to make the trabecular dentine in many other vertebrates much harder, thus becoming sclerotic as the organic matrix is infilled with high levels of mineral. In these instances, cell bodies of odontoblasts are enclosed within the mineralising matrix residing as odontocytes, postulated to still be active in this trabecular dentine. This creates a very hard tissue also seen in the dental plates of several holocephalan chondrichthyans (Smith et al. 2019).

5.3 DEVELOPMENT OF ODONTOBLASTS WITHIN A TOOTH MODULE

The main concept for the interpretation of all odontoblast activity is that cells are segregated into the tooth module (odontode) and are constantly regenerated from both ectomesenchyme and stem-cell odontoblasts (i.e. cells committed to become functional from stem odontoblasts). Smith and Sansom (2000, pp. 78–79) emphasise the separate but modular development of jaws and oral odontodes with the ability to modify cellular potential (Smith and Hall 1993; Smith and Coates 1998).

5.3.1 Loss of Potential to Make Tooth Germs in Extant Holocephalans

One example of this modification may be within the holocephalans, who through evolution have lost teeth, although fossil holocephalans, such as *Helodus* (Devonian to Permian), have revealed that holocephalans ancestrally possessed teeth organised within a replacement tooth whorl, similar to other chondrichthyans (Moy-Thomas 1936; Patterson 1965; Stahl 1999; Coates et al. 2021; Johanson et al. 2021). Despite having lost teeth, extant holocephalans still make a dentine shell first as a mould for a continuously growing dentine plate (e.g. *Harriotta*, Smith et al. 2019), indicating that they have retained odontoblasts, but that these cells instead form an internal dentine scaffold attached to the first outer shell, with spaces in the scaffold that are subsequently infilled with hypermineralised dentine. This hypermineralised dentine can be visualised in CT scans. For example in an embryo of the extant holocephalan, *Chimera* (Figure 5.2D, E), the two dentine plates form the upper dentition, and one dentine plate forms the lower dentition. The hypermineralised dentine is easily segmented from the surrounding dentine scaffold (false coloured red in Figure 5.2E). Hypermineralised dentine takes a variety of forms (Smith et al. 2019; Johanson et al. 2020, 2021), but teeth or tooth germs are never observed. It is interesting to propose that extant holocephalans have only retained the ectomesenchyme dental-stem cell population and, instead, specialised odontoblasts produce the hypermineralised dentine, whose arrangement shapes the functional surface of the dental plate instead of teeth. This transformed function of odontoblasts is discussed further later in the chapter.

5.3.2 Early Stages of Cell Differentiation in the Tooth Module

Chondrichthyan dentitions, as represented by the embryonic catshark (*Scyliorhinus canicula*; Carcharhiniformes), show a regulated temporo-spatial set of developing replacement teeth (tl—t3, Figure 5.2A), which are repeated along the jaw (Smith et al. 2009a), each in the form of a tooth whorl (Figure 5.2B, F). That is, tooth regeneration is always present in shark jaws (Tucker and Fraser 2014; Fraser et al. 2019, 2020), and in the Elasmobranchii in general, including skates and rays (Rasch et al. 2020). Thus, the first tooth to have developed and the most mature in a developmental series (t3; Figure 5.2A) has a space occupied by highly mineralised enameloid (in de-mineralised, stained sections). Also representing the cellular tooth module is one yet to develop any mineralised tissues (t1) but has segregated the ectomesenchyme within the dental papilla from the surrounding ectomesenchymal tissue as an undifferentiated tooth bud. A new potential bud has formed as dental epithelium, later to make the next replacement tooth, newest in the series (Figure 5.2A, dp, ect).

By contrast, Figure 5.2C is an example of one tiny early tooth, representing a single first-generation tooth in the embryonic lungfish jaw (*Neoceratodus*, Figure 5.2C, dp, ba; also Kundrát et al. 2008), a dentine cone (red) with only a single odontoblast in the dental papilla, and a separate bone tissue (red) attaching the tooth to the jaw, as in all osteichthyans (e.g. Rosa et al. 2021). The dental module has formed both the dentine cone and its related 'bone of attachment', also the initiated bud of stem cells to make the next tooth in the dentary row (Figure 5.2C, black arrow). It is probable that chondrichthyans do not form the 'bone of attachment', having lost this cell population from the tooth module (Smith and Hall 1993), following this model of evolution, the tooth base forms the attachment body, translated as osteodentine formed by odontoblasts and attached by fibres to the jaw cartilage.

5.3.3 New Tooth Modules that Form Continuously in Adult Jaws

In chondrichthyans, teeth are continually regenerated, thus the adult tooth whorl has replacement teeth, and the latest in development provides an opportunity to study the development of the earliest mineralised tissues. As seen in the ground sections, the first preserved tooth (Figure 5.2B, F, t1; *Lamna* [Lamniformes]) has hard tissues, well represented. This earliest mineralised tooth is simply an enameloid cone with barely any infilling of dentine near the apex of the pulp, such that the earliest mineralisation to have occurred is that of the enameloid. The next and older tooth (t2) is infilled with osteodentine, but a gap, presumed to retain a vasculo-cellular soft tissue (i.e. the pulp tissue, which is never seen in hard tissue sections),

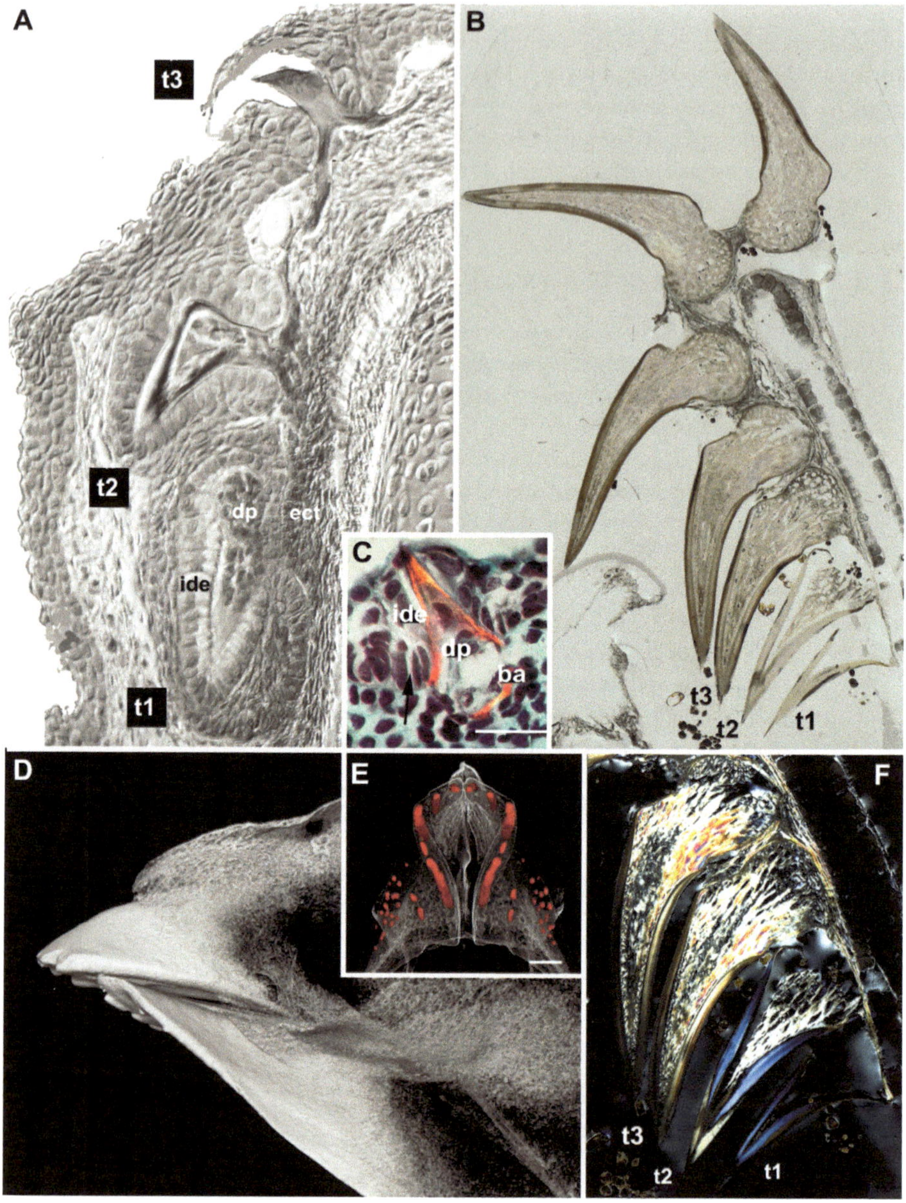

FIGURE 5.2 Chondrichthyan and dipnoan dentitions, extant. (A) Embryonic catshark dentition (*Scyliorhinus canicula*; Scyliorhinidae; Elasmobranchii), decalcified paraffin section, first three tooth germs, t1–t3. (B, F) *Lamna nasus*, ground section, stitched, transmitted light image of a complete tooth whorl. (F) Same three teeth in tooth whorl, as in B, but under cross-polarised light, t1 shows first enameloid, no formed osteodentine, t2 with two layers in enameloid, and osteodentine, t3, enameloid as still a thin layer, denteones feature in pulpal and basal osteodentine. (C) first dentary tooth in the Queensland lungfish *Neoceratodus forsteri* (Dipteridae; Sarcopterygii), histology as indicated, black arrow indicating the next tooth at bud stage. (D, E) Rendered CT scan of *Chimaera monstrosa* embryo, lateral view of two

upper dental plates, in situ; E, ventral view rendered CT scan density-dissected to reveal hypermineralised dentine, first whitlockin to be made (false coloured red). *Abbreviations*: ba, bone of attachment; dp, dental papilla; ect, ectomesenchyme; ide, inner dental epithelium; t1–t3, youngest to older teeth in tooth family, od, osteodentine; scale bar in C, E = 1 mm. Image taken from Smith et al. 2009c, figure 4C (for Figure 5.2A); Smith et al. 2016, figure 2A; Smith et al. 2009b, text-fig. 1A (for Figure 5.2C) and Smith et al. 2020, fig. 1D (for Figure 5.2E).

---

is present between the osteodentine and enameloid (Figure 5.2B). In polarised light with transmission optics (Figure 5.2F), the teeth (t1, t2) have strongly birefringent enameloid as the outer layer and again, osteodentine separately infilling the pulp cavity (t2). We discuss this secretory activity of odontoblasts later in these extant examples and as shown in similar hard tissue sections of a tooth of the fossil shark *Jaekelotodus* (Lamniformes, Morocco, Cretaceous). Comparable odontoblast activity is shown between chondrichthyan teeth in a whorl and those in extant osteichthyan tooth families, but exceptions are described further later in the chapter, such as lamniform shark osteodentine and the 'toothless' but occluding holocephalan dental plates in the former and durophagous fishes such as pufferfish in the latter.

Smith and Sansom (2000, figures 5 and 6) suggested that there are two cell types of odontoblasts in the pulpal layer, which show plurality from the beginning of tooth development, with different roles in the production of the orthodentine and of the enameloid and its maintenance. Here, we emphasise the odontoblast's role in enameloid formation and its maintenance in chondrichthyan teeth, with the contrast and compliance of odontoblast roles in shaping and providing excessive mineralisation. With respect to the latter, we have examples of high levels of mineralisation in the chondrichthyan holocephalans, of a non-apatite-based tissue we called 'whitlockin', in their dental plates to compensate for loss of the tooth module and enameloid. In this, we developed a model that explained how the odontoblasts would differentiate into specialised cells, 'whitloblasts', that ensured the initiation of the mineral phase in a novel way not only by forming small vesicles from the dentine tubules, but also by controlling its increasing levels of mineralisation from a tight layer of the same cells at the vascular surface of tubate dentine (Smith et al. 2019).

5.4 EVOLUTION OF DENTINE TISSUES AND ODONTOBLAST PLURALITY

Some of the earliest microvertebrate skeletal remains, referred to 'chondrichthyan-like' taxa, have been recovered from Late Ordovician deposits (Stairway Sandstone, Australia; Harding Sandstone, North America) and are represented by dermal denticles (odontodes; Sansom et al. 1996; Young 1997; Johanson et al. 2008; Sansom et al. 2012; Andreev et al. 2015), made from dentine. These odontodes present various dentine types in which the odontoblasts form tubules within the deposited dentine that are arranged as all parallel (Figure 5.1A, orthodentine) or as enclosed cells that have many fine branching tubules (mesodentine; Smith 1991; Smith and Sansom 2000, figure 8; Figure 5.1D). Previously, these dentine types were thought to be an evolutionary series (Ørvig 1967; see Donoghue et al. 2006). To vary the arrangement of tubules and to divide the pulp into canals were part of the odontoblast's repertoire that evolved very early in geological time in the Ordovician and Silurian (Smith and Sansom 2000; Andreev et al. 2020).

One group of early vertebrates of interest in this regard is the Thelodonti (Ordovician to Devonian) and their dermal denticles, which, like the Ordovician taxa described earlier, show a range of dentine types (Žigaitė et al. 2013). This includes *Thelodus* denticles (Figure 5.1A, B), which have a single pulp cavity, assumed to be lined with odontoblast cell bodies, as multiple tubules derived from the cell body migration can be seen in the dentine. Starting from the more mineralised outer layer, nearly all branches of the terminal tubules converge and then continue through the dentine to the pulpal cavity. Another type is seen in *Trimerolepis* sp. denticles (Figure 5.1C, odont, pc), which must have had odontoblasts arranged within multiple pulpo-vascular canals that could translate into 'tubate dentine', as in examples described later (e.g. in the toothless extant holocephalan dental plates). In these thelodonts (Figure 5.1A–C, E), fine diameter, splayed tubules extend into the enameloid-like surface (also *Thelodus laevis* [Gross 1967, pl. 1A–K]), although in the real photomicrograph of *Trimerolepis* (Figure 5.1E), they are not resolved in the very translucent, highly mineralised outer layer. In Figure 5.1D, the tissues made by odontoblasts have many branching tubules as an arboriform dentine (mesodentine), where tubules also extend into the translucent tissue of the surface, but, in the dentine, the cell bodies are embedded within as odontocytes. The tubule extensions into the enameloid are often not resolved in the mature tissue without exceptional optics (*Trimerolepis*, Figure 5.1E; Žigaitė et al. 2013, figure 4a–d). The base in many of these denticles had also been made by odontoblasts of the dental papilla, with branching tubules and presumed soft tissue fibres attached as partially mineralised Sharpey's fiber bundles (see base of Figure 5.1D; also Wilson and Märss 2009, figure 3). One diagram (Figure 5.1A, odont) imposes the cells onto the forming surface as free from matrix and able to migrate centripetally, as if a joined layer of cells. We return to this topic in our discussion of the formation of extra hard tissues within the holocephalan dental plates that are compensating for the lack of separate teeth but reveal evolutionary retention of the odontoblast's signals in their hard tissue histology that could demonstrate their putative repertoire in these diverse end products.

5.5 ENAMELOID PRODUCTION BY ODONTOBLASTS WITH VARIATION IN SHARKS AND RAYS (ELASMOBRANCHII)

5.5.1 Shark Age Series in a Tooth Whorl

As described before, the seven teeth illustrated in the *Lamna* tooth whorl (Figure 5.2B, F) each show a different stage of development, from the earliest mineralised crown (t1) through to the adult teeth in the upper part of the whorl. In t1, the earliest mineralised tissues derived from ectomesenchymal odontoblasts have shaped the crown, forming the layer of enameloid. In the catshark embryo tooth whorl with t1–3 in oldest to newest, the odontoblasts are active in the papilla (thin enameloid layer, Figure 5.2A (t3)) and the dentine tubules reach right to the edge of the outer surface (bifurcated first, Figure 5.3H, I) as the enameloid layer is built while they migrate away (centripetally). Changes in development from t1 to t3 in *Lamna* (Figure 5.2) show an increasing amount of osteodentine deposited, as is described for lamniform teeth in general (Jambura et al. 2019, 2020), and this seems to form separately and fills the central pulp but starts in the tooth tips (Figure 5.2F, t1). There is a distinct

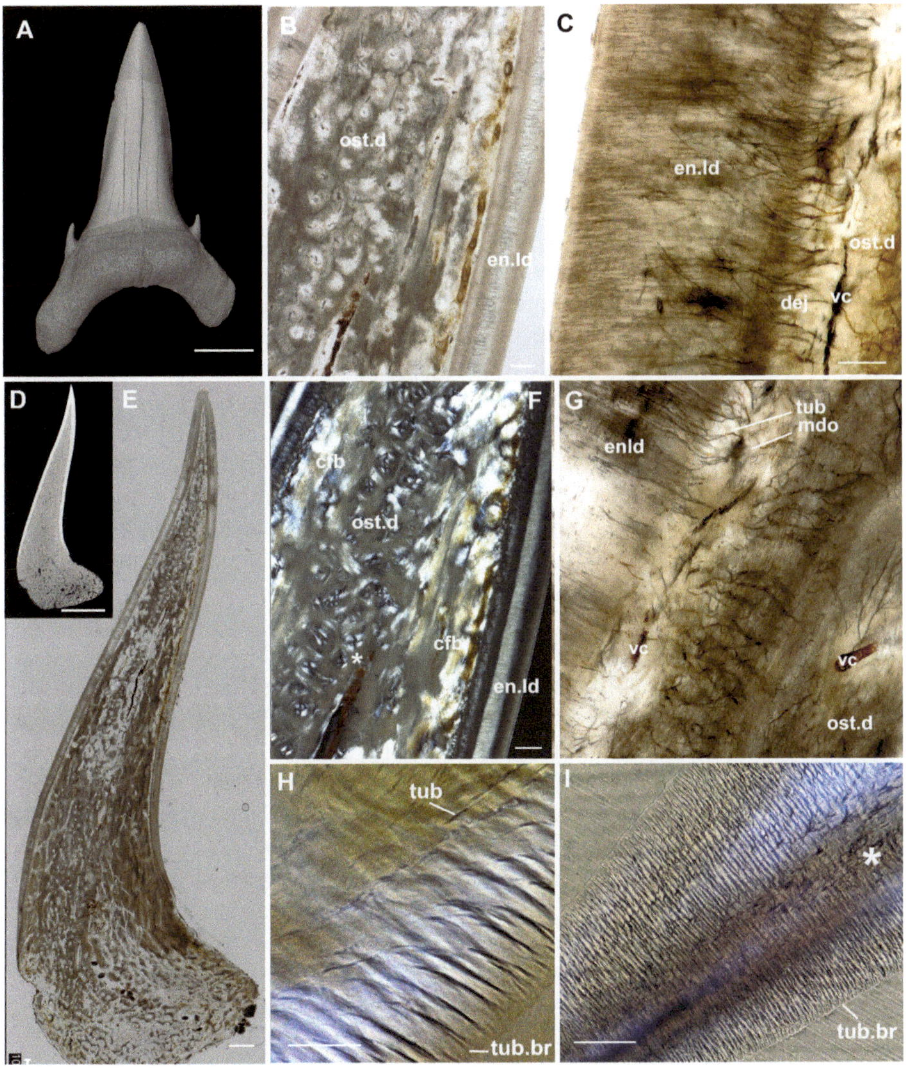

FIGURE 5.3 *Jaekelotodus robustus* (Lamniformes, Elasmobranchii), Cretaceous Morocco. (A) 3D rendering from CT scan of tooth. (D) Virtual section, LS; B, C, E to G, serial LS sections through tooth. (B, F) Images near worn edge of enameloid in transmitted (B) and cross-polarised light (F), showing osteodentine and 'Maltese cross' birefringent pattern (F, white asterisk). (C, G) Closeup of enameloid with microvascular layer (self-stained brown) and osteodentine under mixed transmitted and coaxial light. (E) Stitched image of a complete tooth section in transmitted light. (H, I) *Lamna nasus* (Lamniformes, Elasmobranchii, extant), showing extension of odontoblast tubules through the enameloid to the tooth margin under coaxial, cross-polarised light; white asterisk in I indicates region shown in H. *Abbreviations*: cfb, crystal fibre bundles; dej, dento-enameloid junction; en.ld, enameloid; ost.d, osteodentine; mdo, microdenteones; ost.d, osteodentine; tub, dentine tubules tub.br, branching tubules. Scale bars: B, F = 200 μm, C, I = 50 μm, D, E = 1,000 μm, H = 25 μm.

junction between the enameloid and the osteodentine by the next growth stage, the former including two layers which grade into each other (t3, *Lamna*, Figure 5.2B, F, *Jaekelotodus*, Figure 5.3B, C, F; *Schizorhiza*, Figure 5.5A). Odontoblast tubules in this first formed tissue terminate exactly at the outer enameloid surface, often with a bifurcation (*Lamna*, Figure 5.3H, I); they have the smallest diameter but branch from wider diameter tubules that all have horizontal fine tubules linking across with adjacent ones. The surface of the pulp is where the tubules are widest and is the assumed final location of the odontoblast cells during migration in the development of the enameloid, possibly in two stages.

The complete odontoblast repertoire is illustrated in the adult fossil tooth of *Jaekelotodus*. Tissues are produced in a developmental order; hypermineralised enameloid is present in the whole tooth (longitudinal section [LS], Figure 5.3A, D, E; transverse section [TS], Figure 5.4A, B), in all sections (Figures 5.3B, C, F, G, 5.4B, C) osteodentine is below the enameloid (Figure 5.3C, G; TS, Figure 5.4B, C). In the mature tooth, this osteodentine is a mineralised, dentinous vascular tissue that fills the central pulp. Details from the orientation of the tubules demonstrate the developmental paths of the migratory odontoblasts and their secretory activity, as their repertoire unfolds in the hard tissues of both fossil and extant forms.

5.5.2 Enameloid as a Product of the Odontoblasts

The enameloid is largely formed by the odontoblasts and from the beginning of development (Sasagawa 1989, 1998; Sasagawa and Akai 1992) as already described for the extant forms *Squalus* and *Lamna* (Figures 5.2A, B, F and 5.3E, H, I), it is as hard as enamel and as dense in CT reconstructions, optically translucent but with a multitude of fine and broad branching tubules linking with the odontoblast cell bodies, almost certainly housed in the walls of the vascular spaces (Figures 5.3F, G and 5.4C, G, vc). At the start of enameloid formation in the earliest mineralised tooth germs (t1), regular, closely spaced tubules extend to the pulp surface where they are wider, while at the enameloid outer margin, they display ends as multiple branches in a spray formation (Figures 5.1A–C and 5.3H, I). In the mature tissue, the enameloid from these cells shows the 'parallel to surface', horizontal crystal fibre bundles in polarised light (Figure 5.4C, cfb).

5.5.3 Osteodentine as a Product of the Odontoblasts

In a vertical section through the tooth, the trabecular arrangement is observed to be formed as dentine with a network of coarse crystal fibre bundles, each with oriented crystallites, as seen in polarised light (cfb; Figures 5.3F and 5.4E). The different orientations of these bundles seem to act as a joined-together fibrous sling linking to the thick fibre bundle layer that is parallel to the enameloid junction (Figures 5.3F and 5.4C [black asterisk], 5B). The latter figure reveals the different orientation at each side, at right angles to the dento-enameloid junction, by yellow and blue signs of birefringence. Infilling between these coarse bundles (part of the interstitial dentine) are many denteons where the fine, closely spaced fibres of circumvascular dentine, with Maltese cross birefringence patterns, surround the many branching vascular canals (Figures 5.3B, F and 5.4B, D, E, den, cvdent, vc). Compared to the interstitial dentine, the circumvascular dentine is often quite hypermineralised, as fully mineralised denteones are translucent (Figures 5.3F and 5.4E, asterisk). The same denteons can be compared using different optical configurations (Figure 5.4D, E), which reveal that they form among an interstitial matrix of thicker, crystal-fibre bundles crossing each other, part of a trabecular arrangement of the dentine, as mentioned before (Figure 5.4E, cfb). Multiple fine branching odontoblast tubules appear throughout this tissue, radiating from the vascular centre and through the circumvascular dentine (Figure 5.4D, cvdent) to the margins of the denteon and beyond (Figure 5.4D).

We re-emphasise the contribution of the odontoblasts to enameloid, with their tubules incorporated from the start of the secreted dentinal tissue as demonstrated in the earliest teeth to mineralise in the tooth whorl of adult *Lamna* (Figure 5.3H, I). Tasked with passing a large amount of mineral ions into the forming enameloid, a massive vascular system is located just below the dento-enameloid junction (Figures 5.3B, C and 5.4B, C, dej, vas, mdo) where there is no pallial dentine layer. This may well be a reflection of the concept of two cell types at the pulpal front as mentioned earlier, with odontoblasts remaining there to contribute to mineral excess in the enameloid. In both LS and TS sections of the adult fossil tooth, there are additional smaller diameter vascular capillaries, just below the enameloid, here termed microdenteones, because of the size difference with those in osteonal dentine and from which many branches of tubules extend into the enameloid (Figure 5.3B, C, G, mdo, tub).

From the TS tooth view, this outermost layer of the osteodentine is enhanced by postmortem fossil infilling of the vascular spaces, also in the enameloid tubules that are connected and run normal to the surface (Figure 5.4B, C) suggesting an intimate connection, and thus a very rich supply of mineral ions to create the hypermineralised enameloid. In polarised light, the tubules contrast with the surface-parallel crystal fibre bundles (Figure 5.4B, C, cfb) in the inner layer of enameloid. In TS, apart from the central blood vessels in the osteodentine, there are major vessels entering via the outer edges of the tooth in a regular arrangement on both mesiodistal and labio-lingual sides (Figure 5.4A, black asterisk, B). These would supply the large denteonal vessels and also the smaller microdenteons immediately under the enameloid junction with osteodentine (Figure 5.4B, mdo).

5.5.4 Odontoblast Production in Dermal Saw Teeth

The fossil ray *Schizorhiza* (Cretaceous, Batoidea) is characterised by an elongate rostrum (Smith et al. 2015), with batteries of 'saw teeth' along either side. New saw teeth develop in close association with the previous, stacked below each root (Figure 5.5C, sw.t). In the crown of the exposed functional saw tooth, as seen in Figure 5.5C (asterisk), the dentine independent of the denteones (interstitial; Figure 5.5A, B, D) reveals the same arrangement of the odontoblast tubules as they traverse the dento-enameloid junction and penetrate deeply into the thick layer of enameloid (Figure 5.5A, asterisk). No pallial dentine layer is present, but instead osteodentine is deposited beneath the dento-enameloid junction, where denteones fill the entire

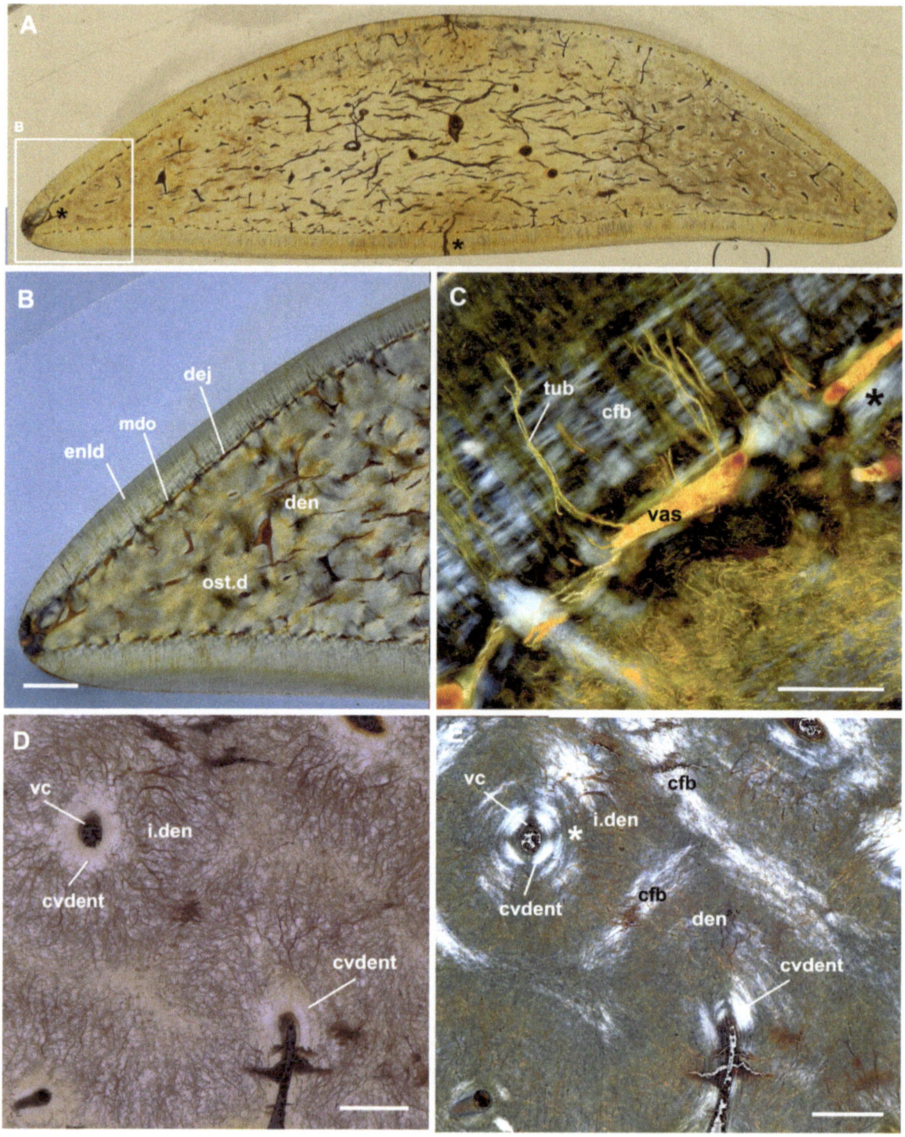

FIGURE 5.4 *Jaekelotodus robustus* (Lamniformes, Elasmobranchii), Cretaceous Morocco. (A) Tooth in transverse section under transmitted light, box shows area in B, black asterisks indicate entry of major vessels. (B) Higher magnification of outer enameloid, with vascular canal entry, linking with microdenteones and vascular system in osteodentine (under partial cross-polarised light). (C) High magnification of outer enameloid (same orientation as B, crossed polars), showing boundary with osteodentine, coincident with microdenteonal layer, birefringent horizontal coarse fibre bundles part of enameloid matrix with vertical fibre bundles crossing the junction, black asterisk indicates thick cfb layer parallel to the enameloid junction. (D, E) High magnification of identical field of osteodentine, with D, a z-stacked image of denteones with major display of tubules (interdenteonal) and a denteone of circumvascular dentine (coaxial light). (E) Same image under cross-polarised coaxial light (asterisk, concentric fibre orientation) of fine fibred composition with maltese cross as signature display. *Abbreviations*: cfb, crystal fibre bundles; cvdent, circumvascular dentine; dej, dento-enameloid junction; den, denteone; enld, enameloid; i.den, interdenteonal dentine; mdo, microdenteones; ost.d, osteodentine; tub, dentine tubules; vas, vascular system at the dej; vc, vascular canals. Scale bars: A, B = 500 μm, C to E = 100 μm.

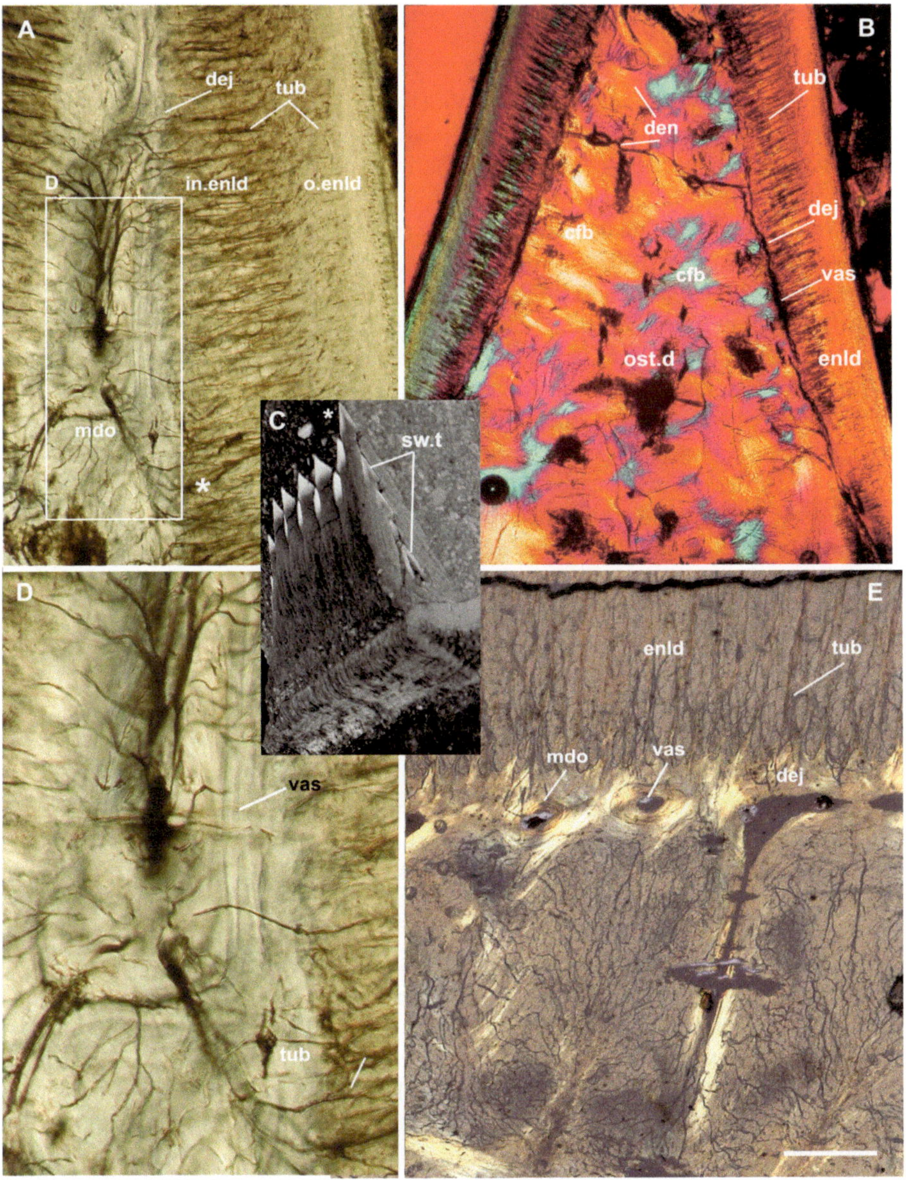

FIGURE 5.5 *Schizorhiza stromeri* (Batoidea, Elasmobranchii), Cretaceous Morocco. NHMUK PV P. 73626, tooth from the rostrum in A, B, D, longitudinal section. (C) Virtual section through segmented CT scan of the rostrum showing stacked teeth, asterisk marks tooth crown shown in A–D (Zeiss Nomarsky optics) as a region of tooth apex, high magnification of enameloid as two layers (B) and narrow vascular, tubule-rich osteodentine A, D (box in A) high magnification region in D; A is the centre of apex where the blood vessels run adjacent to the dento-enameloid junction, and self-stained canals at the centre of a denteon give rise to tubules with many fine branches in D, where the translucent canal of a blood vessel lies parallel to odontoblast canal, B, lower magnification shows entire pulp cavity filled with osteodentine between thick cover of enameloid, yellow and blue signs of birefringence (crossed polars with 45^0 gypsum filter) reveal opposite directions of coarse fibre bundles; E, *Jaekelotodus robustus* transverse section, from section in Figure 5.4, high magnification of enameloid and dentine, prominent microdenteonal layer with small dentone (vas) where circumvascular layers are clear, under cross-polarised coaxial light. *Abbreviations*: cfb, crystal fibre bundles; dej, dento-enameloid junction; den, denteones; enld, enameloid; in.enld, inner enameloid layer; mdo, microdenteones; ost.d, osteodentine; o.enld, outer enameloid layer; sw.t, saw teeth; tub, dentine tubules; vas, vascular system at the dej. Scale bars: E = 100 μm. Images taken from Smith et al. (2015), figure 5.1e, i (B, C).

pulp cavity into the apex of the saw tooth (Figure 5.5B, ost.d, vas) as do the blood vessels. Dominance of osteodentine and absence of a pallial layer of orthodentine also characterise the teeth of lamniform sharks (see Section 5.5.1, also Jambura et al. 2019, 2020).

A feature seen at the enameloid junction with osteodentine of the saw tooth is a microdenteonal layer, with the central blood vessel shown in phase contrast in longitudinal view as translucent canals that are situated parallel to the dento-enameloid junction (Figure 5.5A, D, vas). By comparison, they are dark self-stained spaces in transmitted polarised light (Figure 5.5B, vas). The microdenteonal region is also present in *Jaekelotodus* seen in transverse section with a Z-stacked image (Figure 5.5E, mdo), where each microdenteone is associated with a rich vasculature but is of small diameter. Here, each blood vessel space is reduced by an ordered circumvascular dentine observed in polarised light by a Maltese cross (as also in the osteodentine, Figure 5.4E). Fine tubule branches run through this tissue, indicating that it is also a product of the odontoblasts; note that the density of tubules is almost the same in enameloid as in the dentine (Figure 5.5E).

In Figure 5.5B, a longitudinal section of osteodentine has coarse fibre bundles oriented as broad swathes of the inter-denteonal matrix, as if a sling, demonstrated in polarised light with a gypsum filter (yellow indicates orientation −45°, top right of image, blue indicates orientation +45°), within the denteones the circumvascular dentine matrix is fine-fibered. Large numbers of odontoblast processes run from the vascular spaces in the microdenteonal dentine, with many branches that penetrate the enameloid layer (Figure 5.5A–D), ending in finer diameters in the outer enameloid (Figure 5.5A, o.enld). By direct histological comparison, the saw tooth is very similar in its histological components to the oral teeth of *Lamna* and *Jaekelotodus*, representing a convergent pattern within the chondrichthyan odontoblast repertoire, present in *Schizorhiza* (Batoidea; rays) and the lamniforms *Lamna* and *Jaekelotodus*, but absent in most other sharks, which include orthodentine in their teeth (Jambura et al. 2019, 2020). Despite this convergence of the dominance of osteodentine within the tooth pulp cavity, we emphasise the commonality of odontoblast structures, for example the fine tubules, increasingly branching and equally numerous in both outer and inner enameloid (Figure 5.5A). Also crossing taxonomic boundaries are the microdenteones of the outermost dentine, the orientation of the tubules running in a normal direction through the enameloid, and the link with the vascular spaces at the dento-enameloid junction.

5.5.5 Rays Age Series in Tooth Whorls of *Rhinoptera* and *Rhinobatos*

These tooth whorls allow a detailed study of development and maturation of the teeth (Figure 5.6A, C) that can be compared with sharks and that without embryos or hatchlings in these two taxa is not possible in any other way. Teeth develop as closely packed units with interlocked bases in *Rhinoptera* (Figure 5.6A), but teeth in *Rhinobatos* only overlap their crowns, and the bases are not joined (Figure 5.6C).

The microstructure of the *Rhinobatos* dentine (Figure 5.6C–E) is also significantly different from that in *Rhinoptera* (Figure 5.6A, B), this being more shark-like with very fine tubules in the outer osteodentine transitioning into wider tubules

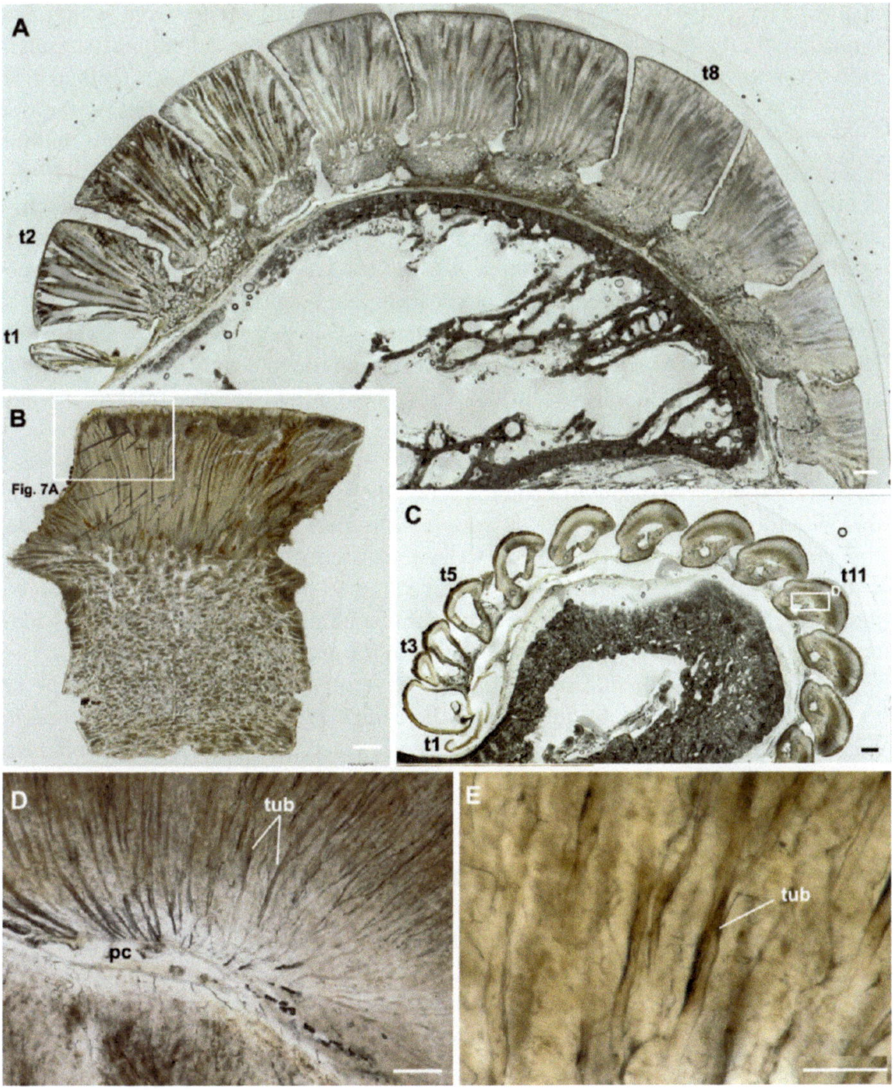

FIGURE 5.6 Chondrichthyan dentitions (Batoidea), *Rhinoptera* (A, B), *Rhinobatos* (C–E). (A) Stitched image of a *Rhinoptera* (extant) section through whole width of tooth whorl (extant) under transmitted light from entire tooth family, t1 to t8 show an increase in age of teeth, wear beginning at t8 on thin enamleoid layer. (B) *Rhinoptera* tooth section (fossil, Cretaceous, Morocco), box indicates area shown in Figure 5.7A, B. (C) Stitched image of *Rhinobatos* (extant), section through whole tooth whorl, left youngest (e.g. t3), to oldest (towards t10) out of the bite, t1 is barely formed tooth with only the outer enameloid present, box indicates region shown in D. (D) High magnification region in t10, closed pulp cavity into which wide tubules connect (coaxial light). (E) High magnification of tubules in D under coaxial light with very fine side branches. *Abbreviations*: pc, pulp cavity; t, teeth in tooth whorl, lower numbers represent the newer, developing teeth; tub, dentine tubules. Scale bars: A, B, C = 1,000 μm, D = 100 μm, E = 50 μm.

nearer to the pulp. Notably, *Rhinobatos* has a large pulp cavity, from which more typical, radially arranged, tubular dentine forms from the odontoblasts, starting early from the margin with the enameloid (from t1 compare with later teeth, e.g. t11, Figure 5.6C). Here, cell bodies forming this dentine would be located on the open pulpal surface for its entire formation, migrating centripetally into the pulp chamber, leaving the tubular space for the cell processes (Figure 5.6D). The advancing front of dentine closes down pulpal space (Figure 5.6C), and many finer tubules branch off the main tubules, intermingling in the interstitial dentine (Figure 5.6E). A separate small space of vascular tissue remains only in the base where osteodentine forms but the crown odontoblasts themselves never form osteodentine. The first mineralised tooth germ to form in the tooth whorl is at almost 45 degrees to the previous ones (Figure 5.6C, t1 versus t3) and is a shell of first enameloid only, marked out as the shape of the tooth with a tiny central cusp but no dentine, nor basal tissue, the latter forms separately later.

By contrast, *Rhinoptera* has 'tubate dentine' (following Ørvig 1985), a very vascularised component of the myliobatid ray type. These vascular tubes are distinct within the adult tooth, predicted to include odontoblasts, and separated by inter-tubate dentine (Figure 5.7, see also Figure 5.14A). A separate fossil tooth of *Rhinoptera* from a fully formed but only partially worn stage (and so predicted to represent approximately t8, Figure 5.6A) was studied at higher resolution (Figure 5.7). We also had the opportunity to study very early development in *Rhinoptera* from the tooth whorl sections. In this developmental series, the vasculo-pulpal canals are wide and the first to form (t1–2, Figure 5.6A) but without the central pulp cavity seen in *Rhinobatos* (t3, Figure 5.6C, D, pc). There is no evidence of osteodentine forming here in either genus but only later and in the separate basal tissue.

In *Rhinoptera*, multiple and numerous branches from the odontoblasts start their journey from the outer surface of the forming enameloid as fine tubules, joining together as a fan-like arrangement, forming only below the thin enameloid (Figure 5.7A–D). This is part of the vascular tubate dentine, leaving many tubules in the inter-tubate dentine between the multiple pulpal canals (e.g. asterisks, Figure 5.7B). This is a tissue zone that is also very well mineralised in extant holocephalan dental plates (Smith et al. 2019). The massive number of fine, branching tubules are apparent in all views, i.e. in TS as small circles (Figure 5.7B, perhaps best seen in Figure 5.7E, F, relative to the vascular canals), suggesting that all odontoblast cell bodies must be in the walls of the vascular canals to make these structures, without a single, large pulp cavity. These features are all indicative of a hypermineralised dentine in many of the studied dentines.

The details of odontoblast activity in *Rhinoptera* are reflected by the pattern of tubules displayed in section (Figure 5.7), for example, from a tooth showing a small amount of superficial wear and so presumed to be just coming into functional position on the tooth whorl (Figures 5.6B and 5.7A). The creation of each new tooth is timed for odontoblast activity making enameloid first, then dentine in a regulated way (i.e. t1, 2, Figure 5.6A). Normally the most mineralised tissue, enameloid is extremely thin and made of two layers (Figure 5.7A, B, D). Perhaps best seen in Figure 5.7D (red box), the inner layer shows bunched tubules from the wider canals below, reflective in incident light (Figure 5.7B, D). Huge numbers of thin branches

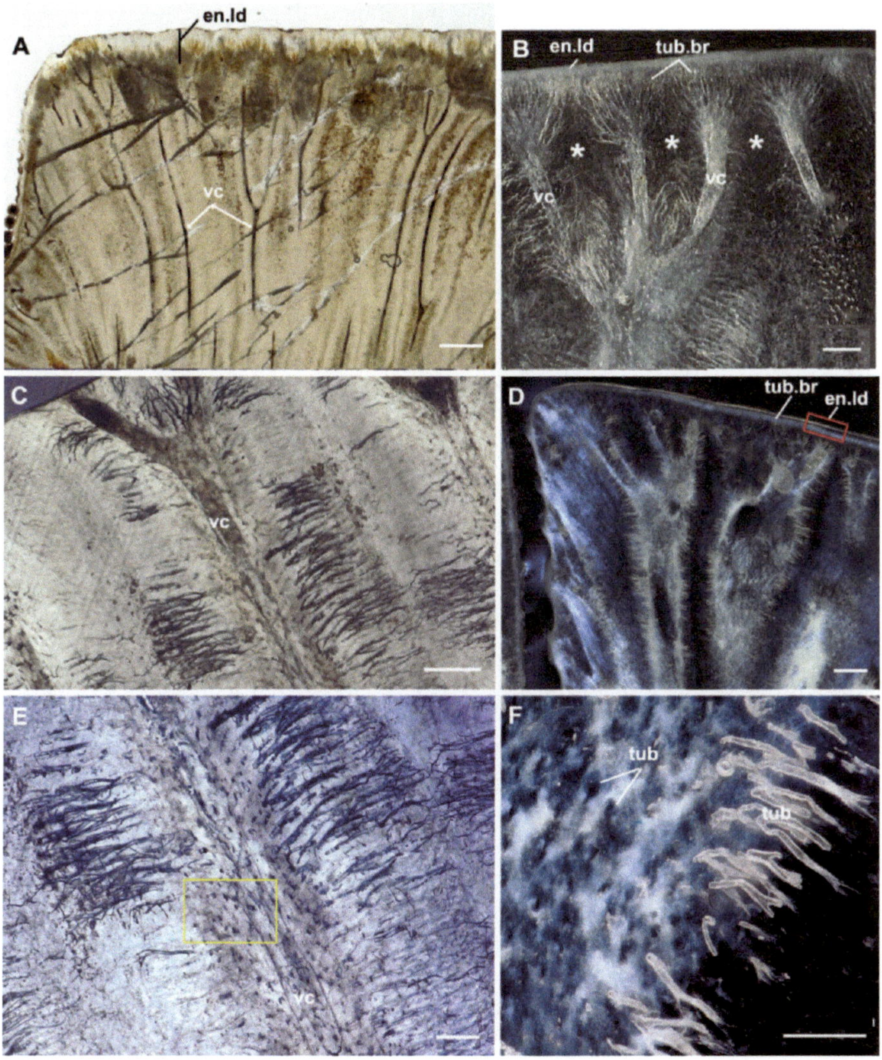

FIGURE 5.7 Chondrichthyan dentition (Batoidea, Elasmobranchii), Cretaceous, Morocco, single tooth. (A) *Rhinoptera*, tooth corner, also box, Figure 5.6B, showing vascular canals as in tubate dentine, with branches near the surface (transmitted light), double branch at the worn surface shown in C. (B) Closeup of enameloid layer, with more finely branching dentine tubules, radiating as sprays towards the enameloid as thin layer at the surface, asterisks indicate inter-tubate dentine (incident light). (C, E) Vascular canal with enormous number of lateral tubules and openings into the vascular canal, under coaxial light; yellow box in E indicates smaller tubules oriented perpendicular to others (appearing as dots). (D) Vascular canal branching below thin enameloid, red box indicates double layered enameloid (cross-polarised coaxial light). (F) Similar region to E, yellow box, inside surface of pulp canal openings for tubules as those projecting into circumvascular tissue. *Abbreviations*: en.ld, enameloid; tub, dentine tubules; tub.br, branching dentine tubules; vc, vascular canals. Scale bars: A = 500 μm, B, C = 100 μm, D = 200 μm, E, F = 50 μm.

radiate back into the broader branching pulp canals (Figure 5.7B—F). The density of tubules is enormous, and many are perpendicular to the radial ones in the plane of the section (in z-shift) seen as small holes (Figure 5.7C, E [yellow box], F) as they emerge from the tube walls, occupied by the cell bodies in life. This type of dental tissue is very similar in the way it is built, and the putative odontoblast activity involved, to dentine referred to as 'tubate dentine' in holocephalans (Smith et al. 2019), described in the next section, for comparison with other chondrichthyans.

5.6 HYPERMINERALISED DENTINE IN HOLOCEPHALANS WITHOUT TEETH

5.6.1 Extinct Holocephalans with Teeth

Teeth do occur in stem-group holocephalans, although all extant holocephalans lack teeth and thus they appear to be lost in evolutionary history (Patterson 1965; Coates et al. 2021; Johanson et al. 2021). These stem-group holocephalans include taxa whose phylogenetic resolution has been controversial; for example they have previously been resolved as stem-taxa relative to the elasmobranchs (e.g. Pradel et al. 2011; Maisey et al. 2020) although most recently they are placed on the holocephalan stem (Coates et al. 2017, 2018; Frey et al. 2019). This includes taxa such as *Cladoselache* and *Helodus*, two taxa with separate teeth held within a tooth family, similar to other chondrichthyans (Moy-Thomas 1936; Williams 2001; Coates et al. 2021; Johanson et al. 2020, 2021; Figure 5.8A, C).

The internal structure of the *Helodus* tooth is dominated by wide tubes running normal to the surface and identified as vascular canals (Stahl 1999; Johanson et al. 2020; Figure 5.8A, C, vc); as such, this tissue can be identified as the 'tubate dentine' described before for the elasmobranch ray *Rhinoptera*, but with enameloid being absent. This dentine dominates the tooth in the stem-holocephalan *Helodus*, but during holocephalan evolution, the tubate dentine may occupy only a proportion of the dental plate in extant holocephalans (crown-group) such as *Callorhinchus* and *Harriotta* (Stahl 1999; Smith et al. 2019, 2020; Johanson et al. 2020, 2021; Figure 5.9C, F, tri). The morphological regions of tubate dentine take the form of what are known as 'tritoral pads', with the bulk of the dental plate formed from trabecular dentine (Smith et al. 2019).

5.6.2 Extant Forms without Teeth and New Hypermineralised Tissue Type

In extant holocephalans, the tubate dentine that is hypermineralised becomes restricted to structures of varying morphologies, including rods, ovoids, and tritoral pads; the rods appearing earlier in ontogeny (Smith et al. 2019; Johanson et al. 2020, 2021; Figures 5.8D–G and 5.9C all segmented and coloured false red in Figure 5.9F). The ovoids form as a developing series stacked vertically from the aboral surface of the dental plate, the most mature ovoids being found towards the oral surface (Figure 5.8G) with very high levels of mineralisation. In these ovoids, the odontoblasts (as specialised cells, 'whitloblasts') are located in the vasculo-fibrous capsule around each structure and part of the trabecular dentine (Smith et al. 2019; Figure 5.8D, blv). The direction of the tubules leads from the outside vascular region of the capsule of

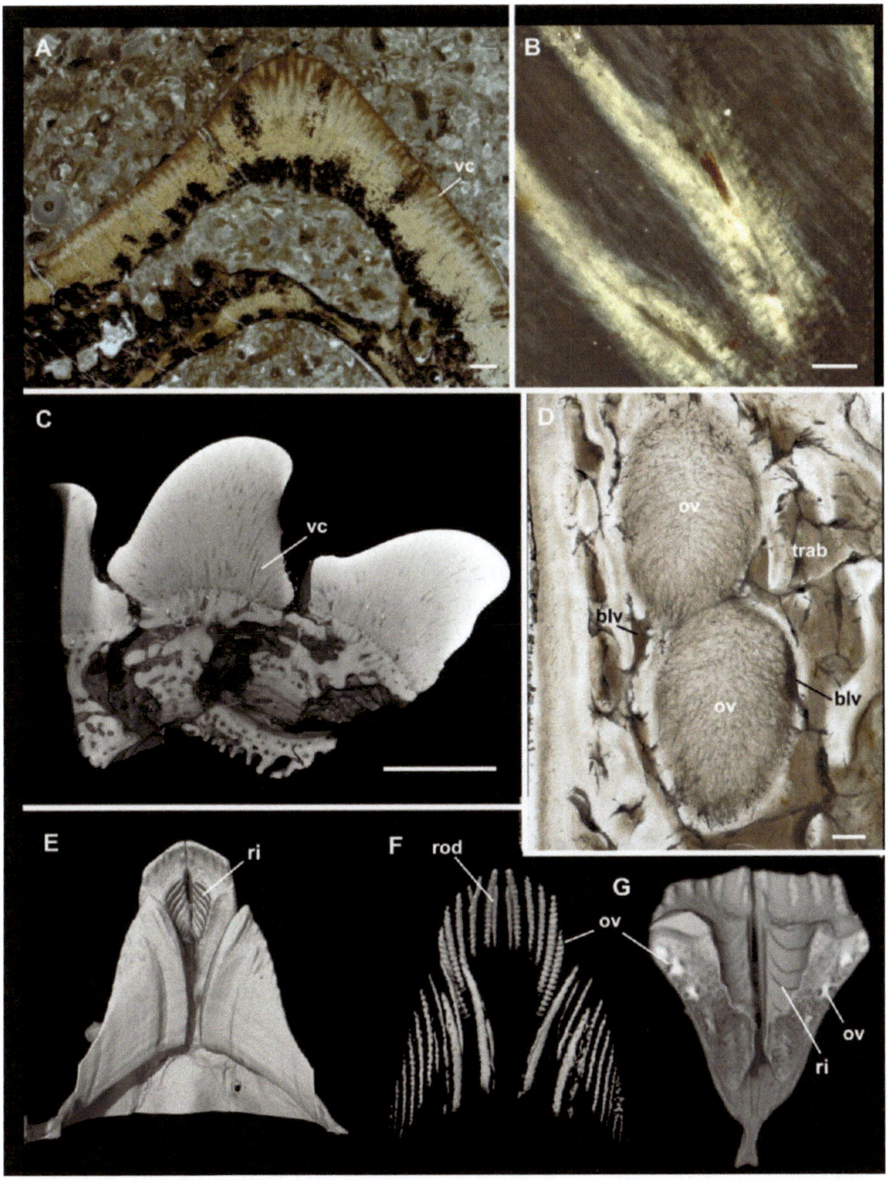

FIGURE 5.8 Holocephali (Chondrichthyes) dentition. (A) NHMUK PV P1432 *Helodus didymus* (Lower Carboniferous, Ireland), ground section, A, conical tooth with large pulp cavity below tubate dentine (transmitted light); (B) polarised light, high birefringence of two vascular canals, many fine tubules branch laterally from these, high magnification of tooth in A (see tubate dentine model, Figure 5.14A); (C) NHMUK PV P8216, *Helodus simplex*, CT scan, virtual section through individual teeth in whorl (Lower Carboniferous, Ireland), narrow vascular canals as in A, wider in dentine base; (D) *Harriotta raleighana* (Family Rhinochimaeridae, extant), ground section of two hypermineralised ovoids, enormous network of very fine dentine tubules linked with peripheral blood vessels, large numbers surround each ovoid (see Figure 5.9F), all held within a less mineralised framework (trabecular dentine), within the dental plate, rich vascular supply (self-stained dark brown), outer dentine shell layer to the left (under mixed transmitted and coaxial light); (E–G) *Chimaera monstrosa* (Family Chimaeridae, extant), E, CT scan of anterior and posterior upper dental plates, anterior plate oral surface with regular spaced ridges of outer dentine, also in G; (F) less mineralised tissue removed by virtual dissection, leaving hypermineralised rods and stacked ovoids (whitlockite); (G) virtual section, cut away revealing structure inside anterior upper tooth plate and patterning of spatial relationship between ovoids and lingual ridges. *Abbreviations*: blv, blood vessel; ov, ovoid; ri, lingual ridges; rod, rods; trab, trabecular dentine; vc, vascular canals. Images taken from Johanson et al. 2021 (C) and Smith et al. 2020, figure 3 (E–G).

the ovoids towards the centre of each ovoid, where mineralisation begins by the same method as in the equivalent developmental zone of the tritors, forming thousands of tubular vesicles (Figure 5.9B, D, E, ves). The tritoral pad develops aborally, forming a block of tubate dentine, punctuated by vascular canals, each lined by odontoblasts (Smith et al. 2019; Figure 5.9C, F). All three hypermineralised structures are made from the products of odontoblasts (in a specialized phase as whitloblasts), differing from regular dentine in the arrangement and role of the tubules, as explained later (Figure 5.14A), the matrix being low in collagen and high in magnesium, forming a calcium phosphate mineral known as whitlockite, with the resulting biological product referred to as whitlockin (Smith et al. 2019).

The specialised cells that secrete this unique type of dentine are the whitloblasts that form a joined-together cellular layer as a membrane along the active dentine surface, functioning to monitor ion exchange (Figure 5.14A), whether ovoid or tritor. Each of these whitloblasts produce massive numbers of tubules (Smith et al. 2019; Figures 5.8D and 5.9A), which in turn give rise to swollen membranous vesicles in enormous numbers (Smith et al. 2019; Figure 5.9B, D, E). It is within these cellular vesicles that crystal formation starts the remarkable process of mineralisation in the absence of any collagen fibres. This is as a shared common process with all morphologies, none are elongate crystals as in hydroxyapatite, but very small, joined together as an irregular mass (Smith et al. 2019, figures 3F and 9G). Importantly, the identity and role of these cells can alternate between whitloblasts and odontoblasts, for example returning to the odontoblasts (identifiable by the presence of fewer tubules) to deposit the less mineralised dentine around the vascular canals of the tritoral pad (Figure 5.14A, PTD; Smith et al. 2019; Johanson et al. 2021, figure 9).

Notably, none of these dental plate structures have developed from anything like a tooth module, but nevertheless clearly involve odontoblasts to make all tissues of the dental plate identified by Smith et al. (2019), including the outer dentine shell, with a linked trabecular dentine framework, and a hypermineralised dentine infilling morphological spaces in the trabecular dentine. From these observations, importantly, we suggested that despite lacking teeth, the neural crest-derived cells (ectomesenchyme) that normally contribute to tooth development were retained during holocephalan evolution, where they make the patterned framework for the location of the whitlockin. Hence, as holocephalans are toothed early in their history (e.g. *Helodus*; Figure 5.8C), they are unique in having retained the cell precursors of odontoblasts without being able to segregate them into tooth modules. These odontoblast stem cells (stem/progenitors [Popa et al. 2019]) in the pulpal tissue can somehow still interact with the cells of the dental epithelium to make all the trabecular dentine, modulating to make the hypermineralised tissue (whitlockin). In support of this, Smith et al. (2020, figures 2–5; Figure 5.8G) observed a correlation between the palatal ridges on the lingual surface of the dental plate (Figure 5.8E, G, ri, ov) and the positions of the whitlockin ovoids forming in the stacks. They suggested a ‘ridge-to-ovoid’ theory of induction, where the ridge at the developing front of the outer dentine shell patterned where the trabecular capsules for the whitlockin were located to initiate the series of ovoids (Figure 5.8E–G).

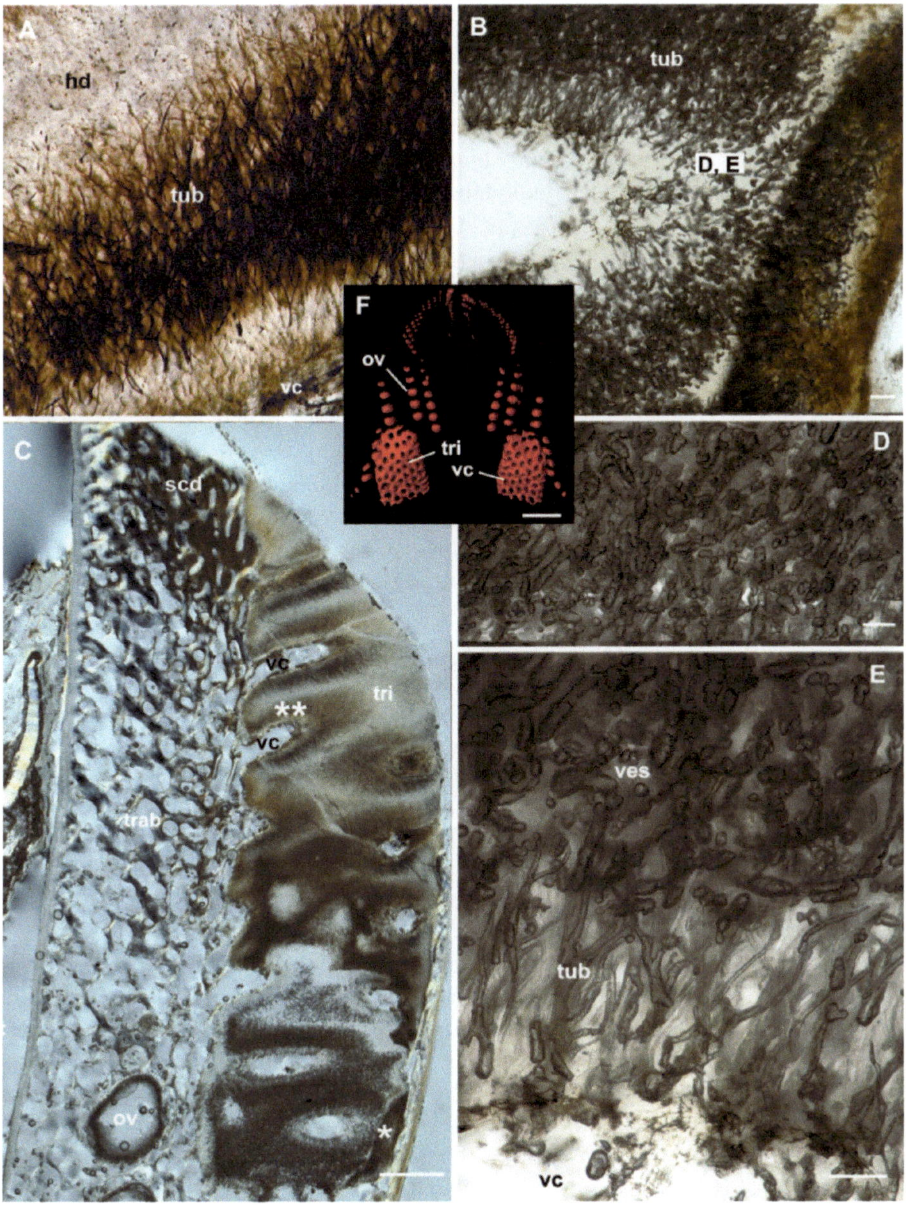

FIGURE 5.9 Holocephali (Chondrichthyes) dentition. *Harriotta raleighana* (Family Rhinochimaeridae, extant) serial sections. (A) high magnification of tubule-forming area in two columns in C indicated by double asterisks (Zeiss Nomarsky optics, self-staining); (B, D, E) increasing magnification of same area in C indicated by single asterisk (transmitted and coaxial light), tubules and tubular vesicles surround the vascular canal (see Figure 5.14B); (C) vertical section through dental plate, sclerotic dentine (scd) below functional wear, inside outer dentine, supporting trabecular dentine, tritoral pad (tri) and a developing ovoid (ov, transmitted, partial cross-polarised light). (F) segmented CT scan of *Harriotta raleighana*, hypermineralised tissues false coloured red, note ovoids (histology shown in Figure 5.8D). *Abbreviations*: hd, hypermineralised dentine; ov; ovoid; scd, sclerotic dentine; trab, trabecular dentine; tri, tritoral pad; tub, dentine tubules; vc, vascular canals; ves, vesicles. Scale bars: B = 100 μm, C = 500 μm, D, E = 25 μm, F = 2.5 mm.

Importantly, the conclusion is that teeth are lost early in the history of the holocephalans, and having been released from the tooth module, or morphogenetic unit, the odontoblasts have become highly pluripotent and able to make different types of dentine structures within the dental plate.

5.7 ODONTOBLASTS IN BONY FISHES (ACTINOPTERYGII), FOSSIL AND EXTANT

The Actinopterygii differ from the chondrichthyan taxa described earlier in having a bony skeleton, including bone that attaches teeth to the jaw (see recent review by Rosa et al. [2021] regarding the bony versus dentinous nature of this attachment). The focus in this section, however, is upon durophagous bony fishes, those that crush prey items during feeding and the contribution of odontoblasts to these specialised dentitions. Under consideration are *Lepidotes* sp. (Ginglymodi; Actinopterygii; Jurassic to Cretaceous) and the pufferfish *Triodon*, *Diodon*, and *Dicotylichthys* (Tetraodontiformes; Actinopterygii; extant).

5.7.1 Odontoblasts Manage the Coronal Enameloid in Crushing Teeth

Lepidotes belongs to the Ginglymodi (order Lepisosteiformes), resolved phylogenetically as a basal group of actinopterygians (e.g. López-Arbarello 2012). *Lepidotes* can be characterised by cushion-shaped teeth (e.g. Forey et al. 2011). The bulk of these teeth is formed from enameloid (acrodin), which characteristically dominates the crown of the tooth (Figure 5.10A, en.ld). A collar of thin hypermineralised tissue is also added incrementally by apposition to the basal dentine surface. This is not formed by odontoblasts but is enamel of a ganoine type, as previously discussed (Figure 5.10A–C, en; Richter and Smith 1995, figure 19).

The coronal enameloid meets the junction with dentine (Figure 5.10C) with very little change in tubule numbers, except that much wider tubules are seen that appear to be continuous with those of the dentine crossing this junction. Whereas those in the dentine are thinner and branching; and those in the enameloid are broader and have fewer fine crossing branches. Nevertheless, the continuity and equivalent arrangement (Figure 5.10C) of the tubules within dentine and enameloid on either side of the junction suggests both originate from the odontoblasts lining the dental papilla. This junction is more distinct from that in chondrichthyans between enameloid and dentine. Along with tubules are swathes of coarse-fibered crystal bundles in the enameloid (Figure 5.10B, cfb), and, in some cases, the dentine tubules follow the courses of these suggesting at least a spatial relationship. Importantly, the direction of odontoblast deposition is from the enameloid and then into the dentine. A major change in the activity of the odontoblast occurs in relation to the orientation of the tubules in the region of the coarse-fibred bundles, from weaving through most of the enameloid to grouped perpendicular to the dento-enameloid junction. At ×1,000 magnification in a stacked image, both tubule direction and size are resolved and vary considerably (tub, Figure 5.10C), often parallel to the bundles in polarised light, and these are shown to run in perpendicular directions (cfb, red and blue, gypsum plate, Figure 5.10B). The grouped parallel organisation at the dento-enameloid

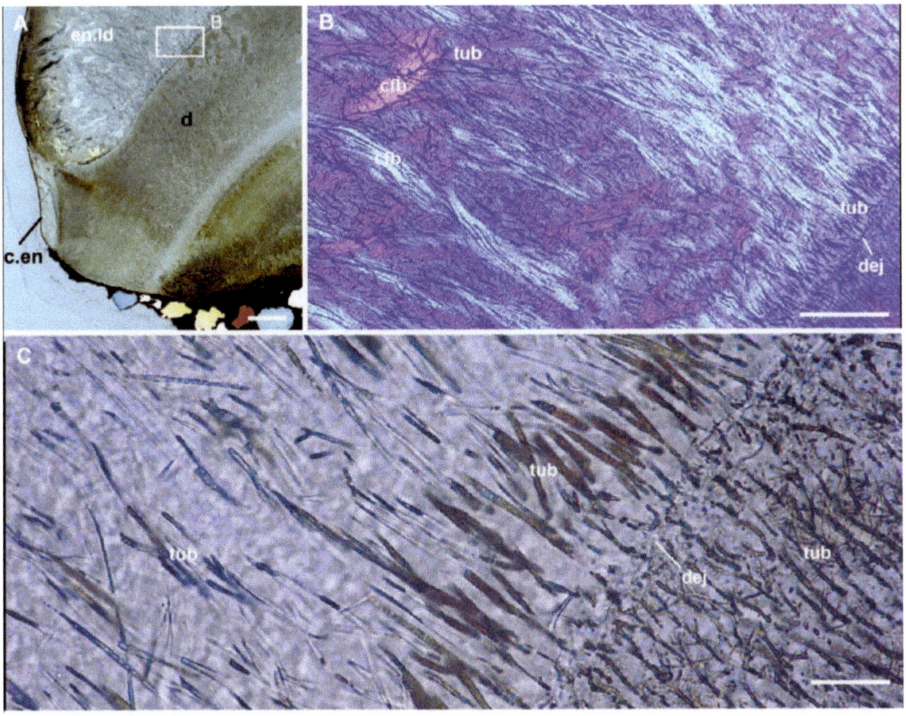

FIGURE 5.10 Actinopterygian tooth, *Lepidotes* sp. NHMUK PV P. 4384 Ginglymodi; Lepisosteiformes, Jurassic to Cretaceous). (A) corner of cushion-shaped tooth, contrast with small layer of collar enamel (c.en), box in A indicates area shown in B; (B) high magnification of thick enameloid and tubules within, continuous with dentine, birefringent crystal fibre bundles in opposite directions red and blue (cfb) (cross-polarised, coaxial light); (C) highest magnification (×1,000) of dento-enameloid junction (dej), broader tubules' contrast with dentine suggest two types of tubules, but continuous across junction (coaxial light). *Abbreviations*: cfb, crystal fibre bundles; dej, dento-enameloid junction; c.en, collar enamel; en.ld, enameloid; tub, dentine/enameloid tubules. Scale bars: A= 500 μm, B = 100 μm, C =25 μm.

junction may reflect the secondary establishment of an organised depositional front at the transition from enameloid to dentine matrix secretion. The microstructure of the tubules also changes again in the dentine. This data supports the pluripotency of odontoblasts, as noted in the introduction, potentially, two cell phenotypes forming the tissues of these crushing teeth.

5.7.2 Odontoblast Activity in the Dentine of In-Group Tetraodontiformes (Neopterygii; Eupercaria)

Tetraodontiformes, or pufferfish, are charismatic ray-finned fish that, through their phylogeny (e.g. Santini et al. 2013), see a change from dentitions comprising separate multiple teeth dominating in early diverging taxa (*Triodon* [Triodontidae], Figure 5.11J–L), with smaller pad-like teeth medially. Later diverging forms are dominated

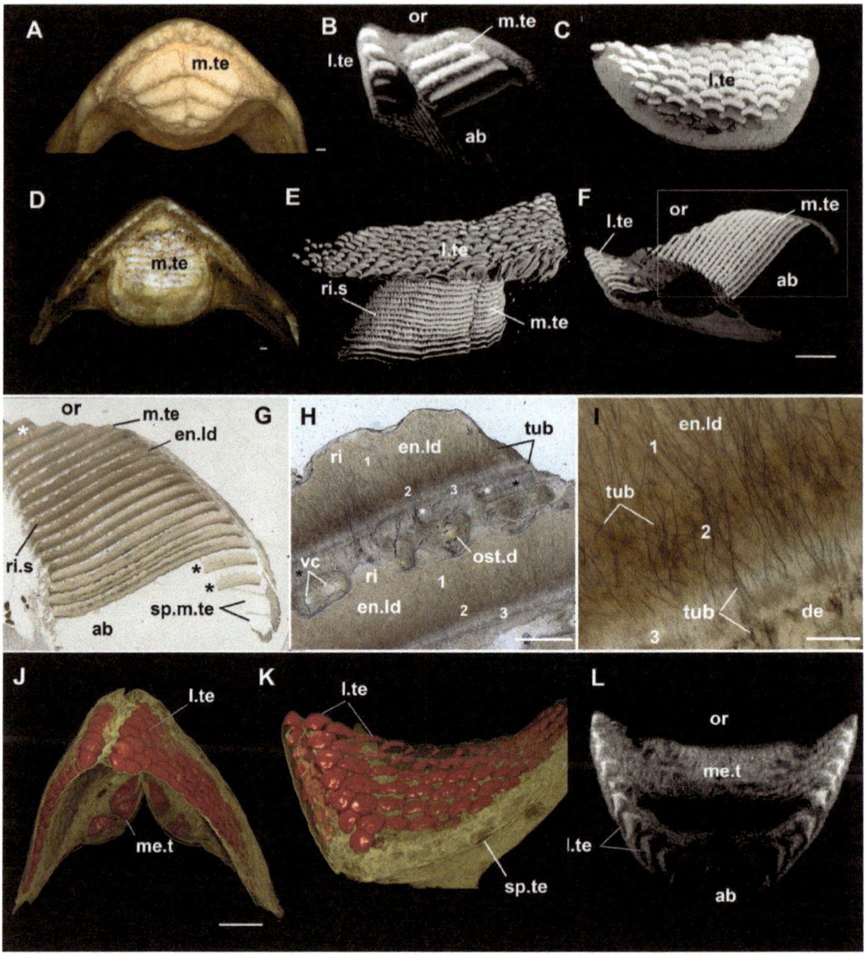

FIGURE 5.11 Actinopterygian dentition, Tetradontiformes (Teleostei). A–C, *Diodon* sp. (Diodontidae), lower jaw, young extant. (A) macrophotograph of the oral surface, dominant broad, medial teeth in series; (B, C) CT scans: B, medial and labial teeth and C, labial in virtual section that reveals how stacks of replacement teeth develop aborally, youngest tooth showing less mineral density; note in C the obvious alternating pattern of large numbers of replacement teeth as also in (J). (D–I) *Dicotylichthys* sp. (Diodontidae), lower jaw extant, processed as ground serial sections as shown in G–I. (D) macrophotograph of the oral surface with ridges exposed on medial teeth, also on lateral teeth, as raised functional surfaces, hypermineralised tissue (enameloid). (E) rendered CT scan lateral view with segmented mineralised tissues reveals medial and lateral teeth with obvious size differences, also alternating development pattern in both and ridged surface on each; (F) CT scan, virtual section through jaw reveals labial and medial teeth and younger teeth with less mineral aborally. (G–I) sections through medial teeth showing (G) stitched image of a stack of teeth and newest teeth aborally (black asterisks), white asterisk indicates area shown in H; (H) low magnification of ridges along functional surface and in forming teeth, with large proportion of enameloid covering only a thin layer of dentine, below which is osteodentine (mixed coaxial, transmitted light); (I) high magnification of enameloid reveals the odontoblast tubules as very fine and numerous also crossing over the dento-enameloid junction (mixed coaxial and transmitted light). (J–L) *Triodon* sp. (Triodontidae), lower jaw, rendered CT scan with mineralised tissues segmented and false coloured red; (J) overview to show much smaller medial teeth than in *Diodon* and *Dicotylichthys*, note size difference along the jaw and symphysis between replacement tooth stacks; (K) labial view showing mineralised teeth, aboral crypts where tooth germs form; (L) virtual section, posterior position through jaw where there is only bone medially (see whole teeth in J), also labial teeth, with diminishing mineral density oral to aboral. *Abbreviations*: ab, aboral; de, dentine; enld, enameloid; l.te, labial teeth; me.t, medial teeth; or, oral; ost.d, osteodentine; ri.s, ridges on medial tooth surface; sp.m.te, space in the jaw where medial tooth will develop; sp.te, space in the jaw where labial tooth will develop; tub, dentine tubules; vc, vascular canal. Scale bars: A, D = 1,000 μm, F = 3.5 mm, H = 250 μm, I = 50 μm; J = 1 mm.

by very broad teeth medially, closely spaced in the palate and lingual lower jaw, with separate small teeth in the labial ridges. All these modified tooth shapes are formed of a hypermineralised tissue, the source of this has been questioned, and here we discuss its histology, especially the role of odontoblasts in forming the enameloid (Figure 5.11H, I).

In the genus *Tetraodon* (Tetraodontidae), a different distribution of the same hypermineralised dentine occurs, but from teeth along the jaw that are only found in the first developmental stages. As demonstrated by Fraser et al. (2012), most tooth positions, except the parasymphyseal ones, lose typical replacement teeth, and instead long bands of hard tissue develop in each quadrant of the upper and lower jaws, derived from only four teeth. Only these four parasympyseal teeth regenerate new tooth germs, expanding their formation of dentine along the jaw where tooth germs have been prevented from regenerating. Hence, the hard tissues become jaw-length bands of the characteristic hypermineralised dentine, and each regenerated product is stacked below the functional band, all expanded to fit jaw growth.

Other taxa such as *Diodon* and *Dicotylichthys* (Diodontidae) show a similar combination of small teeth labially (Figure 5.11B, C, E, F) and very broad teeth lingually, stacked during development in a clear pattern where regeneration is a left–right alternation (Figure 5.11A–F). All these successor teeth are held within the structure of the jaw, forming aborally with the basal ones the least mineralised and below these, soft issue tooth germs not yet mineralised within bony crypts (Figure 5.11B, C, F, G, black asterisks). All teeth become more mineralised towards the oral surface; this is especially well shown by density differences in a virtual CT scan (Figure 5.11B, C, E, F, L). The preformed crypts (bone cavities) form the species-specific patterns into which the dentine and enameloid of the tooth develop (compare Figure 5.11G [*Dicotylichthys*] and Figure 5.11K [*Triodon*]). A comparison can be made with the similar patterning process in holocephalan dental plates.

In the Diodontidae, the major part of the medial broad teeth and labial teeth is formed of enameloid (regions 1–3), each of the former shaped by ridges on the surface domain of enameloid (Figure 5.11H, en.ld, ri.s, ri; Andreucci et al. 1982). The enameloid is characterised by thin elongate tubules that are sparsely arranged orally (region 1), but more aborally are more numerous in the basal enameloid (region 2, Figure 5.11H, I); in region 2, the tubule orientation is random, and less branching occurs in region 3. Also, the tubules here appear thinner (less dark) but still generally organised normal to the surface, and tubules are continuous with those in the region below where there are two different arrangements—one with tubules more aligned (black asterisk) and a second where the tubules are grouped (white asterisk). A very thin layer of dentine is present below the enameloid (Figure 5.11I, de), and a tissue joining the stack of teeth together is where large circular vascular spaces are visible.

Around some vascular spaces, circumvascular dentine is present, creating a denteone (Figure 5.11H, ost.d). The layer of dentine in which tubules are grouped together relates to these denteones, with the vascular canals in this region either supplying potential odontoblasts, or odontoblast cells occupying the canal walls, as described earlier. Andreucci et al. (1982) referred to this as osteodentine. The transition from the enameloid to the osteodentine includes a very narrow layer of dentine (region marked by the white and black asterisks, Figure 5.11H) seen in more detail in Figure 5.11I, where the change in thickness of the tubules between regions is observed to be not dissimilar to that of *Lepidotes*. These observations of the detailed histology represent the paths of odontoblasts as they deposit and then maintain the enameloid. They begin by forming the ridged surface, migrate, and form a transitional dentine layer, either increasing or decreasing in number, as the branches are joined together, and then finally shift to odontoblasts making the vascular, trabecular framework. Therefore, grouped tubules at the dento-enameloid junction (Figure 5.11H, asterisks) signal the end pathway of odontoblasts from the enameloid surface to the vasculature.

As described before for the holocephalans, the tooth module in more phylogenetically derived tetraodontiforms such as *Tetraodon* is modified during development (Fraser et al. 2012), freeing the odontoblast to produce new dental morphologies. In some puffers, the dentine and pulp are lost, leaving the enameloid caps stacked in columns, such as the bands characteristic of *Tetraodon* or the separate teeth and broad, flattened teeth of the family Diodontidae (*Diodon*, *Dicotylicthys*, Figure 5.11A–I). The beak-like jaws consist of osteodentine, instead of bone, again showing a striking similarity to holocephalans. All these tooth morphologies have enameloid as the dominant tissue overlying the dentine, which is in turn underlain by osteodentine (Andreucci et al. 1982). In *Triodon*, the separate teeth dominate. The odontoblasts in the Diodontidae show considerable variability in tubule morphology, pointing to the variability in odontoblast function. This includes increasing/decreasing numbers at the depositional front and odontoblasts associated with the vasculature where both cell and blood vessels have an important role in the extra mineralisation of the enameloid once fully formed.

5.8 ODONTOBLASTS MIGRATE TO REPAIR BONE DAMAGE IN HETEROSTRACI

5.8.1 Dentine Tubercles Renewed and Regenerated by Odontoblasts Making Orthodentine Infills

Psammolepis is a jawless and toothless vertebrate, belonging to the heterostracan family Psammosteidae (Devonian). *Psammolepis* is covered in dermal bony plates, composed of bone covered with orthodentine type tubercles (Figure 5.12A–C).

It has been known early in the historical literature (Tarlo [1964]; Ørvig [1967]; see Keating et al. [2015] for a recent review) that heterostracan bony dermal plates are covered with tubercles attached to the acellular, trabecular bone beneath, surrounded by small ampullae for sensory cells (Figure 5.12A–C, am). Many illustrate bunches of tubules filling the vascular spaces in the bone traversing the trabecular, or spongy, bone (Figure 5.12F, G, J). Whether these tubules originated from those in the primary tubercles was not always clear. Ørvig (1967; Figure 5.12B) referred this reparative dentine to a general category of hypermineralised dentine known as 'pleromin': A collective term meaning infilling of vascular spaces. Here and in other hypermineralised pleromic dentine, it is considered as either a prophylactic against wear of the primary tubules or a reactive response. It was considered appropriate by Denison (1974) to use 'pleromin' only for the secondary dentine infilling vascular spaces to strengthen the spongy bone in response to wear of heterostracan dermal bone.

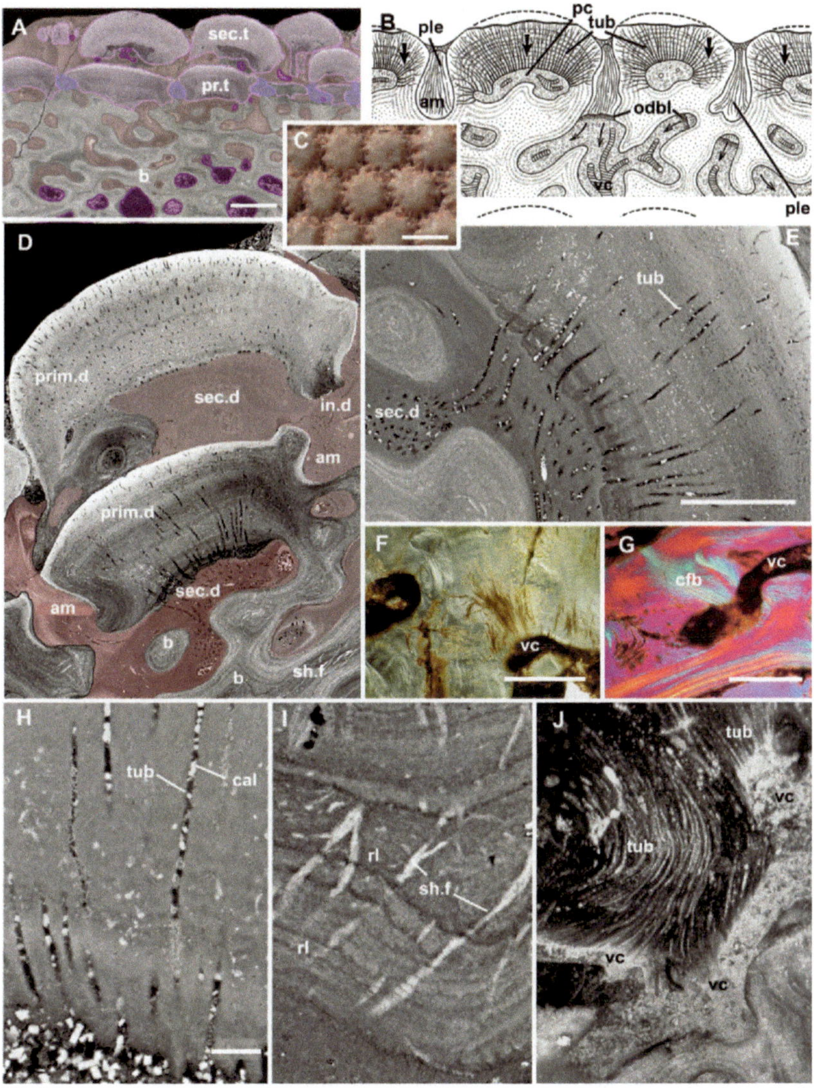

FIGURE 5.12 Wound repair in the jawless vertebrate *Psammolepis alata* (Heterostraci, Middle Devonian, Latvia). GIT 100–4 (Estonian Museum of Natural History), except as noted. (A, B) vertical section through superimposed tubercles and dermal bone of the external skeleton, backscattered electron image, and comparative drawing in B shows pleromic infill of dentine between tubercles (Ørvig 1976, figure 12). (A, D) low power false-coloured to show superimpositional growth of dentine tubercles (pink) and infilling of vascular spaces below (purple): Also ampullae (am), purple assumed vascular spaces (vc); (C) *Psammolepis venykovi* GIT100–7, dorsal view of tubercles, macrophotograph shows ampullar space between the points of each tubercle; (D) higher power than A, superimposed dermal tubercles (primary pr.t, secondary sec.d), mineral dense regions are white, least mineralised are darker, with infilling of spaces around and below them (pink), as secondary dentine (sec.d) fill bases of tubercles; (E) detail of tubercle base showing how dentine tubules of secondary dentine all half filled with calcite continue from primary dentine (bone framework and first dentine contrast, being more electron dense); (F, G) infill below a worn surface; two optical methods show brown infill relates to both dentine and bone (F, Zeiss Nomarsky optics), bone has blue crystal fibre bundles, separate from the layers of growth below (G, blue, using polarised light with gypsum plate for sign of birefringence); (H, I) distinguishing dentine tubules, always filled with calcite crystals, from bone where crystal fibre bundles have high mineral content (white) and cross resting lines in bone growth, all backscattered electron imaging; (J) bending dentine tubules (filled with reflective calcite crystals) show as being grouped together and arising from a vascular canal, z stack in reflective confocal microscopy. *Abbreviations*: am, ampullae; b, bone; cal, calcite crystals; cfb, crystal fibre bundles; in.d, invasive dentine; odbl, putative odontoblast layer; pc, pulp cavity; ple, pleromin; prim.d, primary dentine; pr.t, primary tubercle; rl, resting lines sec.d, secondary dentine; sec.t, secondary tubule; sh.f, Sharpey's fibres; tub, dentine tubules; vc, vascular canal. Scale bars: A = 500 μm; C = 0.5 mm; E = 200 μm; F, G = 1,000 μm; H = 20 μm. Images taken from Johanson et al. (2013).

We identified a piece of bony shield of *Psammolepis* that included both damaged and undamaged regions and using backscattered SEM for higher resolution on polished sectioned surfaces, compared this tissue response (Johanson et al. 2013; Figures 5.12 and 5.13). In the undamaged region, we found that the tubules of the primary dentine continued into secondary dentine that infilled the pulp cavity of the denticle, with the tubules then continuing into bony spaces (Figure 5.12A, D, E, prim.d, sec.d, b, false coloured pink, to purple for vascular spaces). As shown by Ørvig (1967; Figure 5.12B), this infilling property also included dentine tissue within the ampullae next to the denticles (Figure 5.12A, B, D, am). It is clear from the high-resolution SEMs that odontoblast tracks of cell processes are continuous from the tubercle to the vascular space, and also the ampullar dentine contributes to this tertiary dentine (false colour pink added at high resolution SEMs, Figure 5.12D). We show that tubercle dentine contributes as much as the ampullae to the infilling process, not so clearly shown by Ørvig (1967, figures 10–13; Figure 5.12B) who drew the ampullar dentine as the principal process.

It was necessary to identify characters of the infill tissue (response to wear) to be sure of the type, dentine, or bone, which could be responsible for the repair of this damaged bone. These acellular bone regions were seen as filled with a spray of brown self-stained tubules in optical microscopy (Figure 5.12F) but some also were the fibre bundles. The dilemma was resolved in comparing the phase contrast showing brown infill also for the fibre bundle spaces (Figure 5.12F), identified using polarised light and a gypsum plate (Figure 5.12G); here, the blue colour indicated crystal fibre bundles, hence this was the earlier formed bone. It was also evident, using z-stacked images from confocal microscopy, that the tubule bundles curve around the bone to connect with blood vessel canals and with putatively three sets of cell bodies (odontoblasts) located there (Figure 5.12J, vc). This confirms the vascular connection and the invasion of dentine into bone spaces, as a natural event, as depicted in Ørvig's diagram (Figure 5.12B).

We were able to refine tubule details as distinct from Sharpey's fibres using SEM. Dentine tubules were filled unevenly with calcite crystals (Figure 5.12H, tub, cal), whereas the deposition of acellular bone left many cement lines at close intervals, as indicators of normal growth, with dense mineral (white) for Sharpey's fibre bundles crossing these resting lines, two characters that were easy to identify deposition of bone (Figure 5.12I, sh.f, rl). Once certain how to identify invasive dentine growth (Figure 5.12D, E, sec.d) from bone layer growth, we saw how widespread the pleromic dentine was (false colour pink, Figure 5.12D), and that the origins were rationally not only from the secondary dentine of the surface tubercle, but also from within ampullae (Figure 5.12A, B, am).

5.8.2 Response of Odontoblasts to Massive Damage from a Wound to the Armour

After observation of an indentation attack (Figure 5.13A), we suggested that healing of this wound occurred by the natural process of invasive dentine just described, co-opting it in a response to injury, already a prophylactic property of these normal *Psammolepis* odontoblasts in dermal armour (Johanson et al. 2013). Figure 5.13B, F, G shows the relationship among the surface tubercles, dentine tubules and blood

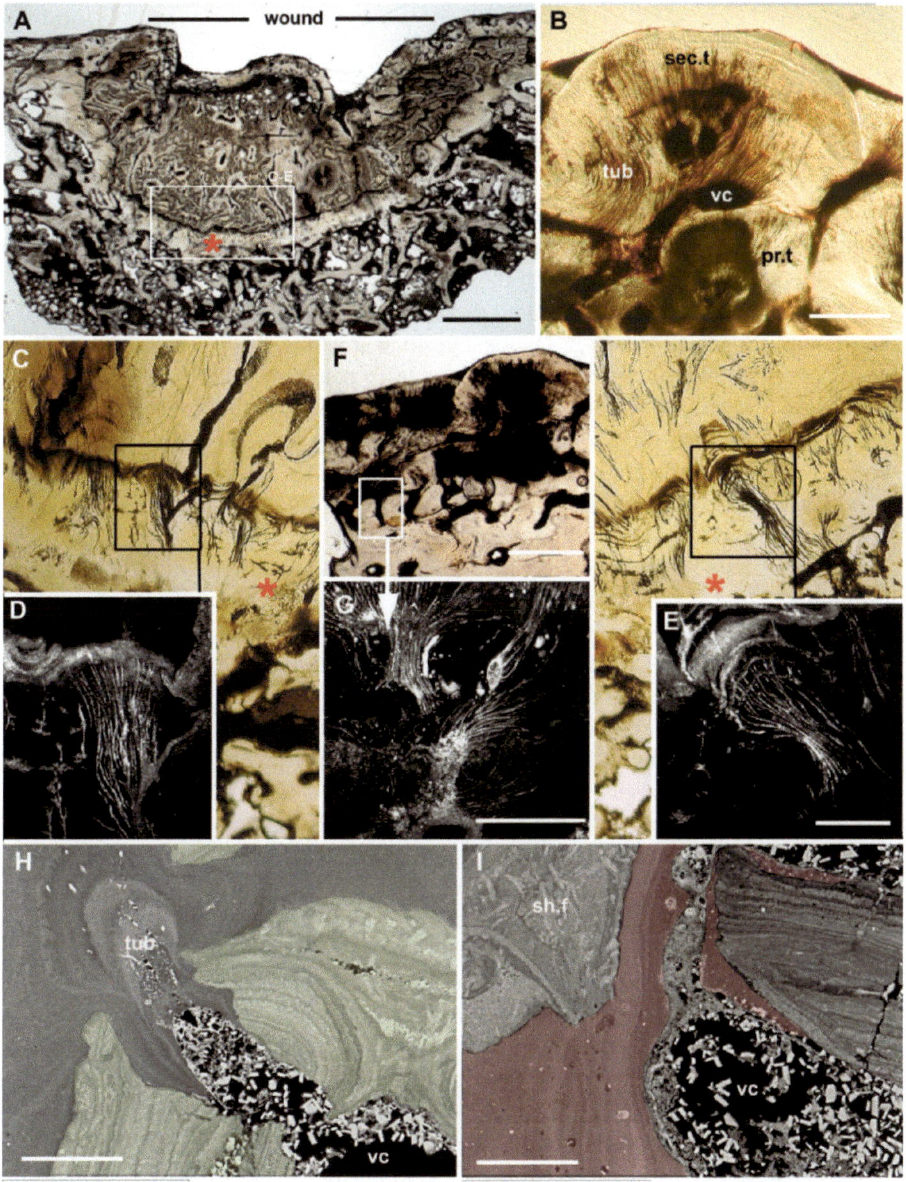

FIGURE 5.13 Wound repair in the jawless vertebrate *Psammolepis alata* (Heterostraci, Middle Devonian, Latvia). GIT 100–4 (Estonian Museum of Natural History). (A) ground section through tuberculate dermal bone and area of damage; box in A indicates region shown in C–E; red asterisk indicates the hypermineralised reactive layer below the repair dentine that first closes off the wound (see also C); (B) slightly worn tubercle at surface (also shown in F), transmitted light with brown infilled tubules; (C) first repair layer below wound damage, with boxes showing positions illustrated in D and E, showing grouped tubules in reflective confocal mode, superimposed on C, all of infill by dentine tubules within this layer, controlled by odontoblast deposition; (F) worn tubercle at surface, transmitted light, with box showing area in G, infilling dentine tubules below this tubercle, emerging from vascular spaces (D, E, G, reflective confocal microscopy); (H, I) details of damaged broken bone (back scattered electron imaging), regions selected to show repair of broken bone by dentine (H bone false coloured green); (I) reparative dentine (false coloured pink) coating the broken bone (Sharpey's fibres) surfaces. *Abbreviations*: pr.t, primary tubercle; sec.t, secondary tubule; sh.f, Sharpey's fibres; tub, dentine tubules; vc, vascular canal. Scale bars: A = 1 mm; B = 100 μm; C = 0.5 mm; E = 200 μm; F = 250 μm; G–I = 50 μm. Images taken from Johanson et al. (2013).

vessels, with the reflective material within the bunched tubules demonstrating the link of new dentine tubules with vascular canals. The wound itself is deep and has been sealed off from the more basal trabecular, or spongy bone, by a solid, newly formed tissue (Figure 5.13A, box in A indicates regions shown in Figure 5.13C, D, E). Figure 5.13D and E shows parts of this layer at higher resolutions with reflective confocal z-stacked images, where bundles of grouped tubular dentine have penetrated this otherwise nearly featureless layer.

This process is demonstrated in all these images that migratory odontoblasts fill in the multiple vascular spaces in response to injury. Along with the solid repair layer, migratory odontoblasts can also deposit reparative dentine onto broken bone surfaces. Because we were able to distinguish repair dentine (tubules in TS or LS) from new bone on the basis of their growth modes (e.g. resting lines and Sharpey's fibres in bone), we could interpret a clear response to broken bone fragments on SEM images in which we false coloured bone in green leaving dentine as shades of grey depending on mineral density (as in the reflected electron mode; Figure 5.13H). The dentine is also identifiable by bunched tubules that were able to change direction of travel through vascular spaces as in dentine invasions from wear (Figure 5.12A, D); reparative dentine was also identified in another coarse-fibered broken site (Figure 5.13I, false pink) on the bone with broken edges that were thus unequivocally coated with tubular dentine. This repair process takes place below the region of dense dentine that separates the main wound area from the damaged bone beneath (red asterisk, Figure 5.13C), including that in the broken bone below the wound (Figure 5.13H, I).

In summary, we suggest that responsive dentine, whether activated due to wear or larger damage, traces not only from the sensory ampullae between the tubercles as shown by Ørvig (1967; Figure 5.12B, D) and by Johanson et al. (2013), but also from the secondary dentine in the tubercles (Figure 5.12E, sec.d; prim.d), i.e. those odontoblasts from the papillary dentine adding to the primary dentine and illustrated as the false pink regions tracing the tubular dentine formed by odontoblast processes (Figure 5.12A, D). These identify odontoblasts and dentine, contrasting with bone in Figure 5.12H, I. Just how far migratory odontoblasts from the primary dentine of the surficial tubercles can migrate via vascular soft tissue can be demonstrated by where tubules end their journey (Figure 5.12J). Here, bright tubule spaces (reflection of crystalline fossil infill) indicate two sets of migratory cells, one curving around a bone surface and the other in a separate part of the vascular space. Critically, these odontoblasts are activated in response to normal wear of the tubercles to form more secondary dentine and join that of the ampullary source by extended odontoblast processes that are far outside the dentine of the papilla. This regular migratory capacity makes them ideal candidates for far-reaching repair.

5.9 DISCUSSION

5.9.1 Interpretations of the Odontoblast Repertoire

Our goal is to establish some of the criteria for tissue features that may allow us to interpret tissue development and growth activity of odontoblast cells, from ground sections with a variety of observational techniques as illustrated herein. This allowed us to compare across not only major vertebrate groups, such as the chondrichthyans and actinopterygians, but also jawless vertebrates (e.g. thelodonts and heterostracans). Indeed, most of the early studies on the microvertebrate dermal skeleton, focused on jawless vertebrates, were only made on ground sections of scales and ornamented bony plates (Gross 1967; Ørvig 1967; Karatajūte-Talimaa 1978), as in Figure 5.1. We pointed out that in these, the odontoblast trail is well-recorded, and two tissue types, both with a central pulp (Figure 5.1A) and a divided one as multivascular dentine (Figure 5.1C), could be transformed during evolution predictably into similar types of orthodentine, and then more specialised tubate dentine, as in the dental plates described here for holocephalans (Figure 5.9C, F).

Additionally, there has been considerable debate surrounding the nature and homology of enameloid in Chondrichthyes and acrodin in Actinopterygii. In both cases, this is the hypermineralised tissue that forms the outer surface and starts to be produced first in the tooth germ at the junction between an epithelial inner layer (inner dental epithelium or in some pre-ameloblasts) and an ectomesenchymal outer papillary layer (pre-odontoblasts) (Figure 5.1A, t1). This debate has also occurred in previous studies, such as those on the early transmission electron microscopy and scanning microscopy of sectioned surfaces, etched to reveal tissue structural details. Sasagawa (1989, 1995, p. 801), in TEM studies of both epithelial and ectomesenchymal cells in the teleost fish *Tilapia nilotica* (Actinopterygii), proposed that "odontoblasts are involved in the formation of a considerable portion of the enameloid matrix including collagen fibrils." Shellis and Miles (1976) using an autoradiographic microscopic study of teleosts (wrasse and eel) to study amino acid proline uptake by, and secretion in, tooth germs were able to show the timing of secretion and activity of both the inner dental epithelium and the odontoblasts. They concluded that the protein secreted by the odontoblasts is collagen and that the matrix of enameloid was first formed at the surface of the papilla where its bulk increased, before any production of dentine. These structural and developmental features are inconsistent with the production of enameloid by ameloblasts.

Odontoblasts also seem to have a major role in the initiation of the enameloid matrix in elasmobranchs, such as the carcharhiniform shark *Mustelus manazo* (Sasagawa 1989), that also includes the process of mineralisation starting at the regions next to the inner layer pre-ameloblasts, i.e. the surface of enameloid. In the batoids, Sasagawa and Akai (1992) studied early tooth formation in embryos of the rays *Dasyatis akajei* and *Urolophus aurantiacus* and from fine details, including TEM and EDX to show levels of calcium and other minerals, demonstrated that the odontoblasts are mainly involved in the formation of the enameloid. Especially, they revealed numerous 'tubular vesicles' at the forefront of enameloid production where the first mineral crystals of the matrix appeared, hence odontoblasts have a role in initiation of mineralisation. One feature that may have a commonality, or not, with both teleosts (Sasagawa 1988, 1995) and elasmobranchs is that mineral crystals are first found inside these single membrane, tubular vesicles that form from the odontoblast process in the outermost part (nearest to ide cells) of the enameloid matrix.

One relatively unreported topic is the process by which ectomesenchymal dental papilla cells are selected first to form the organic enameloid matrix and initiate its mineralisation and then, second, to form dentine. Sasagawa (1995), in the embryonic stingray, referred to odontoblast dimorphism, assigning each cell type to different roles. These roles were to make the two tissues, dentine and enameloid, either as odontoblasts with long cell processes extending into the enameloid matrix and controlling its mineral initiation (Sasagawa 1995, light cells, p. 229: Dark cells, p. 228, figure 13) or as those with only short odontoblast processes and a role of initial dentine matrix formation.

The idea of dimorphism was explored by Smith and Sansom, (2000; section 5.4.1), where they reinforced the concept of these two cell types in the tooth development of the elasmobranch *Squatina japonica* (Squatinidae; personal communication from Ichiro Sasagawa 1998), one retaining a process in the enameloid and the other confined to the dentine, formed via a second process beneath the cap tissue. They further extended this concept to the placoderm *Phlyctaenaspis acadica*, reproducing the drawing that showed coronal odontocytes as 'special coronal cells and the other in the cellular dentine below as unipolar odontocytes' (Gross 1957, figure 9A). Smith and Sansom (2000, p. 72, figure 6) commented that there were at least two phenotypes in the odontoblast repertoire and suggested it was an unexplored area of research requiring molecular markers for future studies.

Another recent example of odontoblast plurality is found in the holocephalan *Chimera phantasma*, which does not have teeth and certainly no enameloid, but was also found to have the same cytoplasmic ability of the cell type involved in early mineralisation of the dental plate tissue of hypermineralised dentine (Ishiyama et al. 1984, 1991; Iijima and Ishiyama 2020). Importantly, the initiation of mineralisation occurred within similar tubular vesicles as identified before in the enameloid of both teleosts and elasmobranchs despite being a different mineral in holocephalans, the magnesium-containing mineral whitlockite. An equivalent feature was recently found in the extant holocephalan *Harriotta raleighana* where new tissue of the hypermineralised dentine (whitlockin) formed in ovoids and tritors (Figure 5.9B, D, E; Smith et al. 2019, figure 9H). Here, the same tubular vesicles are generated in massive numbers in regions with high densities of growing odontoblast tubules that extend into the centre of the mineralisation (Figure 5.14B, tub.br, ves) through extensive ramification and where first mineral crystals form. This is a repertoire where only odontoblasts have the capability to form all components of the dental plate, including the less mineralised framework of the outer shell and trabecular dentine. The latter itself forms the shapes of rods, ovoids, and tritors, within which the whitlockin becomes the super-hard tissue that serves to make higher levels of wear-resistant structures at the functional surface. It is a remarkable property of the odontoblast's repertoire that here in a region of tissue replacement, or regeneration, the morphology of the dental plates is established in the absence of teeth (e.g., Johanson et al. 2021).

In holocephalans, the cells making tritoral dentine in the adult dental plates clearly show a changing morphology, from the regular odontoblasts making the trabecular dentine of the support tissues, to the whitloblasts at the formative front of the whitlockin (model in Figure 5.14A is adapted from Ørvig 1985, figure 33). The model shows an essential phenotype with cells joined as a layer located within the walls of the preformed trabecular dentine spaces, seen as enlarged cell body spaces (Smith et al. 2019, figure 5.4). The odontoblast cells, transformed into whitloblasts, are still connected to the whole of the whitlockin (an intertubate, hypermineralised dentine, Figure 5.14A, ITD) through a massive spread of unrestrained tubules within the tissue. These extensive and numerous tubules are essential to initiate the mineralisation, as they expand to

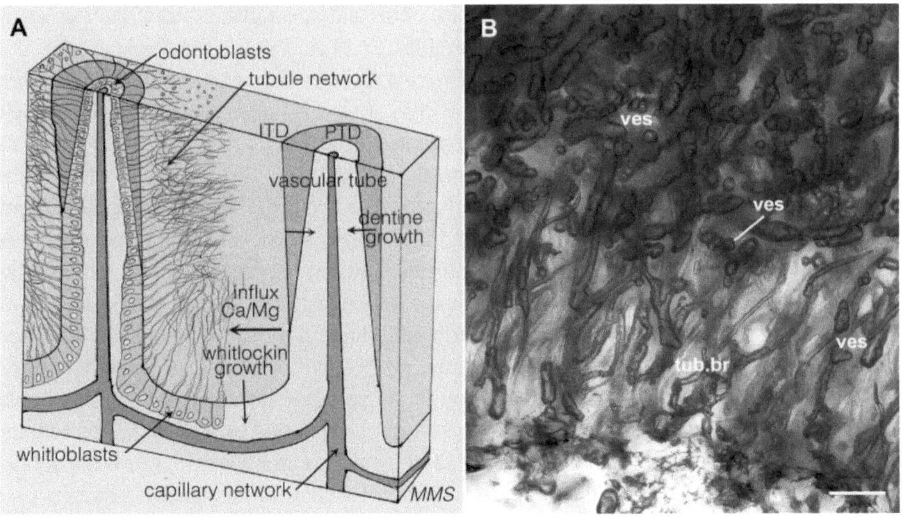

FIGURE 5.14 A schematic of odontoblasts forming tubate dentine, with location of whitloblasts as predicted from our observations on the holocephalan dentition (Figures 5.8, 5.9; also in Smith et al. 2019) cells with massive numbers of tubules intermixing, which produce the hypermineralised tissue whitlockin. Arrows indicate directions of growth, including the regulation by whitloblasts of minerals into the hypermineralised form of dentine. Their phenotype (larger cells) is indicated, with a change back to regular odontoblasts forming the circumvascular (peritubate dentine, PTD) onto the surface of the whitlockin (intertubate dentine, ITD) (Ørvig 1985, figure 33, see also Johanson et al. 2021, figure 9f, including the proposed use of 'whitlockin'); B, maximum optical magnification using Keyence stacked image with transmitted light (see Figure 5.9C, E, F), illustrating details of branching tubules and the creation of tubular vesicles (ves) as separate entities, where mineral crystals are initiated, and later mineral density is increased in the entire matrix to form whitlockin, from the expanded tubules and between the vascular canals (Smith et al. 2019, figures 7–9). *Abbreviations*: tub.br, branching tubules; ves, vesicles.

form saccular vesicles commencing in large amounts, Mg dominating among the ions, to make the whitlockin. This phenotype of the matrix, through an active membrane on the surfaces, shows cellular control of this specialised dentine production via the influx of the massive number of mineral ions. The larger cell bodies reveal exactly how mineralisation is initiated: Their elongated processes expand to discharge huge numbers of separate cell membrane vesicles (Figures 5.9B, D, E and 5.14B) in which granular mineral products first form. These were previously identified in elasmobranch enameloid (Sasagawa 1989, 2002) called tubular vesicles. Once the whitlockin production has completed, this cell reduces in height and is reversed to a regular odontoblast type to form the circumvascular dentine (Figure 5.14A, PTD), one that forms at a lower level of mineralisation and has fewer tubules, but these remain connected to the whitlockin.

This illustrates a remarkable example of the repertoire of odontoblasts in holocephalans in forming dental plates in the absence of tooth germs, but the role in producing dentine with enormous intermingling tubules that give rise to the tubular vesicles to initiate mineralisation may well be universal among elasmobranchs and actinopterygians.

Leaving all versions of enameloid, whitlockin, and specialised oral structures aside, there is another amazing property of odontoblasts reported in the ornamentation of dermal bones in the fossil jawless vertebrate *Psammolepis*, which uses the intrinsic property of the odontoblast cells, migration, in the process of repair. These migratory odontoblasts were proposed to originate from pluripotent neural crest cells at many locations in the dermal plates of this heterostracan (Johanson et al. 2013). We have traced the path of these cells in all locations by the tubules that remain from the *in vivo* cytoplasmic processes, tubules identified by a postmortem infill with extrinsic mineral, as seen in SEM images (Figure 5.12D, E, H–J). This establishes a migratory property of the odontoblast repertoire that is co-opted to effectively repair a wound in the dermal plate, which was quite extensive to include into the acellular bone. The repair of the wound is all-embracing and extensive, recruiting what is a normal property of dentine that the odontoblasts, as free cells or in a layer, will migrate through the vascular spaces in acelluar trabecular bone to infill all, even down into the supporting bone that had broken. It seems remarkable that even spaces created by the breakage of the bone can be repaired through dentine that coats the fragments of bone from the adjacent vascular tissue spaces (Figure 5.13H, I). Notably, these bone fragment surfaces show none of the characteristic scalloped margins that indicate the activity of osteoclasts and bone resorption. As shown recently in the teleost fish *Leedsichthys* (Johanson et al. 2022), the repair of the broken lepidotrichia (formed from dermal bone) required the resorption of the damaged bone margin, prior to the formation and attachment of coarse fibres that anchor new repair bone to this surface. In *Psammolepis*, neither of these mechanisms are necessary, with odontoblasts directly depositing the new repair dentine onto bone. We have given explicit details in this early evolutionary process of wound healing in acellular bone, albeit in a jawless psammolepid, tuberculate head shield, to encourage research on this regenerative process, as an intrinsic property of vascular odontode-bearing dentine. This is a central objective of this exploration into the whole repertoire of the ectomesenchymal odontoblasts.

ACKNOWLEDGMENTS

We thank Tony Wighton and Callum Hatch for slide preparation; Vincent Fernandez and Brett Clark for assistance with computed tomographic scanning (all Science Innovation Platforms, Natural History Museum); and Peter Pileki for assistance with confocal and Keyence microscopy (Center for Oral, Clinical and Translational Sciences, Faculty of Dentistry, Oral and Craniofacial Sciences, King's College London).

REFERENCES

Andreev, P. S., Coates, M. I., Karatajūtė-Talimaa, V., Shelton, R. M., Cooper, P. R., Wang, N.-Z., and Sansom, I. J. 2016. The systematics of the Mongolepidida (Chondrichthyes) and the Ordovician origins of the clade. *PeerJ* 4: e1850.

Andreev, P. S., Coates, M. I., Shelton, R. M., Cooper, P. R., Smith, M. P., and Sansom, I. J. 2015. Upper Ordovician chondrichthyan-like scales from North America. *Palaeontology* 58: 691–704.

Andreev, P. S., Zhao, W., Wang, N.-Z., Smith, M. M., Li, Q., Cui, X., Zhu, M., and Sansom, I. J. 2020. Early Silurian chondrichthyans from the Tarim Basin (Xinjiang, China). *PLoS ONE* 15(2): e0228589.

Andreucci, R. D., Britski, H. A., and Carneiro, J. 1982. Structure and evolution of tetraodontoid teeth: An autoradiographic study (pisces, Tetraodontiformes). *Journal of Morphology* 171(3): 283–292. doi: 10.1002/jmor.1051710304

Coates, M. I., Finarelli, J. A., Sansom, I. J., Andreev, P. S., Criswell, K. E., Tietjen, K., Rivers, M. L., and La Riviere, P. J. 2018. An early chondrichthyan and the evolutionary assembly of a shark body plan. *Proceedings of the Royal Society B, Biological Sciences* 285: 20172418.

Coates, M. I., Gess, R. W., Finarelli, J. A., Criswell, K. E., and Tietjen, K. 2017. A symmoriiform chondrichthyan braincase and the origin of chimaeroid fishes. *Nature* 541: 208–211.

Coates, M. I., Tjetien, K., Johanson, Z., Friedman, M., and Sang, S. 2021. The cranium of *Helodus simplex* (Agassiz, 1838) revised; pp. 193–204 in J. Denton, A. Pradel, and P. Janvier (eds.), *John Maisey Symposium, Ichthyological Explorations of Freshwaters*. Verlag Dr. Friedrich Pfeil, Munich.

Denison, R. H. 1974. The structure and evolution of teeth in lungfishes. *Fieldiana, Geology* 33: 31–58.

Donoghue, P. C. J., Sansom, I. J., and Downs, J. P. 2006. Early evolution of vertebrate skeletal tissues and cellular interactions, and the canalization of skeletal development. *Journal of Experimental Zoology B, Molecular Development and Evolution* 306B: 278–294.

Fincham, A. G., Luo, W., Moradian-Oldak, J., Paine, M. L., Snead, M. L., and Zeichner-David, M. 2000. Enamel biomineralization: The assembly and disassembly of the protein extracellular organic matrix; pp. 37–61 in M. E. Teaford, M. M. Smith, and M. W. J. Ferguson (eds.), *Development, Function and Evolution of Teeth*. Cambridge University Press, Cambridge.

Forey, L., López-Arbarello, A., MacLeod, N. 2011. A New Species of Lepidotes (Actinopterygii: Semionotiformes) from the Cenomanian (Upper Cretaceous) of Morocco. *Palaeontologia Electronica* 14: 1–12.

Fraser, G. J., Britz, R., Hall, A., Johanson, Z., and Smith, M. M. 2012. Replacing the first-generation dentition in pufferfish with a unique beak. *Proceedings of the National Academy of Sciences* 109: 8179–8184.

Fraser, G. J., Hamed, S. S., Martin, K. J., and Hunter, K. D. 2019. Shark tooth regeneration reveals common stem cell characters in both human rested lamina and ameloblastoma. *Science Reports* 9: 15956.

Fraser, G. J., Standing, A., Underwood, C., and Thiery, A. P. 2020. The dental lamina: An essential structure for perpetual tooth regeneration in sharks. *Integrative and Comparative Biology* 60: 644–655.

Frey, L., Coates, M., Ginter, M., Hairapetian, V., Rücklin, M., Jerjen, I., and Klug, C. 2019. The early elasmobranch *Phoebodus*: Phylogenetic relationships, ecomorphology and a new time-scale for shark evolution. *Proceedings of the Royal Society B, Biological Sciences* 286: 20191336.

Gross, W. 1957. Mundzähne und Hautzähne der Acanthodier und Arthrodiren. *Palaeontographica* 109A: 1–40.

Gross, W. 1967. Uber thelodontier-schuppen. *Palaeontographica Abteilung A* 127: 1–6.

Iijima, M., and Ishiyama, M. 2020. A unique mineralization mode of hypermineralized pleromin in the tooth plate of *Chimaera phantasma* contributes to its microhardness. *Science Reports* 10: 18591.

Ishiyama, M., Sasagawa, I., and Akai, J. 1984 The inorganic content of pleromin in tooth plates of the living holocephalan, *Chimaera phantasma*, consists of a crystalline calcium phosphate known as beta-Ca3(PO4)2 (whitlockite). *Archivum Histologicum Japanicum* 47: 89–94.

Ishiyama, M., Yoshie, S., Teraki, Y., and Cooper, E. W. T. 1991. Ultrastructure of pleromin, a highly mineralized tissue comprising crystalline calcium phosphate known as Whitlockite, in holocephalian tooth plates; pp. 453–457 in S. Suga and H. Nakahara (eds.), *Mechanisms and Phylogeny of Mineralization in Biological Systems*. Springer-Verlag, Berlin.

Jambura, P. L., Kindlimann, R., Lopez-Romero, F., Marramà, G., Pfaff, C., Stumpf, S., Türtscher, Underwood C. J., Ward, D. J., and Kriwet, J. 2019. Micro-computed tomography imaging reveals the development of a unique tooth mineralization pattern in mackerel sharks (Chondrichthyes; Lamniformes) in deep time. *Scientific Reports* 9: 9652.

Jambura, P. L., Türtscher, J., Kindlimann, R., Metscher, B., Pfaff, C., Stumpf, S., Weber, G. W., and Kriwet, J. 2020. Evolutionary trajectories of tooth histology patterns in modern sharks (Chondrichthyes, Elasmobranchii). *Journal of Anatomy* 236: 753–771.

Johanson, Z., Liston, J., Davesne, D., Challands, T., and Smith, M. M. 2022. Mechanisms of dermal bone repair after predatory attack in the giant stem-group teleost *Leedsichthys problematicus* Smith Woodward, 1889 (Pachycormiformes). *Journal of Anatomy* 241: 393–406.

Johanson, Z., Manzanares, E., Underwood, C. J., Clark, B., Fernandez, V., and Smith, M. M. 2020. Evolution of the holocephalan dentition. *Integrative and Comparative Anatomy* 60: 630–643.

Johanson, Z., Manzanares, E., Underwood, C. J., Clark, B., Fernandez, V., and Smith, M. M. 2021a. Ontogenetic development of the holocephalan dentition: Morphological transitions of dentine in the absence of teeth. *Journal of Anatomy* 239: 704–719.

Johanson, Z., Smith, M. M., Kearsley, A., Pilecki, P., Mark-Kurik, E., and Howard, C. 2013. Origins of bone repair in the armour of fossil fish: Response to a deep wound by cells depositing dentine instead of dermal bone. *Biology Letters* 9: 20130144.

Johanson, Z., Tanaka, M., Chaplin, N., and Smith, M. M. 2008. Early Palaeozoic dentine and patterned scales in the embryonic catshark tail. *Biology Letters* 4:87–90.

Johanson, Z., Underwood, C. J., Coates, M. I., Fernandez, V., Clark, B., and Smith, M. M. 2021b. The stem-holocephalan *Helodus* (Chondrichthyes; Holocephali) and the evolution of modern chimaeroid dentitions; pp. 205–214 in J. Denton, A. Pradel, and P. Janvier (eds.), *John Maisey Symposium, Ichthyological Explorations of Freshwaters.* Verlag Dr. Friedrich Pfeil, Munich.

Karatajūte-Talimaa, V. 1978. *Silurian and Devonian thelodonts of the SSSR and Spitsbergen.* Mokslas, Vilnius, 334 p. [In Russian, English summary].

Keating, J. N., Marquart, C. L., and Donoghue, P. C. 2015. Histology of the heterostracan dermal skeleton: Insight into the origin of the vertebrate mineralised skeleton. *Journal of Morphology* 276: 657–680.

Kundrát, M., Joss, J. M., and Smith, M. M. 2008. Fate mapping in embryos of *Neoceratodus forsteri* reveals cranial neural crest participation in tooth development is conserved from lungfish to tetrapods. *Evolution and Development* 10: 531–536.

López-Arbarello, A. 2012. Phylogenetic interrelationships of ginglymodian fishes (Actinopterygii: Neopterygii). *PLoS One* 7(7): e39370. doi: 10.1371/journal.pone.0039370

Lumsden, A. G. 1988. Spatial organization of the epithelium and the role of neural crest cells in the initiation of the mammalian tooth germ. *Development* 103(Suppl): 155–169. doi: 10.1242/dev.103

Maisey, J. G., Denton, J. S. S., Burrow, C., and Pradel, A. 2020. Architectural and ultrastructural features of tessellated calcified cartilage in modern and extinct chondrichthyan fishes. *Journal of Fish Biology* 98: 919–941.

Moss, M. L. 1968. The origin of vertebrate calcified tissues; pp. 359–371 in T. Ørvig (ed.), *Current Problems of Lower Vertebrate Phylogeny.* Almquist and Wiskell, Stockholm.

Moy-Thomas, J. A. 1936. The structure and affinities of the fossil elasmobranch fishes from the Lower Carboniferous Rocks of Glencartholm, Eskdale. *Journal of Zoology* 106: 761–788.

Nanci, A. 2017. *Ten Cate's Oral Histology-e-Book: Development, Structure, and Function.* Elsevier Health Sciences, St. Louis, MO.

Ørvig, T. 1967. Phylogeny of tooth tissue: Evolution of some calcified tissues in early vertebrates; pp. 45–110 in A. E. W. Miles (ed.), *Structural and Chemical Organization of Teeth.* Academic Press, New York and London.

Ørvig, T. 1976. The interpretation of pleromin (pleromic hard tissue) in the dermal skeleton of psammosteid heterostracans. *Zoologica Scripta* 5: 35–47.

Ørvig, T. 1985. Histologic studies of ostracoderms, placoderms and fossil elasmobranchs 5. Ptyctodontid tooth plates and their bearing on holocephalan ancestry: The condition of chimaerids. *Zoologica Scripta* 14: 55–79.

Patterson, C. 1965. The phylogeny of the chimaeroids. *Philosophical Transactions of the Royal Society London, Series B, Biological Sciences* 249: 101–219.

Popa, E. M., Buchtova, M., and Tucker, A. S. 2019. Revitalising the rudimentary replacement dentition in the mouse. *Development* 146: dev171363.

Pradel, A., Tafforeau, P., Maisey, J. G., and Janvier, P. 2011. A new Paleozoic Symmoriiformes (Chondrichthyes) from the late Carboniferous of Kansas (USA) and cladistic analysis of early chondrichthyans. *PLoS ONE* 6: e24938.

Rasch, L. J., Cooper, R. L., Underwood, C., Dillard, W. A., Thiery, A. P., and Fraser, G. J. 2020. Development and regeneration of the crushing dentition in skates (Rajidae). *Developmental Biology* 466: 59–72.

Richter, M., and Smith, M. M. 1995. A microstructural study of the ganoine tissue of selected lower vertebrates. *Zoological Journal of the Linnean Society* 114: 173–212.

Rosa, J. T., Witten, P. E., and Huysseune, A. 2021. Cells at the edge: The dentin—bone interface in zebrafish teeth. *Frontiers in Physiology* 12: 723210.

Sansom, I. J., and Andreev, P. 2018. The Ordovician enigma: Fish, first appearances and phylogenetic controversies; pp. 59–70 in Z. Johanson, C. Underwood, and M. Richter (eds.), *Evolution and Development of Fishes.* Cambridge University Press, Cambridge.

Sansom, I. J., Davies, N. S., Coates, M. I., Nicoll, R. S., and Ritchie, A. 2012. Chondrichthyan-like scales from the Middle Devonian of Australia. *Palaeontology* 55: 243–247.

Sansom, I. J., Smith, M. M., and Smith, M. P. 1996. Scales of thelodont and shark-like fishes from the Ordovician of Colorado. *Nature* 379: 628–630.

Santini, F., Sorenson, L., and Alfaro, M. E. 2013. A new phylogeny of tetraodontiform fishes (Tetraodontiformes, Acanthomorpha) based on 22 loci. *Molecular Phylogenetics and Evolution* 69: 177–187.

Sasagawa, I. 1988. The appearance of matrix vesicles and mineralization during tooth development in three teleost fishes with well-developed enameloid and orthodentine. *Archives of Oral Biology* 33: 75–86

Sasagawa, I. 1989. The fine structure of Initial mineralisation during tooth development in the gummy shark, *Mustelus manazo*, Elasmobranchia. *Journal of Anatomy* 164: 175–187.

Sasagawa, I. 1995. Fine structure of tooth germs during the formation of enameloid matrix in *Tilapia nilotica*, a teleost fish. *Archives of Oral Biology* 40: 801–814.

Sasagawa, I. 1998. Mechanisms of mineralization in the enameloid of elasmobranchs and teleosts. *Connective Tissue Research* 39: 207–214.

Sasagawa, I. 2002. Mineralization patterns in elasmobranch fish. *Microscopy Research and Technique* 59: 396–407.

Sasagawa, I., and Akai, J. 1992. The fine structure of the enameloid matrix and initial mineralization during tooth development in the sting rays, *Dasyatis akajei* and *Urolophus aurantiacus*. *Journal of Electron Microscopy* 41: 242–252.

Shellis, R. P., and Miles, A. E. W. 1976. Observations with electron microscope on enameloid formation in common eel (*Anguilla anguilla;* Teleostei). *Proceedings of the Royal Society of London B*, 194: 253–269.

Smith, M. M. 1991. Putative skeletal neural crest cells in early late Ordovician vertebrates from Colorado. *Science* 251: 1–303.

Smith, M. M., and Coates, M. I.1998. Evolutionary origins of the vertebrate dentition: Phylogenetic patterns and developmental evolution. *European Journal of Oral Sciences* 106(Suppl 1): 482–500. doi: 10.1111/j.1600-0722.1998.tb02212.x

Smith, M. M., Fraser, G. J., Chaplin, N., Hobbs, C., and Graham, A. 2009a. Reiterative pattern of sonic hedgehog expression in the catshark dentition reveals a phylogenetic template for jawed vertebrates. *Proceedings of the Royal Society B: Biological Sciences* 276(1660): 1225–1233.

Smith, M. M., Fraser, G. J., and Johanson, Z. 2016. Origin of teeth in jawed vertebrates. *Infocus* 42: 5–17.

Smith, M. M., Fraser, G. J., and Mitsiadis, T. A. 2009c. Dental lamina as source of odontogenic stem cells: Evolutionary origins and developmental control of tooth generation in gnathostomes. *Journal of Experimental Zoology B Molecular Developmental Evolution* 312B: 260–280.

Smith, M. M., and Hall, B. K. 1990. Development and evolutionary origins of vertebrate skeletogenic and odontogenic tissues. *Biological Reviews* 65: 277–373.

Smith, M. M., and Hall, B. K. 1993. A developmental, model for evolution of the vertebrate exoskeleton and teeth: The role of cranial and trunk neural crest. *Evolutionary Biology* 27: 387–448.

Smith, M. M., Manzanares, E., Underwood, C., Healy, C., and Clark, B. 2020. Holocephalan (Chondrichthyes) dental plates with hypermineralized dentine as a substitute for missing teeth through developmental plasticity. *Journal of Fish Biology* 97: 16–27.

Smith, M. M., Okabe, M., and Joss, J. 2009b. Spatial and temporal pattern for the dentition in the Australian lungfish revealed with sonic hedgehog expression profile. *Proceedings of the Royal Society B: Biological Sciences* 276(1657): 623–631.

Smith, M. M., Riley, A., Fraser, G. J., Underwood, C., Welten, M., Kriwet, J., Pfaff, C., and Johanson, Z. 2015. Early development of rostrum saw-teeth in a fossil ray tests classical theories of the evolution of vertebrate dentitions. *Proceedings of the Royal Society B, Biological Sciences* 282: 20151628. http://doi.org/10.1098/rspb.2015.1628

Smith, M. M., and Sansom, I. J. 1997. Exoskeletal micro-remains of an Ordovician fish from the Harding Sandstone of Colorado. *Palaeontology* 40: 645–658.

Smith, M. M., and Sansom, I. J. 2000. Evolutionary origins of dentine in the fossil record of early vertebrates: Diversity, development and function; pp. 65–81 in M. E. Teaford, M. M. Smith, and M. W. J. Ferguson (eds.), *Development, Function and Evolution of Teeth*. Cambridge University Press, Cambridge.

Smith, M. M., Underwood, C., Goral, T., et al. 2019. Growth and mineralogy in dental plates of the holocephalan *Harriotta raleighana* (Chondrichthyes): Novel dentine and conserved patterning combine to create a unique chondrichthyan dentition. *Zoological Letters* 5: 11.

Stahl, B. J. 1999. Chondrichthyes III: Holocephali; pp. 1–164 in H.-P. Schultze (ed.), *Handbook of Paleoichthyology*. Volume 4. Verlag Dr. Friedrich Pfeil, Munich.

Tarlo, L. B. H. 1964. The origin of the bone; pp. 3–17 in H. J. J. Blackwod (ed.), *Bone and Tooth*. Pergamon Press, Oxford.

Tucker, A. S., and Fraser, G. J. 2014. Evolution and developmental diversity of tooth regeneration. *Seminars in Cell and Developmental Biology* 25–26: 71–80.

Williams, M. E. 2001. Tooth retention in cladodont sharks: With a comparison between primitive grasping and swallowing, and modern cutting and gouging feeding mechanisms. *Journal of Vertebrate Paleontology* 21: 214–226.

Wilson, M. V. H., and Märss, T. 2009. Thelodont phylogeny revisited, with inclusion of key scale-based taxa. *Estonian Journal of Earth Sciences* 58: 297–310.

Young, G. C. 1997. Ordovician microvertebrate remains from the Amadeus Basin, central Australia. *Journal of Vertebrate Paleontology* 17: 1–25.

Žigaitė, Ž., Richter, M., Karatajūtė-Talimaa, V., and Smith, M. M. 2013. Tissue diversity and evolutionary trends of the dermal skeleton of Silurian thelodonts. *Historical Biology* 25: 143–154.

6 Shifting Perspectives in the Study of Amniote Tooth Attachment and the Path Forward to Establishing Vertebrate Periodontal Tissue Homology

Aaron R.H. LeBlanc

6.1 INTRODUCTION

Like enamel and dentine, the periodontal (attachment) tissues are essential parts of the tooth, providing the vital connection between a tooth and the bone or cartilage of the jaw. In humans, the periodontal tissues work in concert to maintain proper positioning of each tooth within the mouth, supply mechanosensory feedback for each tooth during chewing, and provide a cushion to dissipate the compressive forces of dental occlusion (Hannam, 1982; Beertsen et al., 1997; Nanci, 2013). Orthodontists carefully manipulate the periodontium of individual teeth to maintain or achieve proper tooth positioning in a human patient's mouth, and periodontists monitor the health of the supporting tissues of a tooth and look for signs of periodontal diseases, which are incredibly common in humans (Pihlstrom et al., 2005). But until recently, we had very few hypotheses to explain where the mammalian periodontal tissues *came* from in the evolutionary sense.

One major obstacle in studying the evolution of the periodontal tissues is the apparent uniqueness of mammalian tooth attachment compared to other vertebrates. A mammal's periodontium is a complex three-tissue system that forms a dynamic interface between a tooth and the jaw. Cementum, a bone-like tissue that coats the roots of each tooth, acts as an anchoring point for an organized network of collagen fiber bundles that form the periodontal ligament (PDL) (Figure 6.1) (Bosshardt and Selvig, 1997). On the other side of the periodontal ligament, the collagen fiber bundles anchor into the surrounding bone. This bone is not an extension of the jaw, however. Along with cementum and the periodontal ligament, each developing tooth forms its own layer of alveolar (socket) bone from which it is suspended (Ten Cate and Mills, 1972; Ten Cate,

DOI: 10.1201/9781003439653-6

1997). When developing teeth are experimentally explanted into different host tissues, they produce a layer of alveolar bone to which they are attached via a PDL and cementum (Ten Cate and Mills, 1972; Cai et al., 2009). A mammal tooth therefore makes its own socket during development. Like enamel and dentine, the periodontal tissues are part of the developmental module of the mammalian tooth; they develop from the dental follicle, a cluster of mesenchymal cells around a tooth bud (Baume, 1956; Ten Cate and Mills, 1972; Beertsen et al., 1997). This integrated system of a tooth and three periodontal tissues (cementum, periodontal ligament, and alveolar bone) is heralded as one of the hallmarks of mammalian dental innovation (Gaengler and Metzler, 1992; Beertsen et al., 1997; McIntosh et al., 2002; Diekwisch, 2016).

The opposite has often been said about the tooth attachment system in other vertebrates. Most comparative anatomy and dentistry textbooks provide a brief description of tooth attachment tissues in reptiles, for example. Reptile teeth are supposedly ankylosed (fused) to the jaws by a bone tissue called "bone of attachment" (Tomes, 1882; Peyer, 1968; Osborn, 1981; Shellis, 1982; Berkovitz and Shellis, 2017), a term first introduced by Tomes (1874) to characterize tooth attachment in snakes, eels, and haddock. As Tomes (1874) noted, this "bone of attachment" tissue developed in association with each tooth, but he was unable to determine how this tissue was equivalent to the periodontium in mammals. Nevertheless, Tomes (1882) later expanded the occurrences of "bone of attachment" in non-mammalian vertebrates and concluded that any vertebrate that ankylosed its teeth to the jaws possessed this singular tissue. The ripple effect of Tomes' work has influenced the study of periodontal tissue evolution ever since.

Researchers have struggled to define "bone of attachment" and identify homologous attachment tissues across vertebrate teeth (Owen, 1840; Tomes, 1874; Smith, 1958; Peyer, 1968; Grady, 1970; Moss, 1977; Reif, 1980; Shellis, 1982; Osborn, 1984; Gaengler and Metzler, 1992; Smith and Hall, 1993; Berkovitz and Shellis, 2017). This has been especially difficult for determining how the attachment tissues of early vertebrate teeth could have given rise to the complex tooth attachment tissue system (the periodontium) in mammals (Noyes, 1902; Smith, 1958; Osborn, 1984). The challenges associated with studying tooth attachment tissue evolution are twofold. First, the disproportionate number of studies of the periodontium of mammals has created a mammal-centric view of attachment tissue development and histology, one that has been partly solved by studying the periodontia of modern crocodylians and other vertebrates with "mammal-like" attachment tissues (Beust, 1938; Miller, 1968; Soule, 1969; Berkovitz and Sloan, 1979; McIntosh et al., 2002; LeBlanc et al., 2017b). The second challenge is the previous lack of comparative histological data from the fossil record needed to trace the evolutionary history of the periodontium and address the historical and literary burdens associated with the "bone of attachment" paradigm.

In this chapter, I focus on the developmental and histological definitions of the attachment tissues in extant and fossil amniotes and provide a framework for identifying homologous tooth attachment tissues in other vertebrates. This chapter necessarily focuses on amniotes because of a tissue taxonomy problem: The names given to the different attachment tissues in mammals are old and well-defined (Owen, 1840; Tomes, 1874, 1882), and the names we give to attachment tissues must first be standardized in order to make testable hypotheses regarding the origin, evolution, and diversification of the vertebrate periodontal tissues.

6.2 DESCRIBING DENTITIONS: TOOTH IMPLANTATION AND ATTACHMENT ARE DIFFERENT

There are several descriptors that anatomists and histologists use to describe how teeth sit on or within a jaw. Of these, two classes should be briefly mentioned and differentiated. The first is tooth implantation, which describes the topology and geometry of how a tooth attaches to a jaw (Gaengler, 2000; Dumont et al., 2016; LeBlanc et al., 2017b, 2021; Bertin et al., 2018). Many reviews and textbooks use three major terms to describe tooth implantation in amniotes and in vertebrates more broadly (Shellis, 1982; Gaengler, 2000) (Figure 6.1). Acrodonty describes a tooth that is attached to the crest of the jaw. Pleurodonty is an asymmetrical form of implantation, where a tooth is attached to the labial wall (the pleura) of the jaw. Finally, thecodonty occurs when a tooth is deeply implanted into the jaw. Tooth implantation

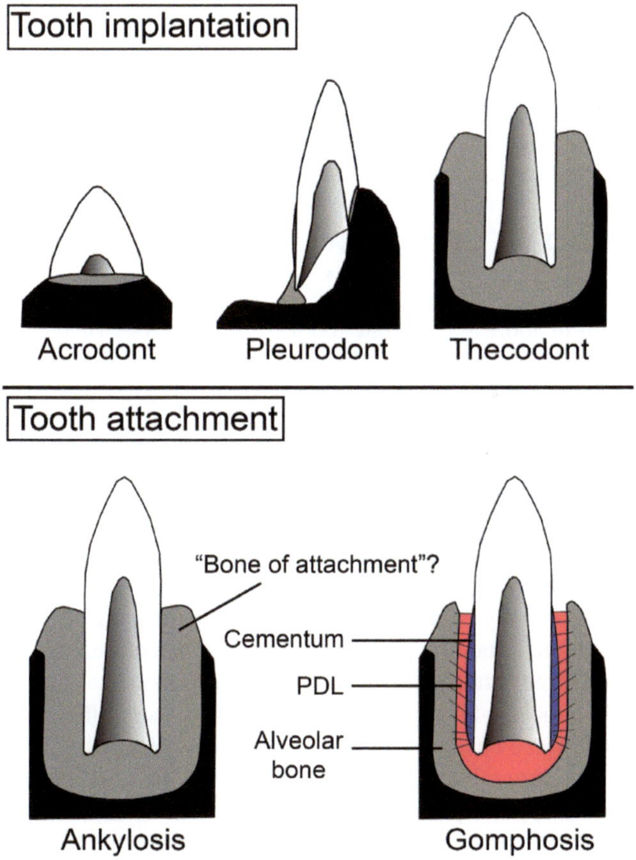

FIGURE 6.1 Tooth implantation defines the geometry of tooth attachment, whereas tooth attachment distinguishes between teeth that are fused to the jaw and those that are suspended in place by a ligament.

Source: Images by the author.

is a subjective and continuous variable—a tooth may be shallowly implanted in the jaws, slightly asymmetrically attached, or sit in a shallow, barely noticeable alveolus (Zaher and Rieppel, 1999; Gaengler, 2000; Bertin et al., 2018; Haridy et al., 2018; LeBlanc et al., 2021). For this reason, many authors have devised additional terms to differentiate between these intermediate conditions (e.g., sub-thecodont, acro-proto-thecodont). Because of this subjectivity, the biological significance of tooth implantation is debated for numerous vertebrate groups, but this should not be confused with the second descriptor that is more clear-cut: Tooth attachment.

Tooth attachment is a discrete variable in dental histology and anatomy. Ankylosis (or synostosis sensu Zaher and Rieppel, 1999) occurs when mineralized tissue fuses a tooth to the jaw. A gomphosis (or syndesmosis sensu Zaher and Rieppel, 1999) occurs when a tooth is held in place by a non-mineralized, fibrous connective tissue (Figure 6.1). The attachment sites for the mineralized or non-mineralized attachment tissues can vary significantly, particularly in bony fish (Owen, 1840; Fink, 1981; Shellis, 1982), but this ankylosis/gomphosis dichotomy holds for every vertebrate group. When initially describing a dentition for the purpose of comparative histology, I advocate for a dual naming system of (1) implantation mode, followed by (2) attachment mode (Zaher and Rieppel, 1999; LeBlanc et al., 2017b; Bertin et al., 2018). However, even if tooth attachment is a more objective and discrete variable, significant challenges remain in identifying and defining the tissues that attach a tooth to the jaw (Figure 6.1).

6.3 WHAT DO WE CALL THE ATTACHMENT TISSUES IN NONMAMMALIAN AMNIOTES?

It may seem odd to address vertebrate tooth tissue homology by focusing on amniote teeth, but this is a top-down problem: We have a discrete set of definitions for the attachment tissues in mammals, thanks to dentistry (Tomes, 1882; Nanci, 2013). Periodontal tissue development in mammals is comparatively well understood (Baume, 1956; Ten Cate and Mills, 1972), and the histological properties of its constituent tissues are well-documented. The challenge lies in identifying homologous attachment tissues outside of Mammalia, a feat that is easier to accomplish by studying amniote tooth tissues first rather than trying to build an evolutionary hypothesis from the earliest fossil vertebrate teeth to those of mammals. There has been significant progress in this regard as more histological analyses of non-mammalian amniote tooth attachment tissues are undertaken and Tomes' "bone of attachment" concept is scrutinized. What follows here are developmental and histological definitions for the periodontal tissues that were first described in mammals, which have been amended for use in the study of non-mammalian amniotes, based on more recent comparative histological analyses (Table 6.1).

Cementum is the bone-like tissue that normally coats the tooth roots. It can be subdivided into several types depending on the origins of the collagen fibers and its cellular content. However, for simplicity's sake, cementum can be divided into two principle types. The first is acellular (or primary) cementum, which in thin sections of human teeth is a thin band of clear tissue located along the dentine surface of the tooth root, external to a distinctive layer of root dentine known as the granular layer of Tomes (Bosshardt and Selvig, 1997; Nanci, 2013; Foster, 2017) (Figure 6.2A, B). The granular layer of Tomes (hereafter simply called "globular zone of dentine", or "globular

TABLE 6.1
Amniote Periodontal Tissues, Their Topologies, and Histological Features.

Tissue Type	Topology	Histology	Mineralization Direction	Unique Identifiers
Acellular cementum	Directly attached to the root dentine. In coronal or sagittal section, extends up the root to the cementum–enamel junction. May extend over the enamel in some mammals.	Acellular band in thin section. Usually thinner than cellular cementum. Sharpey's fibers present in some mammalian acellular cementum. Usually absent in nonmammalian amniotes. May contain incremental growth lines.	Centrifugal (away from tooth root)	Forms light, acellular halo around tooth root in thin section. May also show birefringence in cross-polarized light, distinguishing it from surrounding attachment tissues. In fossils, acellular cementum often breaks away from the dentine, leaving an artificial gap (e.g., Figure 6.6D).
Cellular cementum	External to the acellular cementum. Thicker toward the root apex in mammals and crocodylians. May not entirely cover the acellular cementum near the crown in mammals. Usually covers the entire root surface in other amniotes.	Usually thicker than the acellular cementum and contains cell lacunae (cementocytes). Cementocytes may or may not have canaliculi. Often contains incremental growth lines. Matrix similar to woven bone.	Centrifugal (away from tooth root)	Occasionally has different coloration compared to the surrounding alveolar bone in fossilized samples. Contains Sharpey's fibers from the PDL and often contains incremental growth lines, unlike alveolar bone. Usually avascular, but may contain cementeons (osteocementum).
Periodontal ligament	Usually a soft tissue separating the root cementum and the surrounding alveolar bone.	Unmineralized ligament made of groups of thick collagen fiber bundles. Mineralized ligament made of Sharpey's fibers	Not applicable	Sediment- or mineral-filled space between teeth and socket walls in hard tissue samples and fossils. Sharpey's fibers within cementum and alveolar bone are mineralized ends of the ligament.
Alveolar bone	External-most attachment tissue layer. Separated from the cementum of the tooth root by the periodontal space (unless tooth is ankylosed).	Can be thick and highly vascularized, or thin and avascular. Usually woven bone matrix with or without lamellar bone around the vascular spaces (primary osteons). May be remodeled in older teeth in mammals (secondary osteons).	Centripetal (toward tooth root)	Outer boundaries delineated by a reversal line with the bone of the jaw. Each tooth is encircled by a single generation of alveolar bone in transverse sections.

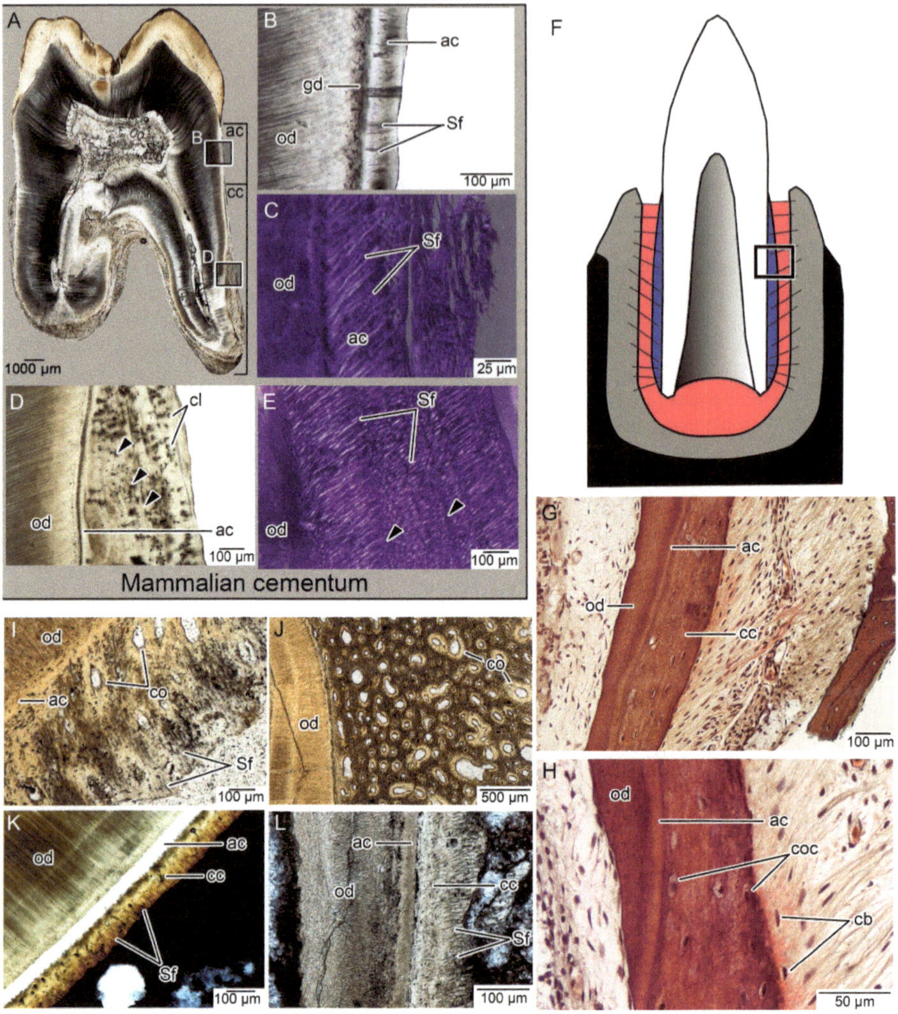

FIGURE 6.2 Cementum types in amniotes. (A) Human tooth section. (B) Closeup of acellular cementum on the human tooth. (C) Cross-polarized light image of acellular cementum on a caribou tooth, showing Sharpey's fibers from the PDL. (D) Closeup of cellular cementum in the human tooth. (E) Cross-polarized light image of cellular cementum in a bear tooth, showing extensive Sharpey's fibers from the PDL. (F) Diagram showing positions of closeup images in G–L. (G) Section of a modern caiman tooth root (H & E staining). (H) Closeup of modern caiman cementum. (I) Osteocementum (vascularized cellular cementum) in a tooth of the dinosaur *Triceratops*. (J) Osteocementum in a mosasaur tooth. (K) Closeup of a tyrannosaurid tooth root. (L) Closeup of the cementum on a tooth of a diadectid (a stem amniote). *Abbreviations*: ac, acellular cementum; cb, cementoblast; cc, cellular cementum; cl, cementocyte lacuna; co, cementeon; coc, cementocyte; gd, globular dentine, od, orthodentine; Sf, Sharpey's fibers. Black arroweads highlight incremental growth lines in cellular cementum.

dentine") is a region of relatively poorly mineralized dentine adjacent to the cementum (Figure 6.2B). Topologically, acellular cementum is the mineralized root tissue located external to the globular zone of dentine, and in humans this can either cover the cervical and mid-portions of the tooth root, or the entire root surface, depending on the tooth position (Bosshardt and Selvig, 1997; Nanci, 2013).

Under a light microscope, the acellular cementum in humans and some other mammals contains incremental growth lines extending parallel to the outer dentine surface and perpendicular Sharpey's fibers (Figure 6.2B, C). These highlight the two most important features of acellular cementum: It grows through apposition of new cementum centrifugally, and it contains the partially mineralized ends of the PDL, making acellular cementum crucial for tooth attachment in many mammals (Bosshardt and Selvig, 1997; Diekwisch, 2001; Nanci, 2013). The cells that produce this tissue are the cementoblasts: Osteoblast-like cells that originate from the dental follicle of the developing tooth (Bosshardt, 2005; Nanci, 2013). In this case, acellular cementum does not contain cell lacunae, indicating that cementoblasts do not become entombed in the matrix that they produce.

Most of our understanding of the development, histology, and function of acellular cementum comes from the study of humans and a select few other mammalian models (Bosshardt and Selvig, 1997; Diekwisch, 2001; Bosshardt, 2005; Nanci, 2013). However, caution should be taken when interpreting the function of acellular cementum in other species. In many instances, even within Mammalia, but especially in reptiles (Berkovitz and Sloan, 1979; McIntosh et al., 2002; LeBlanc et al., 2017b, 2021) and non-mammalian synapsids (LeBlanc et al., 2018), acellular cementum is overlain by the second major category of root-coating hard tissue: The cellular cementum. When this occurs, acellular cementum manifests as a thin, featureless band of mineralized tissue external to a globular zone of root dentine (Figure 6.2). In this form, it often does not contain incremental growth lines and may not even contain Sharpey's fibers. Its function is unclear, but its presence in stem (LeBlanc and Reisz, 2013) and crown amniotes (McIntosh et al., 2002; Caldwell et al., 2003; Luan et al., 2009; Maxwell et al., 2011a; Pretto et al., 2014; LeBlanc and Reisz, 2015; LeBlanc et al., 2017b, 2018, 2021; Mestriner et al., 2021; Spiekman and Klein, 2021) suggests that acellular cementum is an ancestral feature of the amniote tooth attachment system. Based on personal observations, some amniote teeth may lack cellular cementum, but acellular cementum is always present.

Cellular (or secondary) cementum, as the name implies, differs from acellular varieties in the presence of cementocytes (entombed cementoblasts). Cellular cementum is often restricted to the apical third of the tooth roots in many mammals (Nanci, 2013), but it can also be very thick, coating nearly the entire tooth root in other mammals (Figure 6.2D–L), and even coating the enamel in many hypsodont ungulates (Jones and Boyde, 1974; see Section 6.8). In humans, cellular cementum usually lacks the dense Sharpey's fiber networks that are present in acellular cementum, and its restriction to the apical third of the tooth roots suggests that cellular cementum plays an adaptive or reparative role, with only a minor contribution to tooth attachment (Bosshardt and Selvig, 1997; Nanci, 2013). Cellular cementum increases in thickness with age, leaving numerous incremental growth lines and varying numbers of cementocyte lacunae within the growth line intervals (Bosshardt and Selvig, 1997). In humans, cellular cementum contains no vascular spaces, and it is not remodeled (Figure 6.2A, D).

Cellular cementum in several other mammals and reptiles is more variable. For example, cellular cementum in some mammal species does indeed serve a major role in tooth attachment, based on the presence of extensive Sharpey's fiber networks within the cellular cementum (Figure 6.2E). In crocodylians, cellular cementum is the dominant mineralized tissue coating the tooth root (Figure 6.2G, H) (Kvam, 1960; Berkovitz and Sloan, 1979; McIntosh et al., 2002). It completely covers the thinner band of acellular cementum, and it exclusively contains the Sharpey's fibers from the PDL. Cellular cementum in other stem and crown-group amniotes follows a similar pattern. Dinosaurs, toothed birds, most non-mammalian synapsids, marine reptiles, modern squamates, and even diadectids (stem amniotes) have thick, Sharpey's fiber-rich layers of cellular cementum that coat their entire tooth roots (Caldwell et al., 2003; Maxwell et al., 2011a; LeBlanc and Reisz, 2013; Dumont et al., 2016; García and Zurriaguz, 2016; LeBlanc et al., 2018, 2021) (Figure 6.2I–L).

Cellular cementum in non-mammalian amniotes can also be a vascularized tissue. In hadrosaurid and ceratopsid dinosaurs, for example, the cellular cementum can contain abundant vascular spaces and a more haphazard arrangement of cementocyte lacunae, suggesting rapid mineralization of the cementum matrix outwards from the tooth surface (LeBlanc et al., 2016a, 2017b; Bramble et al., 2017). Mosasaurid and ichthyosaur cementum is also notoriously thick along the tooth roots (Owen, 1840; Caldwell et al., 2003; Luan et al., 2009; Maxwell et al., 2011a, 2011b). In these groups, the cementum contains abundant vascular spaces that are mostly oriented parallel to the long axis of the tooth (Figure 6.2I, J). This spongy tissue has been called "bone of attachment" by others (Zaher and Rieppel, 1999; Rieppel and Kearney, 2005); however, its centrifugal growth direction from the dentine of developing mosasaur teeth (Caldwell, 2007), coupled with several other characteristics show that it is true cementum. The vascular spaces in mosasaur cementum are fringed by an acellular halo of mineralized tissue in thin section, creating what Caldwell et al. (2003) termed cementeons. Cementeons are similar to primary osteons in bone, which are vascular spaces fringed by lamellar bone matrix, but in the case of cementeons, cementocyte lacunae are conspicuously absent (Figure 6.2J). This led Caldwell et al. (2003) to define cellular cementum containing these cementeons as osteocementum.

In most amniotes, cementum contains Sharpey's fibers oriented toward the dentine of the tooth root, indicating its primary role in tooth attachment (Figure 6.2I, K, L). In mosasaurs, these are so abundant, that the bulk of the hard tissue component of the cementum is actually composed of Sharpey's fibers from a highly mineralized periodontal ligament (Luan et al., 2009; LeBlanc et al., 2017a).

The periodontal ligament (PDL) is the specialized ligament that connects a tooth root to the socket (Figure 6.3). The PDL plays a pivotal role not only in anchoring a tooth to the jaw, but also in mechanoreception, orthodontic tooth movement, and cushioning during occlusion (Beertsen et al., 1997). Because of its numerous roles in supporting the tooth, the periodontal ligament contains a number of cell types, including fibroblasts, undifferentiated mesenchymal cells, osteoclasts, macrophages, and nerve axons (Berkovitz and Shore, 1982; Hannam, 1982). The PDL also houses the cementoblasts that produce cementum at one end of the ligament, and osteoblasts, which produce alveolar bone at the other (Berkovitz and Shore, 1982). In thin section, the most conspicuous feature of the PDL is a mass of collagen fiber bundles and fibroblasts that span the space between the

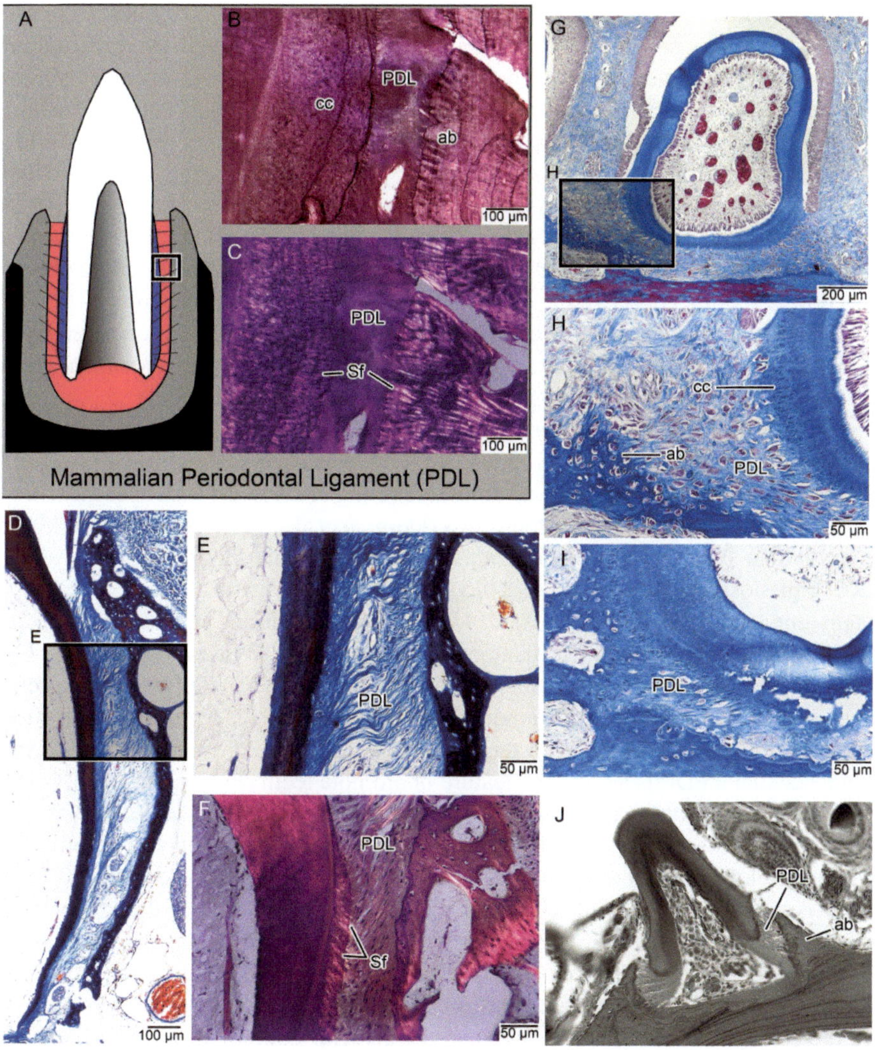

FIGURE 6.3 The Periodontal Ligament (PDL) in extant amniotes. (A) Diagram highlighting positions of closeup images of PDL in B and C. (B) The PDL between the root and alveolar bone of a bear (H & E stain). (C) Same image under cross-polarized light, highlighting Sharpey's fibers of the PDL. (D) Image of the PDL along the tooth root of a caiman (Toluene blue staining). (E) Closeup of the PDL fiber bundles in D. (F) Cross-polarized image of the caiman PDL, highlighting Sharpey's fibers from the PDL in the cementum. (G) Horizontal section through a tooth root of a modern *Iguana* (Masson's trichrome staining). (H) Closeup of the PDL in G, showing the disorganized structure of the PDL prior to ankylosis. (I) Closeup of a slightly older tooth with a more organized PDL prior to complete ankylosis. (J) Palatal tooth (image flipped for comparisons) of the hinge-toothed snake *Xenopeltis*, which has an unmineralized PDL (image courtesy of L. Mahler). *Abbreviations*: ab, alveolar bone; cc, cellular cementum, PDL, periodontal ligament; Sf, Sharpey's fibers.

Source: All other images by the author.

tooth root and the alveolar bone (Figure 6.3). These fiber bundles are typically organized into discrete groups, based on their orientations relative to the tooth root. In mammals, the ligament fiber bundle groups have a complex, overlapping arrangement of mostly oblique and horizontal bundles, which probably resists axial and rotational loads on a tooth (Sloan, 1982). Where the fiber bundles of the PDL meet the cementum and alveolar bone, they are partially mineralized, forming Sharpey's fibers that are embedded within the cementum and alveolar bone matrices (Figures 6.2–6.4).

Crocodylians are the only non-mammalian amniotes for which the PDL has been characterized in any detail. The collagen fiber bundles are mostly oriented obliquely in coronal section (Berkovitz and Sloan, 1979) (Figure 6.3D–F), but may differ in the variation in diameters of the collagen fiber bundles compared to those of mammals. Beyond this, the PDLs of a crocodylian and a mammal are nearly identical (Kvam, 1960; Berkovitz and Sloan, 1979; McIntosh et al., 2002) (Figure 6.3).

It is sometimes erroneously stated that crocodylians are the only non-mammalian amniotes with a PDL. On the contrary, the PDL was clearly present in a growing list of fossil amniote groups, including dinosaurs (Fong et al., 2016; García and Zurriaguz, 2016; LeBlanc et al., 2017b), toothed birds (Dumont et al., 2016; Wang et al., 2023), non-mammalian synapsids (LeBlanc et al., 2018; Whitney and Sidor, 2019; Olroyd et al., 2021), plesiosaurs (Sassoon et al., 2015), ichthyosaurs (Maxwell et al., 2011a, 2011b), stem amniotes (LeBlanc and Reisz, 2013), and even lizards and snakes (Caldwell et al., 2003; Budney et al., 2006; LeBlanc et al., 2017a, 2021). Whereas it is impossible to directly observe the PDL in fossils, the presence of (1) Sharpey's fibers in the cementum of the tooth roots and surrounding alveolar bone and (2) a mineral- or sediment-filled periodontal space between the tooth roots and alveolar margins provide ample evidence that a PDL is a common feature of amniote teeth (Figures 6.2–6.4).

The PDL in extant squamates is often a transient tissue. In *Iguana*, the PDL is a disorganized mass of thick collagen fiber bundles that seem to anchor erupting teeth to the developing alveolar bone along the jaw (Figure 6.3) (LeBlanc et al., 2021). The PDL is quickly mineralized within the growing cementum and alveolar bone layers, and each tooth is eventually completely ankylosed to the jaw (Figure 6.3G–I). A similar condition occurs in snakes and mosasaurs, but here the PDL appears to be more organized, based on the presence of oriented networks of Sharpey's fibers preserved in the osteocementum of mosasaurs and the alveolar bone of snakes (LeBlanc et al., 2017a). In some cases, however, the PDL remains unmineralized. Some shell-crushing mosasaurs retained a soft ligament, because their teeth readily fell out of the jaws post-mortem (Polcyn et al., 2010; LeBlanc et al., 2017a). On the other hand, many modern and fossil snakes have hinged teeth, which are connected to the alveolar bone by a soft PDL (Figure 6.3J) (Savitsky, 1981; Budney et al., 2006; LeBlanc et al., 2017a).

Alveolar bone is the bone that forms the alveolus. This bone tissue is not part of the jaw, but a part of the tooth developmental module. Osteoblasts derived from the dental follicle of a developing tooth form discrete layers of cellular bone along the jaw to which the fiber bundles of the PDL are attached (Ten Cate and Mills, 1972; Ten Cate, 1997). In mammals, the alveolar bone consists of an inner layer of bundle bone, a layer of woven-fibered bone that is perforated by Sharpey's fibers of the PDL (Nanci, 2013). This internal layer may be flanked externally by layers of trabecular bone or dense and highly remodeled bone (Figure 6.4A–C). The histological properties of alveolar bone also change depending on

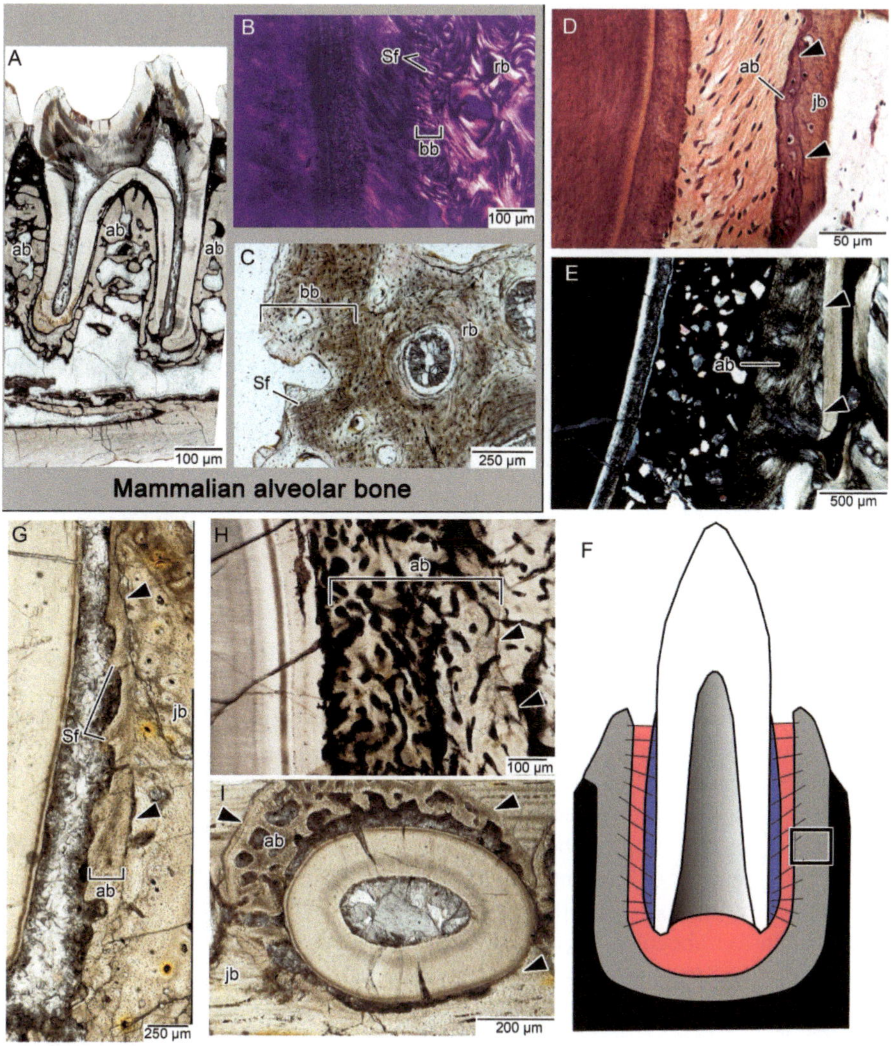

FIGURE 6.4 Alveolar bone in extant and extinct amniotes. (A) Section of a fossil mammal tooth (*Hyopsodus*) showing the arrangement of alveolar bone around the tooth. (B) Cross-polarized light image of alveolar bone in a modern bear tooth, highlighting fibrous bundle bone and more external, remodeled layer. (C) Alveolar bone around a fossil horse tooth, showing fibrous bundle bone layer and remodeled layer. (D) Alveolar bone in a modern caiman (H & E staining). (E) Cross-polarized light image of alveolar bone in a tyrannosaurid dinosaur. (F) An illustration showing the positions of closeup images. (G) Tooth root and socket of a therocephalian tooth, showing thin layer of alveolar bone. (H) Thick, highly vascularized alveolar bone of a dinocephalian tooth. (I) Alveolar bone around a therocephalian tooth. Note the differences in thickness of the alveolar bone, which is due to tooth migration in life (toward the bottom right of the image). *Abbreviations*: ab, alveolar bone; bb, bundle bone layer; jb, bone of the jaw; rb, remodeled bone layer; Sf, Sharpey's fibers. Black arrowheads indicate position of reversal line, which marks the alveolar bone–jawbone boundary.

orthodontic movement (Nanci, 2013). For example, if a tooth is gradually drifting through a jaw, the trailing edge will often consist of highly trabecular bone, whereas the leading edge around the tooth will be resorptive due to the action of osteoclasts that create a resorptive front ahead of the drifting tooth (Saffar et al., 1997) (Figure 6.4I). Alveolar bone is therefore a dynamic bone tissue that supports the tooth and defines the boundaries of the tooth socket. In cases where there is limited or no orthodontic movement, secondary remodeling of older generations of alveolar bone is common (Figure 6.4B, C). Therefore, there are no specific histological properties of the mineralized bone itself that would differentiate it from the bone of the jaw in some cases, especially if it is heavily remodeled and the reversal line between alveolar and jawbone is obscured.

Alveolar bone has similar histological properties in non-mammalian amniotes and is more easily identifiable in polyphyodont species due to a lack of bone remodeling. Alveolar bone in crocodylians was thought to be unique in that it was not resorbed and reformed during tooth replacement (Berkovitz and Sloan, 1979; Shellis, 1982), but an ontogenetic examination of modern and fossil crocodylians suggests that they also reform alveolar bone layers after successive replacement events (LeBlanc et al., 2017b) (Figure 6.4D). Dinosaurs and non-mammalian synapsids make excellent reference points for fossil alveolar bone histology. Because these animals continually replaced their teeth, each tooth developed a new layer of alveolar bone for each tooth generation, leaving a distinct reversal line along its outer fringes, which marks the farthest extent of resorption during tooth replacement, followed by deposition of a new layer of alveolar bone. All the bone internal to this reversal line is alveolar bone that then developed centripetally toward the tooth root (Figure 6.4E–I). In the process, alveolar bone encroaches on the PDL space and forms more extensive networks of Sharpey's fibers within its "bundle bone" layer. Alveolar bone can therefore be identified in fossils by three characteristics: (1) Its centripetal growth direction, (2) the reversal line between the bone of the jaw and the outer region of alveolar bone, and (3) the presence of Sharpey's fibers from the PDL (the bundle bone). Similar to mammals, alveolar bone in other amniotes is histologically highly variable. It can be avascular to trabecular, thick or thin, and the cell lacunae may be small and well-organized or large and randomly distributed depending on the degree of movement of the tooth itself and the rate of growth of the alveolar bone (Figure 6.4).

6.4 PROBLEMATIC TISSUES AND STRUCTURES

"Bone of attachment" is a misleading term for the mineralized tissue that fuses teeth to the jaws in non-mammalian amniotes. It was defined by Tomes (1874, 1882) as performing a function analogous to the socket of a mammal, but it is supposedly the only periodontal tissue present in ankylosed teeth. Tomes also identified four developmental and histological features of "bone of attachment" that made it distinct from cementum and from the bone of the jaw: (1) The bone is coarser in texture compared to jawbone; (2) it is full of irregular vascular spaces and cell lacunae; (3) it is resorbed along with the tooth during a tooth replacement cycle; and (4) it mineralizes from the surface of the jawbone toward the tooth base. "Bone of attachment" and ankylosis became virtually synonymous and continue to be used today to distinguish tooth attachment tissues in mammals from virtually all other vertebrates (Peyer, 1968; Shellis, 1982; Gaengler and

Metzler, 1992; Smith and Hall, 1993; Huysseune and Sire, 1998; Zaher and Rieppel, 1999; Berkovitz and Shellis, 2017). Some authors have subsequently tried to reconstruct the evolutionary transition from "bone of attachment" in early vertebrates to the cementum, PDL, and alveolar bone in mammals (Noyes, 1902; Smith, 1958; Osborn, 1984) but have never found empirical evidence for such a transition either in the fossil record or in modern reptiles. This has led several authors to question if "bone of attachment" is a real entity from which the mammalian compliment of dental tissues evolved (Hughes et al., 1994; Caldwell et al., 2003; Budney et al., 2006; Maxwell et al., 2011b; LeBlanc and Reisz, 2013; Pretto et al., 2014; LeBlanc et al., 2017a, 2021; Rosa et al., 2021).

Tomes' "bone of attachment" paradigm challenged Richard Owen's (1840) hypothesis of homology of dental tissues across toothed vertebrates. Where Owen envisioned all vertebrate teeth as consisting of enamel, dentine, and cementum, Tomes argued that the term cementum should be reserved for teeth that were suspended in a "true" tooth socket by a ligament (Tomes, 1874). Tomes was especially intrigued by the process of ankylosis in reptiles, amphibians, and fish where each tooth is surrounded by a distinct layer of bone that fuses it in place. He disagreed with Owen and recognized that this layer of bone developed very differently from the cementum that coats mammalian tooth roots.

History is important to the story of "bone of attachment", because even though Tomes was correct in distinguishing the bone-like attachment tissues in snakes, amphibians, and fish from cementum or jawbone, his comparisons were based on the science of his day: At the time, mammal teeth were thought to be attached to the jaws via the tooth cementum and a fibrous membrane (later called the periodontal ligament) (Tomes, 1882). While it was known at the time that a mammalian tooth also develops its own socket to which it becomes attached (Tomes, 1874), it was not until later that alveolar bone was described as a distinct periodontal tissue in mammals that developed in association with each tooth (Baume, 1956; Ten Cate and Mills, 1972; Shellis, 1982). As per Tomes' original description, "bone of attachment" is therefore consistent with alveolar bone, which is deposited on top of the jawbone around the developing tooth, and it forms the other end of the attachment of the PDL fibers (Baume, 1956) (Figure 6.4). This is even the case in fully ankylosed teeth in amniotes (see Section 6.5). The largest challenge to the legitimacy of "bone of attachment", however, is the fact that snakes—the original namesakes for "bone of attachment"—can have well-defined alveolar bone and a PDL (Figure 6.3J). This is untenable under the original definition for Tomes' standalone attachment tissue, because snakes should only have one attachment tissue. "Bone of attachment" therefore conflates tooth attachment tissues and represents the largest conceptual roadblock to our understanding of tooth attachment tissue evolution in vertebrates (Caldwell et al., 2003; LeBlanc et al., 2017a, 2021). I strongly suggest avoiding the use of this term, because it obscures any evolutionary link between it and the canonical tooth attachment tissue definitions from mammals and crocodilians.

"Interdental ridges/plates/septa" are standalone ridges or plates of bone along the jaws of many nonmammalian amniotes. These are found either in between adjacent tooth positions or along the lingual surface of a tooth row (Miller, 1968; Zaher and Rieppel, 1999; Caldwell et al., 2003; Budney et al., 2006; Luan et al., 2009; LeBlanc et al., 2017b). These bony plates often have a distinct texture compared to the bone of the jaw and the alveolar bone of the socket (Figure 6.5). Despite considerable debate

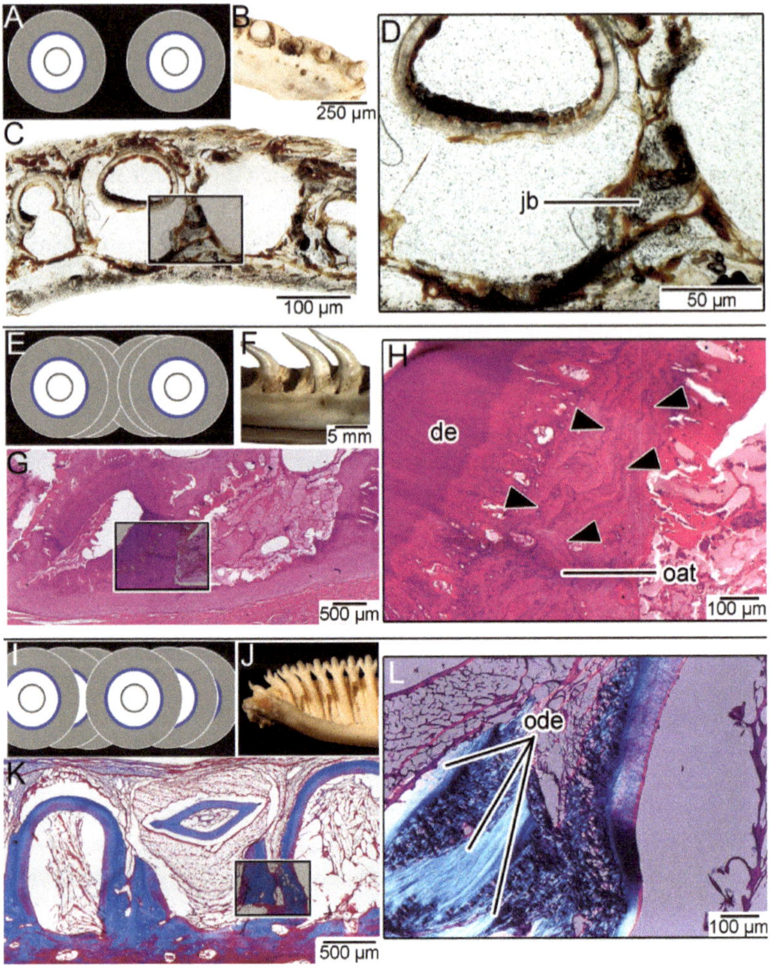

FIGURE 6.5 Comparative histology of interdental structures in modern reptiles. (A) Condition 1: Interdental structure is made of jawbone. (B) Closeup image of a young *Alligator* dentary, showing interdental bone (image courtesy of Diane Scott). (C) Horizontal section of anterior dentary of a young *Alligator*. (D) Closeup of jawbone in between two dentary teeth in *Alligator*. (E) Condition 2: Interdental structure is made of old attachment tissues from two tooth positions. (F) Closeup of a *Python* dentary, showing interdental ridge. (G) Horizontal section through a *Boa* maxilla with well-developed interdental ridge (H & E staining). (H) Closeup of the interdental ridge, showing accumulated layers of attachment tissues from the neighboring tooth positions. (I) Condition 3: Interdental structure is made of dental tissues from a single tooth position. (J) Closeup of a dentary of the marine iguana *Amblyrhynchus* (image courtesy of I. Paparella). (K) Horizontal section through a dentary of *Iguana* (Masson's trichrome staining). (L) Polarized light image of an interdental structure in *Iguana*, highlighting old fragments of dentine from previous tooth generations from the tooth position to the left. *Abbreviations*: jb, jawbone; oat, old attachment tissues; ode, old dentine fragments. Black arrowheads highlight reversal lines.

Source: All other images by the author.

regarding the nature of interdental bone and its homology with the three "mammalian" tooth attachment tissues (e.g., Zaher and Rieppel, 1999; Caldwell et al., 2003; Luan et al., 2009), they are often not proper periodontal tissues. Moreover, the presence of interdental partitions can be ontogenetically variable, particularly in crocodylians and squamates (Miller, 1968; LeBlanc et al., 2021), and their histological composition changes through ontogeny.

For example, in young crocodylians, these structures may be formed of jawbone initially along the anterior region of the dentition (Figure 6.5A–D) and alveolar bone in posterior regions of the jaws (Miller, 1968). In older individuals, these structures are heavily remodeled and consist of old layers of alveolar bone. In other amniotes, these structures often consist of layer-caked arrangements of alveolar bone and even old tooth fragments (LeBlanc and Reisz, 2013; Fong et al., 2016; LeBlanc et al., 2017a, 2018, 2021; Mestriner et al., 2021). Interdental partitions are most commonly the result of the combination of tooth drift and tooth replacement. As the jaw grows in length and the teeth are replaced, old teeth and their alveolar bone layers are only partially resorbed, leaving trails of old tooth generations preserved in between tooth positions (LeBlanc et al., 2017b, 2021). If neighboring tooth positions slowly drift apart as the jaw grows, the interdental region will consist of successive layers of old attachment tissues, all separated by reversal lines, from both tooth positions (Figure 6.5E–H). If the direction of tooth drift is uniform, then the plates of "bone" in between will consist of old dental tissues from one tooth position only (Figure 6.5I–L).

6.5 ANKYLOSIS, GOMPHOSIS, AND THE VARIABLY MINERALIZED PDL: LESSONS FROM SYNAPSIDS AND ARCHOSAURS

Armed with these criteria for identifying cementum, PDL, and alveolar bone in non-mammalian amniotes (Table 6.1), it is possible to reveal major patterns in periodontal tissue evolution and overcome the conceptual "bone of attachment" barrier. Superficially, ankylosed teeth appear to be fused in place by a single tissue (Figure 6.1). This was the key observation Tomes had made to create the concept of "bone of attachment" and was fundamental to the idea that mammals have a more complex periodontium (Peyer, 1968; Osborn, 1984). Indeed, without histological data, it can be almost impossible to determine how each tooth has been fused to the jaw. However, detailed histological studies of Paleozoic and Mesozoic amniotes are changing our perceptions of the nonmammalian tooth attachment tissues and are revealing the hidden complexity of the periodontium, even in ankylosed teeth. Two distantly related amniote groups illustrate these shifting perspectives, thanks in large part to an increasing amount of histological data for representative fossil taxa. On the one hand are the stem mammals: Carboniferous to Triassic "pelycosaurs" and therapsids, which encompass the evolutionary origins of the Synapsida to the origin of mammals (Hopson, 1969; Sidor and Hopson, 1998; Kemp, 2006); and on the other hand are the silesaurids, dinosaurs, and crocodylians, collectively forming a significant portion of the Archosauria (Brusatte et al., 2010).

Synapsida are a large grouping of amniotes, which includes all mammals, and they have an over 300-million-year evolutionary history (Reisz, 1997). "Pelycosaurs" are a paraphyletic grade of the earliest synapsids that originated in the Carboniferous

Period (Hopson, 1969) and include the iconic sail-backed predator *Dimetrodon*, among many others. The phylogenetic relationships of the various "pelycosaur" lineages are under continued revision and debate (Laurin and Reisz, 1995; Benson, 2012; Brocklehurst et al., 2016; Ford and Benson, 2020), but the general pattern of tooth attachment within the earliest synapsid groups is remarkably consistent. All "pelycosaurs", regardless of their presumed ecological roles as carnivores, omnivores, or herbivores, have ankylosed teeth (Edmund, 1960; Peyer, 1968; Osborn, 1984; LeBlanc et al., 2018). However, when viewed in thin section, these ankylosed teeth are surrounded by a complex—albeit fully mineralized—suite of attachment tissues (Figure 6.6A, B).

Each tooth root is coated in thin bands of acellular tissue, which is consistent with acellular cementum (Figure 6.6A, B). This acellular band is often coated in turn by concentric layers of cellular bone-like tissue with occasional growth lines. This tissue is consistent with cellular cementum. Directly adhered to this is the surrounding mass of well-vascularized bone that is often called "bone of attachment" (Peyer, 1968; Osborn, 1984). The most striking feature of these teeth, however, is the abundance of Sharpey's fibers radiating around the tooth roots along the junctions between the surrounding bone and the cementum. These Sharpey's fibers are in the middle of the mineralized periodontal tissues, and they can only be interpreted as the anchoring points for a ligament that was completely encased in the surrounding bone and cementum (Figure 6.6A). LeBlanc et al. (2016b, 2018) interpreted these Sharpey's fibers as the mineralized remnants of the PDL in "pelycosaurs", which must have first suspended the tooth in its socket before becoming completely entombed in the surrounding bone—which was re-interpreted as true alveolar bone. This process of complete mineralization of the PDL was reported before, but in unrelated groups: The giant marine mosasaurid lizards of the Cretaceous period (Caldwell et al., 2003; Luan et al., 2009) and Permo-Carboniferous diadectids (LeBlanc and Reisz, 2013).

"Pelycosaur" tooth histology provides compelling evidence for the presence of cementum and a completely mineralized PDL already in the Carboniferous and early Permian, but the later-evolving therapsids reveal the process of dental ankylosis in much more detail. By cladistic definition, Therapsida includes mammals (Sidor and Hopson, 1998), but, for the sake of this summary, I am referring mainly to the diverse but extinct nonmammalian lineages, including dinocephalians, anomodonts, therocephalians, gorgonopsians, and cynodonts (which include mammals). Osborn (1984) had hypothesized that at some point in the therapsid-to-mammal transition that the periodontium had acquired its mammal-like state of three periodontal tissues from the ankylosed state of earlier synapsids. Osborn even predicted an intermediate condition, whereby "bone of attachment" might fuse one part of a tooth root to the jaw, and at other points along the same tooth, one would find cementum, a PDL, and alveolar bone (Osborn, 1984: figure 5c). The histological data paint a different picture.

As with the earlier "pelycosaurs", many therapsids have ankylosed teeth, and they show evidence of cementum and a mineralized PDL (Sharpey's fibers around the tooth roots) (LeBlanc et al., 2016b, 2018). However, many therapsid teeth are preserved in intermediate stages of periodontal tissue development, presumably

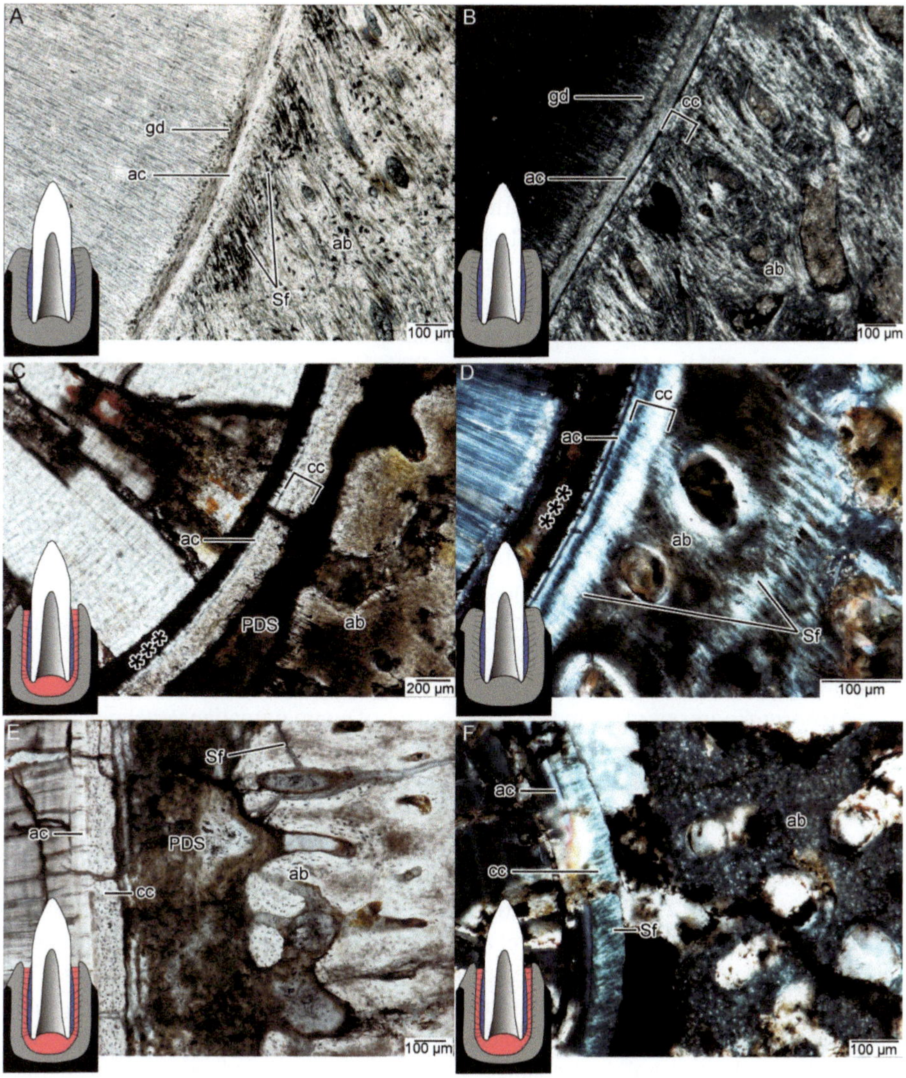

FIGURE 6.6 Periodontal tissues in fossil synapsids. (A) Tooth root section of the "pelycosaur" *Dimetrodon* showing not only ankylosis, but also the presence of Sharpey's fibers from a completely mineralized PDL. (B) Cross-polarized image of the same tooth, highlighting fibrous texture of the alveolar bone and the cementum. (C) Tooth root section of a therocephalian showing gomphosis. (D) Ankylosed tooth in the same therocephalian specimen, showing that both conditions occurred in the same individual. (E) Tooth root section of the cynodont *Cynognathus* showing a gomphosis. (F) Cross-polarized light image of a *Cynognathus* tooth root, highlighting Sharpey's fibers within the root cementum. *Abbreviations*: ab, alveolar bone; ac, acellular cementum; cc, cellular cementum; gd, globular dentine; PDS, periodontal ligament space Sf, Sharpey's fibers. Asterisks indicate taphonomic splitting of acellular cementum from the dentine, which creates an artificial space.

Source: Images by the author.

because the process of ankylosis occurred more slowly in these groups (LeBlanc et al., 2018). One of these intermediate stages is a full gomphosis, where a tooth root is coated in layers of cementum, which incorporated Sharpey's fibers from a PDL, a sediment- or mineral-filled gap between the tooth and the socket (formerly occupied by the soft parts of the PDL), and a discrete layer of alveolar bone surrounding the entire periodontium (Figure 6.6C). This stage progressed in many therapsids to full ankylosis, where the only traces of the PDL are the Sharpey's fibers now completely encased in alveolar bone and, to a lesser degree, in cementum (Figure 6.6D).

Therapsid dental histology has revealed a further insight into the evolution of the mammalian periodontium: Having a gomphosis was not a uniquely mammalian feature, even among therapsids (LeBlanc et al., 2018; Whitney and Sidor, 2019; Olroyd et al., 2021). Many taxa retain the PDL in its earlier developmental state as an unmineralized ligament that would have bridged the gap between the tooth root and the alveolar bone (Figure 6.6E, F). Based on the taxonomic and histological sampling available to date, it appears that this retention of a permanent gomphosis is present in several therapsid groups, including some species of dinocephalians, gorgonopsians, therocephalians, anomodonts, non-mammalian cynodonts, and all mammals (LeBlanc et al., 2018; Whitney and Sidor, 2019; Olroyd et al., 2021). This means that the mammalian gomphosis, this hallmark of dental tissue complexity, is not a product of the evolutionary accumulation of new tissues, but of heterochrony—in this case, the evolutionary truncation of periodontal tissue development and the retention of a non-mineralized ligament.

Archosauria are one of the most studied groups of nonmammalian vertebrates in terms of dental histology (Owen, 1840; Tomes, 1882; Kvam, 1960; Miller, 1968; Berkovitz and Sloan, 1979; McIntosh et al., 2002; Luan et al., 2009; Berkovitz and Shellis, 2017). Part of this is not only because modern crocodylians have the same tooth attachment mode as mammals, but also because of the immense academic interest in dental histology of extinct dinosaurs and toothed birds (e.g., Owen, 1840; Edmund, 1960; Erickson, 1996; Reid, 1996; Sander, 1999; Hwang, 2011; Erickson et al., 2012, 2015, 2017; Brink et al., 2015; Schwarz et al., 2015; Dumont et al., 2016; Fong et al., 2016; García and Zurriaguz, 2016; Bramble et al., 2017; LeBlanc et al., 2017b; Chen et al., 2018; Reisz et al., 2020; Wang et al., 2023). While there are minor differences in the amounts and arrangements of cementum and alveolar bone (McIntosh et al., 2002), these differences pale in comparison to the striking overall resemblance of the crocodylian periodontium to that in mammals (Figures 6.2–6.4), the evolutionary significance of which is almost never discussed. The most recent common ancestor of modern crocodylians and mammals existed over 300 million years ago. Is this similarity in tooth attachment tissue histology convergent or is it evidence of homology between archosaurs and mammals?

Comparing tooth attachment tissue histology between crocodylians and other extinct archosaurs has revealed that cementum, PDL, and alveolar bone are deeply conserved in Archosauria (Figure 6.7). Like crocodylians and mammals, dinosaurs also show clear evidence of the three periodontal tissues (LeBlanc et al., 2017b) (Figure 6.7E, F). From the Triassic theropod *Coelophysis*, to the titanosaurs, hadrosaurids, and tyrannosaurids of the Cretaceous period, all dinosaurs examined to date show evidence of a soft PDL that suspended each tooth within its alveolus. Even toothed birds show the evidence of the three periodontal tissues (Dumont et al., 2016; Wang et al., 2023).

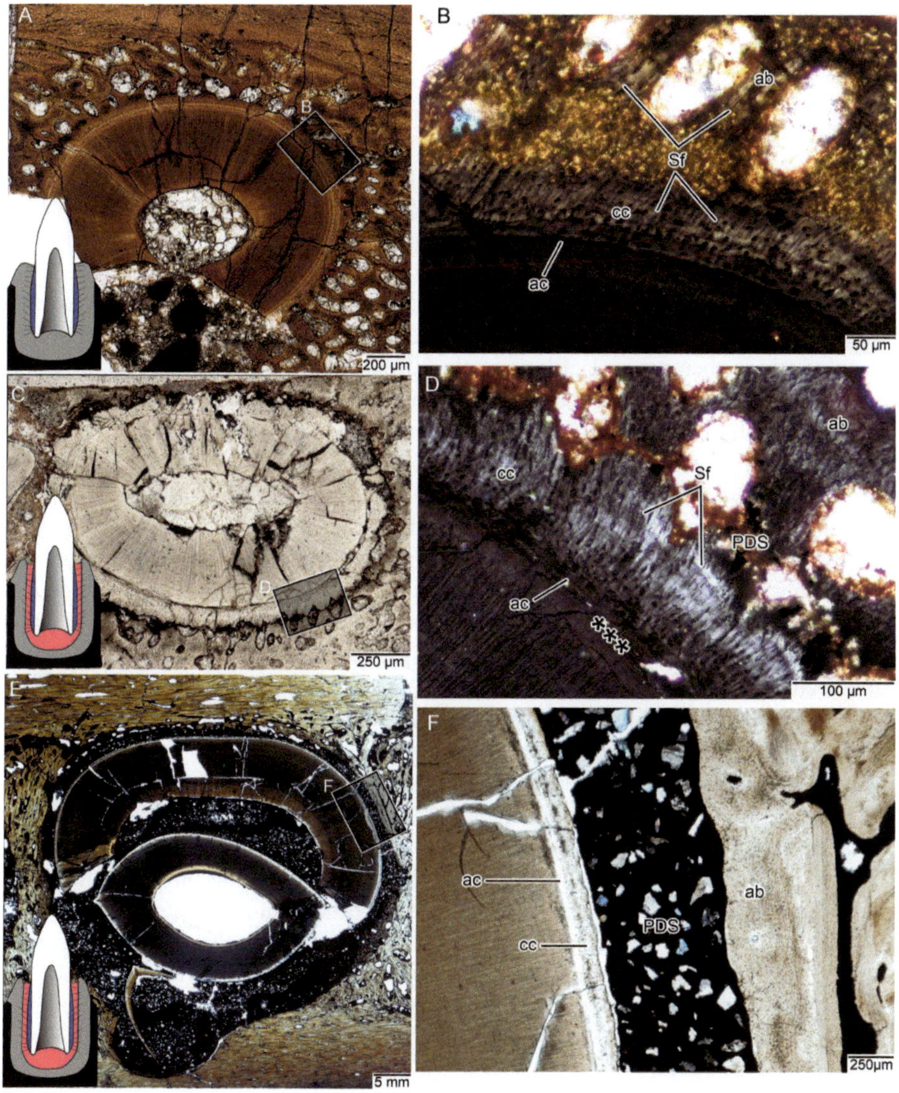

FIGURE 6.7 Periodontal tissue histology in fossil archosaurs. (A) Horizontal section through the ankylosed tooth of the silesaurid *Eucoelophysis* (image courtesy of G. Mestriner). (B) Closeup of tooth attachment tissues in *Eucoelophysis* under cross-polarized light, showing the union between cellular cementum and alveolar bone, as well as Sharpey's fibers from a completely mineralized PDL (image courtesy of G. Mestriner). (C) Horizontal section through an indeterminate silesaurid tooth, showing the gomphosis state (image courtesy of G. Mestriner). (D) Closeup of attachment tissues in C under cross-polarized light, showing the periodontal space and Sharpey's fibers of the PDL (image courtesy of G. Mestriner). (E) Horizontal section of a tyrannosaurid dinosaur tooth root exhibiting a gomphosis. (F) Closeup of the periodontal tissues in E. *Abbreviations*: ab, alveolar bone; ac, acellular cementum; cc, cellular cementum; PDS, periodontal ligament space; Sf, Sharpey's fibers. Asterisks show post-mortem separation of acellular cementum from the dentine.-

The presence of a gomphosis in dinosaurs and crocodylians may not be a shared ancestral feature of these two archosaur groups, however. Mestriner et al. (2021) examined the tooth attachment tissues in several species silesaurids, a key group of Triassic archosaurs that are closely related to dinosaurs. Silesauridae are interpreted either as the closest relatives of the earliest dinosaurs (Nesbitt et al., 2010; Novas et al., 2021), or as early members of the Ornithischia (Muller and Garcia, 2020), which would make them early members of the Dinosauria. These differences in phylogenetic position have a major impact on our understanding of periodontal tissue evolution in dinosaurs, because silesaurids have fully ankylosed teeth (Mestriner et al., 2021).

Mestriner et al. (2021) described teeth at different stages of periodontal tissue development. These sections revealed a developmental sequence that mirrored the process of ankylosis in therapsids (LeBlanc et al., 2018). Each tooth erupted into the mouth in a gomphosis state, where the root was coated in acellular and cellular cementum, a PDL with Sharpey's fibers for anchorage into the surrounding mineralized tissues, and a distinct layer of alveolar bone (Figure 6.7C, D). From there, the surrounding alveolar bone would continue to mineralize the PDL toward the tooth root, eventually meeting the cellular cementum, and fusing the tooth in place (Figure 6.7A, B). Even in fully ankylosed teeth, the Sharpey's fibers of the PDL are still visible despite being completely encased in alveolar bone and cementum (Figure 6.7B). These results indicated that the silesaurid PDL spent enough time in a partially unmineralized state so that some of the teeth could be caught in the gomphosis and mineralization stages in thin sections of partial jaws (Mestriner et al., 2021).

Comparing the timing of periodontal tissue development at equivalent stages between dinosaurs and silesaurids reveals the same pattern as in therapsids and mammals: The dinosaur periodontium is not histologically more complex than that of silesaurids; it is a product of delaying and preventing the onset of ankylosis (Figures 6.7 and 6.8). This truncation of dental ontogeny to produce the permanent gomphosis must have evolved at least twice within archosaurs, depending on the phylogenetic position of silesaurids: It either evolved twice within dinosaurs (Ornithischia and Saurischia), or it evolved once in dinosaurs and again in the crocodile-line archosaurs (Mestriner et al., 2021). It is also possible that this switching between ankylosis and gomphosis evolved even more times, but that will require further histological investigations in other archosaur groups.

6.6 HETEROCHRONY AND AMNIOTE TOOTH ATTACHMENT TISSUE EVOLUTION

Studies of synapsid and archosaur dental histology show that there is no need to invoke the evolution of an increasingly complex periodontium over time in either group. Instead, cementum, PDL, and alveolar bone are homologous and plesiomorphic features of synapsids, archosaurs, and amniotes more generally (Caldwell et al., 2003; LeBlanc and Reisz, 2013). With a recent increase in taxonomic sampling, it seems that these periodontal tissues are present in all major amniote groups (Caldwell et al., 2003; Maxwell et al., 2011b; Pretto et al., 2014; LeBlanc and Reisz, 2015; Snyder et al., 2020; LeBlanc et al., 2021; Spiekman and Klein, 2021) (Figure 6.8A). There is also no need to invoke the existence of "bone of attachment" in fully ankylosed teeth in these groups, given that the three periodontal tissues are present in a completely mineralized state (Figures 6.6 and 6.7). Instead, the rate and extent of mineralization of cementum and alveolar bone dictate the mode of tooth attachment in a given amniote (Figure 6.8B). Observing the intermediate stages in tooth attachment tissue development is key. There are two sites of periodontal tissue mineralization in amniotes: An outer fringe of alveolar bone that interacts with the bone of the jaw and that mineralizes toward the tooth; and cementum, which coats the tooth roots and mineralizes outwards toward the alveolar bone (Figure 6.8B). This occurs at the expense of the PDL, entombing Sharpey's fibers within the matrices of alveolar bone and cementum.

Gomphosis and ankylosis are better considered as two ends of a single, overarching ontogenetic series in amniote teeth (LeBlanc et al., 2016b, 2017a, 2018; Mestriner et al., 2021) The plesiomorphic condition in amniotes is for teeth to pass through a gomphosis stage and to proceed to complete ankylosis, which is often (but not always) accomplished through more extensive alveolar bone growth (Figures 6.6 and 6.7). In some cases, cementum can be the principle mineralized tissue, as in the

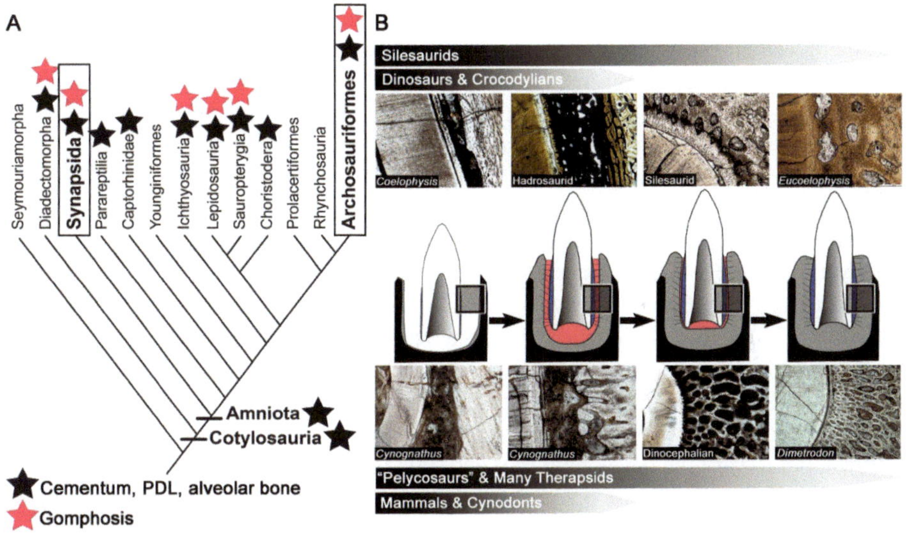

FIGURE 6.8 Shifting perspectives in the prevalence and evolution of "mammal-like" tooth attachment tissues across cotylosaurs (amniotes + diadectomorphs). (A) Reports of cementum, PDL, alveolar bone, and a gomphosis across major stem- and crown-amniote clades [cladogram modified from LeBlanc and Reisz (2013)]. The boxes indicate the groups mentioned in section 6.5. (B) Heterochrony and the evolution of a gomphosis in archosaurs (top images) and synapsids (bottom images). In any given amniote, teeth may pass through a gomphosis stage to complete ankylosis, or dental ontogeny may be truncated, which leads to a retention of teeth in the gomphosis state. Under this hypothesis, mammals, cynodonts, crocodylians, and dinosaurs are paedomorphic relative to their ancestors, which had ankylosed teeth. Images of silesaurid and *Eucoelophysis*, courtesy of G. Mestriner.

Source: All other images by the author.

extensive osteocementum of mosasaurs, ichthyosaurs, and some modern squamates (Caldwell, 2007; Maxwell et al., 2011a; LeBlanc et al., 2021).

In mammals, dinosaurs, crocodylians, and many others, the repeated evolution of a gomphosis is the result of a truncation of this ancestral ontogenetic series, where teeth are retained in an earlier stage of tooth development (Figure 6.8B). These groups are paedomorphic relative to their ancestors that have ankylosed teeth, rather than having independently evolved cementum, PDL, and alveolar bone from an ancestral single-tissue attachment system (contra Tomes, 1874; Osborn, 1984). Based on this increasing taxonomic sampling of tooth attachment tissue histology, it seems that heterochrony has played a major role in the evolution of the diverse forms of tooth attachment that we see across Amniota.

6.7 DEVELOPMENT OF THE PERIODONTAL TISSUES AND HERS, AND THEIR RELATIONSHIP WITH TOOTH IMPLANTATION

The heterochrony model proposed before explains how tooth attachment may vary within an evolutionary lineage; however, this model does not explain how and why variation exists in the geometry of this attachment (tooth implantation). For this, we must look at the development of the periodontal tissues in extant vertebrates. Periodontal tissue formation is well studied in mammalian models (Ten Cate and Mills, 1972; Beertsen et al., 1997; Ten Cate, 1997; Diekwisch, 2001), but it has also received significant attention in crocodylians, because of their histological and developmental similarity to the mammalian condition (Kvam, 1960; Miller, 1968; Berkovitz and Sloan, 1979; McIntosh et al., 2002; Luan et al., 2006). In both cases, the formation of the periodontium begins at an early stage in dental development. Given that the periodontal tissues are all most likely derived from the follicle (Figure 6.9A), all of them are ectomesenchymal (mesenchyme and migratory neural crest cells) in origin, similar to odontoblasts and their secretory product: The dentine. The dental follicle is a mass of these ectomesenchymal cells, which surrounds the developing tooth bud. In mammals and crocodylians, tooth root development begins by the elongation of the epithelium that coats the outer surface of the tooth. This epithelium is a rootward extension of the same epithelial tissue that shapes the tooth crown, but toward the developing root end, this tissue collapses into a structure known as the cervical loop (Figure 6.9A). As the cells of the cervical loop proliferate, the outline of the tooth root elongates; however, the cells of the surrounding dental follicle are unable to contact this root surface, because of an intervening wall of epithelium, known as Hertwig's epithelial root sheath (Hers) (Figure 6.9A). Only after Hers disintegrates into smaller clusters of epithelial cells, known as the epithelial rests of Malassez, can the surrounding dental follicle cells contact the tooth root and begin forming cementum and the periodontal ligament (Figure 6.9A, B).

Periodontal tissue development in crocodylians and mammals is remarkably similar, but outside of these two groups, different arrangements of Hers and the dental follicle can occur. In many reptiles, amphibians, and fish, teeth follow the same early developmental sequence as mentioned previously; however, Hers remains intact (Luan et al., 2006; Zahradnicek et al., 2012) (Figure 6.9C). When this occurs, Hers acts as a permanent physical barrier to periodontal tissue formation, separating the cells of the dental follicle from the surface of the tooth root (Figure 6.9D). Over

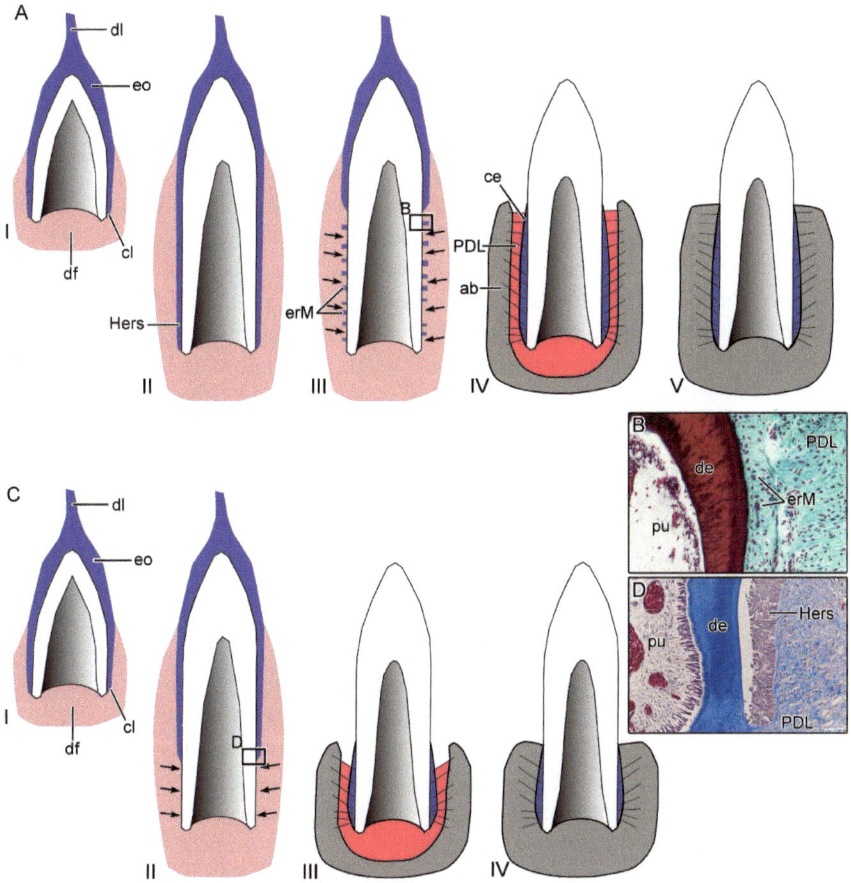

FIGURE 6.9 Periodontal tissue development in crocodylians and mammals, compared with other vertebrates. (A) Stages of periodontal tissue development in mammals and crocodylians, which have a transient Hertwig's epithelial root sheath (Hers). The dental follicle surrounds the basal end of the developing tooth crown, but epithelium (blue) prevents the dental follicle cells (pink) from contacting the surface of the tooth crown (I). Through elongation of the cervical loop, the root of the tooth elongates, but the newly formed Hers still prevents dental follicle cells from contacting the root surface (II). Eventually, Hers dissociates into small clusters of epithelial rests of Malassez (erM), allowing dental follicle cells to contact the root surface and begin forming cementum, PDL, and alveolar bone, which may then go on to completely mineralize, as mentioned above (III–V). (B) Closeup of a section through a tooth root of the crocodylian *Caiman sclerops* showing clusters of erM. (C) Stages of periodontal tissue development in vertebrates with a permanent Hers. Initial tooth development is similar to A (I). As the dentine of the tooth root continues to elongate, Hers remains intact and never dissociates, but the basal end of the tooth is exposed to the surrounding dental follicle (II). This leads to the formation of different topological arrangements of the periodontal tissues (III–IV). (D) Closeup of the permanent Hers along the tooth root of *Iguana iguana*, which prevents the attachment of the surrounding periodontal tissues. *Abbreviations*: ab, alveolar bone; ce, cementum; cl, cervical loop; de, dentine; df, dental follicle; dl, dental lamina; eo, enamel organ; erM, epithelial rests of Malassez; Hers, Hertwig's epithelial root sheath; PDL, periodontal ligament; pu, pulp.

time, some portions of the dentine extend past this epithelial covering (Figure 6.9C). These exposed regions do develop periodontal tissue attachment sites around the tooth, allowing it to form the connection with the jaw. From here, periodontal tissues may completely mineralize and ankylose a tooth to the jaw, or remain partially ligamentous, as is the case for some snakes as well as most cartilaginous fish (Luan et al., 2006, 2009; LeBlanc et al., 2017a).

Regardless, whether Hers remains intact or dissociates into clusters of epithelial cells, I would argue this difference is not indicative of the homology (or non-homology) of the attachment tissues across vertebrate groups. The dental follicle remains the source for the periodontal tissues in these examples, and the relationship of the attachment tissues to the developing tooth root is dictated by the extent of the barrier between the two: Hertwig's epithelial root sheath. The nature and transience of Hers therefore influence the geometry of tooth attachment (i.e., the tooth implantation mode) in a taxon, whereas the degree of mineralization of the periodontal tissues following their initial development defines the tooth attachment mode. These are two distinct processes, with one always preceding the other (Figure 6.9A, C). The extreme example of this occurs in pleurodont squamates, where Hers forms a long epithelial boundary along the lingual surfaces of the teeth thus preventing tooth attachment tissue formation, but a short one on the labial side, allowing teeth to develop extremely asymmetrical forms of implantation (Zahradnicek et al., 2012; LeBlanc et al., 2021).

By comparison, the thecodont implantation in early synapsids and silesaurids is likely the result of the dissociation of Hers into the rests of Malassez, similar to periodontal tissue development in modern mammals and crocodylians. This can be inferred in these fossil groups, because the periodontal tissues connect to all of the surfaces of the tooth below the crown, which would require that the epithelial barrier be broken around the entire tooth root. However, following the formation of these tissues, many synapsid and silesaurid groups possessed teeth that would then completely fuse to the jaw, following the stages mentioned previously (Figures 6.8 and 6.9A). These examples highlight the influence that Hers development and periodontal tissue mineralization have on the diversity of tooth attachment and implantation modes in vertebrates.

6.8 CO-OPTING CEMENTUM: MORE SHIFTS IN DEVELOPMENTAL TIMING TO PRODUCE COMPLEX AND CONTINUALLY ERUPTING TEETH

In most cases, cementum remains a root-bound tissue. However, in some exceptional instances, cementum has been co-opted through different evolutionary processes to extend on to the occlusal surfaces of teeth. These processes highlight the variability in developmental timing of teeth, which have allowed many amniote groups to adopt novel tooth shapes and functions. The most striking of these evolutionary shifts are in the hyposodont (high-crowned) and ever-growing teeth of not only many mammals, but also occasionally in reptiles. Because of its bone-like qualities, cementum is a relatively soft tissue compared to enamel, and its extension onto the occlusal surface is associated with complex grinding surfaces or self-sharpening and continuously erupting teeth (Figure 6.10).

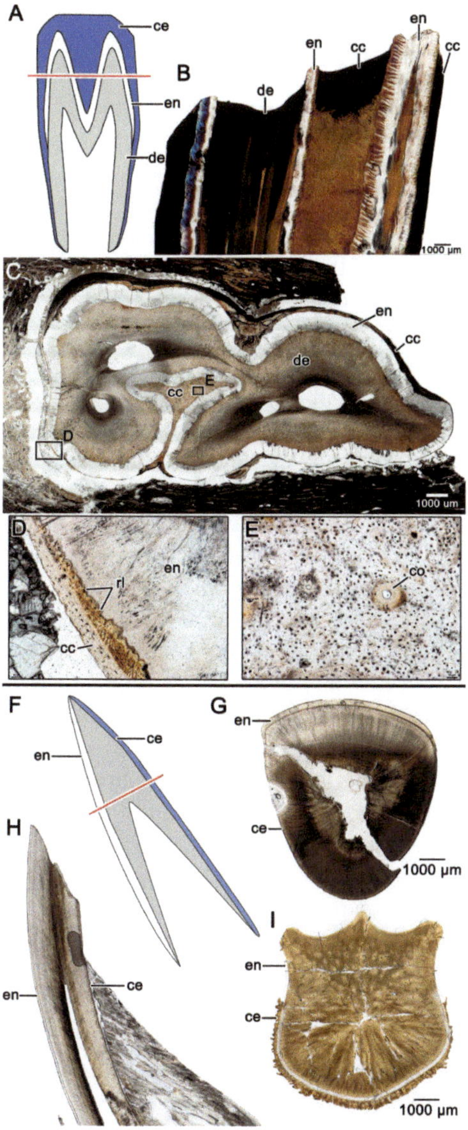

FIGURE 6.10 Examples of coronal cementum and continually erupting teeth. (A) Illustration of the arrangement of the cementum over the tooth crown in an unworn horse tooth. (B) Thin section of a worn tooth of a fossil horse. (C) Transverse section of a horse tooth, showing enamel, dentine, and cementum infoldings. (D) Closeup of coronal cementum in C, which overlaps the enamel and is separated by a scalloped reversal line. (E) Closeup of the vascularized cellular cementum in one of the infolded regions. (F) Illustration of an ever-growing tooth with separate enamel and cementum faces. (G) Transverse section of a beaver incisor, showing enamel and cementum faces (Image courtesy of K. Bramble). (H) Parasagittal section of a rabbit incisor showing separate enamel and cementum-covered regions (Image courtesy of K. Bramble). (I) Transverse section of a hadrosaur tooth showing similar, separate enamel and cementum regions on the tooth.

Coronal cementum is one such modification where cementum overlays the outer enamel surface (Owen, 1840) (Figure 6.10A–E). This unusual arrangement occurs as a result of an extra step in tooth crown development, where cementoblasts are able to deposit cementum along the outer surface of the crown following maturation of the enamel (Jones and Boyde, 1974; Kilic et al., 1997; Sahara, 2014). Scanning electron microscope and histological studies have shown that as horse cheek teeth form below the gingiva, the enamel-producing cells are replaced by a population of tooth-resorbing cells—the odontoclasts—that resorb and scallop the external enamel surface (Jones and Boyde, 1974; Sahara, 2014). Cementoblasts from the surrounding dental follicle then form successive layers of cementum on top of this partially eroded surface, leaving a scalloped reversal line at the junction between the two tissues (Figure 6.10D). Horses provide one of the most complex arrangements of coronal cementum, and many other herbivorous mammals do not display the same level of histological complexity. Some mammals, including lagomorphs, possess acellular coronal cementum that coats their molariform teeth, without any evidence of prior scalloping of the enamel surface (Sahara, 2014). Regardless, the coronal cementum coating the outer crown surfaces is the result of a late-stage extension of cementum onto enamel, and in mammals it appears to serve two roles: Its abrasion and erosion help maintain complex occlusal surfaces with the adjacent enamel and dentine, and it provides an anchoring point for the periodontal ligament in high-crowned teeth of many herbivorous mammals.

In horses, cementum deposition and development appear to be even more complex. Cellular cementum not only coats the external surfaces of the tooth crowns, it also fills in the depressions between layers of infolded enamel (Figure 6.10B, C). This form of coronal cementum is characteristically devoid of any role in tooth attachment, on the basis of its position in the centre of the crown and the lack of Sharpey's fibers within its matrix. This cementum is also vascularized, containing vascular canals fringed with acellular matrix, similar to the cementeons described in marine reptile cementum (Figure 6.2I, J). Cellular cementum of this kind is so modified that it no longer serves a role in tooth attachment and solely serves to infill the troughs in between enamel ridges and to contribute to the complexity of the occlusal surface over continual tooth wear and eruption (Sahara, 2014).

Cementum can extend on to the occlusal surface through another mechanism. In many mammals, and even some extinct reptiles, crown and root development occur simultaneously, allowing a tooth to maintain a crown-domain on one face of a tooth and a root-domain on the other (Figure 6.10F–H). A well-known example is in the ever-growing, self-sharpening incisors of rodents and lagomorphs (Juuri et al., 2013). Here, the labial faces of the teeth are coated in enamel, formed by an ameloblast layer at the base of the tooth, and the lingual faces are coated in cementum, which forms in the same fashion as a typical mammalian tooth root. This arrangement provides a similar dual purpose as coronal cementum: The enamel face wears more slowly than other regions of the tooth, maintaining a sharpened edge of the incisor (Figure 6.10H), whereas the cementum face provides the attachment site for the PDL and maintains a continuously erupting tooth (Juuri et al., 2013).

However, a distinction should be made here between the cementum on a continuously erupting incisor and the coronal cementum on the cheek tooth of an herbivorous mammal. The developmental mechanisms that extend cementum on to the occlusal surface are entirely different. On the one hand, cementogenesis is extended on to the enamel of the developed tooth crown following near-completion of the enamel layers (Figure 6.10A), and, on the other, cementum and enamel form simultaneously because of a topological shift in the crown–root boundary (Figure 6.10F). This distinction is important for interpreting the development of novel tooth morphologies in fossil groups. For example, Erickson et al. (2012, 2015) interpreted the complexity of the occlusal surfaces of dental batteries in hadrosaurid and ceratopsid dinosaurs as the result of the de novo and convergent evolution of coronal cementum in these dinosaur groups, similar to that of modern hypsodont mammals. However, closer inspection and comparisons to other dinosaur groups have revealed that these teeth exhibit the same histological properties as an ever-growing tooth in a rodent (Figure 6.10G, I): Enamel and cementum occupy different, nonoverlapping regions of each tooth. In hadrosaurids and ceratopsids, the crown–root boundary has shifted, allowing cementum to be exposed simultaneously with the enamel as each tooth erupted onto the occlusal surfaces (LeBlanc et al., 2016a, 2017b; Bramble et al., 2017). Cementum is found in all dinosaur groups (see previous sections), and it is its position along the tooth that has shifted, rather than these dinosaurs having independently evolved a novel form of coronal cementum.

6.9 THE PATH FORWARD: WHAT DO WE CALL THE ATTACHMENT TISSUES IN OTHER VERTEBRATES?

Invoking heterochrony in cementum and alveolar bone mineralization provides a robust and testable hypothesis for the evolution and diversification of tooth attachment modes across amniotes, both living and extinct. However, the nature and evolutionary history of the periodontium outside of Cotylosauria (Diadectomorpha + Amniota) remain obscure because the comparative histological data collected thus far address tooth attachment tissue congruence in crown amniotes. So how has tooth attachment been assessed outside of Amniota?

"Bone of attachment" is a far-reaching concept in dental histology, and it should be no surprise that the debates surrounding the nature of the attachment tissues in other vertebrates are old and ongoing (Noyes, 1902; Beust, 1938; Peyer, 1968; Soule, 1969; Shellis, 1982; Hughes et al., 1994; Luan et al., 2006, 2009; Berkovitz and Shellis, 2017; Rosa et al., 2021). However, unlike in amniotes, an additional reason for these discrepancies in histological terminology is the staggering diversity of tooth attachment modes present in non-amniote vertebrates (Figure 6.11).

Two recurring themes occur here that parallel the history of amniote tooth attachment research: The first follows Tomes' (1874, 1882) interpretation of "bone of attachment" in amphibians and fish; and the second explicitly homologizes tooth attachment tissues in other vertebrate lineages with those in mammals, echoing recent studies in amniotes. For example, Gillette's (1955) account of tooth attachment tissue formation in the frog *Rana pipiens* describes cellular cementum along the tooth bases. Howes (1978, 1987) disagreed with Gillette's interpretations of cementum in *R. pipiens*, arguing that the attachment tissues in amphibians are likely not homologous with those in mammals and considered frog teeth to be attached to the jaws by "bone of attachment".

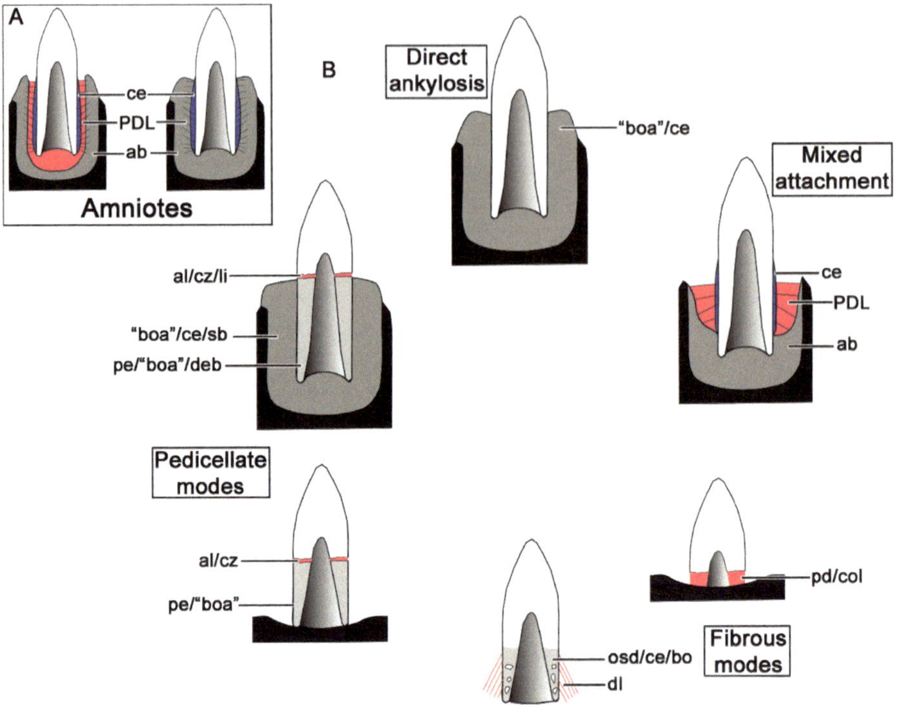

FIGURE 6.11 Examples of the diversity of tooth attachment modes in vertebrates and their associated tissue taxonomies (all taken from the references within the text). (A) The hypothesized homology of tooth attachment tissues between the ankyloses and gomphoses seen in amniotes. (B) The diversity of non-amniote tooth attachment modes and published histological terms used for the attachment tissues. Direct ankylosis is a common condition in some bony fish (Fink, 1981). Fibrous modes are common in cartilaginous (Grady, 1970) and some bony fish (Fink, 1981), and pedicellate modes are very common in lissamphibians and teleost fish (Fink, 1981; Hughes et al., 1994; Davit-Béal et al., 2007). Mixed forms of tooth attachment have been reported in triggerfish and Tautoga (Soule, 1969; Hayes, 1974). *Abbreviations*: al, annular ligament; bo, bone; "boa", "bone of attachment"; ce, cementum; col, collagen; cz, collar zone; deb, dentinous bone; dl, dental ligament; osd, osteodentine; pd, predentine; pe, pedicel; sb, spongy bone.

Source: All illustrations by the author.

Bony fish (osteichthyans) display the most diverse tooth attachment modes among any group of vertebrates. Fish teeth can exhibit complete ankylosis, ligamentous hinges and pedicelly, or mixed tooth attachment modes (Kerr, 1960; Fink, 1981; Berkovitz and Shellis, 2017) (Figure 6.11B). The debates over the nature of the attachment tissues in pedicellate fish teeth in particular parallel the history of amphibian tooth attachment. Tomes (1874) considered the pedicel in eels and haddock to be "bone of attachment", but subsequent studies have challenged this, arguing that the pedicel is formed from intermediate types of dentine and bone and an unmineralized predentine collar zone between the crown and the tooth base (Kerr, 1960; Hughes et al., 1994; Rosa et al., 2021) (Figure 6.11B). Whereas some have disagreed with the tooth pedicel being called "bone of attachment", some of these same studies have equated the cellular, vascularized bone tissue that fuses the pedicel to the jaws in fish to Tomes' "bone of attachment" (e.g., Hughes et al., 1994).

In ankylosed fish teeth, the basal attachment tissue is interpreted as cementum by some researchers and "bone of attachment" by others (e.g., Miller and Hobdell, 1968; Shellis and Poole, 1978; Fink, 1981; Berkovitz and Shellis, 2017). However, others have interpreted the mineralized attachment tissues in fish differently. Beust (1938) examined tooth development and ankylosis in sea breams [*Sargus* (= *Diplodus*)] and barracuda (*Sphyraena*) and noted that during and immediately following tooth eruption, the teeth in both fish were held in place by a fibrous membrane, similar to the PDL in mammals and crocodylians. The ligament was then replaced by bone in later stages, forming a complete ankylosis. Beust (1938) considered this as evidence of mammal-like tooth attachment in the early phases of dental ontogeny in fish, which is later replaced by complete ankylosis. This ontogenetic shift from gomphosis to ankylosis is reminiscent of tooth attachment tissue formation in amniotes (LeBlanc et al., 2016b, 2018; Mestriner et al., 2021). Others have been more explicit in equating tooth attachment tissues in bony fish with those in mammals. Soule (1969) and Hayes (1974) identified acellular cementum, a well-defined PDL, and alveolar bone in triggerfish and blackfish (*Tautoga*), respectively (Figure 6.11B). These authors did not consider the evolutionary implications of this striking similarity to the mammalian condition, and, moreover, Shellis (1982) would later argue that there were still fundamental differences between tooth attachment tissues in these teleost fish and those of mammals.

Chondrichthyans have also played a prominent role in our understanding of tooth origins and development. Excluding ratfishes and their kin (holocephalans), which have highly modified tooth plates (Johanson et al., 2020), most chondrichthyan teeth form whorls of successive tooth generations that are all attached to a fibrous tissue referred to as the dental ligament. The fibers of the dental ligament attach to the base of each tooth, the developmental and histological nature of which has been extensively debated (James, 1953; Grady, 1970; Reif, 1980; Huysseune and Sire, 1998). This "basal plate" on each chondrichthyan tooth has been variably classified as a form of osteodentine, bone, or cementum (James, 1953; Grady, 1970; Reif, 1980; Luan et al., 2006) (Figure 6.11B). The presence of ligamentous tooth attachment in chondrichthyans also begs the question of whether the presence of a PDL or PDL-like soft tissue is also an ancestral feature of the vertebrate tooth attachment tissue system; however, as in other vertebrate groups, the quagmire of histological terminology must first be addressed with detailed comparative histology between fossil and extant chondrichthyans, as well as bony fish.

Smith and Hall (1993) first proposed that the vertebrate dental module consists not only of cap ectoderm (enamel/enameloid-producing cells) and dental papilla (dentine and pulp-producing cells), but also a follicle (attachment tissue-producing cells). The prediction would be that, like enamel, enameloid, and dentine, the attachment tissues in vertebrate teeth share a deep homology that extends into the fossil record of early vertebrates. However, the debates over extant vertebrate tooth tissue homology highlight the need for new approaches to the study of vertebrate tooth attachment tissue evolution, in light of the recent findings in amniotes. The first is to overcome the conceptual barrier of "bone of attachment" and acknowledge that it does not exist. This will be an uncomfortable step for some, given the history of its use in vertebrate comparative dental anatomy, but it is essential for interrogating the development and histology of dental tissues in vertebrates that are farther and farther removed phylogenetically from the periodontium of mammals. Attachment tissue taxonomy is grounded in the study of mammalian teeth, and it is imperative that we default to it before invoking different tissue names. Using different terminology for attachment tissues in non-mammalian vertebrates based on historical, pre-cladistic convention (Tomes, 1874) raises more questions than it answers. Even if a researcher is uncertain of the identity of the attaching tissue in an ankylosed tooth, it makes more sense to communicate that uncertainty rather than to use "bone of attachment" without considering its evolutionary implications (or lack thereof).

The second approach to vertebrate attachment tissue homology is to continue to work backwards from amniotes. While we may not currently know how cementum, PDL, and alveolar bone have changed or remained the same over the immense evolutionary history of teeth and odontodes, only by expanding out to progressively more distantly related vertebrate groups can we establish testable hypotheses of homology. The development, histology, and evolution of periodontal tissues are now best documented in fossil and extant amniotes, which provide the basis for future comparisons in progressively more distantly related vertebrates.

The third approach is to consider dental tissue development and ontogeny in describing vertebrate tooth attachment tissues (Figures 6.8–6.10). Some researchers have considered this before (Beust, 1938), and it has proven to be a useful practice in revealing the hidden complexity of the vertebrate tooth attachment system. Periodontal tissues can undergo significant changes during the life of a single tooth, and, only by describing these in careful detail, will we eventually be able to unravel the history of the periodontium, from mammals to early vertebrates.

REFERENCES

Baume, L. J. 1956. Tooth and investing bone: A developmental entity. *Oral Surgery, Oral Medicine, Oral Pathology* 9:736–741.

Beertsen, W., C. A. McCulloch, and J. Sodek. 1997. The periodontal ligament: A unique, multifunctional connective tissue. *Periodontology 2000* 13:20–40.

Benson, R. B. J. 2012. Interrelationships of basal synapsids: Cranial and postcranial morphological partitions suggest different topologies. *Journal of Systematic Palaeontology* 10:601–624.

Berkovitz, B. K. B., and P. Shellis. 2017. *The Teeth of Non-Mammalian Vertebrates*. Academic Press, London, 342 pp.

Berkovitz, B. K. B., and R. C. Shore. 1982. Cells of the periodontal ligament; pp. 25–50 in B. K. B. Berkovitz, B. J. Moxham, and H. N. Newman (eds.), *The Periodontal Ligament in Health & Disease*, 1st ed. Pergamon Press, Oxford.

Berkovitz, B. K. B., and P. Sloan. 1979. Attachment tissues of the teeth in Caiman sclerops (Crocodilia). *Journal of Zoology* 187:179–194.

Bertin, J. C. T., B. Thivichon-Prince, A. R. H. LeBlanc, M. W. Caldwell, and L. Viriot. 2018. Current perspectives on tooth implantation, attachment, and replacement in Amniota. *Frontiers in Physiology* 9:1630.

Beust, T. B. 1938. Genesis of the periodontium. *The Journal of the American Dental Association and the Dental Cosmos* 25:114–118.

Bosshardt, D. D. 2005. Are cementoblasts a subpopulation of osteoblasts or a unique phenotype? *Journal of Dental Research* 84:390–406.

Bosshardt, D. D., and K. A. Selvig. 1997. Dental cementum: The dynamic tissue covering of the root. *Periodontology 2000* 13:41–75.

Bramble, K., A. R. H. LeBlanc, D. O. Lamoureux, M. Wosik, and P. J. Currie. 2017. Histological evidence for a dynamic dental battery in hadrosaurid dinosaurs. *Scientific Reports* 7.

Brink, K. S., R. R. Reisz, A. R. H. LeBlanc, R. S. Chang, Y. C. Lee, C. C. Chiang, T. Huang, and D. C. Evans. 2015. Developmental and evolutionary novelty in the serrated teeth of theropod dinosaurs. *Scientific Reports* 5:12338.

Brocklehurst, N., R. R. Reisz, V. Fernandez, and J. Frobisch. 2016. A re-description of "Mycterosaurus" smithae, an early Permian eothyridid, and its impact on the phylogeny of pelycosaurian-grade synapsids. *PloS ONE* 11:e0156810.

Brusatte, S. L., M. J. Benton, J. B. Desojo, and M. C. Langer. 2010. The higher-level phylogeny of Archosauria (Tetrapoda: Diapsida). *Journal of Systematic Palaeontology* 8:3–47.

Budney, L. A., M. W. Caldwell, and A. Albino. 2006. Tooth socket histology in the Cretaceous snake Dinilysia, with a review of Amniote dental attachment tissues. *Journal of Vertebrate Paleontology* 26:138–145.

Cai, J., S.-W. Cho, M. Ishiyama, M. Mikami, A. Hosoya, Y. Kozawa, H. Ohshima, and H.-S. Jung. 2009. Chick tooth induction revisited. *Journal of Experimental Zoology Part B: Molecular and Developmental Evolution* 312B:465–472.

Caldwell, M. W. 2007. Ontogeny, anatomy and attachment of the dentition in mosasaurs (Mosasauridae: Squamata). *Zoological Journal of the Linnean Society* 149:687–700.

Caldwell, M. W., L. A. Budney, and D. O. Lamoureux. 2003. Histology of tooth attachment tissues in the late Cretaceous mosasaurid Platecarpus. *Journal of Vertebrate Paleontology* 23:622–630.

Chen, J., A. R. H. LeBlanc, L. Jin, T. Huang, and R. R. Reisz. 2018. Tooth development, histology, and enamel microstructure in Changchunsaurus parvus: Implications for dental evolution in ornithopod dinosaurs. *PloS ONE* 13:e0205206.

Davit-Béal, T., H. Chisaka, S. Delgado, and J.-Y. Sire. 2007. Amphibian teeth: Current knowledge, unanswered questions, and some directions for future research. *Biological Reviews* 82:49–81.

Diekwisch, T. G. H. 2001. The developmental biology of cementum. *International Journal of Developmental Biology* 45:695–706.

Diekwisch, T. G. H. 2016. Our periodontal tissue: A masterpiece of evolution. *Journal of Clinical Periodontology* 43:320–322.

Dumont, M., P. Tafforeau, T. Bertin, B.-A. Bhullar, D. Field, A. Schulp, B. Strilisky, B. Thivichon-Prince, L. Viriot, and A. Louchart. 2016. Synchrotron imaging of dentition provides insights into the biology of Hesperornis and Ichthyornis, the "last" toothed birds. *BMC Evolutionary Biology* 16.

Edmund, A. G. 1960. Tooth replacement phenomena in the lower vertebrates. *Royal Ontario Museum, Life Sciences Division, Contribution* 52:1–190.

Erickson, G. M. 1996. Incremental lines of von Ebner in dinosaurs and the assessment of tooth replacement rates using growth line counts. *Proceedings of the National Academy of Sciences* 93:14623–14627.

Erickson, G. M., B. A. Krick, M. Hamilton, G. R. Bourne, M. A. Norell, E. Lilleodden, and W. G. Sawyer. 2012. Complex dental structure and wear biomechanics in hadrosaurid dinosaurs. *Science* 338:98–101.

Erickson, G. M., M. A. Sidebottom, D. I. Kay, K. T. Turner, N. Ip, M. A. Norell, W. G. Sawyer, and B. A. Krick. 2015. Wear biomechanics in the slicing dentition of the giant horned dinosaur Triceratops. *Science Advances* 1:e1500055.

Erickson, G. M., D. K. Zelenitsky, D. I. Kay, and M. A. Norell. 2017. Dinosaur incubation periods directly determined from growth-line counts in embryonic teeth show reptilian-grade development. *Proceedings of the National Academy of Sciences* 114:540–545.

Fink, W. L. 1981. Ontogeny and phylogeny of tooth attachment modes in actinopterygian fishes. *Journal of Morphology* 167:167–184.

Fong, R. K. M., A. R. H. LeBlanc, D. S. Berman, and R. R. Reisz. 2016. Dental histology of *Coelophysis bauri* and the evolution of tooth attachment tissues in early dinosaurs: Dinosaur Dental Histology. *Journal of Morphology* 277:916–924.

Ford, D. P., and R. B. J. Benson. 2020. The phylogeny of early amniotes and the affinities of Parareptilia and Varanopidae. *Nature Ecology and Evolution* 4:57–65.

Foster, B. L. 2017. On the discovery of cementum. *Journal of Periodontal Research* 52:666–685.

Gaengler, P. 2000. Evolution of tooth attachment in lower vertebrates to tetrapods; pp. 173–185 in M. F. Teaford, M. M. Smith, and M. W. J. Ferguson (eds.), *Development, Function and Evolution of Teeth*. Cambridge University Press, Cambridge.

Gaengler, P., and E. Metzler. 1992. The periodontal differentiation in the phylogeny of teeth-an overview. *Journal of Periodontal Research* 27:214–225.

García, R. A., and V. Zurriaguz. 2016. Histology of teeth and tooth attachment in titanosaurs (Dinosauria; Sauropoda). *Cretaceous Research* 57:248–256.

Gillette, R. 1955. The dynamics of continuous succession of teeth in the frog (Rana pipiens). *American Journal of Anatomy* 96:1–36.

Grady, J. E. 1970. Tooth development in sharks. *Archives of Oral Biology* 15:613–619.

Hannam, A. G. 1982. The innervation of the periodontal ligament; pp. 173–196 in B. K. B. Berkovitz, B. J. Moxham, and N. H. Newman (eds.), *The Periodontal Ligament in Health & Disease*. Pergamon Press, Oxford.

Haridy, Y., A. R. H. LeBlanc, and R. R. Reisz. 2018. The Permian reptile Opisthodontosaurus carrolli: A model for acrodont tooth replacement and dental ontogeny. *Journal of Anatomy* 232:371–382.

Hayes, R. L. 1974. Histological study of the dentition of Tautoga onitis. *Journal of Dental Research* 53:1235–1239.

Hopson, J. A. 1969. The origin and adaptive radiation of mammal-like reptiles and nontherian mammals. *Annals of the New York Academy of Sciences* 167:199–216.

Howes, R. I. 1978. Root formation in ectopically transplanted teeth of the frog, Rana pipiens. II. Comparative aspects of the root tissues. *Acta Anatomica* 109:461–470.

Howes, R. I. 1987. Surface changes in the naturally ankylosed teeth of the frog Rana pipiens during growth and maturation: An SEM study. *Journal of Zoology, London* 212:177–189.

Hughes, D. R., J. R. Bassett, and L. A. Moffat. 1994. Structure and origin of the tooth pedicel (the so-called bone of attachment) and dental-ridge bone in the mandibles of the sea breams Acanthopagrus australis, Pagrus auratus and Rhabdosargus sarba (Sparidae, Perciformes, Teleostei). *Anatamy and Embryology* 189:51–69.

Huysseune, A., and J.-Y. Sire. 1998. Evolution of patterns and processes in teeth and tooth-related tissues in non-mammalian vertebrates. *European Journal of Oral Sciences* 106:437–481.

Hwang, S. H. 2011. The evolution of dinosaur tooth enamel microstructure. *Biological Reviews* 86:183–216.

James, W. W. 1953. The succession of teeth in elasmobranchs. *Proceedings of the Zoological Society of London* 123:419–475.

Johanson, Z., E. Manzanares, C. Underwood, B. Clark, V. Fernandez, and M. Smith. 2020. Evolution of the dentition in holocephalans (Chondrichthyes) through tissue disparity. *Integrative and Comparative Biology* 60:630–643.

Jones, S. H., and A. Boyde. 1974. Coronal cementogenesis in the horse. *Archives of Oral Biology* 19:605–614.

Juuri, E., K. Saito, S. Lefebvre, and F. Michon. 2013. Establishment of crown-root domain borders in mouse incisor. *Gene Expression Patterns* 13:255–264.

Kemp, T. S. 2006. The origin and early radiation of the therapsid mammal-like reptiles: A palaeobiological hypothesis. *Journal of Evolutionary Biology* 19:1231–1247.

Kerr, T. 1960. Development and structure of some actinopterygian and urodele teeth. *Proceedings of the Zoological Society of London* 133:401–421.

Kilic, S., P. M. Dixon, and S. A. Kempson. 1997. A light microscopic and ultrastructural examination of calcified dental tissues of horses: 4. Cementum and the amelocemental junction. *Equine Veterinary Journal* 29:213–219.

Kvam, T. 1960. The teeth of Alligator mississippiensis (Daud) VI. Periodontium. *Acta Odontologica Scandinavica* 18:67–82.

Laurin, M., and R. R. Reisz. 1995. A reevaluation of early amniote phylogeny. *Zoological Journal of the Linnean Society* 113:165–223.

LeBlanc, A. R. H., K. S. Brink, T. M. Cullen, and R. R. Reisz. 2017b. Evolutionary implications of tooth attachment versus tooth implantation: A case study using dinosaur, crocodilian, and mammal teeth. *Journal of Vertebrate Paleontology*:e1354006.

LeBlanc, A. R. H., K. S. Brink, M. R. Whitney, F. Abdala, and R. R. Reisz. 2018. Dental ontogeny in extinct synapsids reveals a complex evolutionary history of the mammalian tooth attachment system. *Proceedings of the Royal Society B: Biological Sciences* 285:20181792.

LeBlanc, A. R. H., D. O. Lamoureux, and M. W. Caldwell. 2017a. Mosasaurs and snakes have a periodontal ligament: Timing and extent of calcification, not tissue complexity, determines tooth attachment mode in reptiles. *Journal of Anatomy* 231:869–885.

LeBlanc, A. R. H., I. Paparella, D. O. Lamoureux, M. R. Doschak, and M. W. Caldwell. 2021. Tooth attachment and pleurodont implantation in lizards: Histology, development, and evolution. *Journal of Anatomy* 238:1156–1178.

LeBlanc, A. R. H., and R. R. Reisz. 2013. Periodontal ligament, cementum, and alveolar bone in the oldest herbivorous tetrapods, and their evolutionary significance. *PLoS ONE* 8:e74697.

LeBlanc, A. R. H., and R. R. Reisz. 2015. Patterns of tooth development and replacement in captorhinid reptiles: A comparative approach for understanding the origin of multiple tooth rows. *Journal of Vertebrate Paleontology*:e919928.

LeBlanc, A. R. H., R. R. Reisz, K. S. Brink, and F. Abdala. 2016b. Mineralized periodontia in extinct relatives of mammals shed light on the evolutionary history of mineral homeostasis in periodontal tissue maintenance. *Journal of Clinical Periodontology* 43:323–332.

LeBlanc, A. R. H., R. R. Reisz, D. C. Evans, and A. M. Bailleul. 2016a. Ontogeny reveals function and evolution of the hadrosaurid dinosaur dental battery. *BMC Evolutionary Biology* 16.

Luan, X., Y. Ito, and T. G. H. Diekwisch. 2006. Evolution and development of Hertwig's epithelial root sheath. *Developmental Dynamics* 235:1167–1180.

Luan, X., C. Walker, S. Dangaria, Y. Ito, R. Druzinsky, K. Jarosius, H. Lesot, and O. Rieppel. 2009. The mosasaur tooth attachment apparatus as paradigm for the evolution of the gnathostome periodontium. *Evolution & Development* 11:247–259.

Maxwell, E. E., M. W. Caldwell, and D. O. Lamoureux. 2011a. Tooth histology, attachment, and replacement in the Ichthyopterygia reviewed in an evolutionary context. *Paläontologische Zeitschrift* 86:1–14.

Maxwell, E. E., M. W. Caldwell, and D. O. Lamoureux. 2011b. Tooth histology in the cretaceous ichthyosaur Platypterygius australis, and its significance for the conservation and divergence of mineralized tooth tissues in amniotes. *Journal of Morphology* 272:129–135.

McIntosh, J. E., X. Anderton, L. Flores-De-Jacoby, D. S. Carlson, C. F. Shuler, and T. G. H. Diekwisch. 2002. Caiman periodontium as an intermediate between basal vertebrate ankylosis-type attachment and mammalian ?true? periodontium. *Microscopy Research and Technique* 59:449–459.

Mestriner, G., A. R. H. LeBlanc, S. J. Nesbitt, J. C. A. Marsola, R. B. Irmis, A. A. S. Da-Rosa, A. M. Ribeiro, J. Ferigolo, and M. Langer. 2021. Histological analysis of ankylothecodonty in Silesauridae (Archosauria: Dinosauriformes) and its implications for the evolution of dinosaur tooth attachment. *The Anatomical Record*:1–31.

Miller, W. A. 1968. Periodontal attachment apparatus in the young Caiman sclerops. *Archives of Oral Biology* 13:735–743.

Miller, W. A., and M. H. Hobdell. 1968. Preliminary report on the histology of the dental and paradental tissues of Latimeria chalumnae (Smith) with a note on tooth replacement. *Archives of Oral Biology* 13:1289–1291.

Moss, M. L. 1977. Skeletal tissues in sharks. *American Zoologist* 17:335–342.

Muller, R. T., and M. S. Garcia. 2020. A paraphyletic "Silesauridae" as an alternative hypothesis for the initial radiation of ornithischian dinosaurs. *Biology Letters* 16:20200417.

Nanci, A. 2013. *Ten Cate's Oral Histology: Development, Structure, and Function*, 8th ed. Elsevier Mosby, St. Louis, MO, 379 pp.

Nesbitt, S. J., C. A. Sidor, R. B. Irmis, K. D. Angielczyk, R. M. H. Smith, and L. A. Tsuji. 2010. Ecologically distinct dinosaurian sister group shows early diversification of Ornithodira. *Nature* 464:95–98.

Novas, F. E., F. L. Agnolin, M. D. Ezcurra, R. T. Muller, A. G. Martinelli, and M. C. Langer. 2021. Review of the fossil record of early dinosaurs from South America, and its phylogenetic implications. *Journal of South American Earth Sciences* 110:103341.

Noyes, F. P. 1902. A comparative study of the attachment of the teeth. *Journal of the American Medical Association* 5:242–244.

Olroyd, S. L., A. R. H. LeBlanc, R. Araújo, K. D. Angielczyk, A. Duhamel, J. Benoit, and M. Amaral. 2021. Histology and μCT reveal the unique evolution and development of multiple tooth rows in the synapsid Endothiodon. *Scientific Reports* 11:16875.

Osborn, J. W. 1981. *Dental Anatomy and Embryology* (A. H. R. Rowe and R. B. Johns, eds.). Blackwell Scientific Publications, Oxford, London, Edinburgh, Boston, Melbourne, 447 pp.

Osborn, J. W. 1984. From reptile to mammal: Evolutionary considerations of the dentition with emphasis on tooth attachment. *Symposium of the Zoological Society of London* 52:549–574.

Owen, R. 1840. *Odontography; or, a Treatise on the Comparative Anatomy of the Teeth; Their Physiological Relations, Mode of Development, and Microscopic Structure, in the Vertebrate Animals.* Hippolyte Bailliere, London, 655 pp.

Peyer, B. 1968. *Comparative Odontology.* The University of Chicago Press, Chicago, IL, 347 pp.

Pihlstrom, B. L., B. S. Michalowicz, and N. W. Johnaon. 2005. Periodontal diseases. *The Lancet* 366:1809–1820.

Polcyn, M. J., L. L. Jacobs, A. S. Schulp, and O. Mateus. 2010. The North African Mosasaur *Globidens phosphaticus* from the Maastrichtian of Angola. *Historical Biology* 22:175–185.

Pretto, F., S. F. Cabreira, and C. L. Schultz. 2014. Tooth microstructure of the Early Permian aquatic predator Stereosternum tumidum and paleobiological implications. *Acta Palaeontologica Polonica* 59:125–133.

Reid, R. E. H. 1996. Bone histology of the Cleveland-Lloyd dinosaurs and of dinosaurs in general, Part I: Introduction: Introduction to bone tissues. *Brigham Young University Geology Studies* 41:25–72.

Reif, W.-E. 1980. Development of dentition and dermal skeleton in embryonic Scyliorhinus canicula. *Journal of Morphology* 166:275–288.

Reisz, R. R. 1997. The origin and early evolutionary history of amniotes. *Trends in Ecology and Evolution* 12:218–222.

Reisz, R. R., A. R. H. LeBlanc, H. C. Maddin, T. W. Dudgeon, D. Scott, T. Huang, C. Jun, C. Chuan-Mu, and S. Zhong. 2020. Early Jurassic dinosaur fetal dental development and its significance for the evolution of sauropod dentition. *Nature Communications* 11:2240.

Rieppel, O., and M. Kearney. 2005. Tooth replacement in the late Cretaceous Mosasaur Clidastes. *Journal of Herpetology* 39:688–692.

Rosa, J. T., P. E. Witten, and A. Huysseune. 2021. Cells at the edge: The dentin-bone interface in zebrafish teeth. *Frontiers in Physiology* 12:723210.

Saffar, J.-L., J.-J. Lasfargues, and M. Cherruau. 1997. Alveolar bone and the alveolar process: The socket that is never stable. *Periodontology 2000* 13:76–90.

Sahara, N. 2014. Development of coronal cementum in hypsodont horse cheek teeth. *The Anatomical Record* 297:716–730.

Sander, P. M. 1999. The microstructure of reptilian tooth enamel: Terminology, function, and phylogeny. *München Geowissenschaftliche Abhandlungen* 38:1–102.

Sassoon, J., D. Foffa, and R. Marek. 2015. Dental ontogeny and replacement in Pliosauridae. *Royal Society Open Science* 2:150384.

Savitsky, A. H. 1981. Hinged teeth in snakes: An adaptation for swallowing hard-bodied prey. *Science* 212:346–349.

Schwarz, D., J. C. D. Kosch, G. Fritsch, and T. Hildebrandt. 2015. Dentition and tooth replacement of *Dicraeosaurus hansemanni* (Dinosauria, Sauropoda, Diplodocoidea) from the Tendaguru Formation of Tanzania. *Journal of Vertebrate Paleontology*:e1008134.

Shellis, R. P. 1982. Comparative anatomy of tooth attachment; pp. 3–24 in *The Periodontal Ligament in Health & Disease*. Pergamon Press, Oxford.

Shellis, R. P., and D. F. G. Poole. 1978. The structure of the dental hard tissues of the coelacanthid fish Latimeria chalumnae Smith. *Archives of Oral Biology* 23:1105–1113.

Sidor, C. A., and J. A. Hopson. 1998. Ghost lineages and "mammalness": Assessing the temporal pattern of character acquisition in the Synapsida. *Paleobiology* 24:254–273.

Sloan, P. 1982. Structural organization of the fibres of the periodontal ligament; pp. 51–72 in B. K. B. Berkovitz, B. J. Moxham, and H. N. Newman (eds.), *The Periodontal Ligament in Health & Disease*. Pergamon Press, Oxford.

Smith, H. M. 1958. Evolutionary lines in tooth attachment and replacement in reptiles: Their possible significance in mammalian dentition. *Transactions of the Kansas Academy of Science* 1903:216–225.

Smith, M. M., and B. K. Hall. 1993. A developmental model for evolution of the vertebrate exoskeleton and teeth: The role of cranial and trunk neural crest. *Journal of Evolutionary Biology* 27:387–448.

Snyder, A. J., A. R. H. LeBlanc, C. Jun, J. J. Bevitt, and R. R. Reisz. 2020. Thecodont tooth attachment and replacement in bolosaurid parareptiles. *PeerJ* 8.

Soule, J. D. 1969. Tooth attachment by means of a periodontium in the trigger-fish (Balistidae). *Journal of Morphology* 127:1–6.

Spiekman, S. N. F., and N. Klein. 2021. An enigmatic lower jaw from the Lower Muschelkalk (Anisian, Middle Triassic) of Winterswijk provides insights into dental configuration, tooth replacement and histology. *Netherlands Journal of Geosciences* 100:e17.

Ten Cate, A. R. 1997. The development of the periodontium—a largely ectomesenchymally derived unit. *Periodontology 2000* 13:9–19.

Ten Cate, A. R., and C. Mills. 1972. The development of the periodontium: The origin of alveolar bone. *The Anatomical Record* 173:69–77.

Tomes, C. S. 1874. Studies upon the attachment of teeth. *Transactions of the Odontological Society of Great Britain* 7:41–58.

Tomes, C. S. 1882. *A Manual of Dental Anatomy: Human and Comparative*, 2nd ed. Presley Blakiston, Philadelphia, 440 pp.

Wang, W., Z. Li, C-. C. Wang, A. M. Bailleul, M. Wang, J. O'Connor, J. Li, X. Zheng, R. Pei, and F. Teng. 2023. Comparative microstructural study on the teeth of Mesozoic birds and non-avian dinosaurs. *Royal Society Open Science* 10:230147.

Whitney, M. R., and C. A. Sidor. 2019. Histological and developmental insights into the herbivorous dentition of tapinocephalid therapsids. *PLoS ONE* 14:e0223860.

Zaher, H., and O. Rieppel. 1999. Tooth implantation and replacement in squamates, with special reference to mosasaur lizards and snakes. *American Museum Novitates*:1–19.

Zahradnicek, O., I. Horacek, and A. S. Tucker. 2012. Tooth development in a model reptile: Functional and null generation teeth in the gecko Paroedura picta. *Journal of Anatomy* 221:195–208.

7 Initiation and Periodic Patterning of Vertebrate Dentitions

Alexa Sadier and Vladimír Soukup

7.1 INTRODUCING THE PATTERNED DENTITIONS

Dentitions have diversified enormously since their first appearance in the Paleozoic vertebrates (Rücklin et al., 2012; Chen et al., 2016, 2020; Vaškaninová et al., 2020), a diversity reflected in many respects, ranging from the position within the oropharyngeal cavity to tooth shape diversity. A central part of this astonishing diversity is, however, reflected in the way teeth are arranged into dental patterns: for example, the conveyor-belt system of replacing teeth of sharks, the pharyngeal jaw-bound toothlets of cypriniforms, or the single-rowed heterodont dentition of mammals. Each group of vertebrates, major or minor, has its own typical dental pattern that identifies and defines it. No wonder the question of how to produce these patterns has been a leitmotif of comparative odontology related to the establishment of the primary and replacing dentitions (Butler, 1939; Edmund, 1960; Osborn, 1978). In this respect, two independent scientific streams occurred during the second half of the 20th century, and, for the sake of this introduction, we will call them "empirical", because it relied on a series of developmental and paleontological data, and "theoretical", because it stemmed from mathematical modeling and simulations.

In a series of papers in the 1970s, Osborn (1970, 1971, 1973, 1974, 1978) criticized the then popular "Zahnreihen" (aka oblique rows) theory of Edmund (1960), which proposed that a dispersal of a tooth-inducing stimulus along the length of the jaw successively initiates tooth germs at predefined focal positions; the oblique tooth rows (especially notable e.g. in sharks or reptiles) being a reflection of the presence of such a stimulus. As an alternative to this contested hypothesis, Osborn instead proposed a model, where teeth are added into the growing jaw with each tooth germ surrounded by a "zone of inhibition", which prevents the initiation of new tooth germs (Osborn, 1971, 1974, 1978). In contrast to "Zahnreihen", positions of teeth are not predetermined, and new tooth germs can be initiated only when the tooth-forming tissues become released from the inhibitory zone, which can occur through growth or maturation of teeth that no longer produce the zone of inhibition. Presence of zones of inhibition and initiation of teeth outside these zones would then account for the regular spacing of repeatedly patterned dentitions. The application of the concept of the zones of inhibition to the developing shark dentitions (Reif, 1976) in combination with the employment of odontodes as the principal building units of

DOI: 10.1201/9781003439653-7

the vertebrate integument (Ørvig, 1967, 1977) next culminated in the proposal of the "odontode regulation theory" (Reif, 1982) that sought for the identification of basic properties and evolutionary transformations of dermal denticles and teeth.

In parallel to studies on the developmental and evolutionary origins of dermal denticles and teeth, an independent line of thoughts occurred to explain the emergence of spatially organized patterns of repeated structures. Originally proposed by Turing (Turing, 1952) and later elaborated by Gierer and Meinhardt (Gierer and Meinhardt, 1972), theoretical models comprising instabilities within a system were designed, where molecular components react among each other and diffuse to surrounding tissues of a competent field. If the variables of this reaction–diffusion (RD) system are set correctly, the instabilities lead to a spontaneous generation of periodically repeated patterns. Within the defined range of variables, the models may return spots with interspot spaces comparable to Osborn's zones of inhibition.

While both the "empirical" and "theoretical" streams were elaborated independently, only the former was intimately connected to the biology of establishment of tooth patterns. The first contact of these two scientific streams occurred no earlier than in the 1990s. Taking advantage of a detailed developmental series by Westergaard and Ferguson (1986, 1987, 1990) and an application of RD kinematics, Kulesa and co-workers (reviewed in Murray, 2003) suggested that the successive appearance of teeth in the alligator dentition follows the behavior of an RD mechanism. Although, in this case, the RD molecular players have never been identified, this work suggested for the first time that the development of vertebrate dentitions may follow the behavior of RD systems. Unfortunately, it has not been until recently that we have been able to fully appreciate the role of RD mechanisms in building vertebrate dentitions, and something that we could call a synthesis of the two streams has not been met. Thus, the general consensus on the role of RD mechanisms taking place during the establishment of vertebrate dentitions has, not been fully distinguished, despite considerations of RD mechanisms taking place in building other biological systems or even structures that bear initial developmental resemblance to teeth, such as hair, feathers, or scales (Marcon and Sharpe, 2012).

The current chapter does not have the intention or ambition of representing the desired synthesis of "empirical" and "theoretical" streams, nor does it represent a comprehensive review of all the possible dental patterns of individual vertebrate groups, for which the reader is referred to elsewhere (see e.g. Peyer, 1968; Berkovitz and Shellis, 2017). Its intention is to evaluate current knowledge on periodic patterning as a process taking place during the development of dentitions. We start with how RD mechanisms regulate the distribution of body cover organs such as the mammalian hair or bird plumage and then turn our attention to dentitions. We review how teeth become initiated and organized and evaluate the role of RD mechanisms in patterning the dentitions in mammals and non-mammalian vertebrates. Last, we propose and discuss current topics related to the patterning of dentitions and the application of RD mechanisms in this process. Implicitly, the ideas in this chapter should be taken as a review and a conceptual framework rather than an exhaustive overview of the topic as, indeed, our knowledge in this respect is (to a greater extent than elsewhere) vastly limited and dependent on the discoveries over the upcoming years.

7.2 REACTION–DIFFUSION MECHANISMS AND PERIODIC PATTERNING OF SKIN DERIVATIVES

7.2.1 Basics of Reaction–Diffusion Mechanisms

Periodic patterning is a process that, in most cases, works based on the action of Turing-like reaction–diffusion (RD) mechanisms. In the simplest form, they consist of a self-activating activator and its feedback inhibitor. In cases when the diffusion rate of the inhibitor is higher than that of the activator, the interaction of these two agents may lead to the establishment of periodically repeated patterns (Figure 7.1). As a premise of a model proposed by Turing, initial concentrations of the activator and inhibitor within a given tissue show a homogeneous distribution; however, noisy

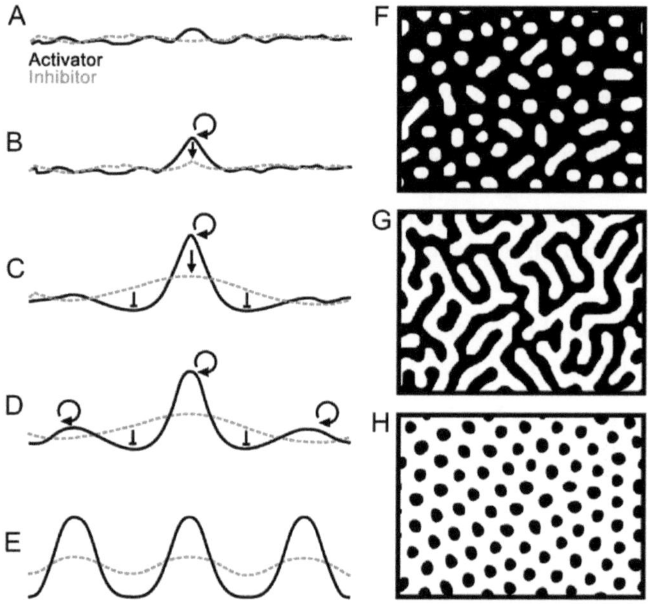

FIGURE 7.1 Spontaneous establishment of patterns by short-range auto-activation and long-range inhibition. (A–E) Distribution of activator (solid lines) and inhibitor (dashed lines) concentrations in space and time as a result of RD mechanics of these interacting agents. (A) The model proposes an initial homogeneous distribution of the activator and the inhibitor; however, minute molecular fluctuations may lead to a locally higher concentration of the activator. (B) The autocatalytic production (circular arrow) generates a local peak of the activator; at the same time, the activator triggers production of the inhibitor (arrow). (C) Continuous production and short-range diffusion of the activator stabilizes the peak, while long-range diffusion of the inhibitor prevents production of new peaks in adjacent areas (bars). (D) New activator peaks (circular arrows) are generated at a distance out of the reach of the inhibitor. (E) The result is a repeating pattern of higher activator and inhibitor concentrations. (F–H) Planar views of the possible pattern outcomes of the RD mechanics.

Source: Adapted from Dalle Nogare and Chitnis (2017) (CC BY license).

molecular fluctuations of these agents can lead to discontinuities and local minute peaks of higher activator concentration (Turing, 1952). Thanks to self-activation, these minute activator peaks are enhanced, leading to subsequent local production of the inhibitor. Due to the higher diffusion rate, the expanding inhibitor cannot suppress the local concentration peak of the activator. However, the influx of the inhibitor to the surrounding regions prevents the production of new activator peaks in the vicinity of the original peak. As the concentration of the inhibitor diminishes spatially, new minute peaks of activators can be produced at a certain distance from the original peak. The result is a set of foci of high activator concentration surrounded by inhibited zones. Cells of the activated zone acquire a certain fate, while cells of the inhibited zone acquire a different fate. Ultimately, the differential fate decision may then turn into a regular morphological pattern of periodically repeated structures. Interestingly, the RD mechanisms represent a self-organizing system, in which each activator peak is produced the same way as any other one without prior asymmetries in a given field of competence. In a planar system, such as the embryonic skin, these RD mechanisms may result in the production of spots, dots, stripes, or mazes of e.g. the various body coating color-patterns or the regular distribution of dots of primordia of integumental organs, such as hair, feathers, scales, or denticles. Basic characteristics of the resulting pattern, i.e. the type of pattern, periodicity, or spacing, stem from the general properties of the system, such as the strength of interactions of the activator and inhibitor and the rates of diffusion of both interacting agents (Kondo and Miura, 2010; Painter et al., 2012).

7.2.2 Reaction–Diffusion Mechanisms in Patterning the Mammalian Hair Follicles

In mammals, hair covers the body surface homogeneously, with individual hair follicles being spaced regularly. During mouse embryonic development, gene expression patterns repeating periodically in the naive epidermis (Figure 7.2A–C) are reflected by cellular condensations in the underlying dermal mesenchyme, and these early events prefigure the nascent appearance of the regularly interspaced hair follicles (Ahtiainen et al., 2014; Glover et al., 2017; Shyer et al., 2017). In a two-step process, primary follicles are specified at a first inductive wave, and secondary follicles are formed in interfollicular regions at a second inductive wave. Past studies have proposed molecular interactions during hair follicle development, qualifying as likely players in the Turing-like RD mechanisms.

Sick et al. (2006) identified a Turing pair, where Wnt/β-catenin signaling pathway acts as the activator and the promoter of follicular fate, and its feedback inhibitor Dkk as the regulator of interfollicular spacing. By modulating the Dkk levels, they were able to reduce the follicular density, while upregulation of Wnt signaling was previously shown sufficient to induce supernumerary follicles (Gat et al., 1998). Computer simulation of induction of different follicular waves modeled and supported the observed experimentally induced phenotypes and provided evidence on the RD mechanisms acting to pattern the mammalian hair (Sick et al., 2006).

In parallel, Mou et al. (2006) suggested a different Turing-like mechanism, where the ectodysplasin receptor (Edar) in the epithelium would act as the activator and

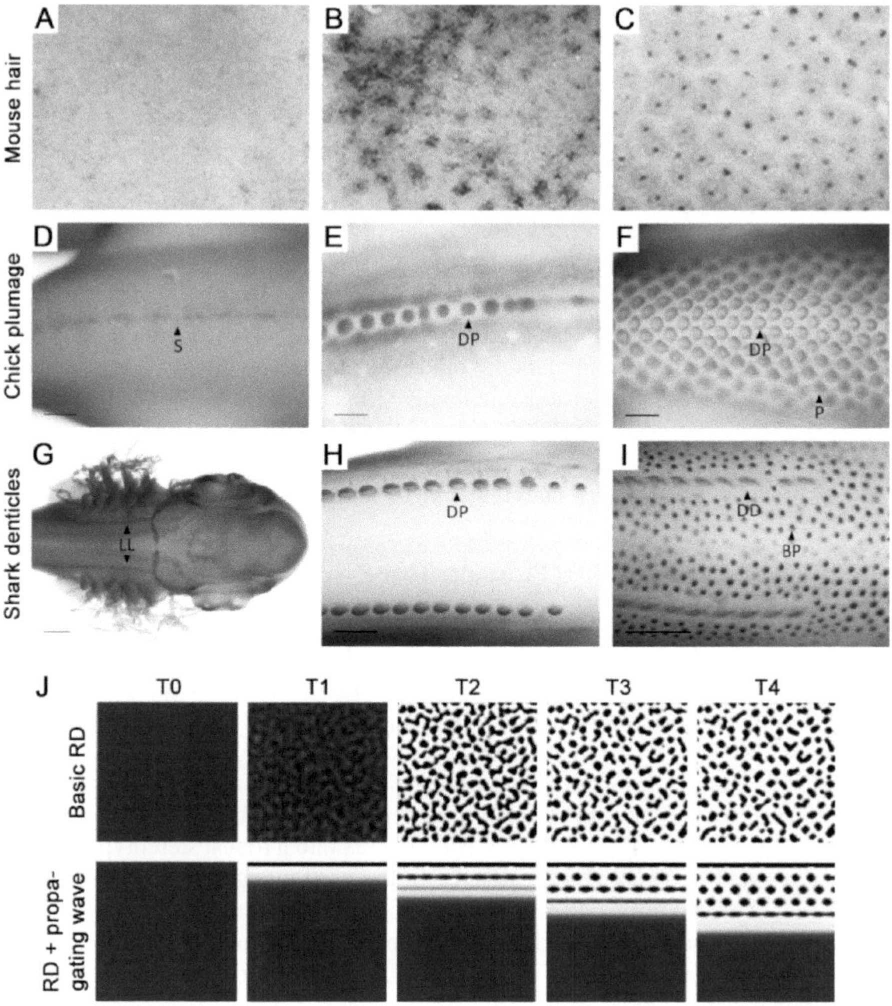

FIGURE 7.2 Periodic patterning of skin derivatives. (A–C) Expression of *Dkk4* shows progressive appearance of nondirectional primary hair pattern on mouse skin explants. (D–I) Expression of *β-catenin* in the chick and shark body surfaces shows directional pattern establishment of chick plumage and shark denticles. (D–F) In chick, the initial median stripe of competence (S) breaks into individual dermal placodes (DP) and further placodes (P) arise bilaterally to cover the dorsal skin with feather buds. (G–I) In shark, two lateral lines (LL) break into dermal placodes (DP) marking the positions of prospective large dermal denticle (DD) rows. During development, minute BPs giving rise to prospective minor dermal denticles arise in the vicinity of the main dermal placodes. (J) Time course of RD models of Ho et al. (2019) to simulate pattern establishment of bird plumage. A basic RD model yields an irregular non-stereotyped array resembling the flightless (ratite) bird plumage pattern and is reminiscent of the mammalian hair pattern (upper row). The same model, but involving a propagating wave, returns a highly stereotyped hexagonal array reflecting the flighted bird plumage pattern development (lower row). Figures A to C are adapted from Glover et al. (2017), D to I from Cooper et al. (2018), and J is adapted from Ho et al. (2019) (CC BY license).

Bmp2 and Bmp7 in the mesenchyme as the repressors of follicular fate. *Eda* and its receptor *Edar* are expressed homogeneously prior to the induction of the first follicular wave, and their transcripts next become confined to follicular foci (Mou et al., 2006). On the other hand, Bmps, which act downstream of the Eda pathway, are expressed focally and have a repressive role on *Edar* expression. While Edar, as a transmembrane protein, exhibits a cell autonomous activity, Bmp ligands may diffuse and act at a distance as a classical Turing-like inhibitor. Moreover, the application of Eda protein to *Eda*-deficient skin explants is able to restore the follicular pattern, and the increasing concentration of the Eda protein leads to the transformation of the spotty pattern into spot fusions and labyrinthine patterns, amid the behavior of a classical RD model (Gierer and Meinhardt, 1972; Headon and Painter, 2009). Although a consequent computational modeling questioned the system proposed by Mou et al. (2006) as being unstable in longer terms (Klika et al., 2012), the outcome of works provided by Sick et al. (2006) and Mou et al. (2006) gave a first glimpse on connecting theoretical RD mechanisms with specific molecular circuitries playing role in periodic patterning of mammalian hair follicles.

7.2.3 Patterning the Bird Plumage: Turing with and without a Wave

The self-organizing properties of the RD systems during the development of the mammalian hair generate locally organized, although globally stochastic, non-stereotyped patterns. However, the distribution of many integumentary organs, such as avian feathers and footpads, fish scales or teeth, displays a much lower degree of pattern stochasticity (Aman et al., 2018; Bailleul et al., 2019; Cooper et al., 2019; Ho et al., 2019). The "higher order" patterning of these organs suggests the presence of a globally acting mechanism on top of the proposed basic Turing-like RD instability to operate in concert and organize individual units into a robust stereotyped fashion.

Past research in chick has suggested that the positioning of feather follicles is a result of RD mechanisms where Fgf and Wnt signaling supposedly act as the activators and Bmp signaling as the inhibitor of feather follicle formation (Song et al., 1996, 2004; Widelitz et al., 1996; Jung et al., 1998; Noramly and Morgan, 1998; Chang et al., 2004; Michon et al., 2008; Wells et al., 2012). *In vitro*, feather follicles of the cultured chick skin explants emerge autonomously by the spontaneous appearance of mesenchymal condensations, suggesting that feather follicles have an innate ability to self-organize (Jiang et al., 1999). *In vivo*, feather follicles develop from competent regions called feather tracts situated at various positions on the body (Neguer and Manceau, 2017). In the dorsal skin, a continuous stripe of midline expression of factors such as β-catenin is among the first signs of feather follicle development (Figure 7.2D, Jiang et al., 1999). During the embryonic development, the β-catenin expression spreads bilaterally toward the flank region, thus gradually expanding the zone competent for feather bud production (Figure 7.2E, Noramly et al., 1999; Ho et al., 2019). The initial compact midline expression becomes compartmentalized into foci of high β-catenin expression, and the lack of β-catenin expression in the direct vicinity to these foci mirrors individualization of positions of the future midline feather row. As the β-catenin expression spreads toward the flank, new rows of feather follicles are initiated laterally from the median follicle row; each new row develops in a staggered arrangement, and this results in the final highly ordered hexagonal array of follicles (Figure 7.2F).

Interestingly, the progressive addition of successive follicular rows is not the sole result of the innate ability to self-organize as this, alone, does not lead to the precise honeycomb lattice arrangement (Jiang et al., 1999; Glover et al., 2017). Ho et al. (2019) have recently found that the precision in the honeycomb feather follicle positioning is substantiated during the successive appearance of feather follicle rows due to the presence of a propagating wave of *Eda* expression. The propagation of this wave from the dorsal midline over the β-catenin positive skin acts on the local innate self-organizing ability of the skin mesenchyme (Jiang et al., 1999) to promote the non-stereotyped pattern into the "higher-order" hexagonal arrangement (Figure 7.2J). This regular hexagonal arrangement seems to be an adaptive trait connected to the flight ability, because flightless birds, such as the ostrich or emu, develop a randomized follicular pattern much alike that of the mammalian hair (Bailleul et al., 2019; Ho et al., 2019).

Assembling the body cover organs into regular patterns is widespread among vertebrates. For example, in extant sharks and rays, skin denticles are arranged into one or two rows running longitudinally on the dorsum (Cooper et al., 2018). These rows trigger the subsequent development of minute denticles that develop bilaterally from the initial rows and progressively cover the entire body surface (Figure 7.2G–I). Experiments with beads soaked in the inhibitor of Fgf signaling together with mathematical modeling have suggested that the development of minute denticles in the vicinity of the original denticle rows proceeds based on the Turing-like mechanism, with Fgf and Shh signaling acting as the activators and Bmp signaling acting as the inhibitor of the RD dynamics (Cooper et al., 2018). Unlike in the volant birds, however, the subsequent minute denticles are not arrayed into a precise hexagonal lattice suggestive of an absence of any "higher order" patterning mechanism similar to that of the Eda propagating wave. Nevertheless, the presence of the initiator row and the progressive addition of single denticles suggest that the process of body cover patterning in sharks and rays is directional, in contrast to, *e.g.*, hair follicle patterning.

7.2.4 From Body Covers to Dentitions

The various examples of body cover development show that diverse periodic pattern generators may act on the competent environment leading to disparate arrangements and contributing to the overall diversity in periodically repeated skin appendages. In some cases, a basic RD circuitry in the absence of prior tissue asymmetries results in an almost abrupt synchronous appearance of the pattern, in others, the presence of the triggering cue initiates the innate RD circuitry and provides a propagating directional generation of the pattern (Inaba et al., 2019). As the directionality and the propagating patterned addition of individual placodes are features of many developing and actively replacing dentitions, it was suggested that the periodic pattern generators playing a role in skin appendage patterning likely control the patterning of vertebrate dentitions as well. However, whereas the abovementioned skin derivatives are usually built to cover vast areas of body surface, teeth are commonly placed in specialized and locally restricted regions of the oropharyngeal cavity. And even in cases where the whole of the oropharyngeal cavity is virtually covered in teeth, such as that of bichirs, gars, bowfin, or coelacanths (Nelson, 1969), these do not represent a single dental entity but are allocated to individual dentate regions with developmentally separate tooth patterns.

The evolutionary origin of teeth is sought in the enigmatic odontode, a structural entity that, in the form of dermal denticles, covers external body surface of extinct Paleozoic fishes and extant sharks and, in the form of teeth, augments the oropharyngeal surfaces of jawed vertebrates (Reif, 1982). Developing at the epithelial–mesenchymal interface and composed of the dentine cone and optionally an enameloid cap (Ørvig, 1967, 1977), a great deal of discussion has struggled to explain how the pattern of the jaw-bound odontodes (teeth) may have evolved from the "scattered" non-tooth odontodes (denticles) (reviewed in Donoghue and Rücklin, 2016). One of the main challenges of these evolutionary scenarios is to answer the question of the precision with which patterns in dentitions versus external denticles arise, often coming to conclusions that these two systems may be fundamentally disparate and unrelated (Fraser et al., 2010; Fraser and Smith, 2011). We do not want to review the hypotheses and their argumentations on possible evolutionary scenarios of dental translocations, although, indeed, this topic involves patterning as well. We just want to emphasize that, stepping beyond the broad polarity between dermal odontodes (denticles) and oral odontodes (teeth), it is important to consider the existence of disparities at the level of patterning dentitions themselves and body cover organs themselves. Importantly, some of the disparity may, however, be explained by a relatively easy mechanism that changes one pattern into another, such as in case of the abovementioned patterning of feather follicles in volant versus flightless birds.

Indeed, as in the case of the plumage pattern of volant birds, the precise patterning of vertebrate dentitions points to the functional demands, *i.e.*, food processing in this case. Although currently unknown to a large extent, the evolutionary stability of dentition patterns in individual vertebrate lineages suggests the presence of conserved mechanisms regulating the formation of dentitions. Generally, the development of a dentition comprises several periods, including the specification of the zone competent for odontogenesis, the initiation of tooth development within this zone, tooth addition, and establishment of the pattern and retainment or alterations in the pattern. In the next section, we aim to put forward topics regarding the development of dental patterns in vertebrates related to Turing-like RD mechanisms. How does the region competent for tooth development is established? How does the dentition itself initiate? How do the individual tooth germs arrayed into more-or-less stereotyped patterns? All these are important questions related to the establishment and sustainment of tooth patterns both in evolutionary and developmental terms.

7.3 REACTION–DIFFUSION MECHANISMS AND PERIODIC PATTERNING OF TEETH

7.3.1 Specification of the Region Committed for Tooth Development

As noted before, unlike the mammalian hair, for which the zone competent for hair placode development is basically the entire body surface, the region competent for tooth development is spatially restricted, thus prefiguring the nascent toothed area. This region is called the primary dental lamina in mammals and reptiles or the odontogenic band in fishes. Reflecting the diversity of vertebrate dentitions and the functional demand to develop them at defined regions used for food processing, the primary dental laminae/odontogenic bands may be allocated to various places across the whole oropharyngeal cavity.

Prior to the first morphological sign of tooth development, the tooth-competent zone is marked by the early epithelial expression of *Pitx2* and *Shh*. Although mutual spatial relations of the expression patterns of *Pitx2* and *Shh* may vary, research in the plethora of vertebrate species has postulated that these two factors generally represent the earliest markers of odontogenesis and factors responsible for the progression to successive stages of tooth development. In most vertebrates, *Pitx2* and *Shh* are co-expressed in a horseshoe-shaped pattern running along jaw primordia, thus prefiguring the nascent jaw-bound dentition (Keränen et al., 1999; Fraser et al., 2004, 2008; Stock et al., 2006; Debiais-Thibaud et al., 2015; Rasch et al., 2016; Pospisilova et al., 2022). In the zebrafish, *Pitx2* and *Shh* expression patterns in the oral region do not colocalize, and this lack of co-expression has been postulated as the proximate reason for the inability of this and other cypriniform species to develop oral teeth (Stock et al., 2006). Similarly, the lack of co-expression of these two factors was reasoned as the cause of the lack of initiation of the third tooth row in the two-rowed dentition in the cichlid fish *Cynotilapia afra*, while *Pitx2/Shh* co-expression and development of subsequent tooth rows do occur in other cichlid species with multirowed dentitions (Fraser et al., 2008).

Indeed, the tooth-competent region is where the patterning processes occur and where individual teeth are progressively added. Therefore, its shape reflects the resulting appearance of the toothed region. A long and narrow strand-shaped region of competence leads to the development of a tooth row, such as that present in mammals or *e.g.* oral teeth in the sterlet sturgeon (Keränen et al., 1999; Pospisilova et al., 2022), while a broad tooth-competent region leads to the development of multirowed dentitions or tooth patches, such as the dentitions of sharks (Rasch et al., 2016). Variations in the extent of *Pitx2/Shh* expression may even be ascribed to the diversity of dental phenotypes in closely related species. For example, in cichlids, broader *Pitx2/Shh* expression patterns are found in species with multiple tooth rows in the jaws, while narrow expression is present in species with only two tooth rows (Fraser et al., 2008). Interestingly, in the Mexican axolotl, *Pitx2* expression pattern demarks a competent region for multiple tooth fields: the initially broad *Pitx2* expression pattern becomes broken up to individual toothed regions on the premaxillary, vomerine, and palatine on the roof, and dentary and coronoid on the floor of the oral cavity (Soukup et al., 2021). However, how the compartmentalization of the *Pitx2* expression domain is established to build multiple tooth fields in the axolotl (versus the lack of it in other vertebrates) remains unknown.

Following the initial specification of the tooth-competent region, the broad *Pitx2/Shh* expression becomes progressively restricted to individual tooth germs to regulate further steps in odontogenesis.

7.3.2 Specification of Tooth Competence in the Mouse

Despite many decades of research on tooth development in various vertebrate species and the extensive body of research done in mice, we still have a limited understanding of the initial steps that control the specification of individual toothed regions (Balic, 2019). As for other vertebrates, *Pitx2* and *Shh* are expressed in a horseshoe-shaped pattern along the mouse jaw primordia, and this broad expression becomes gradually confined to the nascent incisor and first molar areas (Dassule and McMahon, 1998; St. Amand et al., 2000). Additionally, other transcription factors and signaling molecules such as *Foxi3*, *Sox2*, *Irx1*, *Bmp2*, *Wnt10a*, and *Wnt10b* are expressed specifically within the primary dental lamina. This restricted expression distinguishes molecularly the region specified for odontogenesis from the surrounding non-dental epithelium and further plays a role in progressive differentiation of individual dental lamina cell types (Dassule and McMahon, 1998; Juuri et al., 2013; Shirokova et al., 2013). The broad *Pitx2* expression is regulated by signaling molecules that specify the polarity of the jaw primordia: by the proximally expressed *Fgf8* in a positive feedback loop and by the distally expressed *Bmp4* in a negative feedback loop (St. Amand et al., 2000; Liu et al., 2003). Apart from these molecular interactions, *Pitx2* is also a downstream transcription factor of the Wnt/β-catenin pathway (Kioussi et al., 2002; Briata et al., 2003).

The lack of information on the earliest events of mouse odontogenesis has, however, been filled by a recent study that uses single-cell transcriptional profiling of mouse mandibular epithelium (Ye et al., 2022). This work confirmed the fundamental role of Pitx2 in restricting the dental epithelial fate and further helped in the identification of other previously proposed odontogenic markers (such as Irx1), as well as in the identification of novel and previously unidentified factors likely playing a role in tooth initiation (such as Ntrk2). The study further confirmed the progressive restriction and gradual sharpening of odontogenic fate during development, *i.e.*, the initially broad *Pitx2* expression domain at E9.5 becomes spatially confined and colocalizes with the expression of another primary dental lamina marker gene *Irx1* at E10.5. This observation suggests that the major part of mandibular epithelium is initially specified toward a dental fate, and the future position of the primary dental lamina is due to the specification of the bordering epithelia into non-dental fates. Indeed, the establishment of adjacent zones then goes hand in hand with the sharpening of expression boundaries of primary dental lamina marker genes such as *Pitx2* (Ye et al., 2022).

Within the early dental lamina, Pitx2 both regulates the proliferation of $Sox2^+$ dental epithelial progenitor cells and, activates *Shh* and *Lef1* expression in early tooth signaling centers called initiation knots (Ahtiainen et al., 2016; Yu et al., 2020; Mogollón et al., 2021). Later, Pitx2 is also responsible for the establishment of the second signaling centers of tooth development called the primary enamel knots, and *Pitx2*-deficient dental epithelium fails to form both the initiation and the primary enamel knots, leading to arrested tooth development at the cap stage (Yu et al., 2020). Furthermore, Pitx2-activated *Shh* expression seems to be regulated by *Wnt7b* from the adjacent non-dental epithelium, and this interaction restricts the region fated to become dental epithelium from the oral epithelial lining (Sarkar et al., 2000). Pitx2 thus qualifies as a crucial factor responsible both for the differentiation of epithelial dental progenitors and the establishment of tooth signaling centers and, as such, was proposed to act as a "master regulator" of tooth development in mice (Yu et al., 2020).

7.3.3 Specification of Tooth Competence in Ray-finned Fishes

In zebrafish, *pitx2* is expressed anteriorly along the jaw primordia and posteriorly in the caudal portion of the dorsal and ventral pharynx, although teeth develop only ventrally on the fifth ceratobranchial bone (Aigler et al., 2014). Previous research has found that this pharyngeal *pitx2* expression is downstream of retinoic acid signaling, as the perturbation of this pathway leads to the loss of *pitx2* expression and loss of teeth without impacting the underlying bones (Gibert et al., 2010). Upon exposure to the excess retinoic acid signaling, the *pitx2* expression domain expands anteriorly and leads to an ectopic induction of teeth; additionally, teeth emerge ectopically also in the dorsal pharynx (Seritrakul et al., 2012). In accordance with the role of retinoic acid on the zebrafish tooth development, mutants for *cyp26b1*, a gene encoding a retinoic-acid-degrading enzyme, show induction of ectopic tooth at the margin of the first tooth row (Gibert et al., 2015). Retinoic acid signaling thus regulates the extent of *pitx2* expression and, implicitly, tooth development in zebrafish. However, this regulation might occur only in zebrafish as the dependence of tooth development on the retinoic acid signaling has not been observed in other teleosts such as the medaka or the Mexican tetra (Gibert et al., 2010).

Interestingly, the induction of supernumerary teeth on the fifth ceratobranchial and in the dorsal pharynx in zebrafish may also occur upon overactivation of ectodysplasin (Eda) signaling (Aigler et al., 2014), thus phenocopying the effects of retinoic acid gain-of-function experiments (Seritrakul et al., 2012). However, unlike the zebrafish-specific role of retinoic acid on tooth development, the sufficiency of Eda signaling for tooth induction seems to be more general. Eda overexpression in the Mexican tetra, a characiform unrelated to the zebrafish, leads to the development of ectopic teeth at palatal and pharyngeal regions, and the effect of this gain of function experiment phenocopies the supposed ancestral characiform dentition (Figure 7.3, Jandzik and Stock, 2021). In these cases, the induction of ectopic teeth can be interpreted to be a result of a revival of tooth developmental potential at places where dentitions had once been present in predecessors of these lineages. This potential shows either a conserved or an atavistic ancestral competence for tooth development and contrasts with the zebrafish oral region where such potential has been lost, possibly due to spatial dissociation of *pitx2* and *shh* expression (Stock et al., 2006; Jandzik and Stock, 2021).

Taken together, *Pitx2* seems to be the factor responsible for tooth competence in vertebrates, although its sole presence is not sufficient for the odontogenesis to occur. *Pitx2* also regulates the size of the dental lamina/odontogenic band and is responsible for the advancement into successive stages of tooth development. This initial tooth competent region sets up the extent and shape of the dentition, namely the first-generation teeth in mammals, and the first tooth row or tooth patch in multigeneration dentitions of non-mammalian vertebrates. Further competence for tooth production is provided by the deeply invaginated successional dental lamina or the superficial adjacent epithelium. Together with the underlying mesenchyme, this epithelium provides a morphospace for the patterning mechanisms regulating the arrangement of teeth to take place.

7.3.4 How to Initiate Development of the Dentition: The Role of the Initiator Tooth

Compared to other RD-based patterns, such as in hair, for which multiple signaling centers appear at the same time in a large competent field (Mou et al., 2006), tooth development starts with the apparition of a single signaling center that initiates the

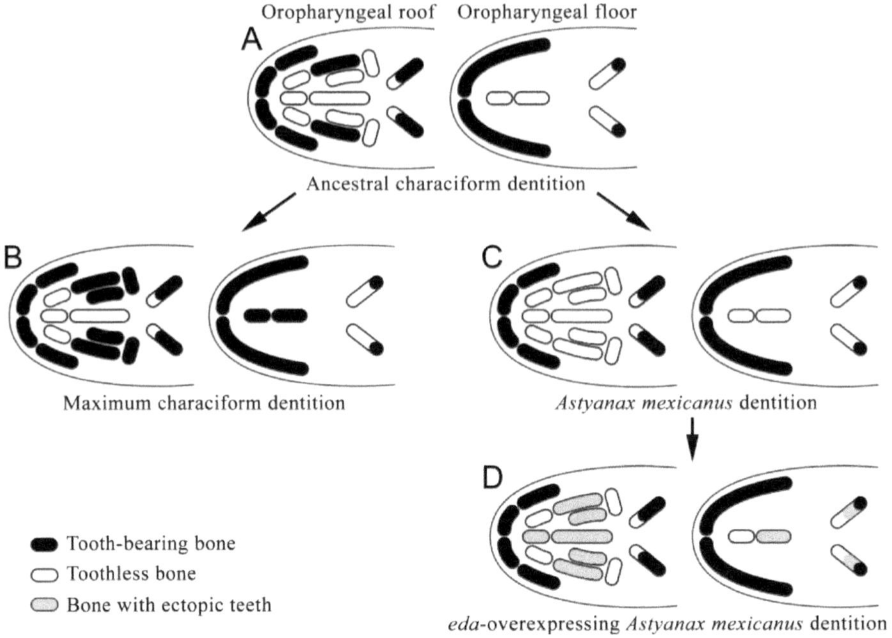

FIGURE 7.3 Gains and losses of dental competence in the characiform oropharyngeal cavities. A cartoon showing distribution of bones constituting the oropharyngeal roof and floor and reflecting (A) the dentition of the common characiform ancestor, (B) the maximum characiform dentition, and (C) the dentition in the Mexican tetra (*Astyanax mexicanus*). (D) Upon overexpression of *eda* in the Mexican tetra, teeth appear on the bones that are not normally tooth-bearing (grey), thus partially phenocopying the maximum extent of the characiform dentition (compare grey regions in D with identical white regions in C and black regions in B).

Source: Adapted from Jandzik and Stock (2021) (permission approved by The Royal Society).

formation of other teeth. The existence of this unique signaling center is due to the intrinsic development of the dental field that expands gradually, either in the jaw or pharyngeal regions, forming new competent regions as the development progresses. This particular aspect of pattern formation through space and time has led researchers to speculate about the existence of the so-called "initiator" tooth, whose role would be to signal the formation of a given tooth row. The initial steps of this first tooth formation have been intensively studied in zebrafish and mice and have not only revealed interesting features regarding how tooth rows are initiated in vertebrates (reviewed in Sadier et al., 2020), but also highlighted differences between clades.

In cypriniforms, the first tooth to form, 4V1, is located on the fifth ceratobranchial arch and corresponds to the fourth tooth of this row in the adult (Van der heyden and Huysseune, 2000; Gibert et al., 2010). This tooth forms early during development and is replaced before the acquisition of chewing, making it intriguingly nonfunctional from a feeding point of view. However, a closer examination of further dental development in zebrafish revealed another functional role of this tooth: instead of being used for chewing food, it triggers the development of the next two teeth of the row, 3V1 and 5V1, before being replaced by 4V2 that will play a role in chewing (Borday-Birraux et al., 2006). This observation led to propose that 4V1 has been maintained during evolution to initiate the dental cascade that triggers the formation of all the teeth of a given row (Van der heyden and Huysseune, 2000; Borday-Birraux et al., 2006). This hypothesis was verified in a series of elegant experiments which blocked 4V1 formation by manipulating FGF or RA signaling (Gibert et al., 2019). From these experiments, came the idea that the first tooth to form is nonfunctional and would only serve to initiate the Turing patterns that cascade successive teeth formation. The results found in other groups support this hypothesis and even suggest that this pattern could represent the ancestral condition (Butler, 1939; Sadier et al., 2020): in actinopterygians and chondrichthyans, each of their multiple rows is initiated one at a time, each row being suspected to initiate its own cascade from an initiator tooth. Interestingly, while a true dental lamina is not always present in these groups (e.g., for the rainbow trout), teeth still develop following a Turing-like cascade. Similarly, in cichlids, the development of the teeth and the different rows has been shown to be dependent on an initiator tooth (Fraser et al., 2008, Fraser et al., 2013). In chondrichthyans, such as the catshark, the first row of teeth has been shown to be nonfunctional and dependent on the apparition of a first initiator tooth. This first tooth and then row, which develop even before the dental lamina invaginates, are supposed to act as a conveyor belt that will generate the multiple tooth rows of these species (Rasch et al., 2016; Tucker and Fraser, 2014).

In other vertebrates, the situation is less clear. Some polyphyodont species such as the bearded dragon (Handrigan et al., 2010; Salomies et al., 2019) or the gecko (Zahradnicek et al., 2012) possesses a nonfunctional initiator tooth. This tooth signals the formation of subsequent teeth both on the caudal and rostral sides that are later replaced by functional teeth. However, in the veiled chameleon (Buchtová et al., 2013), while teeth are also initiated by a single functional initiator tooth, others form at a distance and then intercalate in a manner reminiscent of more classical "hair-type" patterns, making the contribution of this first initiator tooth to the others less clear.

In mammals, most of the studies regarding the initiation of tooth dentition have been done on mice molars, limiting our understanding of the development of the whole dentition. The three mouse molars develop successively one after another, through activation/inhibition mechanisms, as the primary dental lamina grows. These mechanisms have been precisely characterized (Kavanagh et al., 2007): The first tooth of this class to appear is the first molar, which would represent the first functional tooth of the row that initiated the development of the others. However, more recent research considers that the initiation of the first molar could not be separated from complex evolutionary history of muroid dentition during which mice lost their premolars but retained premolar vestigial buds, MS and R2 (Viriot et al., 2002; Prochazka et al., 2010). These buds are recognizable morphologically (Viriot et al., 2002) and express the classical molecular pathways of a tooth bud (Prochazka et al., 2010) but fail to drive a cap transition and abort successively right before the formation of the first molar. As MS and R2 disappear prior to the formation of the mouse first molar, it has been suggested that these premolar vestiges could initiate a whole premolar–molar cascade and act as "non-functional" initiator teeth (Prochazka et al., 2010). This result would also explain the phenotypes of some mutants that exhibit a supernumerary tooth in front of the molars (Charles et al., 2009a), as well as the phenotype of others in which activator/inhibitor pathways are impaired, leading to a "rescue" of R2 (Peterkova et al., 2009; Lochovska et al., 2015). More broadly, this hypothesis implies that premolar buds would have been maintained during evolution because of their fundamental role to odontogenesis to initiate the molar row and maintain a correct tooth positioning. However, more recent studies that have investigated the development of cheek teeth in other mammals contest this view, at least for the mammalian clade as a whole: In opossums, who possess two premolars and three molars, dP3 is the first to develop followed by the M1, C1, and P2. If dP3 initiates a cascade at the origin of cheek teeth, it goes both ways, anteriorly as well as posteriorly. In addition, the canine would be initiated independently (Moustakas et al., 2011), invalidating the existence of a nonfunctional initiator tooth. In noctilionoid bats, premolars and molars have been shown to arise through the development of two independent cascades, and the sequential apparition of the teeth (or their loss) is dependent on growth rate and location (Sadier et al., 2021). The sequence of apparition of teeth in bats is more largely confirmed by the morphological examination of the upper dentition of the fruit bat *Eidolon helvum* (Popa et al., 2016) in which premolars develop sequentially along the posterior–anterior axis (P3 to P1) before M1 initiates and potentially cascades the molar row antero-posteriorly. Together, these results suggest that, while each tooth class develops through an initiator tooth, this one is likely to be functional and initiate the development of the future teeth of the same class. The sequence of apparition of each tooth depends on the space available for other teeth, like in other nondirectional Turing patterns (Sadier et al., 2020, 2021).

7.3.5 The Molecular Basis of Mammalian Dentition Patterning

In mammals, the molecular origin of pattern formation has mostly been studied in detail in mice molars, and some genes playing the role of activators and inhibitors have been found. The initial study by Kavanagh et al. (2007) identified *Bmp4* and *ActivinβA* as activators of tooth formation. Further ones found other activators such as *Fgf20* and *Wnts* (Häärä et al., 2012; Lan et al., 2014) as well as inhibitors *Fgf4*, *Bmp2*, *Shh*, *Sostdc1*, *Follistatin* (Kettunen et al., 2000; Cho et al., 2011; Navarro and Murat Maga, 2018) and some intermediate genes (Cho et al., 2011), including genes of the Eda pathway (Lefebvre and Mikkola, 2014; Sadier et al., 2014). As expected, mutations in the genes of these developmental pathways (FGF, HH, Wnt, BMP and Eda) often modify the size of the tooth germs and/or result in supernumerary teeth due to patterning defects (Kristenová et al., 2002; Charles et al., 2009a, 2009b).

In mutant mice such as in *Edar*-deficient mice, the supernumerary tooth located in front of the first molar was shown to be a rescue of the R2 signaling center, the vestige of the premolar that was lost in the mouse lineage. In the last decade, several studies investigated the molecular basis of R2 rescue and revealed a fairly complex picture regarding the patterning of the vestigial premolars and its impact on the Turing mechanisms that drive molar row development (Peterkova et al., 2009; Ahn et al., 2010; Prochazka et al., 2010; Häärä et al., 2012; Lagronova-Churava et al., 2013). In particular, these findings suggested that the development of vestigial premolars might influence molar development in mice.

To study this idea, a recent work investigated the dynamics of activation waves in the molar field during the patterning of the molar row (Sadier et al., 2019). Using *Edar* expression as a readout of activation level in the molar field, this work highlighted the Turing mechanisms at play in the dental epithelium and revealed that tooth formation starts with a broad activation of the dental field followed by a spatial restriction to a signaling center that marks tooth position (Figure 7.4A). To explain this dynamic pattern and recapitulate the variation of *Edar* expression during molar formation (broad activation followed by a restriction), the authors built a bistable RD system that incorporates an activator and an inhibitor. This bistable RD regime is reiterated as the primary dental lamina grows during the formation of the successive molars as they reach a certain distance from the inhibition field from the previously formed molar (Figure 7.4B). Interestingly, the premolar vestiges, MS and R2, undergo the same *Edar* broadfield activation and restriction before degenerating, which suggests that these bistable RD mechanisms also govern the development of

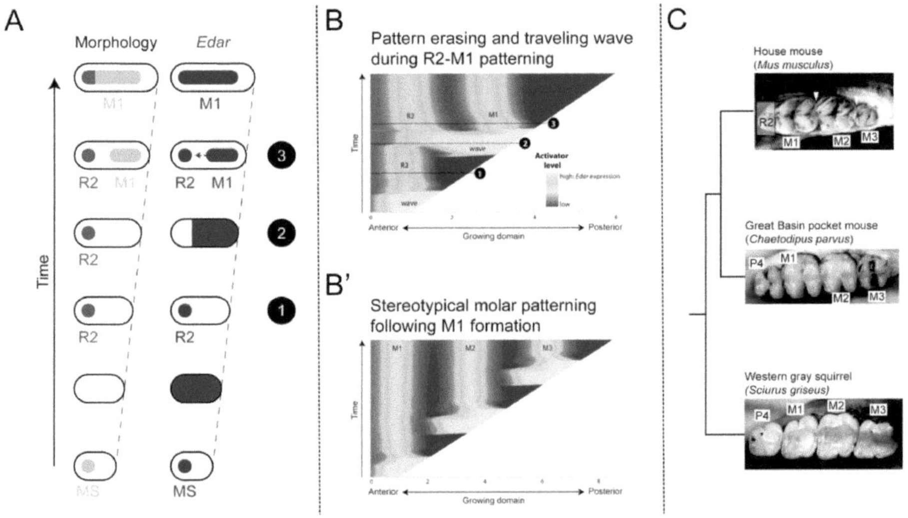

FIGURE 7.4 A developmental palimpsest of alternating reaction–diffusion and bistable regimes regulating the initiation of the mouse molar row. (A) Schematic representation of the apparition of the successive signaling centers (MS, R2, and M1) (left) and Edar expression (right). (B) Model of pattern formation on growing domain involving RD mechanisms, pattern erasing, and traveling waves reflects the observed plasticity during R2 and M1 development (note the corresponding stages in A and B labeled by black circles). (B′) Model of pattern formation on growing domain involving only regular RD mechanism during the successive formation of the M1–M3 molars. (C) Lower jaw cheek teeth of the house mouse, the Great Basin pocket mouse and the Western gray squirrel. The morphology of the mouse M1 with a larger anterior part is supposed to be the consequence of the incorporation of the R2 into M1. The two other rodent lineages that conserve their P4 lack the anteroconid in M1. B and B′ are adapted from Sadier et al. (2019) (CC BY license), C adapted from Prochazka et al. (2010) (permission granted by the NAS) and a courtesy of Dr. Jessica Blois (University of California, Merced).

premolars. Surprisingly, during M1 formation, the broad activation of *Edar* extends anteriorly in the zone of the aborted R2, revealing that 1) the inhibition driven by the aborting signaling centers is not strong enough to compete with the newly developing M1, and 2) newly forming teeth compete with the previously formed ones and can erase previous patterns, revealing that pattern formation is highly dynamic and not direct (Figure 7.4B′). To further explore the interaction between R2 and M1, the authors further studied *Edar* dynamics during M1 formation and revealed that R2 is incorporated into the M1 primary enamel knot, forming a large enamel knot in the lower jaw, but not the upper one, explaining the morphology of the lower molar in mice. Indeed, the mouse exhibits a larger M1 than other rodent lineages that have not lost their premolars (Figure 7.4C). To reconcile this unprecedented behavior with the bistable/RD system in place during molar formation, the authors introduced chemotaxis, which explains how cell migration also contributes to the formation of the large primary enamel knot during lower M1 development. Together, these experiments not only explain in detail how Turing regimes control the dynamics of apparition of successive molars in mice, but also reveal interesting characteristics about the balance between the different signaling centers. Finally, this study also reveals how RD parameters can be fine-tuned during evolution to produce variation in combination with other developmental mechanisms. Interestingly, RD mechanisms are also implicated in the successive apparition of cusps (Jernvall and Thesleff, 2000), and later models, mostly based on RD mechanisms, have revealed that diffusion accounts for a substantial proportion of the variation seen in teeth (Salazar-Ciudad and Jernvall, 2010). Later studies linked the patterning of the cusps with the underlying developmental pathways (Harjunmaa et al., 2012, 2014) and revealed that cusps are patterned by developmental cascades that could originate from the same tooth developmental program. Recent research has revealed that alterations of this program could explain the variation of cusp patterns observed in teeth (Harjunmaa et al., 2014; Couzens et al., 2021).

7.3.6 Periodic Pattern Generators as Assemblers of Multirowed Dentitions

While mammalian dentitions are arranged into a single tooth row, many nonmammalian vertebrates, including chondrichthyans, actinopterygians, amphibians, and reptiles, possess dentitions with several tooth rows erupted and functional at a time. How does a dentition containing many teeth and tooth rows arise? Cichlids represent a wonderful group of fishes, where patterning of multirowed dentitions can be studied thanks to their enormous phenotypic diversity despite a very recent common ancestor of this lineage. This diversity comprises species-specific tooth size and number of tooth rows, so that different species may display different dental phenotypes: Large-toothed species contain few tooth rows and tiny-toothed species many tooth rows. A Turing-like pattern generator was proposed to periodically organize teeth into rows and tooth rows into jaws during the establishment of these diverse dentitions (Figure 7.5, Fraser et al., 2008). This proposed pattern generator contains several factors including Pitx2, Shh, Eda, Edar, and Wnt7b, whose mutual interactions set the basic properties of the nascent dentition. The Wnt7b–Shh interaction, shown to discriminate the fate of oral versus dental epithelium in the mouse (Sarkar et al., 2000), and the Wnt/β-catenin Lef1-mediated Eda–Edar interaction, responsible for tooth and hair placode development (Laurikkala et al., 2001, 2002), supposedly control the positioning of individual tooth germs within the tooth-competent *Pitx2* expressing odontogenic band (Fraser et al., 2008). These molecular interactions further regulate the diameter of tooth germs and the inter-germ zone, i.e., the zone of inhibition of Osborn (1974), and thus spacing among the teeth. The periodically repeating Turing-like mechanism is then responsible for iteratively triggering the subsequent tooth rows, leading to multirowed dental arrangements (Fraser et al., 2008). While this periodic pattern generator is responsible for the establishment of the general multirowed dentition ground plan, subtle molecular tinkering within this groundplan pattern generator is then accountable for the various size, number, and

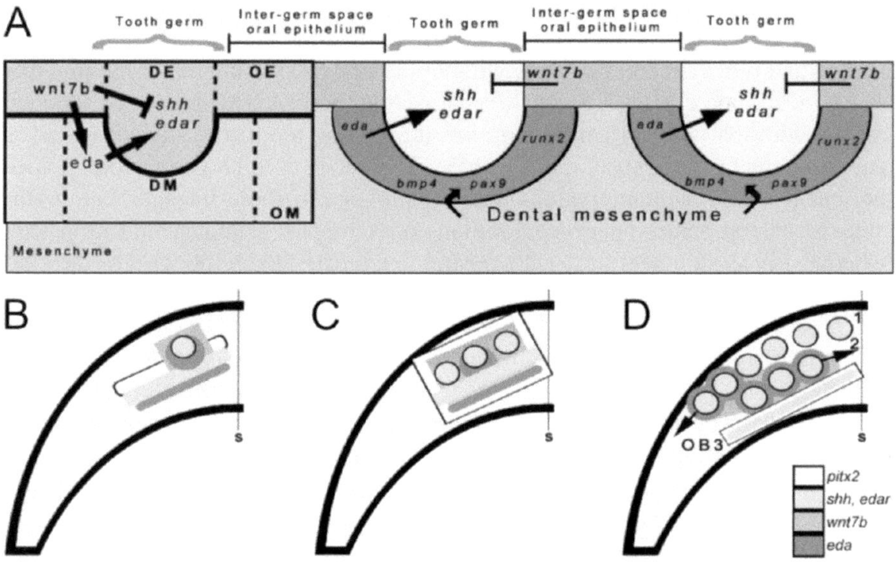

FIGURE 7.5 Periodic pattern generator for the development of multirowed cichlid dentitions. (A) Periodic patterning is facilitated by molecular interactions between the dental epithelium, giving rise to a tooth germ and non-dental (oral) epithelium destined to form the inter-germ space. Molecular mechanisms comprise inductive (wnt7b–eda–edar) and repressive (wnt7b–shh) interactions from oral to dental epithelium. The balance between inductive and repressive interactions regulates the size of tooth germs and the inter-germ spacing and provides a proximate cue for the diversity in dental phenotypes in closely related Lake Malawi cichlid fishes. (B–D) Periodic patterning regulates positioning of tooth germs into the row within the *pitx2*-positive tooth-competent regions. Subsequent tooth rows develop in a similar way from the lingually descending odontogenic band marked by the *pitx2*, *shh*, and *edar* expressions. *Abbreviations*: DE, dental epithelium; DM, dental mesenchyme; OB3, odontogenic band of the third tooth row; OE, oral (non-dental) epithelium; OM, oral (non-dental) mesenchyme; s, mandibular symphysis.

Source: Adapted from Fraser et al. (2008) (CC BY license).

spacing of teeth and tooth rows, thus reflecting the diversity of dental phenotypes among the various cichlid species.

The periodic pattern generator for development of multirowed dentitions proposes a copy-and-paste mechanism of how to organize teeth into rows and tooth rows into jaws, thus conceptually separating these two processes (Fraser et al., 2008). In support of this conceptual separation, development of many multirowed dentitions follows this proposition. For example, in the lesser spotted catshark *Scyliorhinus canicula*, tooth germs are first added from a pair of mesially placed initiator-teeth to fill the extent of the embryonic jaw with a row of five tooth germs, and, next, tooth germs are added in alternate positions to form second and subsequent tooth rows until creating the hexagonal grid (Smith et al., 2009b; Rasch et al., 2016), an arrangement likely plesiomorphic phylogenetically and embryologically for shark dentitions (Underwood et al., 2016; Smith et al., 2018). In other elasmobranchs, however, the embryonic development of dentitions occurs differently. In the rays *Discopyge* and *Myliobatis*, the development proceeds directly from the parasymphyseal initiator-tooth germ present on either side of the jaw by an alternate addition of teeth of the second and other consecutive rows without the prior establishment of the first tooth row (Underwood et al., 2015). In these cases, the toothed area spreads by the incremental addition of tooth germs per each developing tooth row, thus progressively enlarging the field in a splay-mode fashion. This mode of tooth germ addition is also present in many actinopterygians and amphibians. In these lineages, the toothed areas at palatal and/or pharyngeal regions often display arrangement into patches, where tooth development starts with positioning the first tooth followed by the placement of two subsequent teeth lingually, together constituting a triangle. Defined by the position of these three teeth, a continued tooth germ addition into the sector of a circle progressively spreads the toothed area in a splay-mode fashion (Figure 7.6, Nakajima, 1984; Soukup et al., 2021; Pospisilova et al., 2022). Tooth germs of new tooth rows develop in alternate positions, and the splay-mode behavior expands the tooth field similar to the development of *Discopyge* and *Myliobatis* dentitions (Underwood et al., 2015). Either way, both these modes of development of multirowed dentitions create an arrangement into a honeycomb lattice, thus suggesting the presence of an innate pattern generator for hexagonal packing.

How patterning into the row versus patch is regulated can be assessed grossly from the initial shape of the tooth competent zone, although the induction and precise placement of the initiator tooth, from where this zone would be filled up with teeth, are currently unknown. Indeed, studying the development of diverse types of dental arrangements present in a single species, such as the medaka, sturgeon, pike, or axolotl (Debiais-Thibaud et al., 2007; Pospisilova et al., 2019, 2022; Soukup et al., 2021), may help disentangle differences in the molecular regulation of initial steps in dentition patterning (see also Salomies et al., 2019). A more solid dataset in this respect, including functional experiments, is, to a large extent, yet to come.

7.3.7 Dental Stem Cells as the Source of Patterned Replacing Dentitions

A crucial topic of developing dentitions in the majority of nonmammalian vertebrates is how the primary dental pattern established during the initiation of the

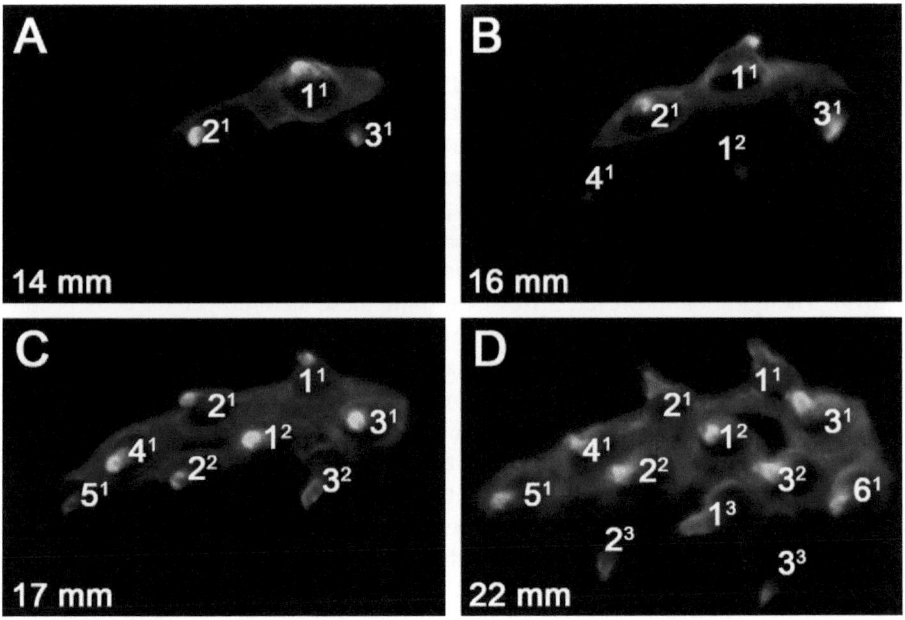

FIGURE 7.6 Splay-mode fashion of tooth addition. Successive stages of development of the left hypobranchial 1 tooth field of the sterlet sturgeon show a splay-mode posterior addition of new tooth germs. Teeth are designated according to their order of appearance (capital) and position within a tooth family (superscript). The initiator-tooth of the field (1^1) is placed anteriorly, and, together with the other two consequently appearing teeth (2^1 and 3^1), form a triangle (A), thus demarking a sector of a circle into which further tooth germs are added (B–D). The addition of tooth germs in alternate positions leads to the enlargement of the field and packing of teeth into a hexagonal lattice (D). Individual stages were assessed according to the total length of the fish. Figures are a courtesy of Dr. Anna Pospisilova (Charles University, Prague).

dentition is maintained at subsequent rounds of tooth replacements. Functional teeth are usually replaced from behind from the successional dental epithelium (Square et al., 2021) either in the form of the deeply invaginated dental lamina (Reif, 1982; Smith et al., 2009a; Whitlock and Richman, 2013) or the adjacent superficial epithelium (Berkovitz, 1977a; Huysseune and Witten, 2008; Vandenplas et al., 2014; Smith et al., 2015; Vandenplas et al., 2016; Pospisilova et al., 2022). The descending successional dental epithelium and the adjacent mesenchyme represent the generative substrate for vertebrate dentitions by providing cellular material for the successive appearance of replacement teeth, and it is these tissues where the pattern set during the initial phases of dental development is maintained or becomes modified (e.g. Huysseune and Witten, 2006; Underwood et al., 2016).

For the continuous tooth replacement to occur, the dentitions must employ epithelial and mesenchymal stem cells that provide the source of progenitor and differentiating tooth-producing cells. The last couple of years has been fruitful in the identification of niches of dental stem cells, mostly in the epithelium (Handrigan et al., 2010;

Wu et al., 2013; Abduweli et al., 2014; Vandenplas et al., 2016; Salomies et al., 2019). For example, in sharks and skates, these epithelial dental stem cells reside in Sox2+ niches found in the basal layer at the lingual aspect of the dental lamina, termed the taste–tooth junction, and these cells contribute superficially to the taste bud-forming epithelium and profoundly to the tooth-forming successional dental lamina (Martin et al., 2016; Rasch et al., 2020). During their progression deeply into the aboral portion of the dental lamina, the stem cells start proliferating, take on a progenitor cell fate, lose their Sox2+ identity, and, thanks to the activated Wnt/β-catenin signaling pathway, differentiate into tooth-producing epithelial cells (Martin et al., 2016; Salomies et al., 2019). The question, however, arises as to how these stem/progenitor cells become employed to maintain the species-specific dental patterns.

In the continuous successional dental lamina of sharks, a cyclical stop/start hypothesis was proposed to explain how to pattern the alternate order of tooth families (Fraser et al., 2020). Here, the successional dental lamina is physically restricted from furrows within the Meckel's and palatoquadrate cartilages, limiting its maximum oral-aboral extent. As teeth progressively migrate toward occlusion in a conveyor-belt-like manner, space becomes available in the aboral part of the successional dental lamina for new tooth germs to be initiated. Cyclical initiation of tooth germs occurs due to the cyclical activity of the Wnt/β-catenin pathway that drives differentiation of Sox2+ progenitors into mature tooth-producing cells. In shark dentitions where tooth families take on a staggered arrangement, a stop phase of the cycle in one family prevents initiation of a new tooth, while a start phase in adjacent families promotes initiation of a new tooth. Alternating repetitive cycles of stop/start phases in adjacent tooth families would then account for the development of the stereotypic hexagonal tooth grid (Fraser et al., 2020).

The multirowed dentitions of sharks and rays represent wonderful systems for the study of tooth patterning. Besides the growing amount of developmental data, there are also examples of anomalous dentitions that may help unravel the underlying patterning mechanisms. The dental anomalies range from insertions of a single tooth, insertions of a new tooth family, splitting of a tooth family, to defects in an overall replacement pattern (Gudger, 1933, 1937; Reif, 1976; Hovestadt and Hovestadt-Euler, 2013; Delpiani, 2016). These anomalies are most likely reflections of early abnormalities in local control of employment of tooth progenitors and, unlike the tooth pattern changes in the progressively degrading dentitions, such as that of sturgeons (Pospisilova et al., 2022), may be informative in the identification of molecular pathways and morphogenetic mechanisms standing behind the replacement of patterned dentitions.

7.4 PROSPECTS AND IDEAS FOR PERIODIC TOOTH PATTERNING

7.4.1 Mathematical Modeling for Periodic Tooth Patterning

As with other integumentary organs, vertebrate dentitions show patterned arrangements. However, while the development of integumental patterns has readily been linked to the dynamics of Turing-like reaction–diffusion (RD) models, those of developing dentitions have not, or at least not explicitly. An association of tooth development with Turing-like mechanisms has long been proposed (Kulesa et al., 1995, 1996a,

1996b; Stock, 2001; Donoghue, 2002), yet a solid connection of tooth patterning to RD kinetics has been rather scarce (Cho et al., 2011; Sadier et al., 2019, 2021). This may, from a superficial point of view, seem unexpected, given the ever-growing knowledge of molecular underpinnings regulating tooth development, but the causation for this discrepancy is that at least two requirements are about to be provided and fulfilled for the activity of Turing-like RD mechanisms during the development of vertebrate dentitions to be evaluated. One of them comprises clear identification of molecular players behaving according to the RD mechanisms. The other is the application of a suitable computational model with RD kinetics capturing the sequential phases of formation of the respective toothed region from initiation to completion of the pattern. Here, we capture several theoretical examples in this respect.

1) Models simulating the establishment of spotted patterns on the basis of basic RD kinetics within a restricted competent field, such as those used to mimic pattern establishment of mammalian hair follicles (Mou et al., 2006; Sick et al., 2006), may help in understanding the basic properties of dentitions, such as the tooth size and inter-tooth spacing. These models do not propose spatial directionality in the pattern establishment, a feature typical for developing dentitions, but rather lead to non-stereotyped patterns (Figure 7.7A). Despite this, they may still be helpful in modeling the development of real dentitions, namely those, where the pattern directionality is missing or at least is not proposed to take place. This comprises pharyngeal tooth plates and ornamented or shagreen regions on the palates present in many early branching fossils and extant actinopterygians and sarcopterygians, such as the bowfin, bichir, gar, *Eusthenopteron*, and coelacanth or in derived actinopterygians such as the Mexican tetra (Jarvik, 1955; Nelson, 1969; Clemen et al., 1998; Grande and Bemis, 1998; Grande, 2010; Jandzik and Stock, 2021). These regions generally serve to roughen the oropharyngeal surfaces or hold the caught prey during its transfer to the esophagus. To our knowledge, no detailed study has been performed on the development of these dentitions, but once relevant studies are performed, the basic RD models may help provide theoretical background for the establishment of a shagreen-type pattern.

2) The vast majority of toothed regions, however, displays directional tooth germ addition and spatial expansion during development. Application of models of pattern establishment on growing domains may help provide a mechanistic background for the successive addition of teeth into a row, such as those on the jaws. The advantage of using models incorporating growing versus non-growing domains is that very regular patterns are formed if patterning occurs during the growth (Krause et al., 2019). In isotropically growing domains, the tissue expansion proceeds more or less uniformly across the extent of the competent zone. Thus, *e.g.* due to the jaw growth, new tooth germs become progressively added in between the already formed teeth (Figure 7.7B). As an example, a model comprising dynamic RD system involving agents acting as an activator, inhibitor, and substrate on the isotropically growing domain reproduces the spatial and temporal sequence of tooth germ appearance in the alligator jaw suggesting that a Turing-based RD mechanism is at play during the development of the dentition in this species (Kulesa and Murray, 1995; Kulesa et al., 1996a, 1996b; Murray and Kulesa, 1996, see Murray, 2003). In anisotropically growing domains, on the other hand, periodically repeating units form from a propagating

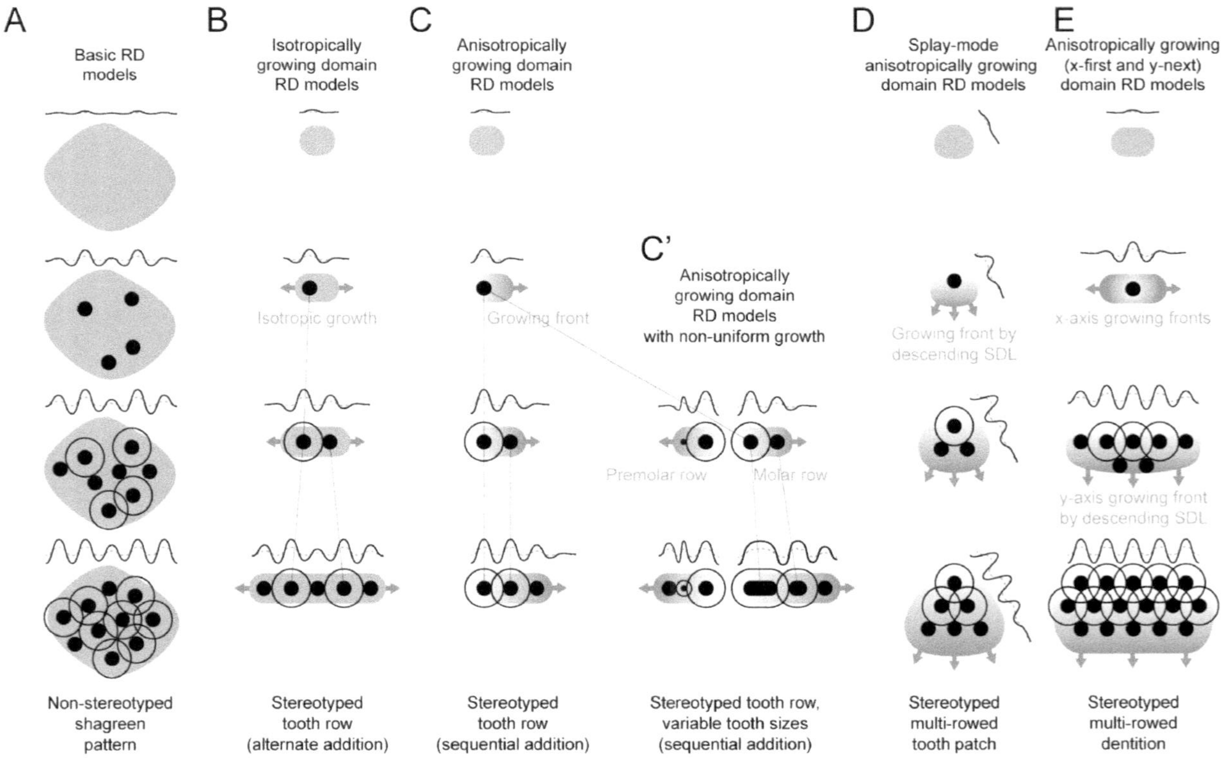

FIGURE 7.7 Models of reaction–diffusion mechanisms applicable to developing dentitions. (A) Basic RD models on a static field of competence result in non-stereotyped patterns best comparable to shagreen-type dentitions. (B) RD models on isotropically growing domains lead to an alternate order of tooth-germ appearance and to stereotyped tooth row patterns (e.g., the development of the alligator tooth row). (C) RD models on anisotropically growing domains lead to a sequential order of tooth germ addition forming stereotyped tooth row patterns. (C′) Modifications to the RD models on anisotropically growing domains (such as a nonuniform growth) may lead to variable tooth sizes (e.g., the development of the noctilionid bat premolars and molar cascades). (D) RD models on splay-mode anisotropically growing domains lead to stereotyped tooth patch patterns by sequential addition of new tooth rows in continuously growing successional dental lamina (SDL). (E) RD models combining growth of domains in x-axis (mesiodistal) first and y-axis (labio-lingual) next may lead to stereotyped multirowed dentitions through sequential addition of successive tooth rows.

front sequentially one after the other (Figure 7.7C). This is for example the case of sequential addition of digit phalanges or mammalian cheek teeth (Kavanagh et al., 2007; Suzuki et al., 2008; Gaete et al., 2015; Scoones and Hiscock, 2020). Interestingly, application of mathematical models of RD mechanisms on differentially growing tooth-forming domains in noctilionid bats has recently revealed that simple modulation of the anisotropic growth rate of the jaw may explain variability in the number and size of teeth (Sadier et al., 2021, Figure 7.7C′). Both sequential and alternate tooth germ addition is common among extant vertebrates allowing broader inclusion of models of RD mechanisms on isotropically and anisotropically growing domains to provide theoretical background on dental patterning.

3) Many multirowed dentitions develop by the sequential addition of tooth germs into a sector of a circle, and this splay-mode addition is due to initiation within the propagating successional dental lamina or odontogenic band (Figure 7.7D). Interestingly, the progressive increase in the number of tooth germs during the growth of the respective tooth field resembles, *e.g.*, the increasing number of cartilaginous elements during the distal outgrowth of the tetrapod limb. Here, models of RD mechanisms on anisotropically growing domains do simulate the observed sequential patterns of condensing cartilaginous elements not only in limbs of various vertebrates but also mutant or experimentally abrogated limbs (Miura et al., 2006; Zhu et al., 2010; Onimaru et al., 2016; Glimm et al., 2020). These models have thus provided a theoretical background for a global mechanism, whose modulations reflect developmental and evolutionary variations of the limb skeletal elements. Analogically to the outgrowing limb, the development of tooth patches, such as those found in amphibians, actinopterygians or some rays, proceeds by the incremental addition of new tooth germs from the progressively expanding propagating front (Figure 7.6). From a modeling perspective, a propagating front allows patterns to form in a sequentially controlled and robust manner, and the system is less prone to pattern defects caused by the noisy biological environment (Woolley et al., 2011; Maini and Woolley, 2019). Models of RD mechanisms on growing domains may thus provide a theoretical background for understanding the incremental addition and reproducible stereotypical patterning of tooth patches.

4) Similar to pattern establishment during the splay-mode tooth germ addition, development of multi-rowed dentitions, such as those present in sharks, may also be linked to patterning on growing domains (Figure 7.7E). Here, the first tooth row is initiated superficially by the compartmentalization of the Shh-positive odontogenic band, and tooth germs of subsequent rows develop from the lingually invaginated dental lamina (Smith et al., 2009b; Rasch et al., 2016), which provides morphospace for ordered tooth germ addition and patterning. While the growth of the dental lamina may be the main cue for the establishment of the hexagonal pattern, an appealing analogy may be a comparison of the sequence of tooth addition in the developing dentition to patterning systems where a propagating wave acts to trigger the innate ability of the tissue to self-organize, such as that present in case of plumage patterning in volant birds (Bailleul et al., 2019; Ho et al., 2019). Both of these systems comprise the initial zone (stripe) of competence (feather tract/odontogenic band) and the sequential addition of new units (feather follicles/tooth germs), where the staggered arrangement of rows takes place after the pattern establishment in

the first (competent stripe) row. Apart from the bi- versus unidirectionality, a major difference is the non-growing (feather follicle appearance sequence) versus growing (dental lamina) property of these two systems. Nonetheless, the propagating character (either by the Eda wave or the growing front) seems to promote precision in the final lattice arrangement, and both systems may thus share a comparable behavior.

7.4.2 Identification of Molecular Players in the Reaction–Diffusion Mechanisms

As noted in the previous section, a crucial premise for the presence of Turing-like mechanisms playing a role in dental patterning is the clear identification of molecular players behaving according to the RD dynamics. This premise is as old as the modeling of RD biological phenomena themselves (Marcon and Sharpe, 2012), but, due to the complexity of gene regulatory networks underlying the development of any periodically repeating structures, teeth notwithstanding, the decisive designation of RD molecular players is a challenging task. While the simplest Turing models involve interactions between two components (*i.e.*, factors or pathways), patterning of periodically repeated structures may contain three- or multi-component relationships. If we, for example, take the case of a simple three-component system, where each one of the players interacts with the other two and with itself, and the interactions may be positive, negative, or neutral, these 3^9 combinations equal 19,683 possible network topologies. Moreover, these interactions may not only be qualitative but also quantitative; and, indeed, their number grows with the growing number of players within the network. A recent approach by Economou et al. (2020) provides a framework to scale down the number of possible network topologies relevant for the establishment of periodically repeated structures. But still, picking the right candidate factors or pathways may be an uneasy task, and the situation may also be complicated by other phenomena acting above or in combination with the potential simple Turing pair, such as growth, cell movements, mechanical morphogenesis, propagating waves, biological noise, alternating Turing regimes, etc. (Ho et al., 2019; Sadier et al., 2019; Scoones and Hiscock, 2020).

This oftentimes non-deconstructable complexity is the reason why simple identification of the Turing pair(s) has not taken place, and many up-to-date theoretical models serve as exploratory frameworks for developing dentitions rather than platforms for directly testing the patterning roles of candidate pathways in this process (Sharpe, 2017). Although, indeed, the exploratory role of these models brings a valuable framework for the evaluation of phenotypic diversity (Sadier et al., 2021). Theoretically, there may be a number of molecules or pathways behaving according to Turing mechanisms. Empirically, examples from integumentary organs comprise Wnt/Dkk and Edar/Bmp/CTGF in patterning the mammalian hair (Mou et al., 2006; Sick et al., 2006) or Fgf4+Shh/Bmp2+Bmp4 and Bmp7/Bmp2 in patterning the feather follicles (Jung et al., 1998; Michon et al., 2008), but more examples of Turing pairs can be found to supposedly play role in the development of other organs (Marcon and Sharpe, 2012). In the case of mouse teeth, a network of activator (Wnt), inhibitor (Sostdc1), and mediator (Shh) was proposed to pattern the molar row (Cho et al., 2011), and a complex modulation of Turing regimes together with cellular chemotaxis was also proposed (Figure 7.4, Sadier et al., 2019). Naturally, the RD candidate molecules or pathways may vary in different systems, and there are examples where the same pathway may act as an activator in one system and an inhibitor in the other (Marcon and Sharpe, 2012). Moreover, and somewhat counterintuitive at first glimpse, not only extracellularly released ligands and their inhibitors, but also other components of the pathway, such as receptors, may be involved in the multi-component RD circuitries (Mou et al., 2006).

As stated in the previous section, only the precise identification of Turing molecules together with the process-simulating model may lead to a robust synergistic relationship between the theory and the experiment; and it is encouraging to see first synergies in the context of periodic tooth patterning appearing progressively.

7.4.3 The Origin of Tooth Classes in Mammals

Periodic patterning of mammalian dentition and the origin of tooth classes has been a topic of interest in classical comparative odontology since the mid-20th century (Butler, 1939; Osborn, 1978), but the implication of Turing mechanisms for tooth patterning has been established only relatively recently (Kavanagh et al., 2007). While the molecular mechanisms controlling this patterning have been deeply studied since then, especially in case of mouse molars (Prochazka et al., 2010; Sadier et al., 2019), they remain relatively under-studied in other mammalian clades, in particular those with a complete dentition, which constitute better models to study the evolution of tooth patterning for other tooth classes. As a result, we have a limited understanding of how other tooth classes are patterned. In particular, the following questions remain to be addressed: Are the molecular signals that drive tooth patterning conserved or divergent among tooth classes? Are all tooth classes patterned by the same type of developmental cascade or are they initiated by different RD mechanisms? Up to recently, the differences between tooth classes were supposed to be established by a homeobox code, a pre-pattern established before the apparition of tooth buds, which delineates the presumptive fields of each tooth class (Sharpe, 1995, reviewed in Catón and Tucker, 2009). Regarding later stages, e.g., tooth initiation and morphogenesis, some studies have revealed an overall conservation of the gene expression pattern among tooth classes with some differences, for example, in the opossum (Moustakas et al., 2011), in the house shrew (Yamanaka et al., 2007), or between incisors and molars in mice (Jheon et al., 2013) without proposing a developmental model to link these observations to the origin of tooth classes. Very recently, a transcriptomic analysis has revealed differences in gene expression between tooth classes in the domestic cat; although this study covered rather late developmental stages (Woodruff et al., 2022). Therefore, more studies remain to be done on species with a complete mammalian dentition to fully understand which molecular mechanisms drive the differences between tooth classes and if these differences play a role in dental patterning at the early stages of tooth development.

While the molecular origin is poorly understood, even less is known about the RD mechanisms responsible for the patterning of each tooth class. Results obtained in mice could suggest a common origin for the patterning of premolars and molars: The vestigial premolar buds develop first, one after another, preceding the lower M1 formation, which suggests that both premolars and molars initiate from a unique premolar center. In addition, the Turing mechanisms that pattern vestigial premolar buds interact with molar patterning (see section 3.5, Prochazka et al., 2010; Sadier et al., 2019), revealing a complex interdependence of these two tooth classes in mice (Figure 7.4B). However, the extension of these results to other mammals remains uncertain: Murine dental patterning is complicated by the loss of canines and premolars, and the dentition is not representative of the ancestral mammalian dentition. Recent results on other clades with a complete dentition could offer a better insight into the developmental differences between tooth classes and the role of Turing mechanisms in establishing them.

In bats, canines are the first teeth to develop, independently from the apparition of postcanine teeth (Popa et al., 2016, personal observations). Then, premolars and molars develop from two distinct buds appearing at the same time, which then progress in two directions (Sadier et al., 2021), suggesting that premolars and molars are patterned by two independent RD mechanisms and/or are influenced by the prepatterning of the field of each tooth class. This idea is in accordance with the distinct patterns of losses between premolars and molars, which are observed in the clade of noctilionoid bats (Sadier et al., 2021) and the sequence of apparition observed in other mammals with a complete dentition such as the house shrew (Yamanaka et al., 2007, 2010) or the ferret (Jussila et al., 2014). While the independence of these RD mechanisms, especially between premolars and molars, has not been characterized molecularly, these results suggest that separate Turing mechanisms initiate the development of each tooth class independently. Finally, the modeling studies based on simulations that incorporate different parameters (e.g., growth, RD mechanisms) have revealed that the modification of multiple developmental mechanisms influences the patterning of the teeth and their cusps (Harjunmaa et al., 2014; Couzens et al., 2021). In the future, more integrative studies will be needed to study how the different RD mechanisms potentially interact and/or compete with each other in more species with a complete dentition.

7.4.4 Is the First Tooth Always Non-functional?

The results obtained in zebrafish suggested that the first tooth could be non-functional and potentially act as an initiator of tooth patterning in all vertebrates. As stated in Section 7.3.4, results obtained in other species revealed that this is far from being a common case and that the presence of a non-functional initiator tooth could depend on taxa, tooth type, or life history traits. In axolotl, which possesses different dental fields, the initiator tooth forms at the ectoderm–endoderm boundary or from the ectodermal region and initiates the formation of the other teeth of the field (Soukup et al., 2021). In the sterlet sturgeon, oral tooth fields are single-rowed, non-renewed, and, while being functional for some time, they are shed quite early during development (Pospisilova et al., 2022), a fate similar to the vomerine and dentary teeth in the lungfish (Smith and Krupina, 2001). The palatal and pharyngeal tooth fields, on the other hand, form multiple tooth rows and are functional for much longer time, although they become eventually lost, too. Thus, while the adult sterlet is completely edentulous, the initiator-teeth of all its temporary tooth fields are functional (Pospisilova et al., 2022). Likewise, the first-formed fetal dentition of viviparous caecilians (Wake, 1976), or the first teeth of oviparous mammals and reptiles, i.e., the egg teeth (Hermyt et al., 2017, 2020; Fons et al., 2020; Fenelon et al., 2021) are functional. On the other hand, the first-generation teeth of the rainbow trout (Berkovitz, 1977b), cichlids (Huysseune and Sire, 1997), gecko (Zahradnicek et al., 2012), or alligators (Westergaard and Ferguson, 1986) are non-functional or barely functional.

In mammals who exhibit a complete dentition, the initiator tooth of each tooth class is likely to be functional (Sadier et al., 2020, 2021), each first tooth initiating the development of all the teeth of the same class. However, this trend seems to be modified in some particular cases of very derived dentition, revealing interesting evolutionary patterns. For example, in elephants, tusks are modified enlarged incisors that grow continuously during the life of the animal (Raubenheimer, 2000). On the contrary to other typical mammalian incisors that either develop directly or are renewed once, tusks develop from a non-functional tooth germ called "tush" that develops, mineralizes, and never erupts but initiates the development of the permanent tusk (Raubenheimer et al., 1995; Raubenheimer, 2000). The tush, which is reabsorbed by the adjacent tissues when the tusk forms, can thus be considered a decidual tusk whose role is to initiate the permanent tusk. The reason for the development of this non-functional tooth is unknown and could be linked to the duration of the pregnancy of the elephant: with a shorter development, this decidual tooth could have erupted before being replaced by the adult dentition. Other tusked animals also possess non-functional teeth that can be linked to differences in sexual selection. For example, in dugongs, tusks that are also a case of modified incisors form in both male and female but only erupt in males and some females older than 40 years upon the exposure to testosterone (Marsh et al., 1984; Burgess et al., 2012). In female dugongs, these tusks could thus be considered as non-functional, whereas their role in older females remains unclear. Altogether, these results suggest that the initiator tooth is far from always being non-functional. Exploration of more clades reveals that life history traits are likely to influence the patterns of tooth formation and that the development of non-functional teeth could be maintained by evolution to initiate development of the permanent dentition.

More broadly, these examples reveal that teeth can be rapidly gained and lost at various stages of their development and across generations in a similar way as in other vestigial structures (Sadier et al., 2021). In edentulous modern birds that have lost teeth 65 Myr ago, vestigial tooth germs are initiated in chick embryo, (Louchart and Viriot, 2011). While some terminal "tooth-producing" genes have been lost, development of teeth can still be reactivated experimentally (Harris et al., 2006). Of note, teeth have been lost multiple times during evolution of the avian lineage (Louchart and Viriot, 2011). This is reminiscent of the situation in anurans, where teeth have also been lost several times either altogether or at certain positions within the mouth but regained in some species (Paluh et al., 2021b), the classic example being the reappearance of teeth on the mandible of *Gastrotheca guentheri* after ~225 mya of anuran toothlessness in this region (Wiens, 2011; Paluh et al., 2021a). However, to what extent vestiges may play a role in the evolutionary reappearance of anuran teeth is unknown. In mammals, vestigial teeth are relatively common. Examples comprise the functionally monophyodont shrews, rabbits, sloths, narwhals, walruses, and bats among others (Järvinen et al., 2008; Yamanaka et al., 2010; Nweeia et al., 2012; Hautier et al., 2016; Kryukova, 2018; Bertonnier-Brouty et al., 2021; Sadier et al., 2021). In addition, within a group, and even within the same tooth class of a species, the degree of reduction of teeth exhibits great variation among loci, suggesting that each locus is to some extent controlled independently. This is exemplified by recent results in bats in which teeth are gained and lost rapidly in relationship with diet at various stages of their development, independently between tooth classes (Sadier et al., 2021), thanks to the independence of the Turing cascades that control their development with the same set of genes. In addition, the sequential addition of teeth during pattern formation on a growing domain could explain why some teeth are lost at specific loci. These mechanisms are likely to be extended in other mammals and could explain why vestigial teeth are so frequently found in this clade and, more broadly, in vertebrates.

7.4.5 Integrating Turing Patterns with Other Developmental Mechanisms

The interaction between developmental mechanisms (such as cell division, biomechanics, diffusion, patterning) is always difficult to assess because it implies a deep understanding of the contribution of each of them to the development of the organ of interest. This explains why, up to recently, the RD mechanisms that drive tooth patterning have often been studied independently of other developmental mechanisms. This lack of integration could explain why some of RD mechanism-derived rules, while being supposedly universal, fail to recapitulate the variation seen in nature. For example, the inhibitory cascade described by Kavanagh et al. (2007) on murine molars was hypothesized to control the development of other repeated organs. This idea was successfully verified on vertebrae and limbs of various clades (Kavanagh et al., 2013; Young et al., 2015) suggesting that the inhibitory cascade model could be seen as a universal rule for the development of repeated organs. However, the exploration of more clades and/or organs challenged its universality (Polly, 2007; Carter and Worthington, 2015; Roseman and Delezene, 2019; Vitek et al., 2020; Sadier et al., 2021). These observations do not invalidate the existence of Turing patterns *per se* but suggest that other mechanisms could interact with Turing mechanisms by changing, modifying, or even erasing patterns (Sadier et al., 2019, 2021). Recent studies that incorporate different mechanisms in addition to Turing-like ones have started to provide some examples of such interactions and how they can modify the patterns that the classical RD mechanisms fail to predict. In mice, the incorporation of the R2 premolar vestige into the lower M1, or the phenotype of some mutants with extra numerous teeth, could be explained by the interaction between the two-regime RD systems and chemotaxis (Sadier et al., 2019, Section 7.3.6), which is in accordance with chemotaxis playing a role in the formation of tooth and hair follicles (Ahtiainen et al., 2014, 2016; Glover et al., 2017). More recently, the variation in growth rate was shown to explain the tremendous diversity of tooth number and size observed in bats by perturbing the apparition of the successive signaling center of premolars and molars (Sadier et al., 2021). Interestingly, premolars and molars exhibit distinct patterns of losses and changes in proportions, suggesting that other parameters such as tissue maturation, chemotaxis, or the mineralization of other teeth might influence the establishment of the successive apparition of signaling centers through RD mechanisms. Finally, another recent study (Renvoisé et al., 2017) revealed the importance of mechanical forces in tooth cusp pattern formation that is also governed by Turing mechanisms, as shown in mouse and rat (Cai et al., 2007). Future studies should focus on identifying how these developmental mechanisms interact together to get a clearer understanding of tooth patterning.

7.5 CONCLUSION

Pattern establishment is a crucial feature of developing dentitions, which provides a proximate source of information on how they would be formed, function, and renew; however, the exact processes that lie behind are, to a vast extent, hazy. We welcome recent connections between mathematical modeling and developmental biology as a synergism of once-separated disciplines, which provides a whole new view on periodical patterning. Upon their successful employment in modeling development of periodically repeating body cover organs, the application of mathematical models using RD mechanisms has been promising in providing a theoretical background for pattern establishment also in diverse dental systems. While we are aware that comparable patterns may be achieved by several disparate models, we see a great deal of potential in Turing's RD mechanisms in building vertebrate dentitions. Whether our outlook is justified is only a matter of upcoming investigations at the edge of biological research and mathematical modeling.

ACKNOWLEDGMENTS

We would like to thank Donglei Chen and Per Erik Ahlberg for their kind invitation to participate in this book. Many thanks further go to Jessica Blois and Anna Pospisilova, who kindly provided data for Figures 7.4 and 7.6, and two reviewers who helped enhance the text. Alexa Sadier is supported by the US National Science Foundation (NSF IOS Award 2017803). Vladimír Soukup is supported by the Czech Science Foundation (GACR 18–04580S) and the Charles University institutional support SVV260571/2022.

REFERENCES

Abduweli, D., O. Baba, M. J. Tabata, K. Higuchi, H. Mitani, and Y. Takano. 2014. Tooth replacement and putative odontogenic stem cell niches in pharyngeal dentition of medaka (*Oryzias latipes*). *Microscopy* 63:141–153.

Ahn, Y., B. W. Sanderson, O. D. Klein, and R. Krumlauf. 2010. Inhibition of Wnt signaling by Wise (Sostdcl) and negative feedback from Shh controls tooth number and patterning. *Development* 137:3221–3231.

Ahtiainen, L., S. Lefebvre, P. H. Lindfors, E. Renvoisé, V. Shirokova, M. K. Vartiainen, I. Thesleff, and M. L. Mikkola. 2014. Directional cell migration, but not proliferation, drives hair placode morphogenesis. *Developmental Cell* 28:588–602.

Ahtiainen, L., I. Uski, I. Thesleff, and M. L. Mikkola. 2016. Early epithelial signaling center governs tooth budding morphogenesis. *Journal of Cell Biology* 214:753–767.

Aigler, S. R., D. Jandzik, K. Hatta, K. Uesugi, and D. W. Stock. 2014. Selection and constraint underlie irreversibility of tooth loss in cypriniform fishes. *Proceedings of the National Academy of Sciences* 111:7707–7712.

Aman, A. J., A. N. Fulbright, and D. M. Parichy. 2018. Wnt/β-catenin regulates an ancient signaling network during zebrafish scale development. *eLife* 7:e37001.

Bailleul, R., C. Curantz, C. Desmarquet-Trin Dinh, M. Hidalgo, J. Touboul, and M. Manceau. 2019. Symmetry breaking in the embryonic skin triggers directional and sequential plumage patterning. *PLOS Biology* 17:e3000448.

Balic, A. 2019. Concise review: Cellular and molecular mechanisms regulation of tooth initiation. *Stem Cells* 37:26–32.

Berkovitz, B. K. B. 1977a. Chronology of tooth development in the rainbow trout (*Salmo gairdneri*). *Journal of Experimental Zoology* 200:65–69.

Berkovitz, B. K. B. 1977b. The order of tooth development and eruption in the rainbow trout (*Salmo gairdneri*). *Journal of Experimental Zoology* 201:221–225.

Berkovitz, B. K. B., and R. P. Shellis. 2017. *The Teeth of Non-Mammalian Vertebrates*. Academic Press is an imprint of Elsevier, London; San Diego, CA, 342 pp.

Bertonnier-Brouty, L., L. Viriot, T. Joly, and C. Charles. 2021. Gene expression patterns associated with dental replacement in the rabbit, a new model for the mammalian dental replacement mechanisms. *Developmental Dynamics* 250:1494–1504.

Borday-Birraux, V., C. heyden, M. Debiais-Thibaud, L. Verreijdt, D. W. Stock, A. Huysseune, and J.-Y. Sire. 2006. Expression of Dlx genes during the development of the zebrafish pharyngeal dentition: Evolutionary implications. *Evolution & Development* 8:130–141.

Briata, P., C. Ilengo, G. Corte, C. Moroni, M. G. Rosenfeld, C.-Y. Chen, and R. Gherzi. 2003. The Wnt/β-Catenin→Pitx2 pathway controls the turnover of Pitx2 and other unstable mRNAs. *Molecular Cell* 12:1201–1211.

Buchtová, M., O. Zahradníček, S. Balková, and A. S. Tucker. 2013. Odontogenesis in the veiled chameleon (*Chamaeleo calyptratus*). *Archives of Oral Biology* 58:118–133.

Burgess, E. A., J. M. Lanyon, and T. Keeley. 2012. Testosterone and tusks: Maturation and seasonal reproductive patterns of live, free-ranging male dugongs (*Dugong dugon*) in a subtropical population. *Reproduction* 143:683–697.

Butler, P. M. 1939. Studies of the mammalian dentition. Differentiation of the post-canine dentition. *Proceedings of the Zoological Society of London* B109:1–36.

Cai, J., S.-W. Cho, J.-Y. Kim, M.-J. Lee, Y.-G. Cha, and H.-S. Jung. 2007. Patterning the size and number of tooth and its cusps. *Developmental Biology* 304:499–507.

Carter, K., and S. Worthington. 2015. Morphologic and demographic predictors of third molar agenesis: A systematic review and meta-analysis. *Journal of Dental Research* 94:886–894.

Catón, J., and A. S. Tucker. 2009. Current knowledge of tooth development: Patterning and mineralization of the murine dentition. *Journal of Anatomy* 214:502–515.

Chang, C.-H., T.-X. Jiang, C.-M. Lin, L. W. Burrus, C.-M. Chuong, and R. Widelitz. 2004. Distinct Wnt members regulate the hierarchical morphogenesis of skin regions (spinal tract) and individual feathers. *Mechanisms of Development* 121:157–171.

Charles, C., S. Pantalacci, R. Peterkova, P. Tafforeau, V. Laudet, and L. Viriot. 2009b. Effect of *eda* loss of function on upper jugal tooth morphology. *The Anatomical Record: Advances in Integrative Anatomy and Evolutionary Biology* 292:299–308.

Charles, C., S. Pantalacci, P. Tafforeau, D. Headon, V. Laudet, and L. Viriot. 2009a. Distinct impacts of Eda and Edar loss of function on the mouse dentition. *PLoS ONE* 4:e4985.

Chen, D., H. Blom, S. Sanchez, P. Tafforeau, and P. E. Ahlberg. 2016. The stem osteichthyan *Andreolepis* and the origin of tooth replacement. *Nature* 539:237–241.

Chen, D., H. Blom, S. Sanchez, P. Tafforeau, T. Märss, and P. E. Ahlberg. 2020. The developmental relationship between teeth and dermal odontodes in the most primitive bony fish *Lophosteus*. *eLife* 9:e60985.

Cho, S.-W., S. Kwak, T. E. Woolley, M.-J. Lee, E.-J. Kim, R. E. Baker, H.-J. Kim, J.-S. Shin, C. Tickle, P. K. Maini, and H.-S. Jung. 2011. Interactions between Shh, Sostdc1 and Wnt signaling and a new feedback loop for spatial patterning of the teeth. *Development* 138:1807–1816.

Clemen, G., P. Bartsch, and K. Wacker. 1998. Dentition and dentigerous bones in juveniles and adults of *Polypterus senegalus* (Cladistia, Actinopterygii). *Annals of Anatomy—Anatomischer Anzeiger* 180:211–221.

Cooper, R. L., V. J. Lloyd, N. Di-Poï, A. G. Fletcher, P. M. Barrett, and G. J. Fraser. 2019. Conserved gene signalling and a derived patterning mechanism underlie the development of avian footpad scales. *EvoDevo* 10:19.

Cooper, R. L., A. P. Thiery, A. G. Fletcher, D. J. Delbarre, L. J. Rasch, and G. J. Fraser. 2018. An ancient Turing-like patterning mechanism regulates skin denticle development in sharks. *Science Advances* 4:eaau5484.

Couzens, A. M. C., K. E. Sears, and M. Rücklin. 2021. Developmental influence on evolutionary rates and the origin of placental mammal tooth complexity. *Proceedings of the National Academy of Sciences* 118:e2019294118.

Dalle Nogare, D., and A. B. Chitnis. 2017. Self-organizing spots get under your skin. *PLoS Biology* 15:e2004412.

Dassule, H. R., and A. P. McMahon. 1998. Analysis of epithelial-mesenchymal interactions in the initial morphogenesis of the mammalian tooth. *Developmental Biology* 202:215–227.

Debiais-Thibaud, M., V. Borday-Birraux, I. Germon, F. Bourrat, C. J. Metcalfe, D. Casane, and P. Laurenti. 2007. Development of oral and pharyngeal teeth in the medaka (*Oryzias latipes*): Comparison of morphology and expression of *eve1* gene. *Journal of Experimental Zoology Part B: Molecular and Developmental Evolution* 308B:693–708.

Debiais-Thibaud, M., R. Chiori, S. Enault, S. Oulion, I. Germon, C. Martinand-Mari, D. Casane, and V. Borday-Birraux. 2015. Tooth and scale morphogenesis in shark: An alternative process to the mammalian enamel knot system. *BMC Evolutionary Biology* 15:292.

Delpiani, G. 2016. Reproductive biology of the southern thorny skate *Amblyraja doellojuradoi* (Chondrichthyes, Rajidae): Reproduction of *Amblyraja doellojuradoi*. *Journal of Fish Biology* 88:1413–1429.

Donoghue, P. C. J. 2002. Evolution of development of the vertebrate dermal and oral skeletons: Unraveling concepts, regulatory theories, and homologies. *Paleobiology* 28:474–507.

Donoghue, P. C. J., and M. Rücklin. 2016. The ins and outs of the evolutionary origin of teeth. *Evolution & Development* 18:19–30.

Economou, A. D., N. A. M. Monk, and J. B. A. Green. 2020. Perturbation analysis of a multimorphogen Turing reaction-diffusion stripe patterning system reveals key regulatory interactions. *Development* dev.190553.

Edmund, A. G. 1960. *Tooth Replacement Phenomena in the Lower Vertebrates*. Royal Ontario Museum, Toronto.

Fenelon, J. C., A. Bennets, N. Anthwal, M. Pyne, S. D. Johnston, A. R. Evans, A. S. Tucker, and M. B. Renfree. 2023. Getting out of a mammalian egg: The egg tooth and caruncle of the echidna. *Developmental Biology* 495:8–18.

Fons, J. M., M. Gaete, O. Zahradnicek, M. Landova, H. Bandali, E. R. Khannoon, J. M. Richman, M. Buchtova, and A. S. Tucker. 2020. Getting out of an egg: Merging of tooth germs to create an egg tooth in the snake. *Developmental Dynamics* 249:199–208.

Fraser, G. J., R. F. Bloomquist, and J. T. Streelman. 2008. A periodic pattern generator for dental diversity. *BMC Biology* 6:32.
Fraser, G. J., R. F. Bloomquist, and J. T. Streelman. 2013. Common developmental pathways link tooth shape to regeneration. *Developmental Biology* 377:399–414.
Fraser, G. J., R. Cerny, V. Soukup, M. Bronner-Fraser, and J. T. Streelman. 2010. The odontode explosion: The origin of tooth-like structures in vertebrates. *BioEssays* 32:808–817.
Fraser, G. J., A. Graham, and M. M. Smith. 2004. Conserved deployment of genes during odontogenesis across osteichthyans. *Proceedings of the Royal Society of London. Series B: Biological Sciences* 271:2311–2317.
Fraser, G. J., and M. M. Smith. 2011. Evolution of developmental pattern for vertebrate dentitions: An oro-pharyngeal specific mechanism. *Journal of Experimental Zoology Part B: Molecular and Developmental Evolution* 316B:99–112.
Fraser, G. J., A. Standing, C. Underwood, and A. P. Thiery. 2020. The dental lamina: An essential structure for perpetual tooth regeneration in sharks. *Integrative and Comparative Biology* 60:644–655.
Gaete, M., J. M. Fons, E. M. Popa, L. Chatzeli, and A. S. Tucker. 2015. Epithelial topography for repetitive tooth formation. *Biology Open* 4:1625–1634.
Gat, U., R. DasGupta, L. Degenstein, and E. Fuchs. 1998. De Novo hair follicle morphogenesis and hair tumors in mice expressing a truncated β-catenin in skin. *Cell* 95:605–614.
Gibert, Y., L. Bernard, M. Debiais-Thibaud, F. Bourrat, J. Joly, K. Pottin, A. Meyer, S. Retaux, D. W. Stock, W. R. Jackman, P. Seritrakul, G. Begemann, and V. Laudet. 2010. Formation of oral and pharyngeal dentition in teleosts depends on differential recruitment of retinoic acid signaling. *The FASEB Journal* 24:3298–3309.
Gibert, Y., E. Samarut, M. K. Ellis, W. R. Jackman, and V. Laudet. 2019. The first formed tooth serves as a signalling centre to induce the formation of the dental row in zebrafish. *Proceedings of the Royal Society B: Biological Sciences* 286:20190401.
Gibert, Y., E. Samarut, E. Pasco-Viel, L. Bernard, V. Borday-Birraux, A. Sadier, C. Labbé, L. Viriot, and V. Laudet. 2015. Altered retinoic acid signalling underpins dentition evolution. *Proceedings of the Royal Society B: Biological Sciences* 282:20142764.
Gierer, A., and H. Meinhardt. 1972. A theory of biological pattern formation. *Kybernetik* 12:30–39.
Glimm, T., R. Bhat, and S. A. Newman. 2020. Multiscale modeling of vertebrate limb development. *WIREs Systems Biology and Medicine* 12:e1485.
Glover, J. D., K. L. Wells, F. Matthäus, K. J. Painter, W. Ho, J. Riddell, J. A. Johansson, M. J. Ford, C. A. B. Jahoda, V. Klika, R. L. Mort, and D. J. Headon. 2017. Hierarchical patterning modes orchestrate hair follicle morphogenesis. *PLoS Biology* 15:e200 2117.
Grande, L. 2010. An empirical synthetic pattern study of gars (Lepisosteiformes) and closely related species, based mostly on skeletal anatomy. *Copeia* 2010:1–871.
Grande, L., and W. E. Bemis. 1998. A comprehensive phylogenetic study of Amiid fishes (Amiidae) based on comparative skeletal anatomy: An empirical search for interconnected patterns of natural history. *Journal of Vertebrate Paleontology* 18:1–696.
Gudger, E. 1933. Abnormal dentition in rays, Batoidei. *Journal of the Elisha Mitchell Scientific Society* 49:57–96.
Gudger, E. 1937. Abnormal dentition in sharks, Selachii. *Bulletin of the American Museum of Natural History* 73:249–280.
Häärä, O., E. Harjunmaa, P. H. Lindfors, S.-H. Huh, I. Fliniaux, T. Åberg, J. Jernvall, D. M. Ornitz, M. L. Mikkola, and I. Thesleff. 2012. Ectodysplasin regulates activator-inhibitor balance in murine tooth development through Fgf20 signaling. *Development* 139:3189–3199.

Handrigan, G. R., K. J. Leung, and J. M. Richman. 2010. Identification of putative dental epithelial stem cells in a lizard with life-long tooth replacement. *Development* 137:3545–3549.

Harjunmaa, E., A. Kallonen, M. Voutilainen, K. Hämäläinen, M. L. Mikkola, and J. Jernvall. 2012. On the difficulty of increasing dental complexity. *Nature* 483:324–327.

Harjunmaa, E., K. Seidel, T. Häkkinen, E. Renvoisé, I. J. Corfe, A. Kallonen, Z.-Q. Zhang, A. R. Evans, M. L. Mikkola, I. Salazar-Ciudad, O. D. Klein, and J. Jernvall. 2014. Replaying evolutionary transitions from the dental fossil record. *Nature* 512:44–48.

Harris, M. P., S. M. Hasso, M. W. J. Ferguson, and J. F. Fallon. 2006. The development of archosaurian first-generation teeth in a chicken mutant. *Current Biology* 16:371–377.

Hautier, L., H. Gomes Rodrigues, G. Billet, and R. J. Asher. 2016. The hidden teeth of sloths: Evolutionary vestiges and the development of a simplified dentition. *Scientific Reports* 6:27763.

Headon, D. J., and K. J. Painter. 2009. Stippling the skin: Generation of anatomical periodicity by reaction-diffusion mechanisms. *Mathematical Modelling of Natural Phenomena* 4:83–102.

Hermyt, M., P. Kaczmarek, M. Kowalska, and W. Rupik. 2017. Development of the egg tooth—The tool facilitating hatching of squamates: Lessons from the grass snake *Natrix natrix*. *Zoologischer Anzeiger* 266:61–70.

Hermyt, M., B. Metscher, and W. Rupik. 2020. Do all geckos hatch in the same way? Histological and 3D studies of egg tooth morphogenesis in the geckos *Eublepharis macularius* Blyth 1854 and *Lepidodactylus lugubris* Duméril & Bibron 1836. *Journal of Morphology* 281:1313–1327.

Ho, W. K. W., L. Freem, D. Zhao, K. J. Painter, T. E. Woolley, E. A. Gaffney, M. J. McGrew, A. Tzika, M. C. Milinkovitch, P. Schneider, A. Drusko, F. Matthäus, J. D. Glover, K. L. Wells, J. A. Johansson, M. G. Davey, H. M. Sang, M. Clinton, and D. J. Headon. 2019. Feather arrays are patterned by interacting signalling and cell density waves. *PLoS Biology* 17:e3000132.

Hovestadt, D. C., and M. Hovestadt-Euler. 2013. Generic assessment and reallocation of Cenozoic Myliobatins based on new information of tooth, tooth plate and caudal spine morphology of extant taxa. *Palaeontos* 24:1–66.

Huysseune, A., and J.-Y. Sire. 1997. Structure and development of first-generation teeth in the cichlid *Hemichromis bimaculatus* (Teleostei, Cichlidae). *Tissue and Cell* 29:679–697.

Huysseune, A., and P. E. Witten. 2006. Developmental mechanisms underlying tooth patterning in continuously replacing osteichthyan dentitions. *Journal of Experimental Zoology Part B: Molecular and Developmental Evolution* 306B:204–215.

Huysseune, A., and P. E. Witten. 2008. An evolutionary view on tooth development and replacement in wild Atlantic salmon (*Salmo salar* L.) *Evolution & Development* 10:6–14.

Inaba, M., H. I.-C. Harn, and C.-M. Chuong. 2019. Turing patterning with and without a global wave. *PLoS Biology* 17:e3000195.

Jandzik, D., and D. W. Stock. 2021. Differences in developmental potential predict the contrasting patterns of dental diversification in characiform and cypriniform fishes. *Proceedings of the Royal Society B: Biological Sciences* 288:20202205.

Jarvik, E. 1955. The oldest tetrapods and their forerunners. *The Scientific Monthly* 80:141–154.

Järvinen, E., K. Vlimki, M. Pummila, I. Thesleff, and J. Jernvall. 2008. The taming of the shrew milk teeth. *Evolution & Development* 10:477–486.

Jernvall, J., and I. Thesleff. 2000. Reiterative signaling and patterning during mammalian tooth morphogenesis. *Mechanisms of Development* 92:19–29.

Jheon, A. H., K. Seidel, B. Biehs, and O. D. Klein. 2013. From molecules to mastication: The development and evolution of teeth. *Wiley Interdisciplinary Reviews: Developmental Biology* 2:165–182.

Jiang, T. X., H. S. Jung, R. B. Widelitz, and C. M. Chuong. 1999. Self-organization of periodic patterns by dissociated feather mesenchymal cells and the regulation of size, number and spacing of primordia. *Development* 126:4997–5009.

Jung, H.-S., P. H. Francis-West, R. B. Widelitz, T.-X. Jiang, S. Ting-Berreth, C. Tickle, L. Wolpert, and C.-M. Chuong. 1998. Local inhibitory action of BMPs and their relationships with activators in feather formation: Implications for periodic patterning. *Developmental Biology* 196:11–23.

Jussila, M., X. Crespo Yanez, and I. Thesleff. 2014. Initiation of teeth from the dental lamina in the ferret. *Differentiation* 87:32–43.

Juuri, E., M. Jussila, K. Seidel, S. Holmes, P. Wu, J. Richman, K. Heikinheimo, C.-M. Chuong, K. Arnold, K. Hochedlinger, O. Klein, F. Michon, and I. Thesleff. 2013. Sox2 marks epithelial competence to generate teeth in mammals and reptiles. *Development* 140:1424–1432.

Kavanagh, K. D., A. R. Evans, and J. Jernvall. 2007. Predicting evolutionary patterns of mammalian teeth from development. *Nature* 449:427–432.

Kavanagh, K. D., O. Shoval, B. B. Winslow, U. Alon, B. P. Leary, A. Kan, and C. J. Tabin. 2013. Developmental bias in the evolution of phalanges. *Proceedings of the National Academy of Sciences* 110:18190–18195.

Keränen, S. V. E., P. Kettunen, T. Åberg, I. Thesleff, and J. Jernvall. 1999. Gene expression patterns associated with suppression of odontogenesis in mouse and vole diastema regions. *Development Genes and Evolution* 209:495–506.

Kettunen, P., J. Laurikkala, P. Itäranta, S. Vainio, N. Itoh, and I. Thesleff. 2000. Associations of FGF-3 and FGF-10 with signaling networks regulating tooth morphogenesis. *Developmental Dynamics* 219:322–332.

Kioussi, C., P. Briata, S. H. Baek, D. W. Rose, N. S. Hamblet, T. Herman, K. A. Ohgi, C. Lin, A. Gleiberman, J. Wang, V. Brault, P. Ruiz-Lozano, H. D. Nguyen, R. Kemler, C. K. Glass, A. Wynshaw-Boris, and M. G. Rosenfeld. 2002. Identification of a Wnt/Dvl/β-Catenin → Pitx2 pathway mediating cell-type-specific proliferation during development. *Cell* 111:673–685.

Klika, V., R. E. Baker, D. Headon, and E. A. Gaffney. 2012. The influence of receptor-mediated interactions on reaction-diffusion mechanisms of cellular self-organisation. *Bulletin of Mathematical Biology* 74:935–957.

Kondo, S., and T. Miura. 2010. Reaction-diffusion model as a framework for understanding biological pattern formation. *Science* 329:1616–1620.

Krause, A. L., M. A. Ellis, and R. A. Van Gorder. 2019. Influence of curvature, growth, and anisotropy on the evolution of Turing patterns on growing manifolds. *Bulletin of Mathematical Biology* 81:759–799.

Kristenová, P., M. Peterka, S. Lisi, J.-L. Gendrault, H. Lesot, and R. Peterková. 2002. Different morphotypes of functional dentition in the lower molar region of tabby (EDA) mice: Post-natal morphotypes of tabby dentition. *Orthodontics & Craniofacial Research* 5:205–214.

Kryukova, N. V. 2018. Vestigial teeth of the Pacific walrus *Odobenus rosmarus divergens* Illiger, 1815 (Carnivora: Odobenidae). *Russian Journal of Theriology* 17:78–84.

Kulesa, P. M., G. C. Cruywagen, S. R. Lubkin, M. W. J. Ferguson, and J. D. Murray. 1996a. Modelling the spatial patterning of teeth primordia in the alligator. *Acta Biotheoretica* 44:153–164.

Kulesa, P. M., G. C. Cruywagen, S. R. Lubkin, P. K. Main, J. Sneyd, M. W. J. Ferguson, and J. D. Murray. 1996b. On a model mechanism for the spatial patterning of teeth primordia in the alligator. *Journal of Theoretical Biology* 180:287–296.

Kulesa, P. M., G. C. Cruywagen, S. R. Lubkin, P. K. Maini, J. Sneyd, and J. D. Murray. 1995. Modelling the spatial patterning of the teeth primordia in the lower jaw of *Alligator mississippiensis*. *Journal of Biological Systems* 3:975–985.

Kulesa, P. M., and J. D. Murray. 1995. Modelling the wave-like initiation of teeth priordia in the alligator. *Forma* 10:259–280.

Lagronova-Churava, S., F. Spoutil, S. Vojtechova, H. Lesot, M. Peterka, O. D. Klein, and R. Peterkova. 2013. The dynamics of supernumerary tooth development are differentially regulated by sprouty genes: Supernumerary tooth development in sprouty mutants. *Journal of Experimental Zoology Part B: Molecular and Developmental Evolution* 320:307–320.

Lan, Y., S. Jia, and R. Jiang. 2014. Molecular patterning of the mammalian dentition. *Seminars in Cell & Developmental Biology* 25–26:61–70.

Laurikkala, J., M. Mikkola, T. Mustonen, T. Åberg, P. Koppinen, J. Pispa, P. Nieminen, J. Galceran, R. Grosschedl, and I. Thesleff. 2001. TNF signaling via the ligand-receptor pair ectodysplasin and edar controls the function of epithelial signaling centers and is regulated by wnt and activin during tooth organogenesis. *Developmental Biology* 229:443–455.

Laurikkala, J., J. Pispa, H.-S. Jung, P. Nieminen, M. Mikkola, X. Wang, U. Saarialho-Kere, J. Galceran, R. Grosschedl, and I. Thesleff. 2002. Regulation of hair follicle development by the TNF signal ectodysplasin and its receptor Edar. *Development* 129:2541–2553.

Lefebvre, S., and M. L. Mikkola. 2014. Ectodysplasin research—where to next? *Seminars in Immunology* 26:220–228.

Liu, W., J. Selever, M.-F. Lu, and J. F. Martin. 2003. Genetic dissection of *Pitx2* in craniofacial development uncovers new functions in branchial arch morphogenesis, late aspects of tooth morphogenesis and cell migration. *Development* 130:6375–6385.

Lochovska, K., R. Peterkova, Z. Pavlikova, and M. Hovorakova. 2015. Sprouty gene dosage influences temporal-spatial dynamics of primary enamel knot formation. *BMC Developmental Biology* 15:21.

Louchart, A., and L. Viriot. 2011. From snout to beak: The loss of teeth in birds. *Trends in Ecology & Evolution* 26:663–673.

Maini, P. K., and T. E. Woolley. 2019. The Turing model for biological pattern formation; pp. 189–204 in A. Bianchi, T. Hillen, M. A. Lewis, and Y. Yi (eds.), *The Dynamics of Biological Systems, Mathematics of Planet Earth vol. 4*. Springer International Publishing, Cham.

Marcon, L., and J. Sharpe. 2012. Turing patterns in development: What about the horse part? *Current Opinion in Genetics & Development* 22:578–584.

Marsh, H., G. E. Heinson, and L. M. Marsh. 1984. Breeding cycle, life history and population dynamics of the dugong, *Dugong dugon* (Sirenia: Dugongidae). *Australian Journal of Zoology* 32:767–788.

Martin, K. J., L. J. Rasch, R. L. Cooper, B. D. Metscher, Z. Johanson, and G. J. Fraser. 2016. Sox2+ progenitors in sharks link taste development with the evolution of regenerative teeth from denticles. *Proceedings of the National Academy of Sciences* 113:14769–14774.

Michon, F., L. Forest, E. Collomb, J. Demongeot, and D. Dhouailly. 2008. BMP2 and BMP7 play antagonistic roles in feather induction. *Development* 135:2797–2805.

Miura, T., K. Shiota, G. Morriss-Kay, and P. K. Maini. 2006. Mixed-mode pattern in *Doublefoot* mutant mouse limb—Turing reaction—diffusion model on a growing domain during limb development. *Journal of Theoretical Biology* 240:562–573.

Mogollón, I., J. E. Moustakas-Verho, M. Niittykoski, and L. Ahtiainen. 2021. The initiation knot is a signaling center required for molar tooth development. *Development* 148:dev194597.

Mou, C., B. Jackson, P. Schneider, P. A. Overbeek, and D. J. Headon. 2006. Generation of the primary hair follicle pattern. *Proceedings of the National Academy of Sciences* 103:9075–9080.

Moustakas, J. E., K. K. Smith, and L. J. Hlusko. 2011. Evolution and development of the mammalian dentition: Insights from the marsupial *Monodelphis domestica*. *Developmental Dynamics* 240:232–239.

Murray, J. D. 2003. *Mathematical Biology. 2: Spatial Models and Biomedical Applications.* Softcover reprint of the hardcover 3rd edition 2003, corrected second printing. Springer-Verlag GmbH, Berlin, Heidelberg, 811 pp.

Murray, J. D., and P. M. Kulesa. 1996. On a dynamic reaction—diffusion mechanism: The spatial patterning of teeth primordia in the alligator. *Journal of the Chemical Society, Faraday Transactions* 92:2927–2932.

Nakajima, T. 1984. Larval vs. adult pharyngeal dentition in some Japanese cyprinid fishes. *Journal of Dental Research* 63:1140–1146.

Navarro, N., and A. Murat Maga. 2018. Genetic mapping of molar size relations identifies inhibitory locus for third molars in mice. *Heredity* 121:1–11.

Neguer, J., and M. Manceau. 2017. Embryonic patterning of the vertebrate skin. *Review in Cell Biology and Molecular Medicine* 3:1.

Nelson, G. J. 1969. Gill arches and the phylogeny of fishes, with notes on the classification of vertebrates. *Bulletin of the American Museum of Natural History* 141:479–552.

Noramly, S., A. Freeman, and B. A. Morgan. 1999. beta-catenin signaling can initiate feather bud development. *Development* 126:3509–3521.

Noramly, S., and B. A. Morgan. 1998. BMPs mediate lateral inhibition at successive stages in feather tract development. *Development* 125:3775–3787.

Nweeia, M. T., F. C. Eichmiller, P. V. Hauschka, E. Tyler, J. G. Mead, C. W. Potter, D. P. Angnatsiak, P. R. Richard, J. R. Orr, and S. R. Black. 2012. Vestigial tooth anatomy and tusk nomenclature for *Monodon monoceros*. *The Anatomical Record: Advances in Integrative Anatomy and Evolutionary Biology* 295:1006–1016.

Onimaru, K., L. Marcon, M. Musy, M. Tanaka, and J. Sharpe. 2016. The fin-to-limb transition as the re-organization of a Turing pattern. *Nature Communications* 7:11582.

Ørvig, T. 1967. Phylogeny of tooth tissues: Evolution of some calcified tissues in early vertebrates; pp. 45–100 in *Structural and Chemical Organization of Teeth.* Academic Press, New York.

Ørvig, T. 1977. A survey of odontodes ('dermal teeth') from developmental, structural, functional and phylogenetic points of view; pp. 53–75 in *Problems in Vertebrate Evolution.* Academic Press, New York.

Osborn, J. W. 1970. New approach to Zahnreihen. *Nature* 225:343–346.

Osborn, J. W. 1971. The ontogeny of tooth succession in *Lacerta vivipara* Jacquin (1787). *Proceedings of the Royal Society of London. Series B. Biological Sciences* 179:261–289.

Osborn, J. W. 1973. The evolution of dentitions. The study of evolution suggests how the development of mammalian dentitions may be controlled. *American Scientist* 61:548–559.

Osborn, J. W. 1974. On the control of tooth replacement in reptiles and its relationship to growth. *Journal of Theoretical Biology* 46:509–527.

Osborn, J. W. 1978. Morphogenetic gradients: Fields versus clones; pp. 171–201 in *Development, Function and Evolution of Teeth.* Academic Press, London.

Painter, K. J., G. S. Hunt, K. L. Wells, J. A. Johansson, and D. J. Headon. 2012. Towards an integrated experimental—theoretical approach for assessing the mechanistic basis of hair and feather morphogenesis. *Interface Focus* 2:433–450.

Paluh, D. J., W. A. Dillard, E. L. Stanley, G. J. Fraser, and D. C. Blackburn. 2021a. Re-evaluating the morphological evidence for the re-evolution of lost mandibular teeth in frogs. *Evolution* 75:3203–3213.

Paluh, D. J., K. Riddell, C. M. Early, M. M. Hantak, G. F. Jongsma, R. M. Keeffe, F. Magalhães Silva, S. V. Nielsen, M. C. Vallejo-Pareja, E. L. Stanley, and D. C. Blackburn. 2021b. Rampant tooth loss across 200 million years of frog evolution. *eLife* 10:e66926.

Peterkova, R., S. Churava, H. Lesot, M. Rothova, J. Prochazka, M. Peterka, and O. D. Klein. 2009. Revitalization of a diastemal tooth primordium in *Spry2* null mice results from increased proliferation and decreased apoptosis. *Journal of Experimental Zoology Part B: Molecular and Developmental Evolution* 312B:292–308.

Peyer, B. 1968. *Comparative Odontology.* The University of Chicago Press, Chicago and London, 347 pp.

Polly, P. D. 2007. Development with a bite. *Nature* 449:413–414.

Popa, E. M., N. Anthwal, and A. S. Tucker. 2016. Complex patterns of tooth replacement revealed in the fruit bat (*Eidolon helvum*). *Journal of Anatomy* 229:847–856.

Pospisilova, A., J. Brejcha, V. Miller, R. Holcman, R. Šanda, and J. Stundl. 2019. Embryonic and larval development of the northern pike: An emerging fish model system for evo-devo research. *Journal of Morphology* 280:1118–1140.

Pospisilova, A., J. Stundl, J. Brejcha, B. D. Metscher, M. Psenicka, R. Cerny, and V. Soukup. 2022. The remarkable dynamics in the establishment, rearrangement, and loss of dentition during the ontogeny of the sterlet sturgeon. *Developmental Dynamics* 251:826–845.

Prochazka, J., S. Pantalacci, S. Churava, M. Rothova, A. Lambert, H. Lesot, O. Klein, M. Peterka, V. Laudet, and R. Peterkova. 2010. Patterning by heritage in mouse molar row development. *Proceedings of the National Academy of Sciences* 107:15497–15502.

Rasch, L. J., R. L. Cooper, C. Underwood, W. A. Dillard, A. P. Thiery, and G. J. Fraser. 2020. Development and regeneration of the crushing dentition in skates (Rajidae). *Developmental Biology* 466:59–72.

Rasch, L. J., K. J. Martin, R. L. Cooper, B. D. Metscher, C. J. Underwood, and G. J. Fraser. 2016. An ancient dental gene set governs development and continuous regeneration of teeth in sharks. *Developmental Biology* 415:347–370.

Raubenheimer, E. J. 2000. Early development of the tush and the tusk of the African elephant (*Loxodonta africana*). *Archives of Oral Biology* 45:983–986.

Raubenheimer, E. J., W. F. P. Van Heerden, P. J. Van Niekerk, V. De Vos, and M. J. Turner. 1995. Morphology of the deciduous tusk (tush) of the African elephant (*Loxodonta africana*). *Archives of Oral Biology* 40:571–576.

Reif, W.-E. 1976. Morphogenesis, pattern formation and function of the dentition of *Heterodontus* (Selachii). *Zoomorphologie* 83:1–47.

Reif, W.-E. 1982. Evolution of dermal skeleton and dentition in vertebrates: The odontode regulation theory; pp. 287–368 in M. K. Hecht, B. Wallace, and G. T. Prance (eds.), *Evolutionary Biology.* Springer, Boston, MA.

Renvoisé, E., K. D. Kavanagh, V. Lazzari, T. J. Häkkinen, R. Rice, S. Pantalacci, I. Salazar-Ciudad, and J. Jernvall. 2017. Mechanical constraint from growing jaw facilitates mammalian dental diversity. *Proceedings of the National Academy of Sciences* 114:9403–9408.

Roseman, C. C., and L. K. Delezene. 2019. The inhibitory cascade model is not a good predictor of molar size covariation. *Evolutionary Biology* 46:229–238.

Rücklin, M., P. C. J. Donoghue, Z. Johanson, K. Trinajstic, F. Marone, and M. Stampanoni. 2012. Development of teeth and jaws in the earliest jawed vertebrates. *Nature* 491:748–751.

Sadier, A., N. Anthwal, A. L. Krause, R. Dessalles, M. Lake, L. Bentolila, R. Haase, N. Nieves, S. Santana, and K. Sears. 2023. Bat teeth illuminate the diversification of mammalian tooth classes. *Nature communications* 14:4687.

Sadier, A., W. R. Jackman, V. Laudet, and Y. Gibert. 2020. The vertebrate tooth row: Is it initiated by a single organizing tooth? *BioEssays* 42:1900229.

Sadier, A., M. Twarogowska, K. Steklikova, L. Hayden, A. Lambert, P. Schneider, V. Laudet, M. Hovorakova, V. Calvez, and S. Pantalacci. 2019. Modeling *Edar* expression reveals the hidden dynamics of tooth signaling center patterning. *PLoS Biology* 17:e3000064.

Sadier, A., L. Viriot, S. Pantalacci, and V. Laudet. 2014. The ectodysplasin pathway: From diseases to adaptations. *Trends in Genetics* 30:24–31.

Salazar-Ciudad, I., and J. Jernvall. 2010. A computational model of teeth and the developmental origins of morphological variation. *Nature* 464:583–586.

Salomies, L., J. Eymann, I. Khan, and N. Di-Poï. 2019. The alternative regenerative strategy of bearded dragon unveils the key processes underlying vertebrate tooth renewal. *eLife* 8:e47702.

Sarkar, L., M. Cobourne, S. Naylor, M. Smalley, T. Dale, and P. T. Sharpe. 2000. Wnt/Shh interactions regulate ectodermal boundary formation during mammalian tooth development. *Proceedings of the National Academy of Sciences* 97:4520–4524.

Scoones, J. C., and T. W. Hiscock. 2020. A dot-stripe Turing model of joint patterning in the tetrapod limb. *Development* 147:dev.183699.

Seritrakul, P., E. Samarut, T. T. S. Lama, Y. Gibert, V. Laudet, and W. R. Jackman. 2012. Retinoic acid expands the evolutionarily reduced dentition of zebrafish. *The FASEB Journal* 26:5014–5024.

Sharpe, J. 2017. Computer modeling in developmental biology: Growing today, essential tomorrow. *Development* 144:4214–4225.

Sharpe, P. T. 1995. Homeobox genes and orofacial development. *Connective Tissue Research* 32:17–25.

Shirokova, V., M. Jussila, M. K. Hytönen, N. Perälä, C. Drögemüller, T. Leeb, H. Lohi, K. Sainio, I. Thesleff, and M. L. Mikkola. 2013. Expression of *Foxi3* is regulated by ectodysplasin in skin appendage placodes: *Foxi3* expression in skin appendages. *Developmental Dynamics* 242:593–603.

Shyer, A. E., A. R. Rodrigues, G. G. Schroeder, E. Kassianidou, S. Kumar, and R. M. Harland. 2017. Emergent cellular self-organization and mechanosensation initiate follicle pattern in the avian skin. *Science* 357:811–815.

Sick, S., S. Reinker, J. Timmer, and T. Schlake. 2006. WNT and DKK determine hair follicle spacing through a reaction-diffusion mechanism. *Science* 314:1447–1450.

Smith, M. M., G. J. Fraser, N. Chaplin, C. Hobbs, and A. Graham. 2009b. Reiterative pattern of *sonic hedgehog* expression in the catshark dentition reveals a phylogenetic template for jawed vertebrates. *Proceedings of the Royal Society B: Biological Sciences* 276:1225–1233.

Smith, M. M., G. J. Fraser, and T. A. Mitsiadis. 2009a. Dental lamina as source of odontogenic stem cells: Evolutionary origins and developmental control of tooth generation in gnathostomes. *Journal of Experimental Zoology Part B: Molecular and Developmental Evolution* 312B:260–280.

Smith, M. M., Z. Johanson, T. Butts, R. Ericsson, M. Modrell, F. J. Tulenko, M. C. Davis, and G. J. Fraser. 2015. Making teeth to order: Conserved genes reveal an ancient molecular pattern in paddlefish (Actinopterygii). *Proceedings of the Royal Society B: Biological Sciences* 282:20142700.

Smith, M. M., and N. I. Krupina. 2001. Conserved developmental processes constrain evolution of lungfish dentitions. *Journal of Anatomy* 199:161–168.

Smith, M. M., C. Underwood, B. Clark, J. Kriwet, and Z. Johanson. 2018. Development and evolution of tooth renewal in neoselachian sharks as a model for transformation in chondrichthyan dentitions. *Journal of Anatomy* 232:891–907.

Song, H.-K., S.-H. Lee, and P. F. Goetinck. 2004. FGF-2 signaling is sufficient to induce dermal condensations during feather development. *Developmental Dynamics* 231:741–749.

Song, H.-K., Y. Wang, and P. F. Goetinck. 1996. Fibroblast growth factor 2 can replace ectodermal signaling for feather development. *Proceedings of the National Academy of Sciences* 93:10246–10249.

Soukup, V., A. Tazaki, Y. Yamazaki, A. Pospisilova, H.-H. Epperlein, E. M. Tanaka, and R. Cerny. 2021. Oral and palatal dentition of axolotl arises from a common tooth-competent zone along the ecto-endodermal boundary. *Frontiers in Cell and Developmental Biology* 8:622308.

Square, T. A., S. Sundaram, E. J. Mackey, and C. T. Miller. 2021. Distinct tooth regeneration systems deploy a conserved battery of genes. *EvoDevo* 12:4.

St. Amand, T. R., Y. Zhang, E. V. Semina, X. Zhao, Y. Hu, L. Nguyen, J. C. Murray, and Y. Chen. 2000. Antagonistic signals between BMP4 and FGF8 define the expression of *Pitx1* and *Pitx2* in mouse tooth-forming anlage. *Developmental Biology* 217:323–332.

Stock, D. W. 2001. The genetic basis of modularity in the development and evolution of the vertebrate dentition. *Philosophical Transactions of the Royal Society of London. Series B: Biological Sciences* 356:1633–1653.

Stock, D. W., W. R. Jackman, and J. Trapani. 2006. Developmental genetic mechanisms of evolutionary tooth loss in cypriniform fishes. *Development* 133:3127–3137.

Suzuki, T., S. M. Hasso, and J. F. Fallon. 2008. Unique SMAD1/5/8 activity at the phalanx-forming region determines digit identity. *Proceedings of the National Academy of Sciences* 105:4185–4190.

Tucker, A. S., and G. J. Fraser. 2014. Evolution and developmental diversity of tooth regeneration. *Seminars in Cell & Developmental Biology* 25–26:71–80.

Turing, A. M. 1952. The chemical basis of morphogenesis. *Bulletin of Mathematical Biology* 52:47–72.

Underwood, C. J., Z. Johanson, and M. M. Smith. 2016. Cutting blade dentitions in squaliform sharks form by modification of inherited alternate tooth ordering patterns. *Royal Society Open Science* 3:160385.

Underwood, C. J., Z. Johanson, M. Welten, B. Metscher, L. J. Rasch, G. J. Fraser, and M. M. Smith. 2015. Development and evolution of dentition pattern and tooth order in the skates and rays (Batoidea; Chondrichthyes). *PLoS ONE* 10:e0122553.

Vandenplas, S., A. De Clercq, and A. Huysseune. 2014. Tooth replacement without a dental lamina: The search for epithelial stem cells in *Polypterus senegalus*. *Journal of Experimental Zoology Part B: Molecular and Developmental Evolution* 322:281–293.

Vandenplas, S., M. Willems, P. E. Witten, T. Hansen, P. G. Fjelldal, and A. Huysseune. 2016. Epithelial label-retaining cells are absent during tooth cycling in *Salmo salar* and *Polypterus senegalus*. *PLoS ONE* 11:e0152870.

Van der heyden, C., and A. Huysseune. 2000. Dynamics of tooth formation and replacement in the zebrafish (*Danio rerio*) (Teleostei, Cyprinidae). *Developmental Dynamics* 219:486–496.

Vaškaninová, V., D. Chen, P. Tafforeau, Z. Johanson, B. Ekrt, H. Blom, and P. E. Ahlberg. 2020. Marginal dentition and multiple dermal jawbones as the ancestral condition of jawed vertebrates. *Science* 369:211–216.

Viriot, L., R. Peterková, M. Peterka, and H. Lesot. 2002. Evolutionary implications of the occurrence of two vestigial tooth germs during early odontogenesis in the mouse lower jaw. *Connective Tissue Research* 43:129–133.

Vitek, N. S., C. C. Roseman, and J. I. Bloch. 2020. Mammal molar size ratios and the inhibitory cascade at the intraspecific scale. *Integrative Organismal Biology* 2:obaa020.

Wake, M. H. 1976. The development and replacement of teeth in viviparous caecilians. *Journal of Morphology* 148:33–63.

Wells, K. L., Y. Hadad, D. Ben-Avraham, J. Hillel, A. Cahaner, and D. J. Headon. 2012. Genome-wide SNP scan of pooled DNA reveals nonsense mutation in FGF20 in the scaleless line of featherless chickens. *BMC Genomics* 13:257.

Westergaard, B., and M. W. J. Ferguson. 1986. Development of the dentition in *Alligator mississippiensis*. Early embryonic development in the lower jaw. *Journal of Zoology* 210:575–597.

Westergaard, B., and M. W. J. Ferguson. 1987. Development of the dentition in *Alligator mississippiensis*. Later development in the lower jaws of embryos, hatchlings and young juveniles. *Journal of Zoology* 212:191–222.

Westergaard, B., and M. W. J. Ferguson. 1990. Development of the dentition in Alligator mississippiensis: Upper jaw dental and craniofacial development in embryos, hatchlings, and young juveniles, with a comparison to lower jaw development. *American Journal of Anatomy* 187:393–421.

Whitlock, J. A., and J. M. Richman. 2013. Biology of tooth replacement in amniotes. *International Journal of Oral Science* 5:66–70.

Widelitz, R. B., T.-X. Jiang, A. Noveen, C.-W. J. Chen, and C.-M. Chuong. 1996. FGF induces new feather buds from developing avian skin. *Journal of Investigative Dermatology* 107:797–803.

Wiens, J. J. 2011. Re-evolution of lost mandibular teeth in frogs after more than 200 million years, and re-evaluating Dollo's law. *Evolution* 65:1283–1296.

Woodruff, E. D., B. K. Kircher, B. A. Armfield, J. K. Levy, J. I. Bloch, and M. J. Cohn. 2022. Domestic cat embryos reveal unique transcriptomes of developing incisor, canine, and premolar teeth. *Journal of Experimental Zoology Part B: Molecular and Developmental Evolution* 338:516–531.

Woolley, T. E., R. E. Baker, E. A. Gaffney, and P. K. Maini. 2011. Stochastic reaction and diffusion on growing domains: Understanding the breakdown of robust pattern formation. *Physical Review E* 84:046216.

Wu, P., X. Wu, T.-X. Jiang, R. M. Elsey, B. L. Temple, S. J. Divers, T. C. Glenn, K. Yuan, M.-H. Chen, R. B. Widelitz, and C.-M. Chuong. 2013. Specialized stem cell niche enables repetitive renewal of alligator teeth. *Proceedings of the National Academy of Sciences* 110:E2009–E2018.

Yamanaka, A., K. Yasui, T. Sonomura, H. Iwai, and M. Uemura. 2010. Development of deciduous and permanent dentitions in the upper jaw of the house shrew (*Suncus murinus*). *Archives of Oral Biology* 55:279–287.

Yamanaka, A., K. Yasui, T. Sonomura, and M. Uemura. 2007. Development of heterodont dentition in house shrew (*Suncus murinus*). *European Journal of Oral Sciences* 115:433–440.

Ye, Q., A. Bhojwani, and J. K. Hu. 2022. Understanding the development of oral epithelial organs through single cell transcriptomic analysis. *Development* 149:dev200539.

Young, N. M., B. Winslow, S. Takkellapati, and K. Kavanagh. 2015. Shared rules of development predict patterns of evolution in vertebrate segmentation. *Nature Communications* 6:6690.

Yu, W., Z. Sun, Y. Sweat, M. Sweat, S. R. Venugopalan, S. Eliason, H. Cao, M. L. Paine, and B. A. Amendt. 2020. *Pitx2-Sox2-Lef-1* interactions specify progenitor oral/dental epithelial cell signaling centers. *Development* 147:dev.186023.

Zahradnicek, O., I. Horacek, and A. S. Tucker. 2012. Tooth development in a model reptile: Functional and null generation teeth in the gecko *Paroedura picta*: Tooth development in a model reptile. *Journal of Anatomy* 221:195–208.

Zhu, J., Y.-T. Zhang, M. S. Alber, and S. A. Newman. 2010. Bare bones pattern formation: A core regulatory network in varying geometries reproduces major features of vertebrate limb development and evolution. *PLoS ONE* 5:e10892.

8 The Selected Deviation

The Acquisition of In-situ Tooth Replacement by Creating a Gap to Fill

Donglei Chen

8.1 INTRODUCTION

8.1.1 Did In-situ Tooth Replacement Evolve De Novo?

In-situ tooth replacement by site-specific hard-tissue resorption is one of the quintessential characteristics of osteichthyans (bony fish and tetrapods). Although some mammals secondarily became functionally monophyodont (single-generation dentition), vestigial deciduous teeth are not uncommonly present (van Nievelt and Smith, 2005; Hautier et al., 2016), evidencing that this tooth replacement mechanism is not modified during the dental reduction occurs in various clades. Even the amniote egg tooth can sometimes show rudimentary replacement tooth buds (Hermyt et al., 2017). The most primitive known bony fish, the stem osteichthyan *Lophosteus* which lived 420 million years ago, had teeth replaced in situ throughout life (Chen et al., 2017, 2020). It is documented by stacks of resorption surfaces (reversal lines in thin sections) similar to those in extant reptiles (e.g., LeBlanc et al., 2017). *Lophosteus* shows a combination of in-situ tooth replacement and the primitive characteristics resembling those of jawed stem gnathostomes, including shark-like transverse organization of teeth and the segregation of the jawbone into short pieces (Vaškaninová et al., 2020). However, in-situ tooth replacement has never been reported in any osteichthyan out-groups, and it seems like an abrupt step of evolution. Are there any intrinsic characters of odontodes, or preadaptations, which allow the evolution of tooth replacement without considerable modifications? This chapter will seek the answer by exploring the process components of odontode ontogeny.

8.1.2 Is Alternation a True Pattern of Dental Development?

Dental development is divided into tooth addition (initiation of new tooth positions) and in-situ tooth replacement (succession at the same tooth position). Plenty of classic theories or models have been proposed for the mechanism of tooth addition and replacement (Table 8.1). When a new one was put forward, it usually incorporated some aspects of the previous ones, even while criticizing them. There are two key ideas shared by most of the theories: 1) The addition of teeth (and odontodes

DOI: 10.1201/9781003439653-8

in general) is controlled by inhibitory zones; and 2) the replacement of teeth is regulated by alternate signals. Over decades, developmental biologists have been searching for the underlying molecular mechanisms and testing the hypotheses. Until recently, gene pathway inhibition experiments, such as those performed on the dermal odontodes of cat sharks (Cooper et al., 2018), supported the idea that Turing's reaction–diffusion system could explain the inhibitory field concept (reviewed by Chapter 7). The zones of inhibition are supposed to prevent teeth being initiating side by side at the same time and can—but do not always—give rise to an alternate pattern.

The alternate tooth replacement has been considered characteristic of nonmammalian dentitions, and the loss of alternate pattern as a key step during the rise of mammals. Early mammals still display back-and-forth reversals between the alternate and sequential replacement among the two or three premolar positions (Luo et al., 2004, fig. 5). Osborn (1977) suggested that the acquisition of the sequential pattern was due to the blocking of the inhibition sphere by the separation of bony crypts in mammals. However, the teeth of Eretmodini cichlids (Huysseune et al., 1999) and bluefish (Bemis et al., 2005), which are also formed in bony crypts, are replaced alternately. While the tooth development of rainbow trout, which takes place in a subepithelial position without a dental lamina (Fraser et al., 2006), can be sequential in the premaxilla, but alternate in the maxilla and dentary (Berkovitz, 1978). Different patterns among tooth-bearing elements are also documented between two closely related cichlids *Astatotilapia elegans* and *A. burtoni* (Huysseune, 1983, 1990). Some squaliform sharks can have the upper dentitions in the alternate pattern but the lower dentitions aligned in parallel rows (Underwood et al., 2016). Replacement series can form from, not only every other tooth, but also every third or every fourth tooth (Miller and Radnor, 1973; Miller and Rowe, 1973; Witten et al., 2005; Huysseune et al., 2007), which suggests that the reversals of the replacement order of the premolars of early mammals may be simply due to developmental plasticity, rather than of phylogenetic significance.

In fact, in fishes and reptiles as well as mammals, the tooth initiation order can be a mix of sequential and alternate addition (Huysseune, 1990). The standard linear mode of osteichthyan dental development along the jaw has been attained in fishes without a dental lamina and can be represented by, for example, the rainbow trout *Salmo gairdneri* (Berkovitz, 1977), the Atlantic salmon *Salmo salar*, and the jewel cichlid *Hemichromis bimaculatus* (Huysseune and Witten, 2006; Huysseune et al., 2007): 1) Starting with a tooth somewhere along the jaw; 2) sequential addition of odd/even positions toward the growing ends of the jaw; 3) filling up of the even/odd positions and establishing alternation; and 4) tooth replacement following the order of tooth addition. Note that the processes of position addition (2–3) continue even after tooth replacement (4) has commenced in older positions, which will further influence the competition for space.

However, this standard mode is not universal. Apart from interspecific and seasonal variations, perturbations in the supplementation or suppression of tooth positions between individuals, between the left and right jaw halves, and between anterior and posterior portions of a jaw half, have been noticed almost by every

TABLE 8.1
Classic Theories or Models of Tooth Addition and Replacement.

Theory and Representative Publication	Animals It Was Initially Based On	True Biological Unit	Summary
Distichy Theory (Bolk, 1912)	Reptiles	Exostichi and endostichi (a pair of alternate longitudinal tooth rows)	The outer tooth row (all the odd-numbered teeth) is replaced by teeth from the inner tooth row (all the even-numbered teeth), giving rise to the alternate replacement. Each row represents a new developmental phase, and teeth in a row are replaced simultaneously.
Field Theory (Butler, 1939)	Mammals	Morpho-genetic field/region	Tooth rudiments are determined to develop in different ways according to their position in the field, which is differentiated into regions (e.g., incisors, canine, and molars) with evolutionary independence.
Inhibition model (Gillette, 1955) (Osborn, 1977)	Frog	Tooth germ	The recently initiated germs prevent the initiation of other germs in their immediate vicinity. The production of an enamel organ uses up the cellular resources of the dental lamina in its immediate vicinity so that a period of recovery is required.
Zahnreihen Theory (Edmund, 1960)	Reptiles	Oblique tooth row	Tooth replacement in all polyphyodont dentitions is regulated by successive stimuli passing along the dental lamina at regular intervals. The teeth produced by a single stimulus form an oblique row, which indicate the replacement waves. The wave that replaces the odd-numbered teeth is followed by the wave that replaces the even-numbered teeth, forming the alternate pattern.
Clone theory (Osborn, 1978)	Reptiles and mammals	Tooth family	A tooth clone is the tissues that produce a tooth family. The dental determinant (initial tooth) has the final shape largely pre-determined, and a group of teeth with similar shape are derived from one clone. The clone cells are self-controlled, always poised to interact in initiating teeth but are prevented by the existence of inhibitory zones.

(Continued)

TABLE 8.1
Continued

Theory and Representative Publication	Animals It Was Initially Based On	True Biological Unit	Summary
Odontode Regulation Theory (Reif, 1982)	Fossil vertebrates and recent sharks	Odontode	Outside the inhibitory field, new dermal elements are automatically induced. The gaps that appear between the scales as the animal grows would be filled with new scales. If the inhibitory field was shorter-lived than an odontode, a new odontode would be induced directly adjacent to the old one. But replacement scales cannot be induced until the old one is shed. The dental lamina allows prefabrication of teeth to take place in order to speed up the replacement process. Dentitions in the proper sense are formed by a dental lamina.
Tooth Position Theory (Westergaard, 1983)	Reptile–mammal transition	Tooth position	Teeth are developed or suppressed at different positions, depending on the differential growth of the jaws. The suppression of some of the tooth positions in reptile dentitions gives rise to mammalian dentitions.
Neuro-osteological Developmental Field (Kjær, 1998)	Human	Nerve pathway	The formation sequence of jaws and teeth follows that of initial peripheral nerve fiber outgrowth. Mandible ossifies before maxilla, but both start around the nerve foramina of mental nerve. Each jaw initially has three never-supply paths, giving rise to three developmental fields: Incisors, canines/premolars, and molars. Each field has its own innervation: The teeth closest to the nerve stem are the first to form, and those farthest are the most affected by agenesis.
Sequential Addition Model (Smith, 2003)	Recent sharks	Generative set (a pair of alternate tooth families)	Tooth initiation progresses in a unidirectional linear pattern, as space becomes available through jaw growth. Reiterative initiation of tooth primordia in each jaw quadrant starts from a putative single primordial tooth-signalling center at the symphysial position. The tooth primordia form the first longitudinal tooth row, and each primordium gives rise to a transverse generative tooth set under autonomous regulatory controls.

Theory and Representative Publication	Animals It Was Initially Based On	True Biological Unit	Summary
Inhibitory Cascade Model (Kavanagh et al., 2007)	Mammals	Individual tooth in a sequence	Whether tooth size will increase or decrease in a sequence depends on the activator and inhibitor ratio, a/i. Proportional relationships among molar sizes: If considering T1 = 1, then T2 = a/i, T3 = 2a/i – 1, T4 = 3a/i – 2, etc. When a/i is consistent throughout the morphospace, if T2 is no more than half of the size of T1, i.e., a/i ≤ 1/2, then T3 cannot be formed; if activation and inhibition are in balance, i.e., a/i = 1, then M1, M2, and M3 are in equal size and may lead to the supernumerary M4.

longitudinal study in various taxa (Gillette, 1955; Kerr, 1960; Lawson et al., 1971; Lawson and Manly, 1973; Berkovitz, 1977; Westergaard and Ferguson, 1986, 1987; Huysseune, 1990; Delgado et al., 2003; Trapani et al., 2005; Huysseune and Witten, 2006; Larionova et al., 2021; Paluh et al., 2021). These irregularities become even more serious with increasing age. Therefore, departures from the regular pattern are actually normal physiological processes, while the perfect regularity postulated by the classic models seems to be an oversimplification of the reality, reflecting a bias arising from exclusive focus on some specialized animal groups. For instance, the conventional view on the reptile condition (Osborn and Crompton, 1973) considered that the primordium tooth locates at Position 9, with teeth posterior to it added sequentially (in the order of Positions 10-11-12, etc.) and teeth anterior to it added alternately first to the odd-numbered positions (in the order of Positions 7-5-3-1) and then the even-numbered positions (in the order of Positions 8-6-4-2). However, this is contradicted by Osborn's own work on the viviparous lizard, *Lacerta vivipara*, in which the initial teeth in the lower jaw are developed at Positions 3, 5, 6, 8, 10, 11, and 13, with Positions 4, 7, 9, and 12 inserted later (Osborn, 1971). Deviations, including abnormalities, as the major sources of natural experiments, are actually more informative in understanding the underlying mechanism of the patterning. A newly evolved character is likely just a selected deviation.

In order to uncover the principle that allows aberrations to happen, the spatio-temporal mapping of the dentitions in all dentate vertebrate groups will be reviewed and reinterpreted carefully in this chapter. Unlike previous reviews that introduced vertebrate dentitions group by group, observations on various extant and extinct animals will in the following sections be compared in terms of their shared ontogenetic process components. Each process component, which describes one aspect of the interrelationship between bone growth and odontode addition (or replacement), is defined in the section heading. Based on this overview, a hypothesis is proposed for how the coordination between bone and tooth development generates the great diversity of observed tooth replacement modes, including the origin of in-situ replacement in osteichthyans. I argue that simple heterochronic modifications of these developmental interactions provide sufficient mechanistic explanations for this remarkable evolutionary radiation.

8.2 PROCESS COMPONENTS OF ODONTODE ONTOGENY

8.2.1 Identical Direction of Tooth Addition and Bone Growth

Unlike hair and feathers, teeth have a functional polarity and show directional growth. Even though the functional spot is the same, such as the jaw margin, dentitions can still be organized longitudinally, transversely, vertically, horizontally, and radially. It is due to the fact that the starting point of the dentition formation is different in different dental elements.

8.2.1.1 The First Tooth Often Locates at the First Ossified Region

The very first tooth germ of a dentition has been called "dental determinant", "primordial tooth-signaling center", "family founder", "pioneer tooth", and "initiator tooth" to emphasize its potential role in the formation of a tooth row or diverging rows (Smith, 2003; Sadier et al., 2020). If the related developmental pathways, such as RA and FGF signaling, are blocked during the induction of the first tooth, after the blocker is washed out, a new first tooth will develop de novo at the same location, rather than at the site of the second tooth, even though the development of the dentition is delayed (Gibert et al., 2019). Intriguingly, the location of the first tooth is usually concomitant with the first bone within the odontogenic competent zone. For instance, in alligators, on both dentary (Westergaard and Ferguson, 1986, PLATE I) and premaxilla (Westergaard and Ferguson, 1990, fig. 3), the first tooth is formed around the jaw anlagen at Position 3 (Figure 8.1C). In axolotl, the dentary tooth row is initiated from Position 1, but from Position 2 on the premaxilla (Soukup et al., 2021). The first initiated tooth usually locates somewhere in the middle of the tooth row in typical nonmammalian osteichthyan dentitions, contradictory to what was assumed by the Zahnreihen Theory that the first tooth anlage of reptiles was always produced at the anterior end of the jaw (Edmund, 1960, fig. 53). But it does form anteriorly at the symphysial or parasymphysial position in modern sharks (Rasch et al., 2016, fig. 8) and at the posterior end of dentigerous jawbones of the stem chondrichthyans ischnacanthids (Ørvig, 1973, fig. 2D). The sequential addition model is based on the elasmobranch condition, in which the formation of the initial dental row progresses from one end to the other of the jaw, but applicable to other conditions with radial tooth rows (Smith, 2003).

8.2.1.2 Radial Tooth Rows as the Vectors of Bone Allometry

Sequential addition can give rise to a radial pattern of tooth arrays according to the shape of the bones and the location of the ossification center. Each array is a tooth family, and all of them radiate from the founder region at the ossification center of the basal bone. The basal bone generally grows concentrically, but has different growth rates in different directions, so the tooth rows extend radially in different length and thus have a different number of teeth (Smith, 1985, fig. 15).

If the ossification center is at one end of the bone, the growing end of a bone will become wider, and the bone will gain a fan shape. When the fanning-out degree of bone is larger than the size increase of successive teeth, there will be more than one tooth row. The ischnacanthid dentigerous jawbones often bear two or three rows (Smith, 2003, fig. 10; Blais et al., 2015, figs. 6, 7, ltr, mtr; Dearden and Giles, 2021). Lungfish have several tooth rows developed on the fan-shaped tooth plates or only two rows on the narrower ones. The initiator tooth can locate at the posterior, anterior, or lateral end of the bone, and the orientation of the tooth rows varies in each tooth plate (Smith and Krupina, 2001, fig. 5).

If the first formed tooth is not at the end of the jawbone, the radial tooth rows will extend in multiple directions. Take the anterior supragnathals, the upper dental plates of stem gnathostomes, as an example: In an acanthothoracid (a jawed stem gnathostome basal to arthrodires) from Arctic Canada, the ossification center is close to the labial margin, and as a consequence teeth are hardly added labially (Vaškaninová et al., 2020); in a buchanosteid (basal arthrodire) from Southeastern Australia, the ossification center is more centrally located, and teeth are added in all directions along a large number of tooth rows (Hu et al., 2019); in the eubrachythoracid (derived arthrodire) *Compagopiscis*, the bone becomes so narrow that teeth are only added posteriorly in two rows and medially in a single short row (Smith, 2003, fig. 5b, asg; Rücklin and Donoghue, 2015, fig. 1f–h).

Actually, all dentitions can be regarded as radial row(s) with one or multiple arrays. The oral lamina of osteichthyan marginal jawbones are typically so slender that tooth positions can only be generated longitudinally in two opposite directions from the primordial tooth at the ossification center, which appears as a single linear tooth row. Generally, combined with insertion of tooth positions (see below), anterior to the primordium, teeth are added posteroanteriorly; posterior to the primordium, teeth are added anteroposteriorly (Osborn and Crompton, 1973; Osborn, 1977, fig. 4A). Therefore, no matter how different are the bone shape and ossification center, the direction of successive addition of teeth is always concordant with that of the appositional growth of bone (Ørvig, 1973, figs. 2D, 4; Smith, 1985, fig. 15).

Note that the radial tooth rows are not always in one plane. Dome-shape dental elements, which increase the depth along with the appositional growth, have the younger tooth rows added at the sides lying at a lower altitude. In the earliest known dentition, the tooth whorl of a stem chondrichthyan *Qianodus* (Andreev et al., 2022), the primary rows are a pair rows of alternate large teeth on the raised medial crest, while teeth of the secondary and tertiary rows on the slope flanking the crest are symmetrical. All tooth rows have the younger teeth with increasing size added lingually. But unlike the classic acanthodian tooth whorl model (Ørvig, 1973, fig. 2A), the growth mode of the whorl base of *Qianodus* is more comparable with that of the tooth cushions of stem osteichthyans in which the labial side of the base also enlarges, even though in a less degree, as shown by the incremental rings (Chen et al., 2017, fig. 1i–j). Accordingly, the initiator teeth of the secondary and tertiary rows of *Qianodus* locate labially to those of the primary rows (Andreev et al., 2022, fig. 2e). In spite of the even smaller size, the initiators of the secondary and tertiary remains controversial, due to the contradictory observations that they are not aligned alternately but symmetrically. The pair primary rows could be regarded as an alternate generative set proposed by the Sequential Addition Model (Smith and Coates, 2001; Smith, 2003) so that the trailing row might also originate from the initiator tooth of progenitor row. But this primary initiator tooth does not directly give rise to the initiators of other rows. The addition of the initiators of different classes of tooth rows occurs asynchronously, but in agreement with the available three-dimensional space on the surface of the tooth whorl.

8.2.2 Differential Timing between Tooth Addition and Bone Growth

The consistency of directional growth does not mean that the initiation of tooth and the mineralization of bone has to be synchronized. The question of whether it is the presence of bone that induces the initiation of teeth, or the other way around, remains controversial, due to contradictory observations.

8.2.2.1 Sequence of Tooth Initiation Is Not Predetermined

The ossification of bone closely follows the calcification of teeth in the tooth patches, as well as the linear tooth row, of Mexican axolotl *Ambystoma mexicanum* (Soukup et al., 2021), sterlet sturgeon *Acipenser ruthenus* (Pospisilova et al., 2021), and three-spined stickleback *Gasterosteus aculeatus* (Ellis et al., 2016). On the pharyngeal tooth plates of stickleback, the first-generation teeth are generally younger peripherally. The extension of the previous row at both ends and the formation of a new row happen in parallel. In other words, tooth addition takes place in multiple directions. There is no standard order for the initiation of tooth positions. Variations among individuals, tooth plates, and even between left and right are more notable in later stages. In other words, there is neither a preset pattern nor a permanent dental lamina to guide the succession, but all depends on the real-time situation. Nevertheless, the addition of a tooth always heralds the formation of the bit of bone beneath it (Ellis et al., 2016, fig. 5). Although this sounds like the former induces the latter, it is equally possible that the irregularity of the tooth patch is due to the allometric growth of the bone. The first dentary teeth (D1–3) of medaka *Oryzias latipes* also emerge irregularly. D2 is formed on either side (lateral/medial) of D1, but D3 is always formed on the opposite side of D2, so the location of D3 depends on that of D2 (Larionova et al., 2021).

8.2.2.2 Offset between the Peak Development of Dentition and Bone

In larval garfish, the premaxillary teeth are initiated anteroposterially, the outward branching of the premaxilla also progresses anteroposterially, but lagging behind. Consequently, teeth have not been attached when they pierce the epidermis (Moy-Thomas, 1934, fig. 3). In the primary dentition of larval bichir, the attachment of teeth falls behind the development of the dentary and prearticular, whereas the coronoids do not expand until attachment sites are needed (De Clercq et al., 2014). In the cichlids *Hemichromis* and *Astatotilapia*, the formation of the first tooth and the dentary is concurrent (Huysseune, 1990). In Mexican tetra, the premaxillae teeth are not mineralized when the entire premaxilla is cartilaginous, but start to calcify when the occluding edge of the premaxilla begins ossification. While the occluding edge is the first region to ossify in the premaxilla of Mexican tetra, it is the last region to ossify in the toothless premaxilla of zebrafish (Hammer et al., 2016). *Fgf10* overexpression in heat-shocked zebrafish larvae delays the tooth development, which is associated with the delayed ossification of the ceratobranchial bone that the teeth will stand on (Jackman et al., 2013).

The first fetal tooth buds of viviparous caecilians are initiated even before the differentiation of jaw cartilage and are thus unattached and nonerupted. Nevertheless, the mesenchyme cells that will form the jawbone have aggregated (Wake, 1976). This implies that the onset of osteogenesis and odontogenesis is synchronized, even though they may not progress simultaneously. But in African clawed frog, odontogenesis does not begin until the jaw quadrants reach a certain length. Then the rate of the extension of the tooth row far exceeds that of the elongation of the jawbones in order to occupy the most length of the jaw rapidly. In spite of the asynchronous growth of teeth and bone, the final length of the tooth row and jaw quadrant shows a correlation (Lamoureux et al., 2018).

Studies on bats inferred that it is the available dental lamina space that determines the initiation order as well as the number of teeth. Bats exhibit an outstanding variation of diet-related skull shape and tooth number, between the long-faced nectar-eating condition that has three premolars (P2–4) and three molars (M1–3), and the short-faced fruit-eating condition that has the middle premolar (P3) and/or the last molar (M3) suppressed (Sadier et al., 2021). Nectar bats show a temporally extended cell proliferation, and the jaw, which begins to elongate at the developmental stage CS22, has enough time to gain a consider length; fruit bats initially have an elevated proliferation rate, leading to an early mature process of cartilage and bone at CS18, at which the canine tooth bud just appears while the jaw growth already slows down (Camacho et al., 2020). After CS19, the postcanine dental lamina develops. Between CS20 and 24, the tooth buds of dP3 and P2 emerge simultaneously in nectar bats because of the continual rapid growth of jaw, but P3 and even dP3 fail to mineralize in fruit bats due to the terminal growth of jaw (Sadier et al., 2021, fig. 2).

8.2.2.3 Orchestration between Dental and Bony Units by Dynamic Rhythms

Not only fruit bats, but also many other mammals show delay of tooth eruption when the jaw grows most rapidly and retardation of jaw growth when the teeth develop. It evidences that rapid bone growth cannot stimulate tooth initiation and is not led by tooth development either. In mouse, no onset of M3 until the retromolar space has been available for a week and no burst of jaw growth around epithelial thickening. Nevertheless, the total jaw length, which jumps from about 5,000 μm to 8,000 μm during the eruption of M1 and M2 (P18–P21), is strongly correlated with the total length of the three molars ($r = 0.90$–0.99), even though the retromolar length that increases steadily is less correlated to molar lengths ($r = 0.79$–0.97) (Ko et al., 2021). It suggests that not the entire retromolar length measured in this study can be stretched through by the dental lamina and the odontogenic competent zone is better represented by the total jaw length. The upper jaws are longer before birth, but are eventually taken over by the lower jaws that have a higher growth rate, which is partially associated with the secondary palate formation (Peterka et al., 2000, fig. 8).

It is concomitant with the molar lengths $M^1 > M_1$, $M^2 \geq M_2$, and $M^3 < M_3$ and the size decrease (M1 > M2 > M3) that is more evident in the upper jaws. Interestingly, regardless of the slower growth rate of the upper jaws and the size differences of molars, the same type of molars are developed synchronously between the upper and lower jaws (Ko et al., 2021). This implies that the developmental timing and growth rate is decoupled between teeth and bone. Furthermore, the growth rate of teeth can be different between the same type of teeth on different jaws, between neighboring teeth on the same jaw, and between different developmental stages of a tooth. M2 is initiated immediately after M1 with their developmental periods largely overlapped, and competition is noticed between them. Initially they share the same growth rate. When M1 accelerates at the start of crown mineralization around E16–18, M2 levels off; when M2 picks up the speed for crown formation from E18 and takes off at birth, M1 drops the spurt (Sofaer, 1977, fig. 1; Peterka et al., 2000, fig. 7; Ko et al., 2021, tab. 3). The mirroring changes of growth rate between M1 and M2 are well-balanced, and the combined increase of the two molars is constant relative to the enlargement of head. It rebuts the hypotheses that the inhibition zone decreases linearly after the initiation of a tooth germ (Osborn, 1971, 1977), or the activator-inhibitor ratio keeps constant through the sequential addition (Kavanagh et al., 2007). Instead, the activation and inhibition, as well as the growth of bone and teeth, are all dynamic processes, which may explain diverging molar proportions departing from the theoretical model (e.g., Billet and Bardin, 2021).

Taken together, competent space provided by jaw elongation is indeed a structural constraint for tooth development and implantation, though unnecessary for tooth initiation; competing for the competent space under the reaction–diffusion mechanism, teeth attain the size and shape adapting to the diet by adjusting the developmental timing.

Odontode and bone, such as tooth and jaw, are certainly two developmental modules (Moy-Thomas, 1934; Smith and Johanson, 2015). Teeth need to stand on a bone or cartilage but may not need the presence of a skeletal support in order to form, and the extent of the supporting skeleton is not necessarily restricted to the tooth-bearing area. It is likely that the formation of both bone and teeth occurs in response to the same developmental signals or growth factors, such as instruction or patterning by nerve tissue that forms before bone and teeth (Kjær, 1998); but that the ossification of bone and the calcification of teeth is influenced by their own developmental rhythm, namely, intrinsic control.

Such intrinsic rhythms may show interspecific, ontogenetic (Wake, 1976; Erickson, 1996), seasonal (Cooper, 1966; Huysseune et al., 2007), and regional variations (e.g., anterior–posterior variations: Lawson and Manly, 1973; Huysseune et al., 2007) and may be related to tooth size (Berkovitz and Moore, 1974, 1975; Wake, 1976; Mateo and López-Jurado, 1997). For example, tooth replacement rate (also referred to "cycle length" and "generation time") varies between tooth families, demonstrating that it is controlled locally at each tooth position/family (Osborn, 1971, 1974; Osborn and Crompton, 1973; Westergaard and Ferguson, 1987; Huysseune and Witten, 2006; Maho et al., 2022). The differential tooth addition rate is clearly seen in the non-shedding dentitions. On a pterygoid tooth plate of the Devonian lungfish *Andreyevichthys epitomus* (Smith and Krupina, 2001), teeth of different rows can align into a curve but mostly not (Figure 8.2A, green and orange lines), with the lateral rows adding smaller teeth more frequently and terminating earlier. But the intrinsic rhythms can be modified by extrinsic factors. In vitro, tooth replacement of cichlid juveniles can be delayed or out of phase between odd- and even-number positions by culturing with the Wnt/β-catenin pathway agonist, LiCl (Fraser et al., 2013). In vivo, the so-called extrinsic factors can simply be a competitive reaction from the neighboring position/family. Similarly, the tooth-bearing elements have to develop in concert with other craniofacial skeletons. Any small change may be enough to make the consequent pattern unpredictable. A highly autonomous coordination among odontodes and between skeletal elements is thus required to make sure that related units work together, even when the pattern has deviated.

8.2.3 Gap-Filling Autonomy during the Initiation of Tooth Position

8.2.3.1 Insertion of Tooth Positions in Single Linear Row Dentitions

Alternate addition is supposed to be the sequential filling of every interspace between teeth of the first round (primary teeth) by a second round (secondary teeth) (Osborn, 1971). To fulfil this, the interspace must be created by the isometric growth of bone. However, allometric growth is more common. For example, osteichthyan mandibles usually grow faster posteriorly than anteriorly. The pattern may thus appear haphazard.

Tracing the tooth initiation in the monophyodont dentition of veiled chameleon, *Chamaeleo calyptratus* (Buchtová et al., 2013, fig. 2, tab. 1), has cast light on the addition of tooth position, without being intervened by tooth replacement. The lower jaw shows a similar initiation sequence (Figure 8.1A) as the upper jaw. In the upper jaw, T1 and T2 are distributed in each half of the initially ossified jawbone, followed by the addition of T3 and T4 at both growing ends of the jaw. Then T5 and T6 are inserted to the wide gaps T1–T2 and T2–T3. After that T7 to T9 are initiated sequentially toward the rostral growing end of the jaw and T10 at the caudal end. T12 and later teeth are all added caudally in sequence. Surprisingly, T11 is inserted into the most recent gap T8–T9. If an inhibition zone created by a tooth germ will reduce as time goes by, and a new tooth germ will be initiated beside once the inhibition goes away (Osborn, 1971, 1977), T11 should have been added to the oldest gap T1–T4, if not anterior to T9 or posterior to T10.

In-vivo charting of 22 tadpoles of African clawed frog, *Xenopus laevis*, displays variable tooth initiation pattern between individuals and even between jaw quadrants (Lamoureux et al., 2018). The first tooth is not initiated at a fix position, and the subsequent teeth are not generated in a definite order. The final length of the tooth row correlates to that of the jaw quadrant, and thus the total number of tooth position varies. Before metamorphism, the first-generation teeth are formed through three rounds of addition, instead of two alternates: The primary, secondary, and tertiary teeth that start at around Stage 54, 56, and 61, respectively (Figure 8.1D). Most of the length of the jaw is quickly occupied by the initial primary teeth with wide interspace, and more primary teeth are added posteriorly as the jaw extends posteriorly at the later stages; then the secondary teeth are inserted into some of the interspace, and as more primary teeth are added posteriorly, more secondary teeth are developed posteriorly

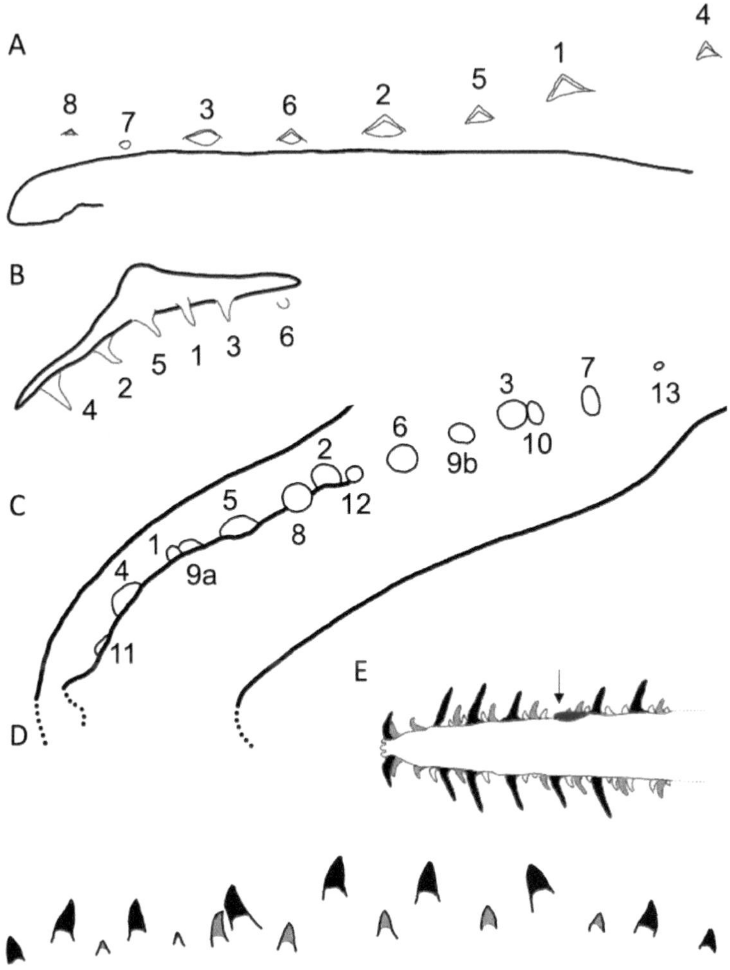

FIGURE 8.1 Tooth addition in a gap-filling manner. Numbering indicates the initiation order of tooth germs. (A) Lower jaw of chameleon embryo (adapted from Buchtová et al., 2013). (B) Premaxilla of Mexican tetra embryo (adapted from Trapani et al., 2005). (C) Lower jaw of alligator embryo (adapted from Westergaard & Ferguson, 1986). (D) Upper jaw of African clawed frog tadpole. Primary tooth germs are colored in black, secondary tooth germs in grey, and tertiary tooth germs shown as open tips (adapted from Lamoureux et al., 2018). (E) Rostrum tip of sawshark adult. Primary, secondary, and tertiary denticles are colored in black, grey, and white, respectively. Replacement denticle that still lies flat is indicated by arrow. (Adapted from Welten et al. 2015).

between the primary teeth; when the initial primary teeth start to be replaced, the tertiary teeth are deposited into some of the gaps between a primary and a secondary teeth, producing a compact tooth row; the initiation of posterior primary teeth, secondary teeth, tertiary teeth, and their replacement teeth can occur in parallel. Between two primary teeth, depending on the width of the interspace, there can be a pair of tertiary teeth separated by a single secondary tooth (triad configuration) or a single tertiary tooth and a single secondary tooth (doublet configuration) or a single secondary tooth (solitary configuration) or no other teeth (null configuration). Such configurations can change constantly during ontogeny. As a result, the odd- and even-numbered tooth positions do not correspond to the primary and secondary teeth. It demonstrates that the initiation of tooth positions is simply a gap-filling process, rather than being programmed by a sequential or alternate pre-pattern. With no metamorphic changes interrupted the observation, four rounds of insertion of first-generation teeth were recorded in alligator embryos (Figure 8.1C), which created up to five interstitial positions (Westergaard and Ferguson, 1987, 1990).

An alternation of the primary, secondary, and tertiary teeth is displayed in extra-oral teeth as well, and the initiation order is reflected by the space-dependent tooth size. In sawsharks, the saw-teeth and the cartilaginous rostrum both mineralize rostrally (Welten et al., 2015). Just like the primary teeth (Figure 8.1) of veiled chameleon, clawed frog, alligator, as well as those of Mexican tetra (Trapani et al., 2005) and wild Atlantic salmon (Huysseune et al., 2007), the primary saw-teeth are widely spread. After the first set of saw-teeth have elevated to the function positions, smaller ones are inserted between them. Later, another set of even smaller size are inserted, if there are gaps between the previous two sets. The tooth size seems not correlated with age but the availability of space. Unlike oral teeth, if no loss, then no replacement; but once a saw-tooth is lost, it will be replaced by a new one of the same size (Figure 8.1E, arrow). Similarly, in the hybodontiform shark *Tribodus limae*, gaps between thorn scales are covered by small scales, but the largest gap is no more than the width of one thorn scale, because once such a large gap appeared, it would have been filled by a thorn scale (Maisey and Denton, 2016, fig. 2A). In this way, the thorn scales maintain a uniform distribution in rows.

8.2.3.2 Splitting and Fusion of Tooth Rows in Multi-row Dentitions

Gap filling is a normal phenomenon even in the modern shark dentitions. A tooth family can split and then fuse back. The splitting varies between the left and right quadrant and is not considered as malformation (Reif, 1980). In the southern thorny skate, *Amblyraja doellojuradoi*, the split rows start or end with a bicuspid teeth, as a transition into a fused row (Delpiani et al., 2012). It is interesting that when a split family carries on the divided status through generations, the pairs of half-teeth are organized alternately as two separated tooth files. This is particularly clear in tiger shark, *Galeocerdo cuvieri*, which has the most asymmetrical sickle-like tooth shape (Grudger, 1937; Reif, 1980). Even in frilled shark, *Chlamydoselachus anguineus*, whose tooth families are normally widely spaced in a single-file pattern, when a double set of teeth are found in the position of a tooth family, the crowded tricuspid teeth automatically form an alternate-file pattern (Grudger, 1937, fig. 5). This implies that the alternate pattern is the result of the space constraint, unlikely requiring a regulation mechanism that is genetically unique to the taxon, and thus not a phylogenetic signal.

A regular pattern will automatically disappear, too, once the size and shape differentiate between neighboring teeth. In the embryos and juveniles of Pacific bullhead shark, *Heterodontus*, both the scales and teeth are needle-like and organized in diagonal rows. As such ordered arrangement is gradually obliterated in the scales by the process of random loss and repair, the diagonal rows also disappear in the teeth when the dentition differentiates from homodont to heterodont. To crush hard shelled prey, the molariform teeth are not separated by gaps. The tooth shape is not identical on the left and right jaw quadrants, and thus the number and position of tooth families are asymmetrical (Grudger, 1937, figs. 6, 7). During the disproportionate growth of the jaw cartilage, a tooth family can be suppressed (Grudger, 1937, fig. 8) or new tooth families would be inserted between preexisting ones (also reported in the bowmouth guitarfish, *Rhina ancylostoma*, Smith et al., 2013, fig. 3E). A gap between tooth families can also be filled by a tooth that has no successors (Reif, 1976, figs. 18, 19, 32, 33b, 34, 38c–e), even though no external disturbance can be detected.

Gap filling is frequently found not only in the linear organization, but also in the radial arrangement in other vertebrate groups. On the supragnathals of a unnamed acanthothoracid (CPW.9; Vaškaninová et al., 2020, fig. S3) and a unnamed buchanosteid arthrodire (ANU V244; Hu et al., 2019, fig. 1C, df.a-c, df.d1–3 and fig. 1G, df.g), since the bone is not in a perfect fan-shape, the array of the tooth families is usually not in a straight line (Figure 8.2B, blue and pink lines). The succession of the elongated ornament-like teeth of CPW.9 marks the former outlines of the bone at the time of tooth formation (Figure 8.2B, green line), and the conical teeth of ANU V244 recurve in the direction of bone growth at that particular position. Teeth to be added at the end of the radial tooth rows will adapt to the differential expansion of the bone margin by slightly adjusting the orientation. Teeth from neighboring tooth families will keep the maximum distance to those on both sides, which makes neighboring tooth families distinguished from each other. Gradually, the inter-family space allows new tooth families to be inserted halfway. Insertion between the main radial tooth rows occurs in lungfish tooth plates as well, especially in the adult forms of basal taxa, but the intermediate teeth are very tiny (Chang and Yu, 1984; Smith and Chang, 1990; Barwick et al., 1997; Smith and Krupina, 2001; Mondéjar-Fernández et al., 2020). As the space between the main rows becomes increasingly wider, the intermediate rows also diverge and appear as V-shaped. Between the divergent rows, another V-shape divergent set is inserted. This happens again and again, forming a fishbone pattern (Figure 8.2A, pink line; Smith and Krupina, 2001, fig. 4A). The main tooth rows also occasionally display misalignment (Barwick et al., 1997, fig. 8.7–8), split (Smith and Chang, 1990, fig. 1B), or merge into a row with a lager tooth size (Smith and Krupina, 2001, fig. 3A).

Pycnodontiforms, a group of Mesozoic actinopterygians, are characterized by the multi-row non-shedding crushing dentitions on the unpaired vomer and paired prearticulars. They also exhibit that not all the tooth rows start from the anterior end, but from halfway where the bone becomes wider. Rows of larger elongated teeth may be intermediated by rows of irregular tiny rounded teeth. While a row of large teeth can diverge into two rows of small teeth, two rows of small teeth can converge to a row of large teeth (Figure 8.2C). More intriguingly, some teeth in a large row are divided into two or three teeth of almost equal size, or into one small tooth and one larger tooth, which are followed by a normal undivided tooth. The divided and undivided teeth both occupy the full width of the tooth row domain. The convergence and divergence are switched back and forth frequently, and all these phenomena are

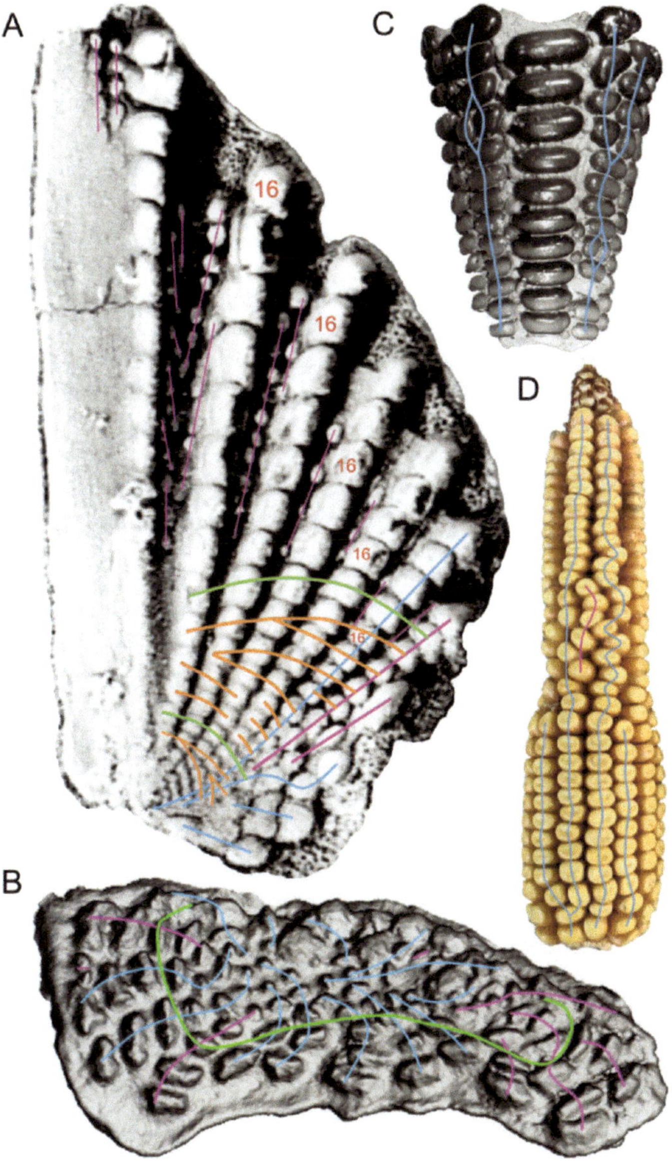

FIGURE 8.2 Insertion, division, and fusion of tooth families in non-shedding dentitions. (A) Pterygoid tooth plate of a lungfish from the Devonian (adapted from Smith and Krupina, 2001). Numbers in red indicate the position of the 16th tooth in each tooth row, reflecting the difference in tooth number. Orange lines indicate the misalignment between tooth rows. (B) Supragnathal of an acanthothoracid fish from the Devonian. Green line indicates interpreted outline of the bone at an earlier developmental stage (adapted from Vaškaninová et al., 2020). (C) Vomer of a juvenile pycnodont fish from the Eocene (adapted from Collins and Underwood, 2021). (D) Modern maize (adapted from u.osu.edu/mastercorn/). Blue line indicates primary row and pink line indicates inserted row.

not uncommon (Longbottom, 1984, fig. 2b, 3–16; Poyato-Ariza and Wenz, 2002, fig. 20A; Kriwet, 2005, figs. 28A-B, 29A, 32, 33B, 36A, 43, 46). Although most of time the tooth rows are well demarcated from each other, and most of species have teeth developed intraosseously it is hard to believe that the tooth addition of each tooth row is guided by a dental lamina. But throughout life, the oral epithelium retains the odontogenic competence and initiates teeth once a gap appears. Even the underlying bone is damaged, multiple small teeth would be initiated with emplacement resorption to repair the broken bit of original large teeth (Collins and Underwood, 2021), in contrast to the one-for-one replacement.

8.2.3.3 Gap-filling in Bony Units

Even the bony pseudoteeth seen in some extinct birds exhibit a similar repetitive sequential size distribution of four ranks, XL-S-M-S-L-S-M-S-XL (Louchart et al., 2018). The smaller a pseudotooth is, the later it is formed. In contrast to the examples given earlier, the jawbone of the beak has been fully grown before the formation of any pseudoteeth. Thus, the process of filling gaps would not be influenced by the regional variations in the growth of bone or cartilage, and the pseudoteeth are organized as neatly as the marked scale on a ruler.

Ordered bony tubercles, just like organized dentinous tubercles, are also not uncommon, such as those on the postbranchial lamina in jawed stem gnathostomes (Johanson and Smith, 2003, 2005). Porolepiform fishes, as well as Devonian lungfishes, are characterized by radial rows of spoon-shaped odontodes anterior to the exposed area of the scales, which become bony tubercles in the derived taxon *Holoptychius*, but the characteristic fan-shaped pattern persists (Ørvig, 1957). A regular pattern produced by gap filling is likely a shared feature across discrete units on dermoskeleton. All of these dermal ornaments can show the insertion of rows and transition between single-file and alternate-file patterns.

8.2.4 Cyclic In-situ Tooth Replacement as a Modification of Columnar Succession

8.2.4.1 Sequential Addition Achieved by Creating Gaps Only at the End of Tooth Row

Modern shark dentitions have been regarded as a perfect model of the strict regulation of dental lamina. Such regularity has been expected in stem chondrichthyans. The non-shedding teeth on the dentigerous jawbones of ischnacanthid acanthodians are initiated sequentially from the posterior to the anterior end, with the next tooth partially overlapping the anterior edge of the previous one (Ørvig, 1973; Rücklin et al., 2021). However, the mechanism of regularity may not be properly understood without studying the abnormality. In a pathological case found in *Taemasacanthus* (Dearden and Giles, 2021), the original penultimate tooth was broken, creating a gap in the sequence. As a result, when a new tooth was formed, it was not laid anteriorly to the ultimate tooth, but above the remnant of the original penultimate tooth. Since the newest tooth is the largest tooth, it also overgrows the posterior edge of the ultimate tooth. As a dental lamina is supposed to extend unidirectionally, this natural experiment demonstrates that dental lamina is not a necessary component for the ischnacanthid dental system to make ordered teeth. Instead, tooth addition may simply be regulated by the anterior extension of the jawbone. The unidirectionality is actually the result of filling gaps at the newly formed bone margin, and it cannot be achieved if gaps appear in the middle of the dentition. Therefore, the pattern of sequential addition, which has been considered as diagnostic of a dental lamina in fossil animals (Smith and Johanson, 2003; Johanson and Smith, 2005; Botella, 2006), is not in any case due to the presence of dental lamina. Dental lamina cannot be considered characteristic of all dentitions of the chondrichthyan total group.

Sequential addition is often seen in ankylosed teeth of statodont dentitions, including the marginal jawbones of acanthothoracids (Vaškaninová et al., 2020), gnathal plates of arthrodires (Ørvig, 1973, 1980; Smith, 2003; Rücklin et al., 2012; Hu et al., 2019), tooth whorls and dentigerous jawbone of acanthodians (Ørvig, 1973; Rücklin et al., 2021), the tooth plates of lungfishes (Smith, 1985), and the monophyodont dentitions of reptiles, such as the later embryonic and postnatal dentition of veiled chameleon (Buchtová et al., 2013, fig. 1F). Because in these dentitions, teeth are neither shed nor able to move through the jaw by a periodontal ligament, interstitial space cannot be generated by adjusting the positions through continuous replacement cycles. As the tooth-bearing bone grows, new space only appears at the growing edge of the bone, i.e., the end of each tooth row.

Polyodontode scales, on which odontodes are non-shedding, also display sequential addition of single- or multi-row odontodes, as well as insertion of odontode rows into the radial pattern, for example, chondrichthyan scales from the Early Silurian (Andreev et al., 2020). In the Silurian stem osteichthyan *Andreolepis* (Qu et al., 2013), the trunk scale expands more ventrally than dorsally. The founder odontode is distinctively trident-shaped and located right above the basal vertical canals, which represent the ossification center of the bony base. From the founder odontode, four increasingly elongated odontodes are laid down sequentially toward the growing ventral end, while only one is added dorsally. When the fifth elongated odontode (Odontode 11 in Qu et al., 2013) is deposited at the ventral end, it considerably overlaps the fourth one (Odontode 5), and the following ones (Odontode 14) will do the same, because the bone growth has started to slow down. A large odontode (Odontode 7) overgrows the short founder odontode, in order to cover the new anterior and posterior margins of the ornamented field of the scale. After that, overgrowing odontodes are added sequentially at the second level toward both the ventral (Odontode 8, 10, and 13) and dorsal (Odontode 9) ends. In the meantime, an even larger odontode of the third level (Odontode 12) overgrows the founder region. Although the shape of overgrowing odontodes becomes irregular due to the existence of previous odontodes, their flattened pulp cavities, which form the so-called horizontal vascular canal system (Qu et al., 2013, fig. 3, in yellow), always fit the gaps between the preexisting odontodes.

8.2.4.2 Tooth Overgrowth as a Sequential Succession in the Vertical Dimension

Bone can grow both appositionally and superpositionally to increase the width and thickness, which allows a dentition to expand horizontally and vertically, respectively. Odontodes tend to take up the new space not only at the end of the row, but also at the top of the bone (Reif, 1982, fig. 9). The sequentially organized dentitions on the jawbones of ischnacanthids do not merely extend in the horizontal dimension. The successors always lie on a level slightly higher than the predecessors like standing on a staircase, matching with the thickening of the basal bone that appears as a slope. The earliest teeth around the ossification center exhibit a more considerable partial overgrowth than the later ones, with the second tooth climbing up almost half the height of the first one (Rücklin et al., 2021), indicating a rapid superpositional growth relative to tooth addition at the early developmental stage. Similarly, in the symphyseal tooth whorls of the stem holocephalan *Edestus*, albeit without the supporting bony tissue, the lingually elongated tooth roots are stacked up (Johanson et al., 2020, fig. 3D–E), appearing as bone laminae, with the tooth crown of tubate dentine sitting at the labial end of each lamina. The fitting of the shorter root of the next tooth onto the grooved root of the previous tooth (Itano 2014, fig. 2) is different from the typical tooth interlocking mode of tooth whorls, in which the next larger tooth has the crown overlaps and the root encompasses the previous tooth (Ørvig 1973, fig. 2A, E). But it exemplifies that stacking between successive teeth can occur in chondrichthyans too.

When the formation of a strip of bone allows the attachment of a new row of teeth, the deposition of a new layer of bone permits the overgrowth of a new generation of teeth. The palatal dentitions of Permian tetrapods *Cacops* and *Captorhinus* display three or four generations of teeth piling up, along with the stacking of new layers of bone (Haridy et al., 2019, fig. S1).

The resultant of the appositional (horizontal) and superpositional (vertical) growth is the oblique stacking of teeth. In the vomer of the Triassic actinopterygian *Boreosomus*, up to eight generations of teeth are stacked sequentially into oblique columns (Ørvig, 1978a). Each column can be regarded as a tooth position, and the tooth positions spread from the ossification center to the growing margin of the bone, as the bone grows wider. The obliquity of the tooth columns also increases toward the bone margin (Ørvig, 1978a), because the margin of the bone expands both appositionally and superpositionally, while the center of the bone only grows superpositionally. Position addition at the bone margin and tooth succession at old positions take place in parallel; hence, the number of tooth generation at a position decreases toward the bone margin.

Superposition of generations of odontodes toward a particular direction has been observed in the dermal elements inside or outside the oral cavity of various jawed and jawless vertebrates (Haridy et al., 2019, fig. 2e, tab. S1). Depending on the taxon, the component odontodes can consist of other types of dentinous tissues like mesodentine and semidentine (Ørvig, 1967), with or without a hypermineralized surface layer. Ontogenetically, successive odontodes may be deposited next to one another or separated by bony tissue, with or without resorption, and orientate from vertical to horizontal (Ørvig, 1978a). Columnar succession of odontodes could be one of the basic growth modes of odontodes primitviely shared by vertebrates, and the evolution of some odontode derivatives might have involved the modification of such a growth mode. For example, in the integument of actinopterygians, such as *Cheirolepis* (Zylberberg et al., 2016), *Moythomasia* (Gardiner, 1984, fig. 144), *Plegmolepis*, *Gyrolepis* (Ørvig, 1978a, figs. 23–29), *Crenilepis*, and *Nephrotus* (Ørvig, 1951, fig. 8D–E), the younger generations of odontodes in the columns gradully lose the dentine layer and eventually turn into superimposed enamel layers, bringing about the evolution of multilayered ganoine.

8.2.4.3 Various Degrees of Resorption within a Replacement Column

A tooth row on the dentigerous jawbones of ischnacanthids or the tooth plates of lungfishes can be considered as a tooth column lying horizontally. The successive teeth only overlap at the edge, and the additon of successor does not induce resorption on the predecessor. Hard-tissue resorption associated with dentitions has occurred in brachythoracid arthrodires (Johanson and Trinajstic, 2014). However, this resorption is due to the systematic remodelling of bone, and a large number of teeth can be removed altogether. It may be comparable with the extensive resorption on the tooth plates of some lungfish (Ahlberg et al., 2006), which does not contribute to the shedding of teeth. The typical osteichthyan in-situ tooth replacement is characterized by cyclic site-specific basal resorption (Chen et al., 2016). Site-specific semi-basal or apical resorption of dentine, which partially removes the preceding odontodes, has commonly occurred in the columnar odontocomplexes of osteichthyans, such as in the sarcopterygians *Glyptolepis* and *Holoptychius* (Bystrow, 1939, fig. 2B, 10E, F), the actinopterygians *Cheirolepis* (Zylberberg et al., 2016, fig. 3J), *Yaomoshania* (Poplin et al., 1991), *Plegmolepis*, *Gyrolepis* (Ørvig, 1978a), and *Scanilepis* (Ørvig, 1973, fig. 3A-C, 1978b, fig. 23), as well as in the tooth columns of both the inner and outer arcades of stem osteichthyan *Lophosteus* (Chen et al., 2017, 2020), the palatal dentition of actinopterygian *Boreosomus* (Ørvig, 1978a), the prearticular dentition of the stem lungfish *powichthys* (King et al., 2021) and tetrapodamorph fish *Eusthenopteron* (Bystrow, 1939, fig. 18B), and the pterygoid of amniote *Captorhinus* (Haridy et al., 2019, fig. S4D).

Great ontogentic plasticity of the successional columns can occur within a single dental element, depending on the relative growth rate between odontodes and bone. It is beautifully illustrated by the marginal jawbones of the oldest and basalmost osteichthyans *Andreolepis* (Chen et al., 2016) and *Lophosteus* (Chen et al., 2020). The marginal jawbone of *Andreolepis* grows very fast initially. Each newly added tooth has enough space to sit at the newly grown lingual margin, forming alternate files, without overlapping the previous teeth. There is thus no need for the teeth to be shed. They function until they are overgrown by dermal odontodes. The long functional life is evidenced by the in-vivo chipped tips. As indicated by the compact lines of growth arrest at the lingual margin, when the oral lamina has been formed, the bone growth suddenly slows down dramatically. New teeth have to sit at roughly the same place as the old ones. The teeth on the oral lamina then undergo basal resorption, and their widely opened pulp cavities are taken over by the new teeth. In this way, teeth are replaced in situ. As indicated by the invasive front of the overgrowing dermal odontodes, the oral-dermal boundary slows down the lingual movement accordingly. When the oral epithelium has produced a new generation of tooth buds, the tooth rows that have not been overgrown will be replaced cyclically. In parallel, when the jaw margin has extended lingually for a certain distance, a new row of first-generation teeth is formed.

In the marginal jawbone of *Lophosteus*, the first (the most labial) row of first-generation teeth is non-shedding, like those of *Andreolepis*, except that the tooth tips are still very sharp, suggesting a shorter functional life. The following rows of the first-generation teeth gradually change from non-shedding to partial shedding, with the most lingual row shed nearly basally. The tooth remnants on a tooth row overlap those of the previous row considerably (Figure 8.3G, H, orange). This implies that the previous tooth must have been shed semi-basally before adding the next. The next tooth that locates more lingually is laid down on a higher level than the previous one, since the bone has grown superpositionally as well as appositionally. It appears as the linguolabial compression of the first-generation tooth files of *Andreolepis*, combined with overgrowth and resorption. Thus, a file of the *Lophosteus* first-generation tooth remnants resembles an odontode column with partial resorption between successive generations. But, different from the odontode column, the semi-basal resorption in *Lophosteus* widely opens up the pulp cavities of the first-generation teeth. The remaining base of each first-generation tooth would not be completely embedded by the next first-generation teeth, instead, replacement teeth can be deposited on top. In this way, each first-generation tooth that has been resorbed semi-basally sets up a tooth position of cyclic replacement. While a first-generation tooth file forms a near horizontal column with dentine remnants, the successive replacement teeth at a tooth position build a vertical column that is also comparable with an odontode column. But the replacement columns that derive from the most lingual first-generation teeth, which represent the marginal row of teeth at any time, can be looked on as a continuation (but a linguolabial compression) of the first-generation tooth file and is thus lying horizontally too. Nevertheless, while the first-generation files flare out lingually like a fan, the marginal replacement columns are roughly parallel (Figure 8.3H–I). These distinct polarities suggest a change of the jawbone shaping and two developmental stages with different signal environment. Like in *Andreolepis*, the resorption of the replacement teeth is usually basal and removes the entire cone of dentine, so that no dentine remnants are found between the resorption surfaces, except remnants of the folding dentine base and bone of attachment. The columns of replacement teeth are stacked up by successive generations of attachment tissues, which is demarcated by a series of basal resorption surfaces (Figure 8.3F, G, J). The stack of attachment tissues and resorption surfaces, as well as the accumulated pulp column, records the ontogenetic trajectory of a tooth position.

8.2.5 The Deposition of Replacement Teeth Requires a Gap to Be Filled

In this section, observations on the marginal jawbone of *Lophosteus* are described as a case study on the gap-filling phenomenon in tooth replacement.

8.2.5.1 Anteroposterior Gradation of Overlap and Shedding

Tracing the stack of resorption surfaces, tooth files can be identified. The successive teeth of a tooth file drift lingually as the jawbone grows. The course of the drift is usually not a straight line, but adapting the available space with neighboring families. In another word, replacement teeth also drift anteriorly or posteriorly (Figure 8.3H, dark green, L1-4). This is due to the inconsistent rate of tooth addition and extent of tooth shedding among anterior and posterior tooth families. The first-generation teeth display a fan-shaped arrangement with an anteroposteriorly differential timing of tooth addition (Figure 8.3H), reminiscent of the mediolateral gradation in a lungfish tooth plate (Figure 8.2A, orange lines). Teeth in the same row are larger anteriorly. The larger size causes a considerable overlap and resorption between the successive teeth in the anterior files. The anterior first-generation tooth files only have the tooth bases retained. The posterior tooth files not only keep more rows

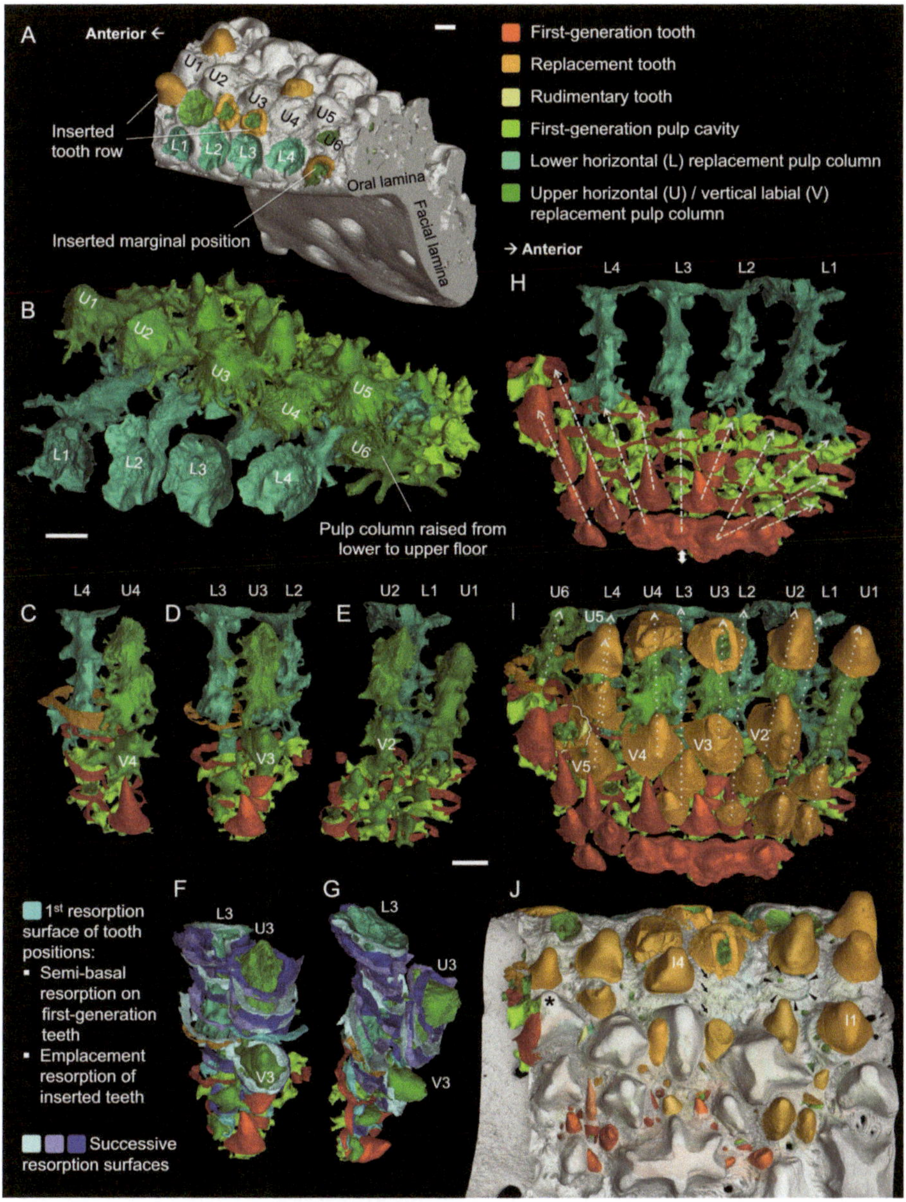

FIGURE 8.3 Tooth insertion and alternation in the marginal jawbone of *Lophosteus*. (A) Linguoposterior view of the modelled section of the jawbone, anterior to the left. The latest replacement teeth of the inserted positions are highlighted in gold. (B) Close-up of the pulp cavities of teeth on the oral lamina, same view as A. (C–F and H–J) Occlusal view of the oral lamina, anterior to the right. (C–E) Dissection of the posterior (C), median (D), and anterior (E) files, showing the division and fusion between files. The latest replacement teeth of the replacement columns are not shown. (F) The stack of resorption surface along the lower and upper replacement columns in D. (G) Posterior view of F. (H) Arrangement of the basal layer

of the dentition, including the first-generation files and the lower horizontal replacement columns. Double arrow below the ridge-like founder odontode marks the ossification center of the jawbone. Dashed arrows show the radiation of the first-generation files from the founder ridge. (I) Upper horizontal and vertical labial replacement columns are shown with the latest replacement teeth, except the inserted positions. Dotted arrows show the extension of the replacement columns. Curves outline the remnant of the first replacement tooth of a non-marginal position. Note that the vestigial successor of V5 is preserved as a rudimentary tooth that only have the tip mineralized. (J) Inserted positions (in gold), overgrowing dermal ornament (not highlighted in colors) and the surface of attachment bone and supporting bone are visible, showing how the labial part of the dentition has been embedded. Asterisk, an example of a dermal odontode overgrowing a first-generation tooth. Arrows, a stack of attachment tissue of successive shed teeth forming the interdental partition. Arrow heads, emplacement resorption pit induced by an upcoming inserted position encroaching the stack of resorption surfaces of the underlying replacement column (U2). Scale bar = 0.1 mm. C–J to the same scale.

FIGURE 8.3 Continued

of first-generation teeth intact (Figure 8.3H), but also have the tooth base of some first replacement teeth (second-generation teeth) retained (compare the retention of replacement teeth, from the tooth base, via the folding tooth root, to attachment tissue without traceable natural edge in Figure 8.3C, D, E, respectively). The first replacement tooth from a non-marginal position (Figure 8.3I, curve) has been shed semi-basally before being overlapped by the first-generation teeth added later to the next row. Such a first replacement tooth would have become the next first-generation tooth of the file if there had been more space for it to stand more lingually. It implies a higher tooth addition rate in a posterior tooth file: At the earlier stage, teeth are added more frequently with a smaller size than the anterior ones, so the labial positions barely overlap and are not shed; At the later stage, a tooth becomes too large and overlaps the predecessor so much that it appears as the first replacement tooth, but its successor establishes a new tooth position as a first-generation tooth, which only requires a semi-basal shedding of the first raplacement tooth. In *Lophosteus*, all teeth are developed from the surface epithelium without the dental lamina and thus always have the options to set up a new position or carry on a preexisting one. According to the availability of space, during the succession of a tooth family, teeth can switch back and forth being the first-generation teeth or replacement teeth, or something in between. There can be no clear demarcation between a position establisher and a position successor.

8.2.5.2 Lower- and Upper-Level Alternation versus Odd- and Even-Number Alternation

As the lingual extension of the jawbone slows down relative to the anteroposterior elongation, the most lingual row of first-generation teeth becomes more anteroposteriorly elongated. Hence, two first-generation files can fuse into one by an extra larger tooth (Figure 8.3C). But they are still able to be carried on by two replacement columns, with one on the lower level (L4) and one on the upper level (U4). The upper one follows the track of the file that is seemingly suppressed, rising from the overgrown labial column (V4) of the file. In the same way, two first-generation files may give rise to three horizontal replacement columns (Figure 8.3D). The inserted column (U3) is positioned above the other two, resorbing the final post-functional tooth of the most lingual vertical replacement column (Figure 8.3F, G, V3) and taking over its pulp cavity after it is overgrown by dermal odontodes. Therefore, the thickening of the oral lamina of the jawbone allows a lower–upper-floor alternation of the marginal replacement columns, which then creates opportunities for the insertion of more replacement columns.

The anterior files shows an even greater difference in the direction of tooth addition between the first-generation and replacement teeth (Figure 8.3H–I, dashed and dotted arrows). A replacement column (Figure 8.3E, V2 and U1) can be based on two first-generation teeth. This then allows one first-generation tooth to split into two replacement columns (L1 and U1). The competition between L1 and U1 pushes L1 to drift posteriorly.

Two neighboring vertical replacement columns can converge into one, making a position straddle two files (Figure 8.3H–I, V5). U5, which derives from V5, is thus drifted anteriorly, almost aligning with L4. This pushes U5 to an even upper lever, higher than U4 (Figure 8.3B). U6, which is supposed to lay on the lower floor, is thus raised to the lever between U5 and L4, at the same height of U4, rearranging the alternation. The upward drift of U6 leaves the jaw margin vacant for a new position to be inserted (Figure 8.3A). These insertions or drift of replacement columns breaks the correspondence between the odd–even-number alternation of the first-generation tooth files and the lower–upper-level alternation of the horizontal replacement columns.

8.2.5.3 Insertion of Tooth Positions Wherever a Gap Appears

As the lower horizontal replacement columns extend lingually, they tilt inward, leaving the upper horizontal and vertical replacement columns of the labial positions far behind (Figure 8.3B). The long replacement history of a horizontal column will produce an interdental septa (see Chapter 6) that is formed from the stacking of attachment bone of the previous generations of shed teeth (Figure 8.3J, arrows). As a large interdental septa may reduce the efficitent of the dentition, once such a gap opens up, a new tooth position (Figure 8.3J, I1 and I4) or a tooth row (Figure 8.3A) is inserted. The inserted position takes root from the nearest replacement columns by inducing emplacement resorption on the interdental partition (Figure 8.3J, arrow heads). Thus, such inserted positions are not established by the first-generation teeth. As the jawbone grows lingually, it will be rotated labially, with the ossification center and the labial part of the oral lamina transported out of the biting area and covered by the skin (Chen et al. 2020, fig. 2). Overgrown by dermal ornament, the labial teeth cannot be shed. Then the development of their replacement teeth is truncated (Figure 8.3I-J, pale yellow) since no gap to fill. The labial positions are no longer replaced. Therefore, the insertion of teeth replenishes active positions and maintains the multi-row property of the dentition. However, the inserted teeth and the overgrown odontodes obscure the alternation pattern of the tooth files.

The self-organization of replacement teeth suggests that, on one hand, successors and predecessors show the same polarity consistent with the bone growth at the time they are initiated; on the other hand, they can be misaligned by spontaneous local interactions triggered by the disturbance, such as the availability of extra space or the overgrowth of dermal odontodes. The offset between successive teeth questions whether tooth families and tooth positions are just arbitrary units to optimize the packing, as has been proposed in the previous studies on urodeles (Reeve, 1962) and alligators (Westergaard and Ferguson, 1987). A tooth family or tooth position may not be recognized until it has been replaced cyclically and the interdental septa is formed. Only the individual tooth is the true developmental unit of a dentition.

8.3 DISCUSSION

8.3.1 Chemical Signals of Activation or Inhibition

The information from this overview of process components allows us to critically evaluate the classic theories about dental patterning. Most of these theories require assumptions about the existence of some mysterious extrinsic factors, which have not yet been tested by experiments. Examples include the successive waves of chemical signals passing through the epithelium surmised by *the Zahnreihen model* (Edmund, 1960) and the gradient of diffusion substances as speculated by *the field model* (Butler, 1939).

The inhibition model proposed that a tooth germ is like a "sink" to "drain" the essential substances from the surroundings, and the depleted substance cannot support the development of a new tooth germ in the vicinity until it is refilled. Thus, the zone of inhibition is supposed to decrease through time (Osborn, 1971), which, however, raises questions about the decreasing rate of the inhibition. Is it the same for each tooth or specific to the tooth family? What factors does it depend on? It has been noticed that new teeth are not always added between the teeth that have the largest sum of age, and the oldest teeth are not always the first to be replaced. This has been documented by the variable lifespan of teeth in embryonic alligator (Westergaard and Ferguson, 1987, fig. 4, 1990, fig. 10) and Mexican tetra (Trapani et al., 2005). And as mentioned earlier, the Tooth 11 of veiled chameleon is not inserted into the gap between Teeth 1 and 4 but into the more recent gap between Teeth 8 and 9 (Buchtová et al., 2013, fig. 2, tab. 1). Another assumption of the model is that the tooth germs may alternatively produce inhibition substances that diffuse into the surroundings. To initiate a new tooth germ, the inhibition-free zone can be created when the preexisting tooth germs have moved away far enough from each other. This may work for the skin teeth of modern sharks that are floating in the dermis (Reif, 1982, figs. 4, 6). But for ankylosed odontodes, it requires the interstitial growth of bone, which is usually not how dermal bones grow, since osteoblasts are lining outside the bone but not inside. In the embryos of viviparous lizard, which Osborn's inhibition model is based on, the tooth rows extend considerably before adding tooth positions. It was interpreted as continuous interstitial growth (Osborn, 1971), but it is actually a result of position drift through tooth replacement. The inhibition-free zones for the addition of interfering tooth families can only be created by position drift, as shown by the trajectory of tooth positions in embryonic alligator (Osborn, 1998, fig. 2).

8.3.2 Dental Lamina and *Sox2*

The odontode regulation theory considers that the replacement of all true teeth, in contrast to the dermal odontodes, has to be regulated by a dental lamina (Reif, 1982). However, the phylogenetically most basal teeth, the dentition of acanthothoracid *Radotina*, demonstrate that teeth originated without forming a fixed tooth family (Vaškaninová et al., 2020). The absence of a dental lamina in *Andreolepis* and *Lophosteus*, the earliest known dentitions with in-situ tooth replacement (Chen et al., 2016, 2017, 2020), and in various clades of actinopterygians (Vandenplas et al., 2014) evidence tooth replacement evolved before the dental lamina. Even in animals with a dental lamina, the earliest teeth have developed before the formation of the dental lamina. Dental lamina is a deep epidermal invagination, but the tooth germs of Teeth 1 and 2 are evaginated in alligator. The earliest trace of a dental lamina forms independently at some lingual distance after the dental epithelial of Teeth 1 and 2 has degenerated. Hence, Teeth 1 and 2 are unconnected to their successors via a dental lamina. Before the formation of Tooth 3, a dental lamina has appeared, but even it can connect to the dental epithelium of Tooth 3, it breaks up into pieces later without giving rise to a successor. As a result, Tooth 3 cannot bring about its own tooth family, and the purported successors assigned to its tooth family are actually located anterior to it. Even after the dental lamina has formed, in some individuals, the family of Tooth 2 is suppressed after one round of replacement, and the successors of the neighboring Tooth 8 drift slightly to occupy the interspace between the positions of Tooth 2 and 8 (Westergaard and Ferguson, 1986, 1987). *The tooth position theory* thus regards tooth classes and tooth families as arbitrary units until isolated by surrounding tissues (Westergaard and Ferguson, 1987). The gap-filling phenomenon on the marginal jawbone of *Lophosteus* challenges the concept of a fixed tooth position as well.

Recently, the expression of the tastebud-linked stem cell factor *Sox2* has been proposed to be a defining character of oral teeth (Smith et al., 2017), differentiating them from the skin and oral denticles. *Sox2* was assumed to ensure the renewal of teeth like the regeneration of taste buds throughout life, on the basis of its lack of expression in the skin denticles of sharks (Atkinson et al., 2016; Martin et al., 2016; Smith et al., 2017). However, whether *Sox2* is expressed in the oral denticles of sharks, which are supposed to form from the same epithelium with the same patterning mechanism as taste buds (Atkinson et al., 2016), is unknown. Instead, *Sox2* is known to regulate the successional formation of the molars (M1–M3) in mouse, which are non-replaced primary teeth (Juuri et al., 2013; Gaete et al., 2015). This implies *Sox2* might be generally related to any sequential addition occurred in both shedding and non-shedding teeth. Positive expression in the oral denticles of sharks is hence not impossible. Even if it were to prove to be negative, this might simply suggest a correlation between *Sox2* and the dental lamina. As a marker gene of dental lamina, *Sox2* has been localized to both primary and successional dental laminae in mammals and reptiles (Gaete and Tucker, 2013; Juuri et al., 2013), including the dental lamina rests in adult humans (Fraser et al., 2019), and even to the belt-like bulges, the indistinctive dental lamina in the pharyngeal dentition of medaka (Abduweli et al., 2014). However, it is negative in the middle dental epithelium, which is the putative functional substitute for a dental lamina, in the jaw teeth of Atlantic salmon and African bichir (Vandenplas et al., 2014, 2016). This challenges the supposed involvement of stem cells in the cyclic tooth replacement when a dental lamina is absent. The middle dental epithelium seems not to be a source of stem cells as speculated. It does not build up until the successor has been initiated and disappears once the successor is attached. Such a spatiotemporal discontinuity between successive generations of middle dental epithelium contrasts with the dental lamina of chondrichthyans, which has the corresponding layer confluent all the way (Huysseune and Witten, 2008, fig. 4). Thus, *Sox2* may not be expressed in any replacing dentitions without a dental lamina, such as coelacanth and bichir teeth (Vandenplas et al., 2014). Put together, the *Sox2*-negative skin denticles evidence the absence of a dental lamina in dermal denticles, but fail to explain the acquisition of true teeth and tooth replacement, which are independent of the convergent evolution of a dental lamina in chondrichthyans and tetrapods (Chen et al., 2016).

8.3.3 Odontogenic Gene Regulatory Network

The sequential addition model presumed that the dental pattern is determined by an ancient regulatory network (Jernvall and Thesleff, 2000; Smith, 2003; Fraser et al. 2004, 2009; Smith et al. 2015; Rasch et al. 2016, 2020) that is repeatedly deployed in multiple organs, including various epithelial appendages such as teeth, scales, feathers, hairs, glands, intestinal crypt villi and sensory organs (see Chapter 3 and 7). This conserved gene network, as has been tested thoroughly in sharks, not only mediates the morphogenesis and patterning of teeth, but also makes dermal odontodes to order (Cooper et al., 2018), as well as regulating the patterned rows of taste buds (Martin et al., 2016) and ampullae of Lorenzini (Cooper et al. 2023). Thus, it can neither explain why teeth have to be added sequentially but dermal odontodes may not, nor the normal phenomenon of tooth insertion or other irregular patterns. If the molecular mechanisms make sure chondrichthyan teeth develop in generative sets, each containing two alternating tooth files from a single primordium (Smith and Coates, 2001), it illuminates the alternation of compacted files, but not the non-alternation of widely spaced files. Meanwhile, the conservation of the gene network poses the conundrum how can the same set of genes give rise to the great diversity of dentitions.

The expression pattern of various molecular signals, such as *Shh*, *Msx1*, *Pitx2*, *Sox2*, *Bmp4*, and *Edar*, during the development of the primary dentitions of catshark (Debiais-Thibaud et al., 2015, fig. 2; Martin et al., 2016), sterlet (Pospisilova et al., 2021), rainbow trout (Smith et al., 2009, fig. 5), Malawi cichlids (Fraser et al., 2008), axolotl (Soukup et al., 2021) and mouse (Jernvall and Thesleff, 2000, fig. 1; Sadier et al., 2019; Ye et al., 2022, fig. 7) all show that the focal expression of individual tooth germs is partitioned from a single broad expression band, as has been reviewed in Chapter 7. Before tooth initiation, the epithelial markers do not exhibit a localized expression but an extensive expression through the range of the initial field of odontogenic competence over all the tooth-bearing bones. It is followed by a compartmentation of the multiple dental fields along the outer and inner arcades and hyoid arches (Soukup et al., 2021) or a regionalization of dental and non-dental epithelia (Ye et al., 2022). When the oral epithelium has been thickened, the expression is confined to the dental epithelia of a specific tooth germ (Smith et al., 2009, fig. 5). The size of the expression sphere is consistent with the initiation order of the tooth germs. But the expression in the second row, if any, is still in the form of a broad band, gradually with protrusions fitting into the gaps between the expression spheres of the first-row positions. No gene expression marks the insertion of a tooth before the gap appears (e.g., the tooth germ 5 to be inserted between germs 1 and 2 or 1 and 3 in Malawi cichlid, Fraser et al., 2008, fig. 2c), demonstrating that the tooth position is not prepatterned. The gene expression associated with taste buds also undergoes a progressive compartmentalization (Pospisilova et al., 2021).

Gene expression at each specific position does not occur synchronically all over the odontogenic area, and the prepattern of tooth distribution presumably determined by the epithelium cannot be visualized and verified before tooth initiation. The gene markers of odontogenesis may indicate and regulate the development of a tooth but cannot determine exactly when and where the tooth will develop. The epithelial induction seems nonspecific: Epithelium may only provide the odontogenic competence across a continuous field, but the specific site of each tooth, as well as the size and number of teeth, may not be determined until the interaction with the underlying mesenchyme. The final determination of the tooth positions in the subsequent rows may be made according to the tooth distribution of the previous row. Whenever space is available, the oral epithelium still has the competence to insert new positions into the preexisting rows, according to the real time circumstance, and may not stick to the "plan". In snakes, the single egg tooth is actually formed from two tooth germs. Before the tooth germs merge, *Edar* and *Shh* are expressed in two distinct domain, rather than a single expression domain, until the tooth germs have been united (Fons et al., 2020). Therefore, the expression of genes serves as a marker of the developmental condition of the tooth germs but cannot determine and preview the final pattern of the dentition.

The tremendously sophisticated signaling feedback pathways of the odontogenic gene regulatory network involve a large number gene families (http://bite-it.helsinki.fi), leaving their application potentials for regenerative therapies remaining to be explored (Yuan and Chai, 2019). In contrast, the key of whatever complicated dental system is space, which can be observed straightforwardly.

8.3.4 Close Packing of Odontodes Coupled with Space Constraint of Skeleton

The alternation pattern may not require any molecular mechanisms, and it is common in teeth and dermal odontodes across vertebrates since it is the result of close packing (Chen et al., 2020). It can be produced by organisms with fundamentally different genetic basis and developmental mechanisms, such as maize kernels. The wild ancestor of maize, teosinte, only has two interleaved rows of kernels that wedge into each other like a single row (Tian et al., 2009), reminiscent of the alternation of odd- and even-number position in a tooth row. Domestic maize can contain more than 20 rows, resembling the multi-row pycnodont dentitions (Figure 8.2C, D). There is a strong tendency for the kernels to align in straight rows: In parallel rows if the kernels press each other into a square shape or in staggered rows if the kernel shape is round. Severe weather or incomplete pollination can cause the abortion of some kernels, and the surviving kernels will become round as the stress of space is gone. Kernels at the base of the ear are the first to be pollinated and thus larger, but the base of the ear tapers abruptly and is where the misalignment often occurs. As analyzed in dental organization, the developmental unit of the kernels is a pair of rows of spikelets, while the evolution of the paired spikelets (Mangelsdorf, 1945) is probably related to the surface area for the spikelets to develop (Jacobs and Pearson, 1991). Therefore, the relationship between the size of the cob and organization of the kernels in maize is comparable with the relationship between bone and odontodes in vertebrates. The same pattern can be produced by organisms whose last common ancestor was a single-cell organism in the Precambrian. Similar to the odontogenic gene regulatory network, the development of kernel rows is regulated by a large number of quantitative trait loci that also control multiple traits of plant architecture, rather than through overall control by a single specific Mendelian gene, and the potential for applying the natural alleles to maize breeding remains elusive (Wang et al., 2019). The high variability of spatial patterns simply follows some universal mathematical models, such as the reaction–diffusion or activator–inhibitor model (Kondo et al., 2010).

The toolkit genes make sure the organs are formed as predicted by the mathematical model: The same toolkit can be used for different jobs, whereas toolkits containing completely different sets of genes can do the same job. For instance, novel dental cell types and dental tissues can emerge by borrowing the gene expression programs of other cell types and tissues (see Chapter 1); chondrichthyans lack the odontode tissue matrix components of osteichthyans but bear similar dental tissues (see Chapter 4). Neither the phenotypic outcome nor the genotypic underpinning has to be homologous: On one hand, the same molecular mechanism can generate different morphology, and the same mutant can cause different aberrant situations between individuals; on the other hand, a trait can be kept even the genotype has drifted (Weiss and Fullerton, 2000). The impressive diversity of dentitions can be achieved by adjusting the parameter values in the equations, as shown by the variability of elasmobranch squamation (Cooper et al., 2018, Table S1). However, how the parameter values are changed across vertebrates, or how variations occurs in signaling molecules, cannot be explained by a steady common gene network.

The inhibitory cascade model (Kavanagh et al., 2007) presents simple formulae for the size sequence of mammalian molars. Exceptions have been reported by many studies and discussed in Chapter 7. The problem of this model is that there are other unknown factors that prevent the a/i from being static; for instance, the evolution of bunodont cusp relief in frugivory anthropoid primates may cause the disproportionately large M2 (Carter and Worthington, 2016). Nevertheless, bearing in mind that a/i is a dynamic parameter, the model can potentially apply to each tooth family of nonmammalian vertebrates, with greater evolutionary versatility. In contrast to the limited number of mammalian molars with more or less reducing size, the indeterminate growth of nonmammalian vertebrates makes sure a/i is greater than 1, giving rise to the constant replacement or a non-shedding row with enlarged teeth. An exception is noticed in the acrodont dentition of the central bearded dragon, *Pogona vitticeps*, in which the extension of the non-shedding tooth row of increasing size ceases once it reaches the coronoid process, with the last two teeth non-ankylosed (Haridy, 2018). A possible explanation is that the inhibition from the coronoid process makes the second last tooth start to reduce size and the last tooth has become too small for further teeth to be initiated, even though the mandible keeps growing posteriorly with the coronoid process being moved away from the fixed tooth row. To put it mathematically, along the length of most of the tooth row a/i is greater than 1, but at the end of the tooth row a/i is no more than 1/2. Therefore, combined with ontogenetic variations, a/i can be spatiotemporally specific, often affected by the growth of the tooth-bearing skeleton. This will bring about a wide variety of deviations and thus a diverse potential of vertebrate dentitions.

A dermal skeleton combining odontodes and bone-like lamellar tissue is present in the earliest known unambiguous vertebrate *Anatolepis* from the Late Cambrian (Smith et al., 1996). Jaw and teeth emerged later, but, by the Late Silurian, the complicated osteichthyan-style dentitions with in-situ replacement had evolved (Chen et al., 2016, 2017, 2020). During such a long evolutionary history, odontode and bone could not have evolved completely independently. Bony tissues, including the lamellar tissue of *Anatolepis*, are generated centrifugally, which allows appositional and superpositional growth, but not interstitial growth; while dentinous tissues, the major component of an odontode, are deposited centripetally, which requires addition of new individual odontodes as the skeleton extends. This makes odontodes interact with bone in a characteristic way. Although a tooth germ can develop in vitro after separating from the jaw tissue, the development of the subsequent teeth will be delayed or ceased, which is likely due to the disruption on the mesenchymal feedback during the activator–inhibitor regulation (Kavanagh et al., 2007). It demonstrates that tooth is a distinct developmental module from bone, but the formation and organization of dentition rely on the response and architecture of bone. A dentate skeletal element should be considered as an integrated evolutionary unit.

8.4 CONCLUSION

1) The consistent co-location of the first tooth of the dentition and the ossification center of bone explains why the first tooth serves as the signaling center for the formation of the entire dentition. Where the first tooth will develop varies among bones and taxa, as the bone ossification center varies. The shape of the tooth-bearing element determines whether the growing front of the bone is a point at the end or a line at the margin and, thus, whether it produces a linear tooth row or a tooth patch.

2) If the initiation of primary/earliest teeth cannot catch up with the growth of bone, the teeth will spread along the bone by being far separated from each other initially, in order to make the dentate area quickly occupy the length of the jaw and leave the gaps to be filled by later teeth.
3) Subsequent teeth will be initiated at any large enough gaps within the odontogenic competent zone, either at the growing ends of the skeleton or between existing teeth. Whether the tooth number will be increased depends on whether the size increase of the teeth can catch up with the growth of the bone. The initiation site does not have to be regulated by a dental lamina or predetermined by a programmer and thus creates opportunities for variations to emerge. The initiation order is nevertheless not just random. Whenever a tooth is ready to initiate, it will be put at the widest gap between the preexisting teeth, but this does not have to be between the oldest teeth.
4) Teeth may not require skeletal support for the initial development, but will do for the final attachment. This allows an offset between the developmental timing of the dentition and the supporting skeleton, including the initiation, rate, and duration of the development of both modules. If tooth addition and bone growth are synchronized, then teeth are non-shedding and display a sequential pattern; if the rate of tooth addition is higher than bone growth, then previous teeth have to be shed at least partially; if the rate of tooth addition is lower than bone appositional growth, existing teeth can migrate to occupy the entire biting area before the mineralization of periodontal ligament; if the rate of tooth addition is lower than bone superpositional growth, the older generations of teeth will be entirely buried by the new bone and then overgrown by the successive generations of teeth. Different types of heterochrony shifts between the modules can occur at different developmental stages of a dentition.
5) A series of replacement teeth can be modified from a column of sequentially added dermal odontodes by introducing site-specific partial-to-entire resorption. The degree of resorption is a continuous variation depends on the extent of appositional and superpositional growth of bone during the intervals between the initiation of successive teeth. In-situ replacing dentitions could have evolved from non-shedding dentitions via increasing the tooth initiation rate relative to the bone growth rate. The origin of in-situ tooth replacement exemplifies the great dental diversity that can be acquired through straightforward spatiotemporal developmental shifts between teeth and the supporting skeleton.

REFERENCES

Abduweli, D., O. Baba, M. J. Tabata, K. Higuchi, H. Mitani, and Y. Takano. 2014. Tooth replacement and putative odontogenic stem cell niches in pharyngeal dentition of medaka (*Oryzias latipes*). *Microscopy* 63:141–153.

Ahlberg, P. E., M. M. Smith, and Z. Johanson. 2006. Developmental plasticity and disparity in early dipnoan (lungfish) dentitions. *Evolution & Development* 8:331–349.

Andreev, P. S., I. J. Sansom, Q. Li, W. Zhao, J. Wang, C. Wang, L. Peng, L. Jia, T. Qiao, and M. Zhu. 2022. The oldest gnathostome teeth. *Nature* 609:964–968.

Andreev, P. S., W. Zhao, N. Wang, M. M. Smith, Q. Li, X. Cui, M. Zhu, and I. J. Sansom. 2020. Early Silurian chondrichthyans from the Tarim Basin (Xinjiang, China). *PLoS ONE* 15:e0228589.

Atkinson, C. J. L., K. J. Martin, G. J. Fraser, and S. P. Collin. 2016. Morphology and distribution of taste papillae and oral denticles in the developing oropharyngeal cavity of the bamboo shark, *Chiloscyllium punctatum*. *Biology Open* 5:1759–1769.

Barwick, R. E., K. S. W. Campbell, and E. Mark-Kurik. 1997. Tarachomylax: A new early Devonian dipnoan from Severnaya Zemlya, and its place in the evolution of the Dipnoi. *Geobios* 30:45–73.

Bemis, W. E., A. Giuliano, and B. McGuire. 2005. Structure, attachment, replacement and growth of teeth in bluefish, *Pomatomus saltatrix* ((Linnaeus, 1766), a teleost with deeply socketed teeth. *Zoology* 108:317–327.

Berkovitz, B. K. B. 1977. The order of tooth development and eruption in the rainbow trout (Salmo gairdneri). *Journal of Experimental Zoology* 201:221–225.

Berkovitz, B. K. B. 1978. Tooth ontogeny in the upper jaw and tongue of the rainbow trout (*Salmo Gairdneri*). *Journal de Biologie Buccale* 6:205–215.

Berkovitz, B. K. B., and M. H. Moore. 1974. A longitudinal study of replacement patterns of teeth on the lower jaw and tongue in the rainbow trout *Salmo gairdneri*. *Archives of Oral Biology* 19:1111–1119.

Berkovitz, B. K. B., and M. H. Moore. 1975. Tooth replacement in the upper jaw of the rainbow trout (*Salmo gairdneri*). *Journal of Experimental Zoology* 193:221–234.

Billet, G., and J. Bardin. 2021. Segmental series and size: Clade-wide investigation of molar proportions reveals a major evolutionary allometry in the dentition of placental mammals. *Systematic Biology* 70:1101–1109.

Blais, S. A., C. R. Hermus, and M. V. H. Wilson. 2015. Four new Early Devonian ischnacanthid acanthodians from the Mackenzie Mountains, Northwest Territories, Canada: An early experiment in dental diversity. *Journal of Vertebrate Paleontology* 35:e948546.

Bolk, L. 1912. On the structure of the dental system of reptiles. *KNAW, Proceedings* 14:950–961.

Botella, H. 2006. The oldest fossil evidence of a dental lamina in sharks. *Journal of Vertebrate Paleontology* 26:1002–1003.

Buchtová, M., O. Zahradníček, S. Balková, and A. S. Tucker. 2013. Odontogenesis in the veiled chameleon (*Veiled Chameleon*). *Archives of Oral Biology* 58:118–133.

Butler, P. M. 1939. Studies of the mammalian dentition. Differentiation of the post-canine dentition. *Journal of Zoology* B109:1–36.

Bystrow, A. P. 1939. Zahnstruktur der Crossopterygier. *Acta Zoologica* 20:283–338.

Camacho, J., R. Moon, S. K. Smith, J. D. Lin, C. Randolph, J. J. Rasweiler, R. R. Behringer, and A. Abzhanov. 2020. Differential cellular proliferation underlies heterochronic generation of cranial diversity in phyllostomid bats. *EvoDevo* 11:1–17.

Carter, K. E., and S. Worthington. 2016. The evolution of anthropoid molar proportions. *BMC Evolutionary Biology* 16:1–18.

Chang, M.-M., and X. Yu. 1984. Structure and phylogenetic significance of *Diabolichthys speratus* gen. et sp. nov., a new dipnoan-like form from the Lower Devonian of eastern Yunnan, China. *Proceedings of the Linnean Society of New South Wales* 107: 171–184.

Chen, D., H. Blom, S. Sanchez, P. Tafforeau, and P. E. Ahlberg. 2016. The stem osteichthyan *Andreolepis* and the origin of tooth replacement. *Nature* 539:237–241.

Chen, D., H. Blom, S. Sanchez, P. Tafforeau, T. Märss, and P. E. Ahlberg. 2017. Development of cyclic shedding teeth from semi-shedding teeth: The inner dental arcade of the stem osteichthyan *Lophosteus*. *Royal Society Open Science* 4:161084.

Chen, D., H. Blom, S. Sanchez, P. Tafforeau, T. Märss, and P. E. Ahlberg. 2020. Dental ontogeny in the most primitive bony fish *Lophosteus* reveals the developmental relationship between teeth and dermal odontodes. *eLife* 9:e60985.

Collins, S. E., and C. J. Underwood. 2021. Unique damage-related, gap-filling tooth replacement in pycnodont fishes. *Palaeontology* 64:489–504.

Cooper, J. S. 1966. Tooth replacement in the slow worm *Anguis fragilis*. *Journal of Zoology* 150:235–248.

Cooper, R. L., A. P. Thiery, A. G. Fletcher, D. J. Delbarre, L. J. Rasch, and G. J. Fraser. 2018. An ancient turing-like patterning mechanism regulates skin denticle development in sharks. *Science Advances* 4:eaau5484.

Cooper, R. L., E. F. Nicklin, L. J. Rasch, and G. J. Fraser. 2023. Teeth outside the mouth: The evolution and development of shark denticles. *Evolution & Development* 25:54–72.

Dearden, R. P., and S. Giles. 2021. Diverse stem-chondrichthyan oral structures and evidence for an independently acquired acanthodid dentition. *Royal Society Open Science* 8:210822.

Debiais-Thibaud, M., R. Chiori, S. Enault, S. Oulion, I. Germon, C. Martinand-Mari, D. Casane, and V. Borday-Birraux. 2015. Tooth and scale morphogenesis in shark: An alternative process to the mammalian enamel knot system. *BMC Evolutionary Biology* 15:292.

De Clercq, A., S. Vandenplas, and A. Huysseune. 2014. A comparison of the larval and juvenile dentition in *Polypterus senegalus*. *Journal of Applied Ichthyology* 30:790–795.

Delgado, S., T. Davit-Beal, and J. Sire. 2003. Dentition and tooth replacement pattern in *Chalcides* (Squamata; Scincidae). *Journal of Morphology* 159:146–159.

Delpiani, G., D. E. Figueroa, and E. Mabragaña. 2012. Dental abnormalities of the southern thorny skate *Amblyraja doellojuradoi* (Chondrichthyes, Rajidae). *Revista de Biología Marina y Oceanografía* 47:135–140.

Edmund, G. A. 1960. *Tooth Replacement Phenomena in the Lower Vertebrates*. The University of Toronto Press, Toronto, 190 pp.

Ellis, N. A., N. N. Donde, and C. T. Miller. 2016. Early development and replacement of the stickleback dentition. *Journal of Morphology* 277:1072–1083.

Erickson, G. M. 1996. Daily deposition of dentine in Juvenile *Alligator* and assessment of tooth replacement rates using incremental line counts. *Journal of Morphology* 228:189–194.

Fons, J. M., M. Gaete, O. Zahradnicek, M. Landova, H. Bandali, E. R. Khannoon, J. M. Richman, M. Buchtova, and A. S. Tucker. 2020. Getting out of an egg: Merging of tooth germs to create an egg tooth in the snake. *Developmental Dynamics* 249:199–208.

Fraser, G. J., A. Graham, and M. M. Smith. 2006. Conserved deployment of genes during odontogenesis across osteichthyans. *Proceedings of the Royal Society B: Biological Sciences* 271:2311–2317.

Fraser, G. J., B. K. Berkovitz, A. Graham, and M. M. Smith. 2006. Gene deployment for tooth replacement in the rainbow trout (*Oncorhynchus mykiss*): A developmental model for evolution of the osteichthyan dentition. *Evolution and Development* 8:446–457.

Fraser, G. J., R. F. Bloomquist, and J. T. Streelman. 2008. A periodic pattern generator for dental diversity. *BMC Biology* 6:32.

Fraser, G. J., C. D. Hulsey, R. F. Bloomquist, K. Uyesugi, N. R. Manley, and J. T. Streelman. 2009. An ancient gene network is co-opted for teeth on old and new jaws. *PLoS biology* 7:233–247.

Fraser, G. J., R. F. Bloomquist, and J. T. Streelman. 2013. Common developmental pathways link tooth shape to regeneration. *Developmental Biology* 377:399–414.

Fraser, G. J., S. S. Hamed, K. J. Martin, and K. D. Hunter. 2019. Shark tooth regeneration reveals common stem cell characters in both human rested lamina and ameloblastoma. *Scientific Reports* 9:1–8.

Gaete, M., J. M. Fons, and E. Ma. 2015. Epithelial topography for repetitive tooth formation. *Biology Open* 4:1625–1634.

Gaete, M., and A. S. Tucker. 2013. Organized emergence of multiple-generations of teeth in snakes is dysregulated by activation of Wnt/Beta-catenin signalling. *PLoS ONE* 8:e74484.

Gardiner, B. G. 1984. The relationships of the palaeoniscid fishes, a review based on new specimens of *Mimia* and *Moythomasia* from the Upper Devonian of Western Australia. *Bulletin of The British Museum (Natural History) Geology* 37:173–428.

Gibert, Y., E. Samarut, M. K. Ellis, W. R. Jackman, and V. Laudet. 2019. The first formed tooth serves as a signalling centre to induce the formation of the dental row in zebrafish. *Proceedings of the Royal Society B: Biological Sciences* 286:20190401.

Gillette, R. 1955. The dynamics of continuous succession of teeth in the frog (*Rana pipiens*). *American Journal of Anatomy* 96:1–36.

Grudger, E. 1937. Abnormnal Dentition on Sharks, Selachii. *Bulletin American Museum of Natural History* 73:249–280.

Hammer, C. L., A. D. S. Atukorala, and T. A. Franz-Odendaal. 2016. What shapes the oral jaws? Accommodation of complex dentition correlates with premaxillary but not mandibular shape. *Mechanisms of Development* 141:100–108.

Haridy, Y. 2018. Histological analysis of post-eruption tooth wear adaptations, and ontogenetic changes in tooth implantation in the acrodontan squamate *Pogona vitticeps*. *PeerJ* 6:e5923.

Haridy, Y., B. M. Gee, F. Witzmann, J. J. Bevitt, and R. R. Reisz. 2019. Retention of fish-like odontode overgrowth in Permian tetrapod dentition supports outside-in theory of tooth origins. *Biology Letters* 15:1–6.

Hautier, L., H. Gomes Rodrigues, G. Billet, and R. J. Asher. 2016. The hidden teeth of sloths: Evolutionary vestiges and the development of a simplified dentition. *Scientific Reports* 6:1–9.

Hermyt, M., P. Kaczmarek, M. Kowalska, and W. Rupik. 2017. Development of the egg tooth—The tool facilitating hatching of squamates: Lessons from the grass snake *Natrix natrix*. *Zoologischer Anzeiger* 266:61–70.

Hu, Y., G. C. Young, C. Burrow, Y. Zhu, and J. Lu. 2019. High resolution XCT scanning reveals complex morphology of gnathal elements in an Early Devonian arthrodire. *Palaeoworld* 28:525–534.

Huysseune, A. 1983. Observations on tooth development and implantation in the upper pharyngeal jaws in *Astatotilapia elegans* (Teleostei, Cichlidae). *Journal of Morphology* 175:217–234.

Huysseune, A. 1990. Development of the anterior part of the mandible and the mandibular dentition in two species of Cichlidae (Teleostei). *Cybium* 14:327–344.

Huysseune, A., B. K. Hall, and P. E. Witten. 2007. Establishment, maintenance and modifications of the lower jaw dentition of wild Atlantic salmon (*Salmo salar* L.) throughout its life cycle. *Journal of Anatomy* 211:471–484.

Huysseune, A., L. Rüber, and E. Verheyen. 1999. A single tooth replacement pattern generates diversity in the dentition in cichlids of the tribe Eretmodini, endemic to Lake Tanganyika (Teleostei: Cichlidae). *Belgian Journal of Zoology* 129:157–173.

Huysseune, A., and P. E. Witten. 2006. Developmental mechanisms underlying tooth patterning in continuously replacing osteichthyan dentitions. *Journal of Experimental Zoology Part B: Molecular and Developmental Evolution* 215:204–215.

Huysseune, A., and P. E. Witten. 2008. An evolutionary view on tooth development and replacement in wild Atlantic salmon (*Salmo salar* L.). *Evolution & Development* 10:6–14.

Itano, W. M. 2014. Edestus, the strangest shark? First report from New Mexico, North American paleobiogeography, and a new hypothesis on its method of predation. *Mountain Geologist* 51:201–221.

Jackman, W. R., S. H. Davies, D. B. Lyons, C. K. Stauder, B. R. Denton-Schneider, A. Jowdry, S. R. Aigler, S. A. Vogel, and D. W. Stock. 2013. Manipulation of Fgf and Bmp signaling in teleost fishes suggests potential pathways for the evolutionary origin of multicuspid teeth. *Evolution and Development* 15:107–118.

Jacobs, B. C., and C. J. Pearson. 1991. Potential yield of maize, determined by rates of growth and development of ears. *Field Crops Research* 27:281–298.

Jernvall, J., and I. Thesleff. 2000. Reiterative signaling and patterning during mammalian tooth morphogenesis. *Mechanisms of Development* 92:19–29.

Johanson, Z., E. Manzanares, C. Underwood, B. Clark, V. Fernandez, and M. Smith. 2020. Evolution of the dentition in holocephalans (Chondrichthyes) through tissue disparity. *Integrative and Comparative Biology* 60:630–643.

Johanson, Z., and M. M. Smith. 2003. Placoderm fishes, pharyngeal denticles, and the vertebrate dentition. *Journal of Morphology* 257:289–307.

Johanson, Z., and M. M. Smith. 2005. Origin and evolution of gnathostome dentitions: A question of teeth and pharyngeal denticles in placoderms. *Biological Reviews of the Cambridge Philosophical Society* 80:303–345.

Johanson, Z., and K. Trinajstic. 2014. Fossilized ontogenies: The contribution of placoderm ontogeny to our understanding of the evolution of early gnathostomes. *Palaeontology* 57:505–516.

Juuri, E., M. Jussila, K. Seidel, S. Holmes, P. Wu, J. Richman, K. Heikinheimo, C.-M. Chuong, K. Arnold, K. Hochedlinger, O. Klein, F. Michon, and I. Thesleff. 2013. Sox2 marks epithelial competence to generate teeth in mammals and reptiles. *Development* 140:1424–1432.

Kavanagh, K. D., A. R. Evans, and J. Jernvall. 2007. Predicting evolutionary patterns of mammalian teeth from development. *Nature* 449:427–432.

Kerr, T. 1960. Development and structure of some actinopterygian and urodele teeth. *Proceedings of the Zoological Society of London* 133:401–422.

King, B., F. Marone, and M. Rücklin. 2021. Tooth development in the Early Devonian sarcopterygian *Powichthys* and the evolution of the crown osteichthyan dentition. *Palaeontology* 64:645–659.

Kjær, I. 1998. Neuro-osteology. *Critical Reviews in Oral Biology & Medicine* 9:224–244.

Ko, D. (Jihyung), T. Kelly, L. Thompson, J. K. Uppal, N. Rostampour, M. A. Webb, N. Zhu, G. Belev, P. Mondal, D. M. L. Cooper, and J. C. Boughner. 2021. Timing of mouse molar formation is independent of jaw length Including retromolar space. *Journal of Developmental Biology* 9:8.

Kondo, S., T. Miura, and T. Turing. 2010. Reaction-diffusion model as a framework of understanding biological pattern formation. *Science* 329:1616–1620.

Kriwet, J. 2005. A comprehensive study of the skull and dentition of pycnodont fishes. *Zitteliana Reihe A: Mitteilungen Der Bayerischen Staatssammlung Fur Palaontologie Und Geologie* 45:135–188.

Lamoureux, D. O., A. R. H. LeBlanc, and M. W. Caldwell. 2018. Tooth germ initiation patterns in a developing dentition: An in vivo study of *Xenopus laevis* tadpoles. *Journal of Morphology* 279:616–625.

Larionova, D., H. Lesot, and A. Huysseune. 2021. Miniaturization: How many cells are needed to build a tooth? *Developmental Dynamics* 250:1021–1035.

Lawson, R., and B. F. J. Manly. 1973. Tooth growth and replacement in *Ctenolucius hujeta*, a neotropical characoid fish. *Journal of Morphology* 141:383–394.

Lawson, R., D. B. Wake, and N. T. Beck. 1971. Tooth replacement in the red-backed salamander, *Plethodon cinereus. Journal of Morphology* 134:259–269.

LeBlanc, A. R. H., D. O. Lamoureux, and M. W. Caldwell. 2017. Mosasaurs and snakes have a periodontal ligament: Timing and extent of calcification, not tissue complexity, determines tooth attachment mode in reptiles. *Journal of Anatomy* 231:869–885.

Longbottom, A. E. 1984. New tertiary pycnodonts from the Tilemsi Valley, Republic of Mali. *Bulletin of the British Museum (Natural History)* 38:1–26.

Louchart, A., V. de Buffrénil, E. Bourdon, M. Dumont, L. Viriot, and J.-Y. Sire. 2018. Bony pseudoteeth of extinct pelagic birds (Aves, Odontopterygiformes) formed through a response of bone cells to tooth-specific epithelial signals under unique conditions. *Scientific Reports* 8:12952.

Luo, Z.-X., Z. Kielan-Jaworowska, and R. L. Cifelli. 2004. Evolution of dental replacement in mammals. *Bulletin of Carnegie Museum of Natural History* 36:159–175.

Maho, T., S. Maho, D. Scott, and R. R. Reisz. 2022. Permian hypercarnivore suggests dental complexity among early amniotes. *Nature Communications* 13:4882.

Maisey, J. G., and J. S. S. Denton. 2016. Dermal denticle patterning in the Cretaceous hybodont shark *Tribodus limae* (Euselachii, Hybodontiformes), and its implications for the evolution of patterning in the chondrichthyan dermal skeleton. *Journal of Vertebrate Paleontology* 1179200:e1179200.

Mangelsdorf, P. C. 1945. The origin and nature of the ear of maize. *Botanical Museum Leaflets* 12:33–81.

Martin, K. J., L. J. Rasch, R. L. Cooper, B. D. Metscher, Z. Johanson, and G. J. Fraser. 2016. Sox2+ stem cells in sharks link taste bud development to the evolution of 'true' regenerative teeth from skin teeth. *PNAS* 113:14769–14774.

Mateo, J. A., and L. F. López-Jurado. 1997. Dental ontogeny in *Lacerta lepida* (Sauria, Lacertidae) and its relationship to diet. *Copeia* 1997:461–463.

Miller, W. A., and C. J. P. Radnor. 1973. Tooth replacement in the bowfin (*Amia calva*—Holostei). *Journal of Morphology* 140:381–395.

Miller, W. A., and D. J. Rowe. 1973. Preliminary investigation of variations in tooth replacement in adult *Necturus maculosus. Journal of Morphology* 140:63–76.

Mondéjar-Fernández, J., T. J. Challands, H. N. Huu, P. T. Hoa, G. Clément, and P. Janvier. 2020. A lungfish (Sarcopterygii, Dipnomorpha) tooth plate from the Lower Devonian of Vietnam and the onset of modern dipnoan dental organization. *Journal of Vertebrate Paleontology* 40:e1772274.

Moy-Thomas, J. A. 1934. On the teeth of the larval *Belone vulgaris*, and the attachment of teeth in fishes. *Journal of Cell Science* 76:481–498.

Ørvig, T. 1951. Histologic studies of Placoderms and fossil Elasmobranchs 1. The endoskeleton, with remarks on the hard tissues of lower vertebrates in general. *Arkiv För Zoologi* 2:321–454.

Ørvig, T. 1957. Remarks on the vertebrate fauna of the Lower Upper Devonian of Escuminac Bay, P.Q., Canada, with special reference to the Porolepiform Crossopterygians. *Arkiv För Zoologi* 10:367–420.

Ørvig, T. 1967. Phylogeny of tooth tissues: Evolution of some calcified tissues in early vertebrates; pp. 45–110 in A. E. W. Miles (ed.), *Structural and Chemical Organisation of Teeth*. Academic Press, New York and London.

Ørvig, T. 1973. Acanthodian dentition and its bearing on the relationships of the group. *Palaeontographica, Abteilung A* 143:119–150.

Ørvig, T. 1978a. Microstructure and growth of the dermal skeleton in fossil actinopterygian fishes: *Boreosomus, Plegmolepis* and *Gyrolepis. Zoologica Scripta* 7:125–144.

Ørvig, T. 1978b. Microstructure and growth of the dermal skeleton in fossil actinopterygian fishes: *Birgeria* and *Scanilepis. Zoologica Scripta* 7:33–56.

Ørvig, T. 1980. Histologic studies of ostracoderms, placoderms and fossil elasmobranchs 3. Structure and growth of the gnathalia of certain arthrodire. *Zoologica Scripta* 9:141–159.

Osborn, J. W. 1971. The ontogeny of tooth succession in *Lacerta vivipara. Proceedings of the Royal Society B* 179:261–289.

Osborn, J. W. 1974. On tooth succession in *Diademodon. Evolution* 28:141–157.

Osborn, J. W. 1977. The interpretation of patterns in dentitions. *Biological Journal of the Linnean Society* 9:217–229.

Osborn, J. W. 1978. Morphogenetic gradients: Fields versus clones; pp. 171–201 in P. M. Butler and K. A. Joysey (eds.), *Development, Function and Evolution of Teeth.* Academic Press, New York.

Osborn, J. W. 1998. Relationship between growth and the pattern of tooth initiation in alligator embryos. *Journal of Dental Research* 77:1730–1738.

Osborn, J. W., and A. Crompton. 1973. The evolution of mammalian from reptilian dentitions. *Breviora* 399:1–18.

Paluh, D. J., W. A. Dillard, E. L. Stanley, G. J. Fraser, and D. C. Blackburn. 2021. Re-evaluating the morphological evidence for the re-evolution of lost mandibular teeth in frogs. *Evolution* 1–11.

Peterka, M., V. J.-L. Vonesch, J.-V. Ruch, Y. Cam, R. Peterkova, and H. Lesot. 2000. Position and growth of upper and lower tooth primordia in prenatal mouse. *Journal of Craniofacial Genetics and Developmental Biology* 20:35–43.

Poplin, C., M. Richter, and M. Smith. 1991. An enigmatic actinopterygian (Pisces: Osteichthyes) from the Upper Permian of China. *Zoological Journal of the Linnean Society* 103:1–20.

Pospisilova, A., J. Stundl, J. Brejcha, B. D. Metscher, M. Psenicka, R. Cerny, and V. Soukup. 2022. The remarkable dynamics in the establishment, rearrangement, and loss of dentition during the ontogeny of the sterlet sturgeon. *Developmental Dynamics* 251:826–845.

Poyato-Ariza, F., and S. Wenz. 2002. A new insight into pycnodontiform fishes. *Geodiversitas* 24:139–248.

Qu, Q., S. Sanchez, H. Blom, P. Tafforeau, and P. E. Ahlberg. 2013. Scales and tooth whorls of ancient fishes challenge distinction between external and oral 'teeth.' *PLoS ONE* 8:e71890.

Rasch, L. J., K. J. Martin, R. L. Cooper, B. D. Metscher, C. J. Underwood, and G. J. Fraser. 2016. An ancient dental gene set governs development and continuous regeneration of teeth in sharks. *Developmental Biology* 415:347–370.

Rasch, L. J., R. L. Cooper, C. J. Underwood, W. A. Dillard, A. P. Thiery, and G. J. Fraser. 2020. Development and regeneration of the crushing dentition in skates (Rajidae). *Developmental Biology* 466:59–72.

Reeve, P. 1962. *The Ontogeny of the Dentitions of Some Urodela.* University of London, London.

Reif, W. E. 1976. Morphogenesis, pattern formation and function of the dentition of *Heterodontus* (Selachii). *Zoomorphologie* 83:1–47.

Reif, W. E. 1980. A mechanism for tooth pattern reversal in sharks: The polarity switch model. *Wilhelm Roux's Archives of Developmental Biology* 188:115–122.

Reif, W. E. 1982. Evolution of dermal skeleton and dentition in vertebrates. *Journal of Evolutionary Biology* 15:287–368.

Rücklin, M., and P. C. J. Donoghue. 2015. *Romundina* and the evolutionary origin of teeth. *Biology Letters* 11:20150326.

Rücklin, M., P. C. J. Donoghue, Z. Johanson, K. Trinajstic, F. Marone, and M. Stampanoni. 2012. Development of teeth and jaws in the earliest jawed vertebrates. *Nature* 491:748–751.

Rücklin, M., B. King, J. A. Cunningham, Z. Johanson, F. Marone, and P. C. J. Donoghue. 2021. Acanthodian dental development and the origin of gnathostome dentitions. *Nature Ecology & Evolution* 5:919–926.

Sadier, A. A., N. Anthwal, A. L. Krause, R. Dessalles, L. Bentolila, R. Haase, N. Nieves, S. Santana, and K. Sears. 2023. Bat teeth illuminate the diversification of mammalian tooth classes. *Nature communications* 14:4687.

Sadier, A., W. R. Jackman, V. Laudet, and Y. Gibert. 2020. The vertebrate tooth row: Is it initiated by a single organizing tooth? *BioEssays* 42:1–7.

Sadier, A., M. Twarogowskaid, K. Steklikova, L. Haydenid, A. Lambert, P. Schneiderid, V. Laudet, M. Hovorakova, V. Calvez, and S. Pantalacci. 2019. Modeling edar expression reveals the hidden dynamics of tooth signaling center patterning. *PLoS Biology* 17:1–28.

Smith, M. M. 1985. The pattern of histogenesis and growth of tooth plates in larval stages of extant lungfish. *Journal of Anatomy* 140:627–643.

Smith, M. M. 2003. Vertebrate dentitions at the origin of jaws: When and how pattern evolved. *Evolution & Development* 413:394–413.

Smith, M. M., and M. Chang. 1990. The dentition of *Diabolepis speratus* Chang and Yu, with further consideration of its relationships and the primitive dipnoan dentition. *Journal of Vertebrate Paleontology* 10:420–433.

Smith, M. M., and M. I. Coates. 2001. The evolution of vertebrate dentitions: Phylogenetic pattern and developmental models; pp. 223–240 in P. E. Ahlberg (ed.), *Major Events in Early Vertebrate Evolution*. Taylor and Francis, London.

Smith, M. M., G. J. Fraser, and T. A. Mitsiadis. 2009. Dental lamina as source of odontogenic stem cells: Evolutionary origins and developmental control of tooth generation in gnathostomes. *Journal of Experimental Zoology Part B: Molecular and Developmental Evolution* 312B:260–280.

Smith, M. M., and Z. Johanson. 2003. Separate evolutionary origins of teeth from evidence in fossil jawed vertebrates. *Science* 299:1235–1236.

Smith, M. M., and Z. Johanson. 2015. Origin of the vertebrate dentition: Teeth transform jaws into a biting force; pp. 9–29 in E. L. B. Kenneth P. Dial, and Neil Shubin (eds.), *Great Transformations in Vertebrate Evolution*. University of Chicago Press, Chicago.

Smith, M. M., Z. Johanson, T. Butts, R. Ericsson, M. Modrell, F. J. Tulenko, M. C. Davis, G. J. Fraser. 2015. Making teeth to order: conserved genes reveal an ancient molecular pattern in paddlefish (Actinopterygii). *Proceedings of the Royal Society B: Biological Sciences* 282: 20142700.

Smith, M. M., Z. Johanson, C. Underwood, and T. G. H. Diekwisch. 2013. Pattern formation in development of chondrichthyan dentitions: A review of an evolutionary model. *Historical Biology* 25:127–142.

Smith, M. M., and N. I. Krupina. 2001. Conserved developmental processes constrain evolution of lungfish dentitions. *Journal of Anatomy* 199:161–168.

Smith, M. M., C. Underwood, and G. Fraser. 2017. How teeth are organised into functional dentitions. *Infocus Magazine*:4–11.

Smith, M. P., I. J. Sansom, and J. E. Repetski. 1996. Histology of the first fish. *Nature* 380:702–704.

Sofaer, J. A. 1977. Coordinated growth of successively initiated tooth germs in the mouse. *Archives of Oral Biology* 22:71–72.

Soukup, V., A. Tazaki, Y. Yamazaki, A. Pospisilova, H. H. Epperlein, E. M. Tanaka, and R. Cerny. 2021. Oral and palatal dentition of axolotl arises from a common tooth-competent zone along the ecto-endodermal boundary. *Frontiers in Cell and Developmental Biology* 8:1–14.

Tian, F., N. M. Stevens, and E. S. Buckler. 2009. Tracking footprints of maize domestication and evidence for a massive selective sweep on chromosome 10; pp. 111–127 in J. C. Avise, and F. J. Ayala (eds.), *In the Light of Evolution III: Two Centuries of Darwin*. The National Academies Press, Washington, DC.

Trapani, J., Y. Yamamoto, and D. W. Stock. 2005. Ontogenetic transition from unicuspid to multicuspid oral dentition in a teleost fish: *Astyanax mexicanus*, the Mexican tetra (Ostariophysi: Characidae). *Zoological Journal of the Linnean Society* 145:523–538.

Underwood, C., Z. Johanson, and M. M. Smith. 2016. Cutting blade dentitions in squaliform sharks form by modification of inherited alternate tooth ordering patterns. *Royal Society Open Science* 3:160385.

Vandenplas, S., A. De Clercq, and A. Huysseune. 2014. Tooth replacement without a dental lamina: The search for epithelial stem cells in *Polypterus senegalus*. *Journal of Experimental Zoology Part B: Molecular and Developmental Evolution* 322:281–293.

Vandenplas, S., M. Willems, P. E. Witten, T. Hansen, P. G. Fjelldal, and A. Huysseune. 2016. Epithelial label-retaining cells are absent during tooth cycling in *Salmo salar* and *Polypterus senegalus*. *PLoS ONE* 11:e0152870.

van Nievelt, A. F. H., and K. K. Smith. 2005. To replace or not to replace: The significance of reduced functional tooth replacement in marsupial and placental mammals. *Paleobiology* 31:324–346.

Vaškaninová, V., D. Chen, P. Tafforeau, Z. Johanson, B. Ekrt, H. Blom, and P. E. Ahlberg. 2020. Marginal dentition and multiple dermal jawbones as the ancestral condition of jawed vertebrates. *Science* 369:211–216.

Wake, M. H. 1976. The development and replacement of teeth in viviparous caecilians. *Journal of Morphology* 148:33–63.

Wang, J., Z. Lin, X. Zhang, H. Liu, L. Zhou, S. Zhong, Y. Li, C. Zhu, and Z. Lin. 2019. Krn1, a major quantitative trait locus for kernel row number in Maize. *New Phytologist* 223:1634–1646.

Weiss, K. M., and S. M. Fullerton. 2000. Phenogenetic drift and the evolution of genotype-phenotype relationships. *Theoretical Population Biology* 57:187–195.

Welten, M., M. M. Smith, C. Underwood, and Z. Johanson. 2015. Evolutionary origins and development of saw-teeth on the sawfish and sawshark rostrum (Elasmobranchii; Chondrichthyes). *Royal Society Open Science* 2:150189.

Westergaard, B. 1983. A new detailed model for mammalian dentitional evolution. *Journal of Zoological Systematics and Evolutionary Research* 21:68–78.

Westergaard, B., and M. W. J. Ferguson. 1986. Development of the dentition in *Alligator mississippiensis*. Early embryonic development in the lower jaw. *Journal of Zoology* 210:575–597.

Westergaard, B., and M. W. J. Ferguson. 1987. Development of the dentition in *Alligator mississippiensis*. Later development in the lower jaws of embryos, hatchlings and young juveniles. *Journal of Zoology* 212:191–222.

Westergaard, B., and M. W. J. Ferguson. 1990. Development of the dentition in *Alligator mississippiensis*: Upper jaw dental and craniofacial development in embryos, hatchlings, and young juveniles, with a comparison to lower jaw development. *American Journal of Anatomy* 187:393–421.

Witten, P. E., B. K. Hall, and A. Huysseune. 2005. Are breeding teeth in Atlantic salmon a component of the drastic alterations of the oral facial skeleton? *Archives of Oral Biology* 50:213–217.

Ye, Q., A. Bhojwani, and J. K. Hu. 2022. Understanding the development of oral epithelial organs through single cell transcriptomic analysis. *Development* 149:dev200539.

Yuan, Y., and Y. Chai. 2019. Regulatory mechanisms of jaw bone and tooth development; pp. 91–118 in B. R. Olsen (ed.), *Vertebrate Skeletal Development.* Academic Press, London.

Zylberberg, L., F. J. Meunier, and M. Laurin. 2016. A microanatomical and histological study of the postcranial dermal skeleton of the Devonian actinopterygian *Cheirolepis canadensis. Acta Palaeontologica Polonica* 61:363–376.

9 Complexity, Networking, and Many-Model Thinking Enhance Understanding of the Patterning, Variation, and Interactions of Human Teeth and Dental Arches

Alan Henry Brook and Matthew Brook O'Donnell

9.1 INTRODUCTION

What is the developmental basis of the dentition and the dental arches and how do variations arise? What are the interactions during development between them, both having been shown to be Complex Adaptive Systems (Brook and O'Donnell, 2011a, 2011b; Patel et al., 2018)? Are changes in these two components of the stomatognathic system coordinated for effective function? In evolution, what are the changes currently occurring and those probable in the near future? Can different models increase our understanding of factors and interactions during development and variation in outcomes?

To address these questions, this chapter will use the pattern outlined in the "Contents". We will move from considering recent advances in methodology and concepts to reviewing evidence of the developmental basis of variation of teeth and dental arches. This leads to examining how tooth and arch development are coordinated and the effect of tooth size variation on dental arch morphology. We will conclude by considering evolutionary trends and the multiple models that can enhance our understanding of process and variation.

Over the past 100 years, advances in knowledge have been stimulated by interactions between the related disciplines. Advances in one discipline have enabled developments in another. From these multiple interdisciplinary interactions, new levels of investigation and knowledge have emerged. Now applying concepts from complexity science, network science, and multiple model thinking enables moving from reviewing current knowledge to understanding the multilevel complex interactive networks that

 DOI: 10.1201/9781003439653-9

influence the patterning of human teeth and the development of tooth variations and the dental arches, together with their coordination. These approaches provide indications of their possible future co-evolution and provide the basis for aetiological models.

9.2 BACKGROUND

9.2.1 Investigating Variation Using Advances in Methodology and Concepts

We begin the exploration of these issues by looking at recent advances in research methodology, complexity science, network science, and modelling (Brook et al., 2014a; Brook and O'Donnell, 2011). The cycle of research (Figure 9.1), which is fundamental, begins with an initial research question that may arise from an observation or from the outcome of previous work. It flows on from the question to formation of a hypothesis and to planning around the available material and methods. Application

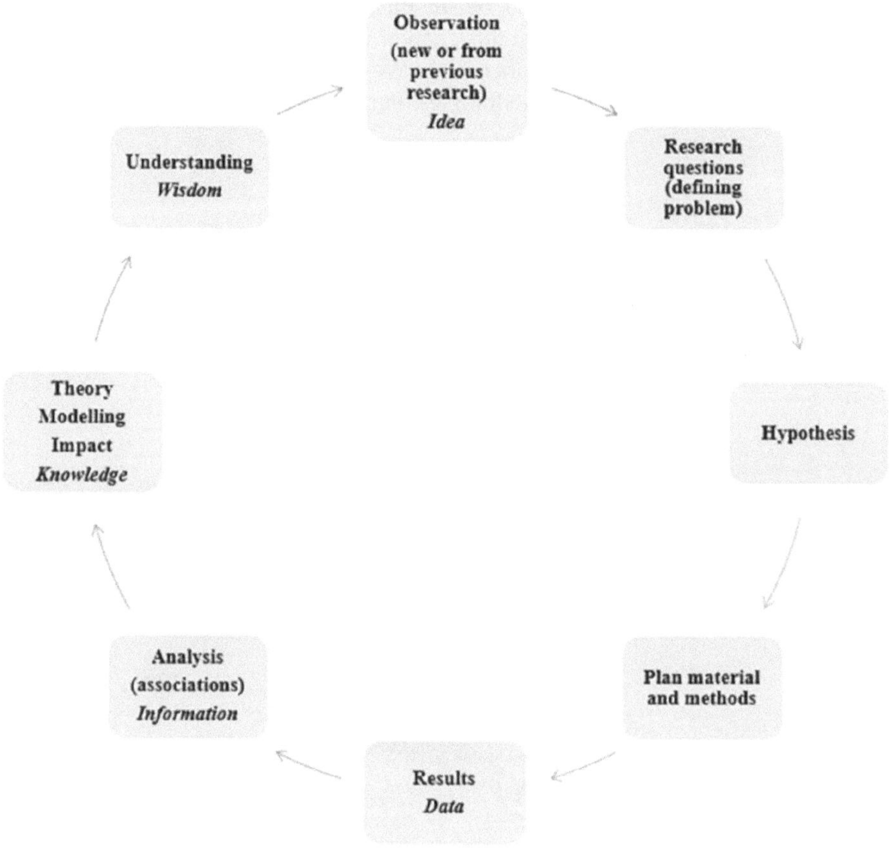

FIGURE 9.1 Cycle of research incorporating the Wisdom Hierarchy given in italics.

of these provides results in the new data. Data of itself lacks structure and meaning, so it is formed into categories to provide information. Analysis of this information enables associations to be explored, providing knowledge. There follows the application of this new knowledge, which allows the construction of models, theories, and impact. This move from knowledge to enhanced understanding is the final step in the Wisdom Hierarchy, a model structuring the relationships between data, information, knowledge, and wisdom. There are multiple origins and variants of this model); here, Page). Incorporating the Wisdom Hierarchy into the cycle of research develops an original interactive diagram that brings together these concepts (Figure 9.1). An example, relevant to the current topic, of repeatedly using the research cycle to progressively advance knowledge has been in studies of tooth number, size, and shape.

The research cycles have moved from visual scoring (Brook, 1984), to 2D measurement (Brook et al., 1999) to 3D measurement (Horrocks et al., 2009). This progress has been made possible by developments in measurement methodology. Further new knowledge and understanding, gained by repeating this cycle of research using new questions and new methodology, lead to cumulative advances that follow an upward spiral of development (Figure 9.2).

These advances are accelerated by an interdisciplinary and multidisciplinary approach. In the field considered in this chapter, there are multiple interacting disciplines, which over the past 100 years have cross-fertilized, stimulating new research questions and providing new methodologies. Parallel advances over this time period

FIGURE 9.2 The spiral of interacting advances in related disciplines over the past 100 years.

have occurred in the related disciplines of measurement methodology, research studies, concept development, and model construction as set out in Table 9.1. Reading horizontally across this table illustrates how advances in one discipline have stimulated development in another. For example the second row describes studies in which measurements rely primarily upon physical (mechanical) measurements and are focused on examining the frequency of variation in the dentition. The conceptual frameworks are informed by consideration of genetic, epigenetic, and environment influence, and explanatory models are framed (Section 9.9).

Over recent decades, the new methodologies listed in Table 9.1 have been developed in science and can be applied to understanding and modelling advances in knowledge concerning teeth and dental arches. Reading down the "Measurement Methodology" column in the table, we observe progression from: (1) Visual inspection of phenomena; to (2) their physical measurement; to (3) the use of x-ray techniques; to (4) specialized

TABLE 9.1
Advances in Related and Interacting Disciplines.

Measurement Methodology	Research Studies	Concept Application	Model Construction	Timeline
Visual	Descriptive	Unifactorial	Single genes	Past 100 years ↓
Mechanical measurement	Prevalence: Mendelian and population genetics	Genetic or environmental	Environmental factors	
Radiographic	Associations	"Normal" teeth and occlusion	Chromosomal	
2D image analysis	Molecular genetics	Quasi-continuous variables	Epithelial–mesenchymal reiterative interactions	
Digital photography software tools	Developmental interactions	Multifactorial	Dentition as Complex Adaptive System	
3D image analysis	Epigenetics	Differential effects	Power Law distributions	
3D printing software tools	Evo/Devo	Evolution: Variation not anomalies	Networks in dental development	
Automated digital measurement	Statistical advances	Complexity science	Multilayer networks	
Machine Learning	Mathematics	Network science	Complex interactions between dentition and dental arches	
Advances in machine learning	Data science	Many-model approach and Wisdom Hierarchy	Multiple models	

digital capture and analysis in bespoke two-dimensional image analysis tools; (5) general developments in digital photography and software-increased accessibility to these methods; (6) recent developments have moved into three-dimensions with specialized image analysis tools and as with 2D imaging; (7) the wide availability of 3D printing tools opens up new approaches; finally, the state-of-the-art is (8) automated measurement tools driven by machine learning. The influences outlined suggest multiple interactions between the different fields, and the following text explores how these occur in Complex Systems and networks.

In this chapter, the approaches of complexity science (Mitchell, 2009), networks (Newman, 2018), and many-model thinking (Page, 2018) are applied to the development and the emergent phenotypic outcome of the teeth and dental arches, their interactions, and coordination.

Complexity science is a unique mix of biology, physics, and social sciences (Mitchell, 2009; Thurner et al., 2018). It sees the world as interconnected with forms and patterns shaped by evolution, history, and context. It explores interrelationships, collective behaviour, dynamic flow, values, unintended consequences, multiple perspectives, the ways factors interact, and the emergence of the unexpected.

Complex Systems are co-evolving multilayer networks (Bianconi, 2018). They are composed of many elements, or nodes, whose interactions can be represented as links or edges. These interactions can change over time as can the elements. They evolve together by influencing each other. Complex Systems are context-dependent; for any dynamic process in a network, the other nodes and links provide the context. Information about the past can be stored in the nodes, the links, or the network structure.

Models are formal structures that help in the understanding of data, information, and knowledge. Many-model thinking (Page, 2018) begins from the realization that no single model can satisfactorily contain and represent the full range of information and knowledge on a complex topic. This is in contrast to the traditional scientific approach in which the adoption or development of a single model that is seen to most fully encompass the observed phenomena is a central goal. Many-model thinking uses the knowledge from multiple models to make sense of complex phenomena. Different models can be seen to be providing complementary and composite explanations rather than being in competition for a unified explanation. It produces wisdom by accentuating different casual forces and develops deeper understanding of processes that overlap and interact. In this context, wisdom is the ability to identify and apply relevant knowledge to develop new understanding of variations in the development of the teeth and the dental arches, their interactions, and possible future evolutionary trends.

9.3 THE DEVELOPMENTAL BASIS FOR VARIATION

9.3.1 Process

Tooth and dental arch development exhibits the general characteristics of Complex Adaptive Systems (Brook and O'Donnell, 2011) in which multiple interacting components at a lower level give rise to higher-level emergent phenomena. The multiple, reiterative interactions at a molecular level produce developmental stages which have further interactions leading to the emergence of the phenotype, the individual tooth. The morphology of these mature teeth varies in accordance with their tooth type and position in the dental arches. Diversity makes substantial contributions; it enhances the robustness of the process, variation in outcome, and the possibility of change (Brook et al., 2014a). The general characteristics of a Complex Adaptive System evident in dental development include self-organization, bottom-up emergence, multitasking, self-adaptation, variation, tipping points, critical phases, and robustness (Brook and O'Donnell, 2011).

The key components and characteristics of dental development are summarized in Table 9.2, which is further expanded from Brook (2009) and Brook et al. (2014a). Dental development passes through multiple levels from molecular to cellular, to

TABLE 9.2
Key Characteristics and Components of Dental Development, Expanded from Brook (2009) and Brook et al. (2014a).

Multi-level					Mineralized tooth
				Tooth germ	
			Tissue		
		Cellular			
	Molecular				
Multi-Dimensional	Spatial		X dimensions		
			Y dimensions		
			Z dimensions		
	Relationships between tooth dimensions		Tooth type series		
			Within arch		
			Between arches		
	Time—Progression		6 weeks in *utero*—22 years of age		
Multifactorial	Genetic code DNA—RNA—micro RNA				
	Epigenetic code	Activate genes	Modify linear DNA structure (DNA methylation)		
		Deactivate genes	Modify DNA structural complexes (histone acetylation)		
	Environmental	Systemic			
		Local			
Multiple interactions	Gene/gene			Signalling pathways	
	Epigenetic/gene methylation			Activation	
	Acetylation			Deactivation	
	Micro RNA				
	Cell/cell				
	Cell/matrix				
	Tooth germ/tooth germ				
	Signalling between tissues—reciprocal—reiterative—progressive				
	Patterns in developmental fields				
	Stochastic variations				

(Continued)

TABLE 9.2 (*Continued*)
Key Characteristics and Components of Dental Development, Expanded from Brook (2009) and Brook et al. (2014a).

Developmental stages	Each tooth	Initiation	Morphogenesis	Differentiation	Biomineralization
Models of developmental patterning	Phenotype quasi-continuous distribution Complex System power law distribution Single-level complex network Multilevel complex interactive network Neuro-osteological fields Vascular supply Morphogenetic fields Clone model Odontogenic homeobox code Genetic interaction concept Inhibitory cascade model				
Progression over time	Each tooth	Initiation Critical phases/thresholds		Mature tooth—eruption Apoptosis	
	Tooth type	Key tooth		Later forming teeth	
	Dentition	Primary Permanent		Overlapping development	

tissue, to tooth germ, to the phenotypic outcome of the erupted mineralized tooth. It occurs in four dimensions—the three spatial dimensions and that of time.

Dental development is multifactorial with interactions between genetic, epigenetic, and environmental factors at multiple stages of the process. It progresses over a substantial period for each individual tooth with critical phases and times within this period (Brook, 2009; Radlanski and Renz, 2005). The time from the start of development of the primary dentition to the completion of the permanent dentition covers a time frame from some 6 weeks in utero to 20–25 years of age.

The multiple interactions listed in Table 9.2 occur at multiple levels and between multiple factors throughout the development of each tooth. Some of these interactions arise from signalling between epithelium and mesenchyme, which is reciprocal, reiterative, and progressive over time.

Each tooth passes through the developmental stages of initiation, morphogenesis, differentiation, and mineralization. As the overall development of the dentition progresses, different teeth are at different stages at a given time. As progression over time occurs within each tooth type, incisor, canine, premolar, and molar, the key tooth forms first. In humans, it is the last tooth in the type which most often varies, and often varies most, in number, size, and shape.

The progressive development of each tooth is illustrated in Figure 9.3. In the horizontal plane, the stages of development are shown, while the vertical presents the progress from genotype to phenotype. The tooth forms by a series of reciprocal, reiterative epithelial–mesenchymal interactions (Steele-Perkins et al., 2003; Thesleff, 2014). During the initiation phase, the position of the tooth germ is established, the earliest morphological indication of being a placode-like thickening of the dental lamina. At this stage, a failure of development will result clinically in the congenital absence of the tooth. Cessation of development from failing to progress at a later critical phase also leads to clinical congenital absence.

In morphogenesis, the size and shape of the soft tissue tooth germ are outlined. The repeated activation and inhibition of signalling between the epithelium and ectomesenchyme, shown by the arrows in the genetic line of Figure 9.3, lead to differential growth and folding within the tooth germ. The primary and secondary enamel knots shown as dots on the tooth germs in the cell/tissues line on Figure 9.3 control the development of crown dimensions and cusp pattern. The enamel knots, specialized signalling centres, do not proliferate, but express growth stimulatory signals that cause the proliferation of adjacent cells (Fleischmannova et al., 2008).

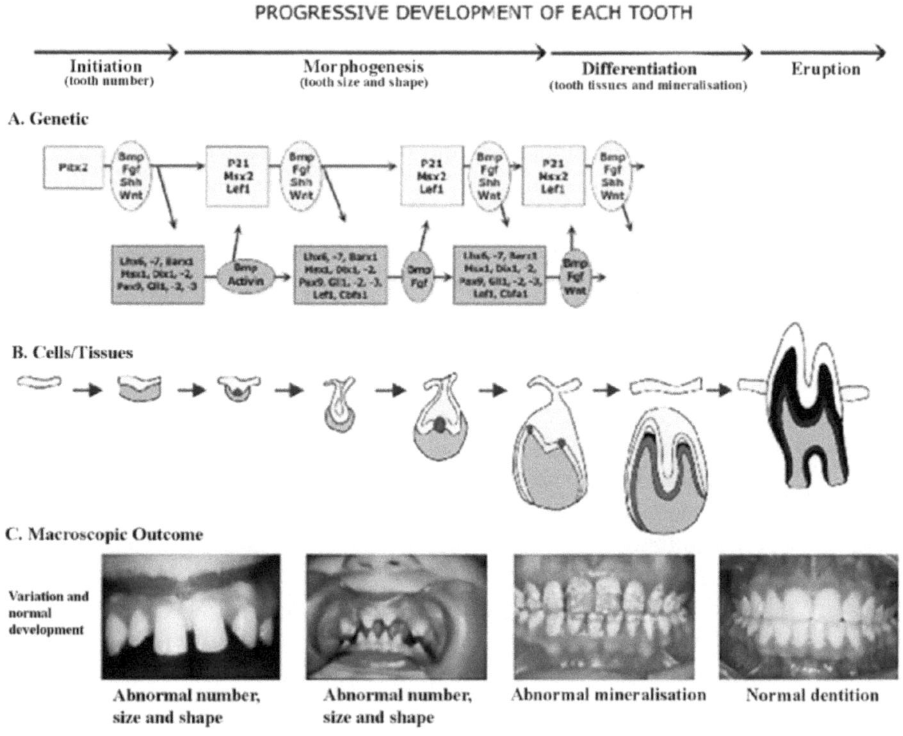

FIGURE 9.3 Progressive development of tooth from genotype to phenotype. Part of the diagram is derived from http://bite-it helsinki.fi//. Diagram further is developed from Brook et al. (2014a).

Source: Genetic signalling factors and transcriptions factors.

The primary enamel knot inhibits the formation of other enamel knots around itself, but when the inner enamel epithelium proliferates beyond its influence, secondary enamel knots develop. This process is fundamental to the inhibitory cascade model (Salazar-Ciudad and Jernvall, 2002, 2010; Riga et al., 2014) (Section 9.9).

Differentiation occurs at the bell stage of the tooth germ with the odontoblasts differentiating. Initially, the odontoblasts secrete the dentine matrix, and then adjacent cells of the epithelium differentiate into ameloblasts and secrete enamel matrix. The odontoblasts and ameloblasts then control the phase of mineralization of enamel and dentine. Once enamel formation is completed, the final external dimensions of the tooth crown are fixed.

With the emergence of the tooth into the mouth, the macroscopic outcome of the process becomes visible and available for study and treatment planning. As the individual teeth develop, they interact with the formation of the surrounding alveolar bone and influence the size and shape of the dental arch (Sections 9.6 and 9.7). To further investigate and understand this complex interactive process, the following sections are organized to review evidence for the factors, interactions, patterns, and outcomes involved in the development of the dentition.

9.3.2 Factors

Throughout development, these interactions are not only between genetic and epigenetic interactions and environmental factors but also between cells, matrix, and the developing tooth germs (Table 9.2). In reviewing the evidence concerning each of these factors, the emphasis is on understanding the process and how variations can arise. Crown size has high heritability in both primary and permanent human teeth, the level of this varying between teeth and different parameters of the same tooth (Hughes et al., 2014).

9.3.2.1 Genetic Factors

These genetic factors are distributed in the epithelium and mesenchyme. Several signalling pathways acting during early tooth development are Bmp, Fgf, Shh, and Wnt (Figure 9.3) (Hardcastle et al., 1998; Tucker et al., 1998; Amand et al., 2000; Dassule et al., 2000; Kettunen et al., 2000; Sarkar et al., 2000). The signalling molecule Bmp4 has been associated with apoptosis that terminates the enamel knot activity and helps to determine tooth size and shape, with different dimensions affected to varying degrees (Brook et al., 2009). Transcription factors in early tooth development include Msx1, Msx2, Dix1, Dix2, and Pax9 (Figure 9.3) (Qiu et al., 1997; Peters et al., 1998; Bei et al., 2000). The Ectodysplasin-A hormone signalling system is involved in late tooth development (Tucker et al., 2000). Shh signalling is deeply involved in tooth formation and has different functions at each stage of tooth development (Hosoya et al., 2020).

9.3.2.2 Epigenetic Factors

The term epigenetics, introduced by Waddington (1959), has been used for all processes through which the phenotype emerges from the genotype (Townsend et al., 2009). Epigenetics broadly refers to the study of changes in gene functions that are mitotically and meiotically heritable and that do not entail a change in DNA sequence (Wu and Morris, 2001). Therefore, epigenetics encompasses a group of acquired or inherited molecular mechanisms that are potentially transgenerational. These mechanisms are influenced by the environment and act on the genome to regulate gene expression (Williams et al., 2014). These epigenetic mechanisms include DNA methylation, histone modification, and mRNA. DNA methylation is involved in X chromosome inactivation in females (Williams et al., 2014). Twin studies offer an opportunity to study how epigenetic factors influence specific phenotypic variations across the same genome (Bell and Spector, 2011; Bell and Saffery, 2012; Hughes et al., 2014).

Hormones have an important role in epigenetic modifications and can signal the type, severity, and duration of the environmental cues to developing tissues. They act to alter gene expression in many different tissues. By producing an epigenome specific to the prevailing condition in utero, hormones act as epigenetic signals in developmental planning (Fowden and Forhead, 2009). Sex- specific changes in epigenetic markers are generated early after fertilization in response to environmental conditions. Sexually dimorphic characteristics in various tissues are influenced in this way, and the differences in sex hormones' expression affect a child's response to environmental conditions (Kurek et al., 2015).

Testosterone levels fluctuate throughout pre and post-natal development associated with variation in tooth size between males and females (Ribeiro et al., 2013). Studies of DZOS (dizygotic opposite-sex) and DZSS (dizygotic same-sex) twins have increased the understanding of how prenatal male sex hormones affect dental development (Dempsey et al., 1999; Ribeiro et al., 2013; Lam et al., 2017). Sexual dimorphism was observed in both primary and permanent dentition, with greater differences in permanent teeth (Taduran et al., 2016). These findings support the Twin Testosterone Transfer Hypothesis that female foetuses that gestate with a male co-twin are affected in development by exposure to pre-natal androgens (Miller and Martin, 1995; Tapp et al., 2011).

Prenatal and perinatal developments are critical periods in the programming of human postnatal life (Hales and Barker, 2001; Wells, 2012). Intrauterine disturbances, including malnutrition, vitamin A and D deficiencies, antimalarial drugs, and antiepileptic drugs, affect human deciduous enamel development (Rugg-Gunn et al., 1998; Mobley and Reifsnider, 2005; Jacobsen et al., 2013).

Low birth weight has also been associated with dental defects particularly in children who were born in less than 32 weeks, were ill in the neonatal period, and required long-term ventilator support (Seow et al., 1984; Fearne et al., 1990). Prenatal and neonatal variables and low birth weight are associated with enamel defects in deciduous teeth (Fearne et al., 1990; Franco et al., 2007).

9.3.2.3 Environmental Factors

Environmental influences have significant generalized effects on the dentition (Proffit et al., 1975; Richards et al., 1990) at any time during dental development from prenatal period through to early childhood. The degree of environmental influence is dependent on the timing (stage of dental development), severity, and duration of the exposure as well as host susceptibility (Brook and Winter, 1975; Brook, 2009).

Water fluoridation at 1 ppm ingested during tooth development possibly has a small effect on tooth size and shape but none on the frequency of hypodontia and supernumerary teeth (Brook, 1974). Localized infection can disrupt the formation of the dental tissues. Pulpal infection of primary teeth can spread to involve the follicle of the underlying developing permanent tooth in some 20–30% of cases causing lesions in the permanent tooth, ranging from enamel defects to arrested development of the permanent tooth germ (Brook and Winter, 1975). Generalized major infections such as rubella can also affect tooth development (Gullikson, 1975), as can recurrent infections (Farwell and Molleson, 1993), drugs, e.g. thalidomide (Axrup et al., 1966), irradiation of the oro-facial tissues (Rushton, 1947), and environmental toxins such as lead, which can also influence the process. Kieser (1998) showed that smoking and higher maternal age significantly increased fluctuating tooth size asymmetry.

9.3.3 Interactions

As illustrated in Figure 9.3, the epithelial–mesenchymal interactions continue throughout the initiation and morphogenesis stages of tooth development. During development, odontoblast differentiation initiates ameloblast differentiation, and there is the possibility of co-regulation (Lesot et al., 2001). Autocrine and paracrine interactions of cells are mediated by signalling factors such as Wnt and Shh and growth factors such as Tgf (Hughes et al., 2014).

Developmental interactions in size occur in maxillary lateral incisors (Sofaer et al., 1971). This suggests compensatory interactions between the tooth germ during development. Tadros et al. (2019) revisited this work using a 2D image analysis system on dental models of Australian twins. They found that developmental variations of the maxillary lateral incisors influenced the size and shape of adjacent teeth and that there was further evidence of compensating genetic, epigenetic, and environmental interactions in the Complex Adaptive System of dental development. For example individuals with one missing maxillary lateral incisor and the other lateral of average dimensions had significantly larger central incisors than the control group.

9.3.4 Patterning

The developmental patterns of the size and shape of tooth type arise during the process outlined earlier. During development, trigeminal nerve fibres produce axonal projections to tooth germs in a spatiotemporally controlled manner, coordinating axon navigation and patterning of tooth formation involving epithelial–mesenchymal interactions. Epithelial Wnt4 and Tgf-β1 regulate expression in the dental mesenchyme (Kettunen et al., 2005). However, the role of innervation in tooth initiation requires further clarification (Pagella et al., 2014).

The genes involved in dental development produce morphological traits indirectly through controlling the rate, timing, and orientation of odontogenetic processes. Epigenetic factors provide the link between genotype and phenotype. Some epigenetic effects follow a tooth specific pattern, e.g. in odontoblast terminal differentiation, which implies tempero-spatially regulated epigenetic signalling (Lesot et al., 2001).

The study of Hlusko et al. (2004) demonstrated the difference in patterning of maxillary and mandibular molar teeth; mandibular morphological homologues in the molar series exhibit complete pleiotropy showing they are influenced by the same suite of genes, while the maxillary molars show incomplete pleiotropic effects.

9.4 VARIATION IN TOOTH NUMBER, SIZE, AND SHAPE

9.4.1 Prevalence

Supernumerary and congenitally missing teeth are found frequently in humans. The prevalence of these variations is greater in the permanent dentition than in the primary. There are also differences in frequency between the genders and between ethnic groups. In the primary dentition, the prevalence is lower for all variations except for double teeth. Supernumerary teeth occur in 0.2–0.8% and congenital absence in 0.1–0.5% of children in the primary dentition. The prevalence in the permanent dentition of supernumerary teeth ranges from 1.5% to 3.8%. The prevalence of congenital absence of third permanent molars is 9–37% in various populations, and 1.6–9.6% for other permanent teeth (Vastardis, 2000; Matalova et al., 2008; De Coster et al., 2009; Huang et al., 2013; Brook et al., 2014a). Therefore, from an evolutionary aspect, these conditions are considered to be variations rather than anomalies. Supernumerary teeth occur more frequently in males, and congenital absence is present more often in females. Megadontia (large teeth) is rare in the primary dentition but was reported in 1.1% of children in the permanent dentition in an epidemiological study in England. The prevalence of microdontia (small teeth) ranges from 0.2% to 0.5% in the primary dentition and from 0.5% to 3.1% in the permanent.

The clinical appearance of the variations in shape includes extra cusps, double teeth, invaginations, and tapering and conical teeth. The shape of supernumerary teeth may be conical, tuberculate, supplemental, or odontoma-like. The formed teeth in hypodontia patients tend to be more rounded, tapering, or conical (Figure 9.4a, b). In patients with supernumeraries, the other teeth may have extra cusps and/or a barrel shape in incisors in the region of anterior maxillary supernumeraries (Figure 9.4c, d).

9.4.2 Factors

Building on the general factors in the development process reviewed in Section 9.3, the following text incorporates specific examples of factors associated with variations in number, size, and shape.

9.4.2.1 Genetic Mutations

There has been a major emphasis on identifying mutations in genes involved in dental development, which are associated with variations in tooth number. Recent advances in sequencing technology have progressed the knowledge of the genes involved. For isolated tooth agenesis, the genes involved include Axin2, Antxrl, Bmp4, Col17A1, Dkk1, Eda, Edar, Edaradd, Fgfrl, Grem2, Irf6, Msx1, Lama3, Lrp6, Ltbp3, Pax9, Smoc2, Wnt10A, and Wnt10B (Williams and Letra, 2018; Yu, Wong et al., 2019).

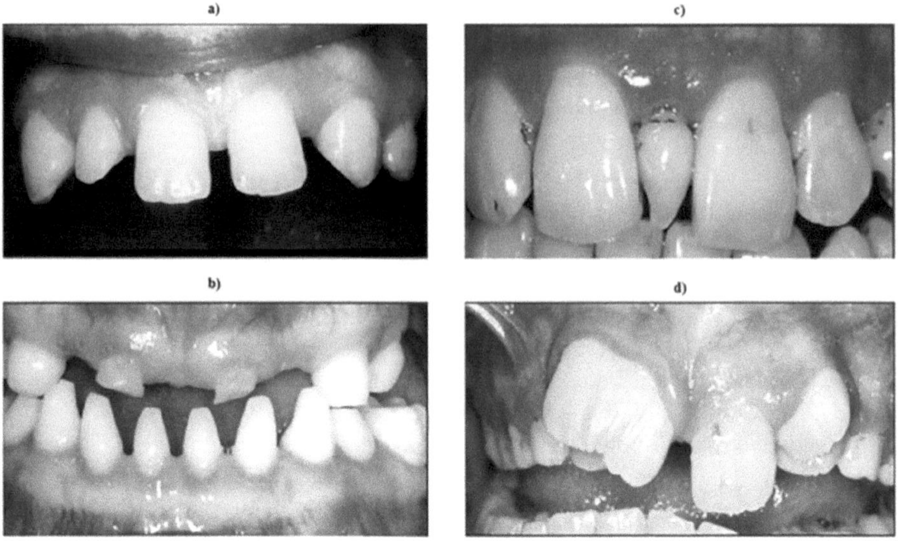

FIGURE 9.4 Clinical photographs of the variation in size and shape of teeth of patients with (a) mild and (b) severe hypodontia and (c and d) anterior maxillary supernumeraries.

A cluster of key genes among this list is estimated to be involved in over 90% of studied cases of hypodontia (Yu, Wong et al., 2019). Findings point to a major role for genes from the Wnt pathway in isolated hypodontia (Arte et al., 2013). Wnt signalling is involved in the patterning, proliferation, and differentiation during embryogenic development. During dental development, the absence of Wnt signalling leads to the disorder of the enamel knot and arrest of tooth development (Yamashiro et al., 2007). A substantial number of the reported cases of tooth agenesis have mutations in the Wnt pathway genes Axin2, Wnt10A, Wnt10B, and Lrp6 (Williams and Letra, 2018).

The transcription factors Msx1 and Pax9 are upstream effectors of Bmp4 expression in dental mesenchyme. Bmp4 is a member of the Tgf-β superfamily and is initially found in the dental epithelium but later appears in the dental mesenchyme. It is an important factor in the bud-to-cap transition and in the formation of the enamel knots. Bmp4 is also essential for the full development of many other organs. Huang et al. (2013) describe tooth agenesis in a family with a Bmp4 pre-domain tooth agenesis.

Further progress has been to demonstrate that different mutations within the same gene give rise to different phenotypes. Yu, Wang, et al. (2019) described different mutations of Bmp4 in five individuals giving rise to a range of variation from one small tooth to many congenitally absent teeth. As these genes are repeatedly active in the sequential stages with reiterative signalling during initiation and morphogenesis, mutations of them can cause variations not only in tooth number but also in tooth size and shape (Figures 9.3 and 9.4).

The effect of the same mutation within different individuals in the same family produces phenotypes with substantial differences. Table 9.3 illustrates the different

TABLE 9.3
Hypodontia Variation in Affected Individuals within the Same Family Affected by the Same Mutation.

	Right quadrants								Left quadrants							
	molars			premolars		can	incisors		incisors		can	premolars		molars		
Affected family members	3	2	1	2	1	1	2	1	1	2	1	1	2	1	2	3
II:3	*	*		*									*		*	*
	*	*		*			*	*	*				*		*	*
II:4							*			*						
III:2	*	*														*
	*															*
III:5	*	*	*		*	*					*	*			*	*
	*							*	*						*	*
III:6	*	*	*		*		*			*		*		*	*	*
	*	*		*				*	*				*		*	*
IV:1	*	*	*	*										*	*	*
	*	*														*

Note: * represents missing teeth.

congenitally absent teeth in individuals in one family with the same mutation of Pax9. The number of absent teeth varies from 2 to 10 in an affected individual, and all tooth types are affected (Brook, Elcock, et al., 2009).

For supernumerary teeth, it is suggested from studies of the mouse dentition that mutations of Fgf, Eda, Osr2, Runx2, Apc, Shh, and β-catenin are associated (Peterkova et al., 2002; Mustonen et al., 2003; Klein et al., 2006; Peterkova et al., 2006; Ohazama et al., 2008).

9.4.2.2 Epigenetic Factors

In monozygotic (MZ) twins, epigenetic events relating to the special arrangement of cells and the timing of the interactive signalling may influence differences in tooth number, size, and shape, as well as dental asymmetry (Townsend and Brook, 2008).

In a study using 2D image analysis, Lam et al. (2017) found that the females from the dizygotic opposite sex (DZOS) cohort had larger tooth crown dimensions for all variables measured than females in the dizygotic same sex (DZSS) cohort. These findings support the twin testosterone transfer hypothesis, confirming the effect of male hormone levels in utero on tooth size and shape in female co-twins. In males, the influence of the testosterone is likely to be in addition to the direct effects of the Y chromosome as detailed by Alvesalo (2009).

Reduced primary and permanent tooth size is seen in low birthweight children (Fearne and Brook, 1993; Harila-Kaera et al., 2001; Harila et al., 2003; Apps et al., 2004). Seow and Wan (2000) measured less than 11% smaller mesiodistal and buccolingual width of low birthweight children compared to a normal birthweight cohort.

Smoking during pregnancy alters tooth development at specific sensitive periods to influence postnatal crown formation (Heikkinen et al., 1992; Heikkinen et al., 1994). They found a 2–4% reduction in primary and permanent crowns in children whose mothers smoked during pregnancy. Al-Ani et al. (2017) report that maternal smoking during pregnancy is associated with hypodontia in this offspring; there were also observed associations between hypodontia and second-hand smoke inhalation as well as with alcohol and caffeine consumption of the mother that did not reach statistical significance in this sample.

9.4.2.3 Environmental Factors

Compared to modern Britons, Romano-Britons had a higher frequency of hypodontia and microdontia and a lower frequency of supernumerary teeth and megadontia (Brook et al., 2006). The difference probably results from the effects of the major environmental results of excess lead ingestion, poor nutrition, and recurrent infections in the Romano-Britons interacting in the developmental process summarized in Figure 9.5 (Brook et al., 2016). Environmental stressors like malnutrition and systemic diseases have a significant effect on molar morphology and generate a developmental response which increases morphological variability (Riga et al., 2014).

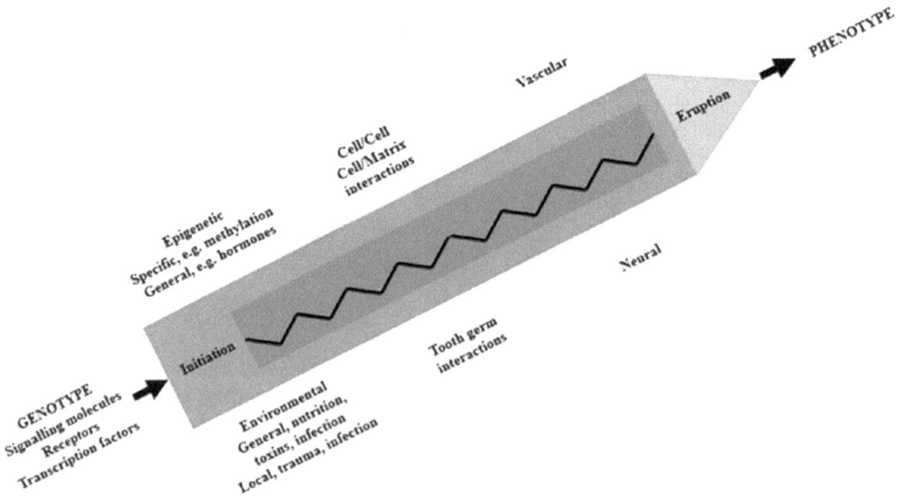

FIGURE 9.5 The multiple factors influencing the development from genotype to phenotype.

Source: Developed from Brook et al. (2016).

9.4.3 Interactions

Kjær et al. (1994) propose that nerve supply, oral mucosa, and supporting tissues are aetiological factors in mandibular hypodontia. These components interact during the development with other genetic, epigenetic, and environmental factors. Interactions between the systemic environmental factors of excess lead ingestion, poor nutrition, and recurrent infections found in the Romano-Britons increase the environmental stress during dental development and produce findings that when compared to modern Britons are unexpectedly in relation to the evolutionary trend of human dental reduction over time (Koh et al., 2016).

9.4.3.1 Phenotypic Outcomes

Some of the clinical phenotypic variations in the outcome of number, tooth size and shape, and mineralization are shown at the bottom of Figures 9.3 and 9.4. As Figure 9.5 illustrates, the genotype interacting with a series of epigenetic and environmental factors as development proceeds leads to variation in the phenotype between individuals. The clinical observations are that the greater the number of missing teeth, the smaller are the formed teeth and the more evident is their different shape. This is shown in the difference between the first and second photographs on the bottom line of Figure 9.3 and Figure 9.4a and b. These clinical observations have been quantified for some tooth parameters by hand measurements and 2D image analysis of dental study models (Brook et al., 2014a).

In Table 9.4, the mean mesiodistal crown dimensions of individuals with varying numbers of missing teeth are given together with those of their relatives who have a full complement of teeth and those of a control group. There are significant increases in tooth size from individuals with severe hypodontia to those with moderate hypodontia to mild hypodontia to unaffected relatives to controls.

Al-Shahrani et al. (2014) applied 3D geometric morphometric principles to assess the tooth shape of subjects with mild, moderate, and severe hypodontia against controls matched for age and sex. Significant shape differences were found between the

TABLE 9.4
Degree of Hypodontia and Tooth Size. Mean Mesiodistal Crown Dimensions in Patients with Different Degrees of Hypodontia, Unaffected Relatives, and Controls.

	Severe	Moderate	Mild	Unaffected Relative	Control
Upper central	7.80*	8.24*	8.43*	8.30*	9.26
Upper first premolar	6.43*	6.44*	6.72*	6.81*	7.37
Lower first premolar	6.63**	**6.72****	6.82**	7.11**	7.56

*P < 0.001, **P < 0.01

hypodontia groups compared to the controls and between sexes with hypodontia. The extent of shape variation was also linked to the severity of the hypodontia. These differences in tooth shape and size in hypodontia are associated with smaller arch dimensions (Section 9.7).

Not only is the tooth perimeter of all teeth reduced in hypodontia patients compared to controls but the number of cusps on posterior teeth is also reduced (Kerekes-Máthé et al., 2015). This has been found for hypodontia of one to five teeth and is more marked for hypodontia of six or more teeth, excluding third molars. Also, different dimensions of individual teeth, e.g. the mesiodistal and buccolingual, are influenced to different degrees in hypodontia patients as compared to controls (Kerekes-Máthé et al., 2022).

Males have both a higher prevalence of supernumerary teeth and larger teeth than females (Brook, 1984). In individuals with supernumerary teeth, the other teeth are larger than those of controls (Khalaf et al., 2005; Brook, Griffin, et al., 2009), and while this difference holds across the whole dentition, there is a gradient effect from the site of the supernumerary. If the supernumerary teeth are in the upper central incisor region, the incisors in both the maxilla and mandible are more affected than the rest of the dentition (Khalaf et al., 2009). Moreover, there are differences with the adjacent maxillary central incisors having a barrel shape when viewed from the labial aspect (Brook et al., 2014a). These findings are compatible with a local field effect (Khalaf et al., 2009). Supernumerary teeth are found associated with megadontia, double teeth, and invaginations (Figures 9.4c,d, and 9.6) (Brook et al., 2014a).

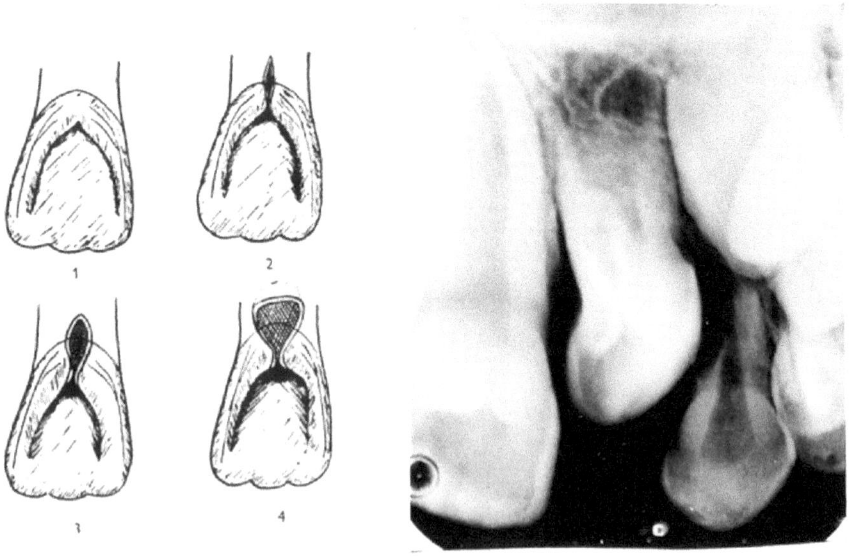

FIGURE 9.6 The diagram illustrates the four grades of invagination seen in radiographs as classified by Hallett (1953). The radiograph shows a grade 3 invagination in the most frequently affected tooth, the maxillary lateral incisor, and a grade 2 invagination in the central incisor.

Similar findings for sex differences in tooth number, size, and shape in human populations were gained in studies of Romano-Britons. Females had smaller teeth than males and had a higher frequency of hypodontia and microdontia; males had larger teeth and a higher frequency of supernumerary teeth and megadontia (Brook and Johns, 1995; Brook et al., 2006).

9.5 DENTAL ARCHES

9.5.1 Development

The dental arch changes with growth and is defined as the curved contour of the dentition or the residual ridge, and each dental arch generally takes the form of a catenary curve (Fleischer-Peters, 1978; Braun et al., 1998; Lee, 1999; Berkovitz et al., 2009). Usually, the maxillary arch is broader than the mandibular arch.

After about 37 days of development, a continuous band of thickened epithelium forms around the mouth in the presumptive upper and lower jaws. These bands are roughly horseshoe-shaped and correspond in position to the future dental arches of the upper and lower jaws (Avery and El Nesk, 2001; Nanci, 2017).

The maxillary and mandibular arches develop intramembranously. Ossification centres appear within mandibular arch at the location of the future mental foremen adjacent to the future lower premolars in week seven in-utero, and bone formation promptly takes place in various directions to form the mandible which continues to develop and increase in size as bone formation continues (Berkovitz et al., 2009). At eight weeks in-utero, the alveolar process develops and grows to surround the developing tooth germ, which is initially surrounded by an eggshell of bone known as the tooth crypt within the developing arches (Berkovitz et al., 2009). The growth and height of the alveolar process continue as increases in the dimensions of each tooth occur (Avery and El Nesk, 2001).

The maxillary ossification centre appears at the eighth week of intrauterine life, nearby to the developing deciduous canine. Growth of the maxilla does not occur in isolation and is influenced by the development of nasal, orbital, and oral cavities due to its position in the developing skull (Berkovitz et al., 2009). The progression in the height of the maxilla is associated with the development of the alveolar process (Berkovitz et al., 2009).

By birth, 45% of craniofacial bone growth is completed, and by seven years of age, 70% is completed (Ferguson and Dean, 2015). A study by Foster et al. (1977) showed that dental arches undergo an irregular growth pattern in size and form up to the age of ten years as the primary teeth are exfoliated, and the permanent teeth erupt rather than showing a steady change.

In adolescence, the maximum growth of the maxilla and mandible occurs concurrently, but growth reduces and discontinues at different times. The mandible continues to mature in length and finishes at a later age than the maxilla, which stops growing between 14 to 16 years of age (Avery and El Nesk, 2001; Berkovitz et al., 2009).

In a study by Bishara et al. (1997), the greatest increases in maxillary and mandibular arch length were seen up to the age of two years. Incremental increases in both maxillary and mandibular arches were seen at 13 years and 8 years, respectively. Decreased increments in both the maxillary and mandibular arches were seen after these ages. However, mandibular growth continues after that in the maxilla.

With regard to intercanine and intermolar widths, in the maxillary arch, the intercanine width increases by about 6.0 mm between 3 and 13 years of age and continues to increase a further 1.7 mm by the age of 45 years. The width between the left and right first permanent molars increases between 8 and 13 years of age by approximately 2.2 mm and, by the age of 45 years, decreases by 1.0 mm (Ferguson and Bishara, 2001). In the mandibular arch, the intercanine width increases by approximately 3.7 mm between 3 and 13 years of age, and then it decreases by 1.22 mm by the age of 45 years. The first permanent intermolar width increases between 8 and 13 years of age by 1.0 mm, and by the age of 45, it decreases by 1.0 mm (Bishara et al., 1997; Ferguson and Bishara, 2001). This decrease may relate to approximal wear between all teeth, with the associated forward drifting of the first molars into narrower part of the curved arch.

9.5.2 Factors

Dental arch development is determined by genetic, epigenetic, and environmental factors. It is influenced by the size and shape of the underlying basal bone. Alveolar bone growth is also influenced by the developing teeth, which have a central role affecting the extent and direction of growth of the dental arches as they grow in all spatial dimensions (Nanci, 2017). This is apparent in hypodontia patients, when the alveolus is deficient in the area of the congenitally absent teeth. The positions and alignment of teeth within the alveolar bone continually change through occlusal development in the primary, mixed, and permanent dentition stages affecting alveolar growth (Brown et al., 1983).

9.5.2.1 Genetic Factors

Genetic factors affect arch morphology (Eguchi et al., 2004; Harris and Smith, 1980; Richards et al., 1990) and have influence on occlusal traits such as overjet, overbite, and crowding (Corruccini and Potter, 1980; Townsend et al., 1988; Weaver et al., 2017). Additionally, the presence of an extra X chromosome affects maxillary and mandibular growth in 47, XXY (Klinefelter syndrome) men with decreased transverse and vertical growth of the maxilla and decreased transverse growth of the mandible (Laine and Alvesalo, 1993).

Variation in the size and shape of the dental arches in the primary dentition is under strong genetic influence and has a high heritability. However, the incisal overbite and overjet and the dental arch asymmetries are influenced by environmental factors and have a low heritability (Hughes et al., 2001). Genes play an important role in the dental arches during the transition from primary to permanent dentition, while environmental factors influence the anterior aspect of the mandible. Studies of dental arch morphology in the permanent dentition of twins have indicated that genetic factors contribute significantly to variation in arch shape but not to other aspects of arch morphology, such as asymmetry (Richards et al., 1990; Kasai et al., 1995).

Candidate Gene analysis has shown that specific genetic pathways are associated with 3D dentoalveolar phenotypic variation (Weaver et al., 2017). Associations are suggested with Pitx2, Snal3, and Fgf8 while fluctuating asymmetry is associated with Bmp3 and Lats1. Decreased alveolar bone formation is reported when Nf1 transcription replication proteins are deficient (Steele-Perkins et al., 2003). Sonic Hedgehog (Shh) is involved in craniofacial patterning and morphogenesis as well as tooth development (Dworkin et al., 2016; Xavier et al., 2016)

Some of the genes involved in the development of teeth and dental arches are also essential for the development of other tissues and organs. Bone morphogenetic proteins initiate bone induction in vertebrate tissues and have other functions in developmental regulation (Thesleff et al., 1995). Mutations in Bmp4, for example, have been associated with congenital anomalies of the kidney and urinary tract (Tabatabaeifar et al., 2009), as well as tooth variations from small teeth to hypodontia (Yu, Wang, et al., 2019). Therefore, a potential for further study is to accurately measure dental arch dimensions in patients with Bmp4 mutations.

Tgf-β super family members signal through a receptor complex to regulate craniofacial development (Zhao et al., 2008). Tgf-β type 1 receptor (Alk5) regulates the fate of cranial neural crest cells during the development of the mandible and the teeth (Zhao et al., 2008).

9.5.2.2 Epigenetic Factors

Disturbances such as hypoxia, smoke exposure, and famine during foetal life and infancy modify the development of craniofacial structures (Lampl et al., 2003; Zadzińska et al., 2013).

Differences are larger in males than females (Sillman, 1964; Bishara et al., 1997; Thilander, 2009). These sexual differences relate to arch size and not arch shape and are thought to be associated with sex differences in the size of the general body, craniofacial structures, and the dentition (Buschang et al., 1987). In females of dizygotic opposite sex twin pairs, there were larger arch dimensions related to the exposure to male sex hormones in utero (Hughes et al., 2001). In Figure 9.7, we propose a model for the process by which twin testosterone transfer leads to increased tooth and dental arch size in females in dizygotic opposite twin pairs. Patel et al. (2018) measured significant differences in arch size in DZOS females compared with either DZSS or MZ females. These differences were present in the primary but not the permanent dentition.

These increased dimensions have occurred due to an epigenetic influence of intrauterine male hormones, which was produced by the male co-twin during the development of the dental arch. No trend was found for arch dimensions in the permanent dentition of DZOS females, suggesting that this epigenetic effect of the male hormone, although organizational, lasted until the age of five to seven years with no or minimal effect on arch dimensions later in life. In contrast, studies on teeth found significantly larger tooth size dimensions in the permanent dentition but not the primary dentition in females from DZOS twin pairs (Dempsey et al., 1999; Ribeiro et al., 2013). This suggests that in the case of tooth dimensions, the epigenetic effect of the male hormones may be indirect and last until a later age. Conversely, no evidence of feminization was observed on tooth dimensions of male DZOS twins,

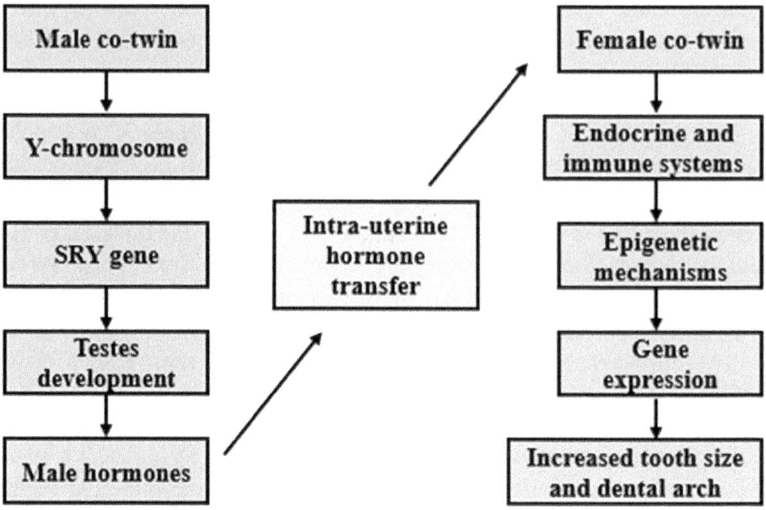

FIGURE 9.7 Proposed model for the process by which testosterone is transferred from the male twin to the female twin in-utero in dizygotic opposite sex twin pairs and the resultant effects.

suggesting that males are generally unaffected or only slightly affected by the presence of the female twin in utero (Ribeiro et al., 2013).

The dentoalveolar process in the primary dentition begins to form just prior to the testosterone surge in utero at around 7–8 weeks post conception and continues to develop in the presence of the male hormone until birth. This exposure to male hormone in utero may explain the effect seen in certain arch dimensions in the primary dentition of DZOS females compared with the other female twin groups. Arch dimensions in the permanent dentition of DZOS females did not display any evidence of a male hormonal effect in utero, possibly because in the permanent dentition, post-natal environmental factors play a more important role as changes in the dental arch occur with the eruption of teeth and masticatory forces which could be associated with dietary consistency (masticatory function) and the occlusal development (Harris and Johnson, 1991).

There have been varying findings concerning sexual dimorphism of arch shape. Kawata et al. (1974) concluded that narrower arch forms were seen in females, but other studies have shown no differences in arch shape between males and females (Richards et al., 1990; Raberin et al., 1993; Henrikson et al., 2001). Additionally, no sex differences in arch asymmetry were seen in a study by Burris and Harris (2000).

9.5.2.3 Environmental Factors

Environmental influences can have significant generalized effects on the dental arches (Proffit et al., 1975; Richards et al., 1990) at any time during development from prenatally through to early childhood. The degree of environmental influence is dependent on the timing (stage of development), severity, and duration of the exposure as well as host susceptibility (Brook, 2009).

The environmental factors of the circumoral musculature and the intraoral functional forces affect the position of the teeth, which influences the shape of the dental arches (Weinstein et al., 1963; Corruccini and Potter, 1980; Harris and Smith, 1980; Brown et al., 1983; Braun et al., 1998).

In regard to nutritive sucking, Warren and Bishara (2002) found that during the first year of life, there was no relationship between dental arch morphology and the length of breast feeding. Non-nutritive sucking, however, has been found in several studies to result in a decreased maxillary arch width, an increased lower arch width, and consequently a greater occurrence of posterior crossbite (Modeer et al., 1982; Lindner and Modeer, 1989; Adair et al., 1992; Øgaard et al., 1994; Adair et al., 1995; Warren and Bishara, 2002). Prolonged pacifier use results in decreased maxillary intercanine arch width and increased mandibular intercanine arch width, resulting in a posterior crossbite (Øgaard et al., 1994; Warren and Bishara, 2002). Decreased maxillary arch width resulted after a duration of use of two years or longer and an increased mandibular arch width resulted after three years of use (Øgaard et al., 1994). Warren and Bishara (2002) also observed that prolonged digit habits were associated with greater maxillary arch depths and narrowed maxillary arch widths.

In a study by Proffit et al. (1975) on Australian aborigines, tongue pressures were inversely related to dental arch dimensions. The larger the tongue pressure, the smaller the arch; the smaller the tongue pressure, the larger the arch.

9.6 RELATIONSHIP AND COORDINATION OF TOOTH AND DENTAL ARCH DEVELOPMENT

9.6.1 Relationship

Foster et al. (1969) suggested that the dimension of the dental arches was related to the jaw size and the position of the teeth but was little affected by the size of the teeth. This was also seen in a study by Moorrees and Reed (1954) in which no correlation was found between tooth size and dental arch size after measurements of the mandibular teeth and dental arch.

In males, the difference in size of permanent and deciduous incisors is much greater than in females (Garn et al., 1967; El-Nofely et al., 1989). El-Nofely et al. (1989) also found that the presence of interdental spaces depends considerably on the mesiodistal crown diameters and the intercanine arch width; they concluded that broad arches and small teeth cause interdental spacing of the deciduous teeth and vice versa.

The dental arches may be susceptible to crowding due to large teeth, very small bony bases of the jaws, or the presence of both (Howe et al., 1983). Arch dimensions have been noted to be smaller in subjects with dental crowding compared to those with no crowding (Howe et al., 1983), and arch perimeter decreases as crowding increases (Lundström, 1951). Howe et al. (1983) also observed asymmetry, irregularity, and frequent narrowing in the arch form of arches with crowded teeth.

There are examples of genes and their products involved in the development of both the teeth and the dental arches. Sonic Hedgehog has an active role in both dental and craniofacial development (Hosoya et al., 2020). The transcription factor Dlx-3 is present in ameloblasts (enamel), odontoblasts (dentine), osteoblasts (bone), and chondroblast (cartilage) and may be involved in matrix deposition and biomineralization in both the teeth and dental arches (Ghoul-Mazgar et al., 2005). Syndecan-1 is a cell surface proteoglycan involved in tooth development and is also found in osteoblasts and osteoclasts in alveolar bone (Filatova et al., 2015).

These findings raise the questions—is there co-ordination of tooth and dental development? If so, how is it achieved?

9.6.2 Coordination

Coordination between tooth and jaw development is important for good mastication of food by an individual. Marked discordance between the dentition and jaws in size and shape leads to less efficiency in chewing. The development, including developmental timing and positioning, of these two Complex Adaptive Systems needs to be closely coordinated. As the individual advances through infancy, childhood, teenage, and the stages of adult life, there are dietary differences at each stage. The variations and changes in each system need to be compatible for the fitness and success of the individual. This will include the capacity of the stomatognathic complex to deal with differences in the environment over time and location.

Integrated variation is important to maintain the function of teeth and jaws when there are large variations in each component. Yet, how is the coordination of jaw and tooth development achieved and maintained? Is it by central regulation, mediated by a common factor, or is it the outcome of selection for coordinated molecular pathways and growth trajectories? What part do local, mechanical, and cellular stimulation play? A further consideration is that maxilla and mandible have different developmental origins with the mandible being derived from the first pharyngeal arch, little affected by cranial and sub-cranial development (Boughner and Hallgrímsson, 2008).

A number of the mutations affecting tooth development also affect other developmental programmes, including hair and skin formation, bone formation, and craniofacial development. These pleiotropic phenotypes may arise from the expression of these genes in migrating neural crest cells (Chai et al., 2000; Ferguson et al., 2000; Steele-Perkins et al., 2003). Loss of the Tgf-β type 1 receptor Alk5 in neural crest tissue results in delayed tooth initiation and development, defects in early mandibular patterning, and altered expression of key patterning genes, including Msx1, Bmp4, Bmp2, Pax9, Alx4, Lhx6/7, and Gsc. Alk5 controls the survival of cranial neural crest cells by regulating the expression of genes in the developing mandible (Zhao et al., 2008).

There is evidence of stasis in teeth with strong functional and developmental interactions such as first permanent molars, while directional evolution is occurring in teeth with marginal roles in mastication, dental arch form, and occlusion. These teeth are temporally and spatially the last in their morphological tooth type. As these evolutionary changes occur in the dentition, the jaws need to undergo complementary change.

9.7 EFFECT OF VARIATIONS OF TOOTH NUMBER, SIZE, AND SHAPE ON DENTAL ARCHES

The large majority of the studies on the dimensions of the dental arches in patients with these tooth variations have been undertaken in cases of hypodontia. There have been differing results in these studies. A few investigators, e.g. Woodworth et al. (1985), have found no differences, but the majority of published studies have found significant differences between hypodontia patients and controls. In light of the varying considering the results of these different investigations, the methodology and study design used in each are important consideration. The studies vary in the severity of hypodontia studied, the landmarks for the parameters measured, the control subjects and whether manual (calliper) techniques or image analysis were employed.

Le Bot and Salmon (1977) found that upper arch width and length were reduced when measured by callipers in 109 male hypodontia patients compared to an age-matched control group. However, no differentiation was made between those with only missing lateral incisors and those with more severe hypodontia. Woodworth et al. (1985) found no difference in upper and lower arch width in 45 patients with bilateral absence of maxillary lateral incisors.

In a carefully structured study using 2D image analysis, Nelson et al. (2001) measured arch dimensions in 40 patients with mild/moderate hypodontia (1–5 congenitally absent teeth), 39 with severe hypodontia (6 or more absent), and 40 controls. In the mild/moderate hypodontia group, the upper arch depth and mean upper arch chords were significantly reduced. The severe hypodontia group showed greater changes with significantly reduced upper intermolar width, upper arch depth, and mean upper arch chords in comparison to the controls. Bu et al. (2008) reported similar findings in severe hypodontia patients.

Kerekes-Máthé et al. (2022) found that patients with mild and moderate hypodontia had significantly smaller arch circumference, arch length, and intercanine width in the upper arch than controls. This agrees with the findings of Nelson et al. (2001) and Patel et al. (2018) for mild and moderate hypodontia.

Overall, the maxillary arch is more affected than the mandibular (Table 9.5) (Patel et al., 2018). The only difference in the lower arch found by Kerekes-Máthé et al. (2022) was a larger intermolar width in the hypodontia group. This had previously been reported by Hobkirk et al. (2010), but not by Fekonja (2013) and Higgins (2017). Table 9.5 records the relevant results from a study of 68 patients with mild or moderate hypodontia and 68 controls matched for age, sex, and ethnicity. There was a highly significant difference in maxillary intercanine width, but not in the other dimensions reported (Patel et al., 2018).

Two possible explanations for a larger mandibular intermolar width in hypodontia patients can be considered. First, in response to the narrower upper arch, there may be increased tongue pressure in the lower molar region arising from the position of

TABLE 9.5
Overall Arch Dimensions in 68 Mild and Moderate Hypodontia Patients and 68 Controls.

Overall Maxillary Effect

	Hypodontia	Control	Difference
IM arch width	51.72	52.56	$P < 0.27$
IC arch width	32.49	35.06	$P < 0.0001$
Arch length	123.86	126.36	$P < 0.14$

Overall mandibular effect

	Hypodontia	Control	Difference
IM arch width	46.12	46.51	$P < 0.68$
IC arch width	25.01	25.14	$P < 0.69$
Arch length	111.49	112.26	$P < 0.49$

Abbreviations: IM, intermolar; IC, intercanine

the tongue. Second, if the lower second premolars are congenitally absent, the lower second primary molars may be retained, preventing the forward movement of the first permanent molars, which are then held back in a wider part of the arch parabola.

The reductions in arch dimensions are greater in male hypodontia patients than in female hypodontia patients (Patel et al., 2018). In the upper arch, male hypodontia patients had highly significant reductions in arch circumference, arch length, and intercanine width, while females had less difference in these three parameters (Kerekes-Máthé et al., 2022). The reduction in upper intercanine width associated with congenitally absent maxillary lateral incisors showed sexual dimorphism; for males, it was statistically significant (Table 9.5) (Patel et al., 2018). In a geometric morphometric study of the dental arches in hypodontia patients, Oeschger et al. (2022) found that the changes in facial profile, with a more retrusive maxilla and a more pronounced mandible as compared to controls, were similar for both sexes but were more marked in males.

The location of the congenitally absent teeth has a significant impact on the dental arch parameters. When the maxillary lateral incisors are congenitally absent, the arch circumference, arch length, and intercanine width are significantly reduced (Patel et al., 2018; Kerekes-Máthé et al., 2022). This suggests that the absence of these teeth has a substantial effect on the development of the anterior segment of the upper arch. No similar change is seen in the posterior area of the maxilla when the maxillary second premolars are congenitally missing. In the mandible, the significant difference is an increase in intermolar width when the mandibular second premolar is congenitally absent (Kerekes-Máthé et al., 2022). In accordance with these results, Oeschger et al. (2022) found that when tooth agenesis was located only in the mandible, both maxilla and mandible were affected. When tooth agenesis was only in the maxilla, no effect was measured in the mandible.

In summary, hypodontia does influence the development of the dental arches. The upper arch is more affected than the lower. Males are more affected than females, and the location of the hypodontia is influential. The more severe the hypodontia, the greater the effect. These changes are evidence of the interactions between two Complex Adaptive Systems that are part of the stomatognathic complex. The influences of hypodontia on dental arch morphology are part of the multiple phenotypic changes associated with congenital absence of teeth.

9.8 EVOLUTIONARY TRENDS

Human teeth and dental arches evolve in a complex scenario, which include constraints that limit dental evolution. These constraints range from developmental interactions between individual teeth to functional constrains that involve both teeth and dental arches in the occlusion. Gómez-Robles and Polly (2012) evaluated tooth shape characteristics using geometric morphometrics and found premolars and molars displayed significant co-variation. Integration was stronger in the mandible than in the maxilla. There was an association between morphological integration and evolution, with stasis observed in teeth with strong functional and developmental interactions. As reviewed before, in modern humans, the later forming teeth in each tooth type are the more variable and have higher prevalences of hypodontia, smaller size, and reduced shape. Clark and Henneberg (2017) suggest that the structural reduction of the anterior dentition and anterior maxilla allowed developments in speech and communication. Less teeth are associated with a flatter profile and decreased facial height (Oeschger et al., 2022). Advances in food availability, extraoral food processing, and cooking have reduced the selective pressures on the dentofacial complex.

Natural selection for a smaller dentition with the benefits of accommodation of molars and reduced generalized crowding has been proposed as a mechanism for the ongoing reduction in the human dentition. With the large number of mutations in many genes involved in dental development leading to hypodontia, the probable mutation effect is a potential explanation for the reduction in tooth number, size, and shape, leading to a smaller dentition. The probable mutations effect and natural selection may both be effective in these changes. The evidence considered in this chapter suggests that these reductions will probably continue in the future.

Substantial morphological changes can occur within humans within a few thousand years in a fast but gradual fashion (Henneberg, 2006). It is probable that there will be an increase in the prevalence of hypodontia, small tooth size, and reductions in shape, particularly of the last forming teeth in each tooth type, i.e. the third molars, the second premolars, the maxillary lateral incisors and the mandibular central incisors. With the increase in hypodontia, there are likely to be changes in the dental arches such as decreased maxillary intercanine width associated with hypodontia of the maxillary lateral incisors. From the laboratory findings of the clinical studies, e.g. Kerekes-Máthé et al. (2022), it is probable that the future changes in the dental arches will vary in the maxilla and the mandible and in different areas of each arch with tooth variations being a major influence. The broader relationship between a smaller number of teeth and overall craniofacial development also has evolutionary implications for facial height and facial convexity (Oeschger et al., 2022).

9.9 COMPLEXITY, NETWORKS, AND MULTIPLE MODELS ENHANCE OUR UNDERSTANDING OF DEVELOPMENT

9.9.1 Complexity and Networks

Throughout this chapter, there has been evidence of multiple interacting factors at multiple locations during the development of the dentition and the dental arches. This is further evidence that Complex Adaptive Systems are functioning during these developmental processes. The mature dentition and dental arches that are the phenotypic o Complex Adaptive Systems (Brook and O'Donnell, 2011).

Within each of these Complex Systems, the multiple factors interact in a single level complex network (Figure 9.8), the active components of which may change as

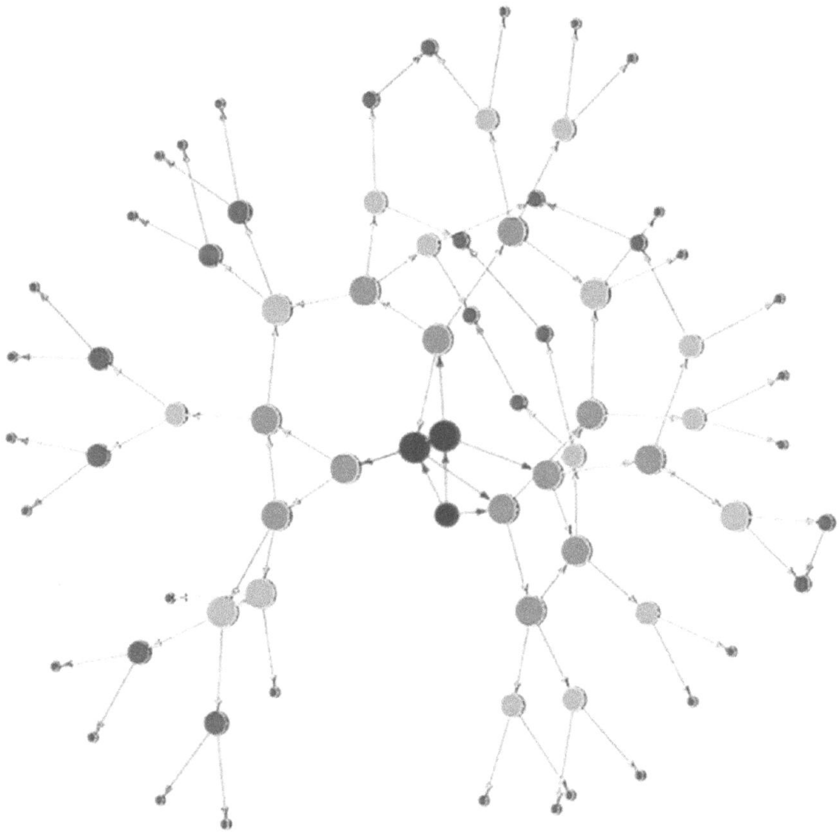

FIGURE 9.8 A complex network. Each individual factor is a "node" represented by a circle and is connected to other nodes by a line or "edge". This edge represents the interaction between nodes. The size of the circle reflects the number of interconnections (or degree) of that node. Nodes with multiple interconnections (high degree) are 'hubs'. If during development there is an error, e.g. the mutation of a gene, the effect is likely to be much greater when that gene is a hub, such as the repetitively activated gene Pax9.

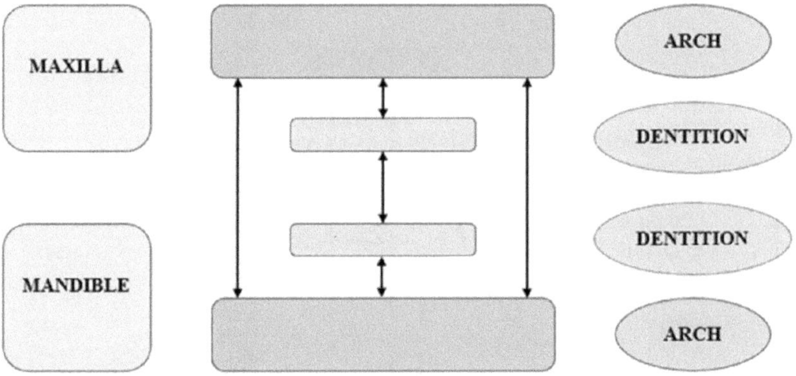

FIGURE 9.9 A multilevel complex interactive network model of the interactions between the maxillary and mandibular teeth and dental arches.

the process progresses in time and space (Brook, 2009; Brook et al., 2014a). The separate single-level complex networks of the dentition and the dental arches interact to form a multilevel complex interactive network producing the dentoalveolar complex (Figure 9.9).

9.9.2 Multiple Models

To investigate and understand this multilevel complex interactive network further, a multi-model approach is necessary. The following text reviews the available models that suggest how the factors, interactions, patterns, and outcomes of the development of the dentition and the dental arches can be interpreted. No single model explains all of the published findings. Therefore, the individual models should be considered as complementary rather than competing explanatory frameworks.

Rather than considering the different models in the historical order that they were proposed, they will be discussed as those arising from phenotypic studies, then those based on pattering in the dentition, and then those constructed from genetic studies.

9.9.2.1 Models Based on Phenotypic Studies

Investigations of the phenotype have the advantage of repeatable, highly accurate measurement, increased by the continuing development of technology and software. Studies have been enhanced by including molecular genetics and family histories to provide genetic, epigenetic, and environmental data. These models also can be investigated by statistical analysis.

The first model is developed from the quasi-continuous model of the normal distribution of a risk in population genetics. This is single tailed with a threshold beyond which individuals are affected by a condition. Rather than a series of separate models for each variation of tooth number, size, and shape, Brook (1984) produced a unified model. It is based on the normal distribution of tooth size, with small teeth and hypodontia at the lower tail and large teeth and supernumeraries at the

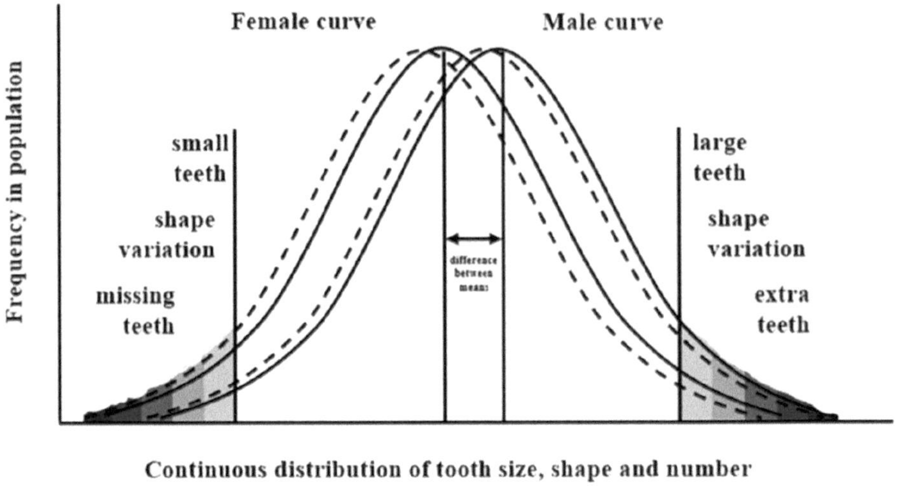

FIGURE 9.10 Model of the normal distribution of tooth number, size, and shape with thresholds beyond which clinically relevant variations occur. The variation in the shading reflects the increasing severity as the tails of the distribution are approached. The dotted curves are for female and male.

upper tail. There are separate distributions for males and females, which reflect the smaller teeth and higher frequency of hypodontia in females and the larger teeth and higher frequency of supernumeraries in males. As additional evidence has emerged from a series of studies, the model has been further developed several times (Brook et al., 2014a; Brook et al., 2014b), and the latest iteration is given here as Figure 9.10. This model is compatible with a random network model and is found for a number of parameters in the process of oral development and the emergent phenotypes. In this threshold model, there is the normal distribution of an underlying continuously varying parameter, i.e. tooth number, size, and shape, on which thresholds are superimposed. The lower tail has a threshold beyond which the individual will have small teeth, variations in tooth shape, and hypodontia. The gradation of the shading beyond the threshold from light to dark as the extreme of the distribution is approached reflects the changes to a smaller size and more rounded shape of the formed teeth, some with a reduced number of cusps, as the number of congenitally missing teeth increases (clinical examples are given in Figures 9.3 and 9.4). At the upper tail beyond the threshold, there are large teeth, variations in shape, and supernumeraries. The dotted curves show the smaller tooth dimensions and different frequencies of variations found in Romano-Britons. The curves for modern Britons and Romano-Britons have similar continuous distributions with the curves for the Romano-Britons shifted to the left, probably due to the combination of severe environmental effects of excess lead ingestion, poor nutrition, and recurrent infections.

While the model in Figure 9.10 incorporates the large majority of the clinical findings, it does not allow for the rare cases when hypodontia and supernumerary

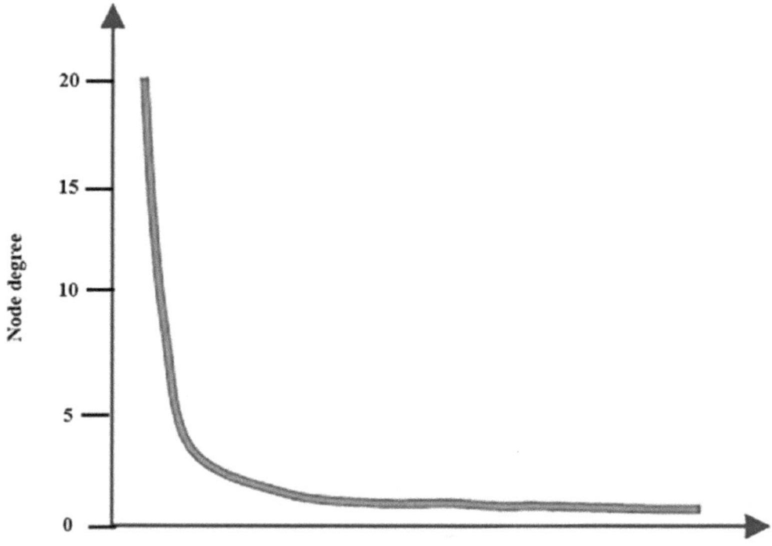

FIGURE 9.11 A power law distribution where a few components with a high node degree have a major effect in a multifactorial system. This was found when analysing the parameters of size and shape of upper and lower permanent incisors (Khalaf et al., 2009, 2022).

tooth are seen in the same dentition. It has now been found that upregulation of the gene Shh leads to supernumerary teeth, while downregulation of Shh gives rise to hypodontia (Peterkova et al., 2006; Salazar-Ciudad and Jernvall, 2010). Both may be happening in the same dentition and even the same tooth type. This reinforces how no single model can incorporate all findings or fully explain a complex biological process.

It is also found that some results do not fit the random network of the first model. For example when examining multiple parameters of human maxillary central and lateral incisors (Khalaf et al., 2009), differential effects of these parameters were discovered, suggesting the presence of a long-tailed or power law distribution (Figure 9.11). Seven factors out of 34 resulting from a Principal Components Analysis of the parameters accounted for 95% of the total variance in the maxillary central incisor measurements and 8 factors for 94% of the variance in measurements of the maxillary lateral incisors. There were similar findings for the mandibular incisors. This suggests that certain factors determining specific dimensions have a greater influence on the shape of the tooth than others. A power law distribution is a frequent finding in Complex Adaptive Systems and therefore fits with this understanding of the background to dental and dental arch development.

The third model, Figure 9.8, enhances the concept that these developmental processes are Complex Adaptive Systems. Brook and Brook O'Donnell (2011) define a Complex Adaptive System as "a dynamic system in which the interacting components at a lower level give rise to higher level emergent, variable phenomena". They show that dental development fulfils the general characteristics of a Complex

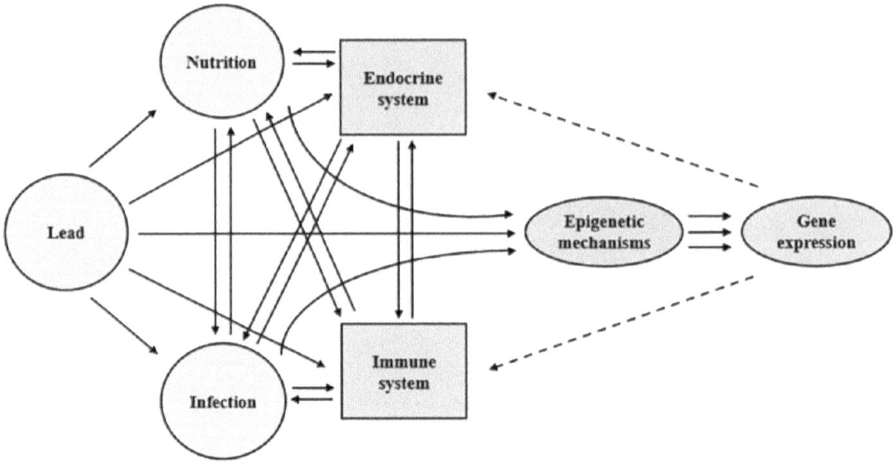

FIGURE 9.12 A multilevel complex interactive network model. The interactive effects of major environmental factors of toxin (lead), infection (generalized), and nutrition (deficient) giving rise to effects on different body systems and to epigenetic and gene expression changes. These give rise to reductions in tooth number, size, and shape in Romano-British sample (Brook et al., 2016).

Adaptive System. Since that time, a complex network of the interacting genetic, epigenetic, and environmental factors has been explored. In Figure 9.8, each factor is a node which interacts with other nodes, shown by connecting line edges. Those nodes with a greater number of connections form the hubs.

The fourth model is a development from the third. For the major environmental factors impacting the development of the dentition in Romano-Britons, the multilevel complex interactive network of effects has been used to construct Figure 9.12. The environmental factors interact with one another, with the endocrine and immune systems and also epigenetic mechanisms influencing gene expression. Another example of this model is when the multiple factors in dental development in the mandible form network of hubs and nodes in a single level. In the dentoalveolar complex, there are also single-level complex networks of factors in development of the maxillary dentition, the maxillary arch, and the mandibular arch. Interactions occur between the four levels, as discussed in Section 9.6. This forms the multilevel complex interactive network as illustrated in Figure 9.9. Further investigating the data, information, and knowledge generated by an ongoing series of studies of the interactions between hypodontia and dental arch morphology will allow an increased understanding of multilevel complex interacting network model for dentoalveolar development.

9.9.2.2 Models Based on Patterning

Butler (1939) proposed the field theory on the basis of the morphogenetic concentration causing a tooth to resemble its neighbour. Differences in morphogenetic concentration lead to variation in tooth morphology. There is a gradient of morphogenetic concentration along a tooth group being less concentrated at the end of the group.

Dahlberg (1945) applied the field theory to human dental development, modifying it to incorporate the patterning observed in the human permanent dentition, giving rise to incisors, canine, premolar and molar fields, and variations in size and morphology. The first tooth in each class is usually the key tooth and the most stable in terms of presence and morphology. Different expressions of morphogenetic gradients leading to variation in tooth size and morphological traits have been described (Townsend and Brown, 1981; Townsend et al., 1990).

The "gradient pre-pattern model" of Van Valen (1970) further explored variation within the same tooth group, suggesting that all developing tooth buds contain an identical pre-pattern on which the gradient causes variation.

Osborne (1978) introduced the clone theory, which proposes every structure is derived from a clone and suggests three different clones—incisor, canine, and molar. As the clone grows distally, over time new primordia arise in the space available. The movement of clone cells create a gradient that determines the shape of the resultant primordia. This is different from the field theory model where the primary gradient induces the shape of the primordia.

A further patterning model is that of neuro-osteological fields (Kjær, 1998). This is based on the production of axonal projections from the trigeminal nerve to tooth germs. It is suggested that there is a coordination between axon navigation and patterning of dental development. There is also coordination between the vascular supply and the developing tooth germs.

9.9.2.3 Models Based on Molecular Genetics

Studies of the molecular genetic basis of tooth development initiated the homeobox model on the basis of findings in mice (Sharpe, 1995). This model proposes that the patterning of teeth is the result of varying expression of several homeobox genes in ectomesenchymal cells. These genes play a significant role in various stages of tooth development as reviewed in Section 9.3.2.1. Two important homeobox genes are Msx1 and Dlx (Sharpe, 1995). This model suggests that overlapping domains of these two genes with others such as Barx give rise to positional information for tooth type morphogenesis (Nanci, 2008). Thus, in this model, region-specific combinatorial homeobox gene expression determines each tooth identity.

Mitsiadis and Smith (2006) proposed a cooperative genetic interaction model which incorporated the homeobox gene model, morphogenetic fields, and the clone theory. According to this concept, every component, including cells, signals, and homeobox genes, is acting to pattern teeth. Any disruption in cell numbers, signals, or mutations will affect the development of tooth position, number, and shape. This combination of models was further explored by Townsend et al. (2009).

The inhibitory cascade model was proposed by Kavanagh et al. (2007) on the basis of studies of mice. It describes an activator–inhibitor mechanism to predict evolutionary size patterns of mammalian teeth, including hominins. The activation mechanism is considered to be mainly of mesenchymal, while the early developing molars produce signalling molecules that inhibit the development of subsequent buds. Kavanagh et al. (2007) propose that a balance between activation and inhibition results in equal sized molars ($M1 = M2 = M3$), and increasing inhibition has a cumulative effect on the posterior teeth giving a distinct ($M1 > M2 > M3$)

pattern. The inhibitory cascade model seems to explain a high proportion of the variation in relative molar size in other mammals as well as mice, but in some studies, particularly in primates, the results have not fitted well (Bernal et al., 2013; Carter and Worthington, 2016; Roseman and Delezene, 2019; Bermúdez de Castro et al., 2021). The inhibitory cascade model was applied to lower primary and permanent molars of different hominin samples by Evans et al. (2016) using the computed crown area, obtained by multiplying the mesiodistal and buccolingual dimensions, and they suggested a tight link between tooth proportions and absolute size in these teeth. However, when Bermúdez de Castro et al. (2021) applied direct measurement of the crown area, instead of the computer crown area, their results suggested a strong relationship between the size of M1 and M2 but a more moderate relationship of size of M3 to M1 and M2. They found that in this modern human sample, the inhibitory cascade model did not fit the separate samples of males or females or the pooled sample. A contributing factor to these different findings may be that Bermúdez de Castro et al. (2021) used individual findings while Evans et al. (2016) used averages. Another consideration is that there are other factors not included in the model affecting molar size in modern humans. Therefore, while this model contributes to understanding dental development, it has its limitations.

In conclusion, reviewing these individual models has indicated that no one model can fully encompass the multiple parameters of the complex process of dentoalveolar development. Rather, the approach is to embrace multiple models, each of which enhances the understanding of the patterning, variations, and interactions of the human dentition and the dental arches.

ACKNOWLEDGMENTS

We are most grateful to Dr. Maryam Hajishafiee and to Angela Gurr for their considerable assistance in the production of this chapter. Their care and commitment are much appreciated. The chapter has been improved by constructive and critical editorial comments. This work was supported by a grant from the Paul Kwok Lee Fund, University of Adelaide.

REFERENCES

Ackoff, R. L. (1989). From data to wisdom. *Journal of Applied Systems Analysis*, *16*(1), 3–9.

Adair, S. M., Milano, M., & Dushku, J. C. (1992). Evaluation of the effects of orthodontic pacifiers on the primary dentitions of 24- to 59-month-old children: Preliminary study. *Pediatric Dentistry Journal*, *14*(1), 13–18.

Adair, S. M., Milano, M., Lorenzo, I., & Russell, C. (1995). Effects of current and former pacifier use on the dentition of 24- to 59-month-old children. *Pediatric Dentistry Journal*, *17*(7), 437–444.

Al-Ani, A., Antoun, J., Thomson, W., Merriman, T., & Farella, M. (2017). Maternal smoking during pregnancy is associated with offspring hypodontia. *Journal of Dental Research*, *96*(9), 1014–1019.

Al-Shahrani, I., Dirks, W., Jepson, N., & Khalaf, K. (2014). 3D-geomorphometrics tooth shape analysis in hypodontia. *Frontiers in Physiology*, *5*, 154.

Alvesalo, L. (2009). Human sex chromosomes in oral and craniofacial growth. *Archives of Oral Biology, 54*, S18–S24.

Amand, T. R. S., Zhang, Y., Semina, E. V., Zhao, X., Hu, Y., Nguyen, L., Murray, J. C., & Chen, Y. (2000). Antagonistic signals between BMP4 and FGF8 define the expression of Pitx1 and Pitx2 in mouse tooth-forming anlage. *Developmental Biology, 217*(2), 323–332.

Apps, M. V., Hughes, T. E., & Townsend, G. C. (2004). The effect of birthweight on tooth-size variability in twins. *Twin Research and Human Genetics, 7*(5), 415–420.

Arte, S., Parmanen, S., Pirinen, S., Alaluusua, S., & Nieminen, P. (2013). Candidate gene analysis of tooth agenesis identifies novel mutations in six genes and suggests significant role for WNT and EDA signaling and allele combinations. *PLoS ONE, 8*(8), e73705.

Avery, J., & El Nesk, N. (2001). *Oral Development and Histology.* 3rd ed. Thieme Medical Publishers, New York.

Axrup, K., d'Avignon, M., Hellgren, K., Henrikson, C.-O., Juhlin, I.-M., Larsson, K., Persson, G., & Welander, E. (1966). Children with thalidomide emrryopathy: Odontological observations and aspects. *Acta Odontologica Scandinavica, 24*(1), 3–21.

Bei, M., Kratochwil, K., & Maas, R. L. (2000). BMP4 rescues a non-cell-autonomous function of Msx1 in tooth development. *Development, 127*(21), 4711–4718.

Bell, J. T., & Saffery, R. (2012). The value of twins in epigenetic epidemiology. *International Journal of Epidemiology, 41*(1), 140–150.

Bell, J. T., & Spector, T. D. (2011). A twin approach to unraveling epigenetics. *Trends in Genetics, 27*(3), 116–125.

Berkovitz, B., Holland, G., & Moxham, B. (2009). Oral anatomy, histology and embryology. *European Journal of Orthodontics, 31*, 457.

Bermúdez de Castro, J. M., Modesto-Mata, M., García-Campos, C., Sarmiento, S., Martín-Francés, L., Martínez de Pinillos, M., & Martinón-Torres, M. (2021). Testing the inhibitory cascade model in a recent human sample. *Journal of Anatomy, 239*(5), 1170–1181.

Bernal, V., Gonzalez, P. N., & Perez, S. I. (2013). Developmental processes, evolvability, and dental diversification of New World monkeys. *Evolutionary Biology, 40*(4), 532–541.

Bianconi, G. (2018). *Multilayer Networks: Structure and Function.* Oxford University Press, Oxford.

Bishara, S. E., Ortho, D., Jakobsen, J. R., Treder, J., & Nowak, A. (1997). Arch width changes from 6 weeks to 45 years of age. *American Journal of Orthodontics and Dentofacial Orthopedics, 111*(4), 401–409.

Boughner, J. C., & Hallgrímsson, B. (2008). Biological spacetime and the temporal integration of functional modules: A case study of dento-gnathic developmental timing. *Developmental Dynamics, 237*(1), 1–17.

Braun, S., Hnat, W. P., Fender, D. E., & Legan, H. L. (1998). The form of the human dental arch. *The Angle Orthodontist, 68*(1), 29–36.

Brook, A. H. (1974). The prevalence of dental anomalies in 11–14 year old children: A comparison of continuous and discontinuous residents of a natural fluoride area. *Proceedings of the British Paedodontic Society, 4*, 7–12.

Brook, A. H. (1984). A unifying aetiological explanation for anomalies of human tooth number and size. *Archives of Oral Biology, 29*(5), 373–378.

Brook, A. H. (2009). Multilevel complex interactions between genetic, epigenetic and environmental factors in the aetiology of anomalies of dental development. *Archives of Oral Biology, 54*(Suppl 1), S3–17.

Brook, A. H., Elcock, C., Aggarwal, M., Lath, D. L., Russell, J. M., Patel, P. I., & Smith, R. N. (2009). Tooth dimensions in hypodontia with a known PAX9 mutation. *Archives of Oral Biology, 54*, S57–S62.

Brook, A. H., Griffin, R. C., Smith, R. N., Townsend, G. C., Kaur, G., Davis, G. R., & Fearne, J. (2009). Tooth size patterns in patients with hypodontia and supernumerary teeth. *Archives of Oral Biology, 54*, S63–S70.

Brook, A. H., Jernvall, J., Smith, R. N., Hughes, T. E., & Townsend, G. C. (2014a). The dentition: The outcomes of morphogenesis leading to variations of tooth number, size and shape. *Australian Dental Journal, 59*(Suppl 1), 131–142.

Brook, A. H., & Johns, C. C. (1995). Dental anomalies of number and size in a Romano-British population. In R. Radlanski, & H. Renz (Eds.), *Proceedings of the 10th International Symposium on Dental Morphology* (pp. 177–180). "M" Marketing Services, Berlin.

Brook, A. H., Koh, K. S. B., & Toh, V. K. L. (2016). Influences in a biologically complex adaptive system: Environmental stress affects dental development in a group of Romano-Britons. *Complex Systems: Fundamentals and Applications, 90*, 239.

Brook, A. H., & O'Donnell, M. B. (2011a). The dentition: A complex system demonstrating self-* principles. 2011 IEEE Fifth International Conference on Self-Adaptive and Self-Organizing Systems. Ann Arbor, MI, pp. 208–209. doi: 10.1109/SASO.2011.41

Brook, A. H., & O'Donnell, M. B. (2011b). Modelling the complexity of the dentition. In G. Townsend, E. Kanazawa, & H. Takayama (Eds.), *New Directions in Dental Anthropology* (pp. 1–9). University of Adelaide Press, Adelaide.

Brook, A. H., O'Donnell, M. B., Hone, A., Hart, E., Hughes, T. E., Smith, R. N., & Townsend, G. C. (2014b). General and craniofacial development are complex adaptive processes influenced by diversity. *Australian Dental Journal, 59*, 13–22.

Brook, A. H., Smith, R. N., Elcock, C., Al-Sharood, M., Shah, A., & Karmo, M. (1999). The measurement of tooth morphology: Development and validation of a new image analysis technique. *Dental Morphology 1998*, 380–387.

Brook, A. H., Underhill, C., Foo, L. K., & Hector, M. (2006). Approximal attrition and permanent tooth crown size in a Romano-British population. *Dental Anthropology Journal, 19*(1), 23–28.

Brook, A. H., & Winter, G. B. (1975). Developmental arrest of permanent tooth germs following pulpal infection of deciduous teeth. *British Dental Journal, 139*(1), 9–11.

Brown, T., Abbott, A. H., & Burgess, V. B. (1983). Age changes in dental arch dimensions of Australian Aboriginals. *American Journal of Physical Anthropology, 62*(3), 291–303.

Bu, X., Khalaf, K., & Hobson, R. S. (2008). Dental arch dimensions in oligodontia patients. *American Journal of Orthodontics and Dentofacial Orthopedics, 134*(6), 768–772.

Burris, B., & Harris, E. (2000). Maxillary arch size and shape in American blacks and whites. *The Angle Orthodontist, 70*(4), 297–302.

Buschang, P. H., Demirjian, A., Tanguay, R., & Charbonneau, C. (1987). Multivariate patterns of variation in dental arch morphology of French-Canadians. *Human Biology*, 911–920.

Butler, P. M. (1939). Studies of the mammalian dentition. Differentiation of the post-canine dentition. *Proceedings of the Zoological Society of London, B109*, 1–36. https://doi.org/10.1111/j.1469-7998.1939.tb00021.x

Carter, K. E., & Worthington, S. (2016). The evolution of anthropoid molar proportions. *BMC Evolutionary Biology, 16*(1), 1–18.

Chai, Y., Jiang, X., Ito, Y., Bringas, P., Han, J., Rowitch, D. H., Soriano, P., McMahon, A. P., & Sucov, H. M. (2000). Fate of the mammalian cranial neural crest during tooth and mandibular morphogenesis. *Development, 127*(8), 1671–1679.

Clark, G., & Henneberg, M. (2017). Ardipithecus ramidus and the evolution of language and singing: An early origin for hominin vocal capability. *Journal of Comparative Human Biology, 68*(2), 101–121.

Corruccini, R. S., & Potter, R. H. Y. (1980). Genetic analysis of occlusal variation in twins. *American Journal of Orthodontics*, *78*(2), 140–154.

Dahlberg, A. A. (1945). The changing dentition of man. *The Journal of the American Dental Association*, *32*(11), 676–690.

Dassule, H. R., Lewis, P., Bei, M., Maas, R., & McMahon, A. P. (2000). Sonic hedgehog regulates growth and morphogenesis of the tooth. *Development*, *127*(22), 4775–4785.

De Coster, P. J., Marks, L. A., Martens, L. C., & Huysseune, A. (2009). Dental agenesis: Genetic and clinical perspectives. *Journal of Oral Pathology and Medicine*, *38*(1), 1–17.

Dempsey, P. J., Townsend, G. C., & Richards, L. C. (1999). Increased tooth crown size in females with twin brothers: Evidence for hormonal diffusion between human twins in utero. *American Journal of Human Biology: The Official Journal of the Human Biology Association*, *11*(5), 577–586.

Dworkin, S., Boglev, Y., Owens, H., & Goldie, S. J. (2016). The role of sonic hedgehog in craniofacial patterning, morphogenesis and cranial neural crest survival. *Journal of Developmental Biology*, *4*(3), 24.

Eguchi, S., Townsend, G. C., Richards, L. C., Hughes, T., & Kasai, K. (2004). Genetic contribution to dental arch size variation in Australian twins. *Archives of Oral Biology*, *49*(12), 1015–1024.

El-Nofely, A., Sadek, L., & Soliman, N. (1989). Spacing in the human deciduous dentition in relation to tooth size and dental arch size. *Archives of Oral Biology*, *34*(6), 437–441.

Evans, A. R., Daly, E. S., Catlett, K. K., Paul, K. S., King, S. J., Skinner, M. M., Nesse, H. P., Hublin, J.-J., Townsend, G. C., & Schwartz, G. T. (2016). A simple rule governs the evolution and development of hominin tooth size. *Nature*, *530*(7591), 477–480.

Farwell, D., & Molleson, T. (1993). Poundbury volume 2: The cemeteries. *Dorchester: Dorset Natural History and Archaeological Society Monograph Series*, *11*.

Fearne, J. M., & Brook, A. H. (1993). Small primary tooth-crown size in low birthweight children. *Early Human Development*, *33*(2), 81–90.

Fearne, J. M., Bryan, E. M., Elliman, A. M., Brook, A. H., & Williams, D. M. (1990). Enamel defects in the primary dentition of children born weighing less than 2000 g. *British Dental Journal*, *168*(11), 433–437.

Fekonja, A. (2013). Comparison of mesiodistal crown dimension and arch width in subjects with and without hypodontia. *Journal of Esthetic and Restorative Dentistry*, *25*(3), 203–210.

Ferguson, C. A., Tucker, A. S., & Sharpe, P. T. (2000). Temporospatial cell interactions regulating mandibular and maxillary arch patterning. *Development*, *127*(2), 403–412.

Ferguson, D., & Bishara, S. (2001). Dental arch development. *Textbook of Orthodontics. Philadelphia: Saunders*, 61–65.

Ferguson, D., & Dean, J. A. (2015). Growth of the face and dental arches. In R. E. McDonald, & D. R. Avery (Eds.), *Dentistry for the Child and Adolescent* (10th ed., pp. 375–389). Eslevier, Moby, Philadelphia.

Filatova, A., Pagella, P., & Mitsiadis, T. A. (2015). Distribution of syndecan-1 protein in developing mouse teeth. *Frontiers in Physiology*, *5*, 518.

Fleischer-Peters, A. (1978). Walther's orthodontic notes. *Fortschritte der Kieferorthopädie*, *39*(2), 179–179.

Fleischmannova, J., Matalova, E., Tucker, A. S., & Sharpe, P. T. (2008). Mouse models of tooth abnormalities. *European Journal of Oral Sciences*, *116*(1), 1–10.

Foster, T., Grundy, M., & Lavelle, C. (1977). A longitudinal study of dental arch growth. *American Journal of Orthodontics*, *72*(3), 309–314.

Foster, T., Hamilton, M., & Lavelle, C. (1969). Dentition and dental arch dimensions in British children at the age of 212 to 3 years. *Archives of Oral Biology*, *14*(9), 1031–1040.

Fowden, A. L., & Forhead, A. J. (2009). Hormones as epigenetic signals in developmental programming. *Experimental Physiology*, *94*(6), 607–625.

Franco, K. M. D., Line, S. R. P., & Moura-Ribeiro, M. V. L. D. (2007). Prenatal and neonatal variables associated with enamel hypoplasia in deciduous teeth in low birth weight preterm infants. *Journal of Applied Oral Science*, *15*(6), 518–523.

Garn, S. M., Lewis, A. B., Swindler, D. R., & Kerewsky, R. S. (1967). Genetic control of sexual dimorphism in tooth size. *Journal of Dental Research*, *46*(5), 963–972.

Ghoul-Mazgar, S., Hotton, D., Lézot, F., Blin-Wakkach, C., Asselin, A., Sautier, J.-M., & Berdal, A. (2005). Expression pattern of Dlx3 during cell differentiation in mineralized tissues. *Bone*, *37*(6), 799–809.

Gómez-Robles, A., & Polly, P. D. (2012). Morphological integration in the hominin dentition: Evolutionary, developmental, and functional factors. *Evolution: International Journal of Organic Evolution*, *66*(4), 1024–1043.

Gullikson, J. (1975). Tooth morphology in rubella syndrome children. *ASDC Journal of Dentistry for Children*, *42*(6), 479–482.

Hales, C. N., & Barker, D. J. (2001). The thrifty phenotype hypothesis: Type 2 diabetes. *British Medical Bulletin*, *60*(1), 5–20.

Hallett, G. (1953). The incidence, nature, and clinical significance of palatal invaginations in the maxillary incisor teeth. *Proceedings of the Royal Society of Medicine*, *46*, 15.

Hardcastle, Z., Mo, R., Hui, C., & Sharpe, P. T. (1998). The Shh signalling pathway in tooth development: Defects in Gli2 and Gli3 mutants. *Development*, *125*(15), 2803–2811.

Harila, V., Heikkinen, T., & Alvesalo, L. (2003). Deciduous tooth crown size in prematurely born children. *Early Human Development*, *75*(1–2), 9–20.

Harila-Kaera, V., Heikkinen, T., Alvesalo, L., & Osborne, R. H. (2001). Permanent tooth crown dimensions in prematurely born children. *Early Human Development*, *62*(2), 131–147.

Harris, E. F., & Johnson, M. G. (1991). Heritability of craniometric and occlusal variables: A longitudinal sib analysis. *American Journal of Orthodontics and Dentofacial Orthopedics*, *99*(3), 258–268.

Harris, E. F., & Smith, R. J. (1980). A study of occlusion and arch widths in families. *American Journal of Orthodontics*, *78*(2), 155–163.

Heikkinen, T., Alvesalo, L., Osborne, R. H., & Pirttiniemi, P. (1992). Maternal smoking and tooth formation in the foetus. I. Tooth crown size in the deciduous dentition. *Early Human Development*, *30*(1), 49–59.

Heikkinen, T., Alvesalo, L., Osborne, R. H., & Tienari, J. (1994). Maternal smoking and tooth formation in the foetus. II. Tooth crown size in the permanent dentition. *Early Human Development*, *40*(1), 73–86.

Henneberg, M. (2006). The rate of human morphological microevolution and taxonomic diversity. *Studies in Historical Anthropology*, *4*(2004), 49–59.

Henrikson, J., Persson, M., & Thilander, B. (2001). Long-term stability of dental arch form in normal occlusion from 13 to 31 years of age. *The European Journal of Orthodontics*, *23*(1), 51–61.

Higgins, S. A. (2017). *Tooth Size and Arch Dimensions in a Hypodontia Patient Group.* M.Sc Thesis, University of Birmingham, Birmingham.

Hlusko, L. J., Maas, M. L., & Mahaney, M. C. (2004). Statistical genetics of molar cusp patterning in pedigreed baboons: Implications for primate dental development and evolution. *Journal of Experimental Zoology Part B: Molecular and Developmental Evolution*, *302*(3), 268–283.

Hobkirk, J. A., Gill, D. S., Jones, S. P., Hemmings, K. W., Bassi, G. S., O'Donnell, A. L., & Goodman, J. R. (2010). *Hypodontia: A Team Approach to Management.* John Wiley & Sons, Chichester.

Horrocks, L. R., Brook, A. H., Alvesalo, L., & Smith, R. N. (2009). A three-dimensional comparison of the modifying effects of familial genetic contribution in Turner syndrome. In *Comparative Dental Morphology* (Vol. 13, pp. 148–152). Karger Publishers, Basel, Switzerland.

Hosoya, A., Shalehin, N., Takebe, H., Shimo, T., & Irie, K. (2020). Sonic hedgehog signaling and tooth development. *International Journal of Molecular Sciences, 21*(5), 1587.

Howe, R. P., McNamara Jr, J. A., & O'connor, K. A. (1983). An examination of dental crowding and its relationship to tooth size and arch dimension. *American Journal of Orthodontics, 83*(5), 363–373.

Huang, Y., Lu, Y., Mues, G., Wang, S., Bonds, J., & D'Souza, R. (2013). Functional evaluation of a novel tooth agenesis-associated bone morphogenetic protein 4 prodomain mutation. *European Journal of Oral Sciences, 121*(4), 313–318.

Hughes, T. E., Richards, L., & Townsend, G. (2001). Dental arch form in young Australian twins. In A. H.Brook (Ed.), *Dental Morphology* (pp. 309–319). Sheffield Academic Press, Sheffield, England.

Hughes, T. E., Townsend, G. C., Pinkerton, S. K., Bockmann, M. R., Seow, W. K., Brook, A. H., Richards, L. C., Mihailidis, S., Ranjitkar, S., & Lekkas, D. (2014). The teeth and faces of twins: Providing insights into dentofacial development and oral health for practising oral health professionals. *Australian Dental Journal, 59*, 101–116.

Jacobsen, P. E., Henriksen, T. B., Haubek, D., & Østergaard, J. R. (2013). Developmental enamel defects in children prenatally exposed to anti-epileptic drugs. *PLoS ONE, 8*(3), e58213.

Kasai, K., Richards, L. C., Townsend, G. C., Kanazawa, E., & Iwasawa, T. (1995). Fourier analysis of dental arch morphology in South Australian twins. *Anthropological Science, 103*(1), 39–48.

Kavanagh, K. D., Evans, A. R., & Jernvall, J. (2007). Predicting evolutionary patterns of mammalian teeth from development. *Nature, 449*(7161), 427–432.

Kawata, T., Nakagawa, H., Kuroda, Y., Kushima, F., & Okada, M. (1974). Statistical and dynamical analysis of dental arch form in adult human with normal occlusion (part II). *The Journal of Osaka University Dental School, 14*, 91–95.

Kerekes-Máthé, B., Brook, A. H., Mártha, K., Székely, M., & Smith, R. N. (2015). Mild hypodontia is associated with smaller tooth dimensions and cusp numbers than in controls. *Archives of Oral Biology, 60*(9), 1442–1449.

Kerekes-Máthé, B., Mártha, K., Brook O'Donnell, M., & Brook, A. H. (2022). Dental arch dimensions in a matched pairs study of hypodontia patients and controls. *Biomedical Journal of Scientific and Technical Research*, In Press.

Kettunen, P. I., Laurikkala, J., Itäranta, P., Vainio, S., Itoh, N., & Thesleff, I. (2000). Associations of FGF-3 and FGF-10 with signaling networks regulating tooth morphogenesis. *Developmental Dynamics: An Official Publication of the American Association of Anatomists, 219*(3), 322–332.

Kettunen, P. I., Løes, S., Furmanek, T., Fjeld, K., Kvinnsland, I. H., Behar, O., Yagi, T., Fujisawa, H., Vainio, S., & Taniguchi, M. (2005). Coordination of trigeminal axon navigation and patterning with tooth organ formation: Epithelial-mesenchymal interactions, and epithelial Wnt4 and Tgfβ1 regulate semaphorin 3a expression in the dental mesenchyme. *Development, 132*(2), 323–334. doi: 10.1242/dev.01541

Khalaf, K., Robinson, D. L., Elcock, C., Smith, R. N., & Brook, A. H. (2005). Tooth size in patients with supernumerary teeth and a control group measured by image analysis system. *Archives of Oral Biology, 50*(2), 243–248.

Khalaf, K., Smith, R. N., & Brook, A. H. (2022). Regional genetic influences around anterior maxillary supernumeraries include variations of the clinical crowns of mandibular incisors. *Genes, 13*(12), 2232. https://doi.org/10.3390/genes13122232

Khalaf, K., Smith, R. N., Elcock, C., & Brook, A. H. (2009). Multiple crown size variables of the upper incisors in patients with supernumerary teeth compared with controls. *Archives of Oral Biology*, *54*, S71–S78.

Kieser, J. A. (1998). Fluctuating dental asymmetry and prenatal exposure to tobacco smoke. *Human Dental Development, Morphology, and Pathology: A Tribute to Albert A.*, 287–297.

Kjær, I. (1998). Neuro-osteology. *Critical Reviews in Oral Biology & Medicine*, *9*(2), 224–244.

Kjær, I., Kocsis, G., Nodal, M., & Christensen, L. R. (1994). Aetiological aspects of mandibular tooth agenesis—focusing on the role of nerve, oral mucosa, and supporting tissues. *The European Journal of Orthodontics*, *16*(5), 371–375.

Klein, O. D., Minowada, G., Peterkova, R., Kangas, A., Benjamin, D. Y., Lesot, H., Peterka, M., Jernvall, J., & Martin, G. R. (2006). Sprouty genes control diastema tooth development via bidirectional antagonism of epithelial-mesenchymal FGF signaling. *Developmental Cell*, *11*(2), 181–190.

Koh, K. S. B., Toh, V. K. L., Brook O'Donnell, M., Ranjitkar, S., & Brook, A. H. (2016). A complex adaptive system in which environmental stress affects gene expression during development. *International Journal of Design and Nature and Ecodynamics*, *11*, 686–695.

Kurek, M., Żądzińska, E., Sitek, A., Borowska-Strugińska, B., Rosset, I., & Lorkiewicz, W. (2015). Prenatal factors associated with the neonatal line thickness in human deciduous incisors. *Homo*, *66*(3), 251–263.

Laine, T., & Alvesalo, L. (1993). Palatal and mandibular arch morphology in 47, XYY men and in other sex-chromosome anomalies. *Archives of Oral Biology*, *38*(2), 101–105.

Lam, F., Yong, R., Ranjitkar, S., Townsend, G. C., & Brook, A. H. (2017). Agents within a developmental complex adaptive system: Intrauterine male hormones influence human tooth size and shape. *Complex Systems: Theory and Applications*, *214*.

Lampl, M., Kuzawa, C. W., & Jeanty, P. (2003). Prenatal smoke exposure alters growth in limb proportions and head shape in the midgestation human fetus. *American Journal of Human Biology*, *15*(4), 533–546.

Le Bot, P., & Salmon, D. (1977). Congenital defects of the upper lateral incisors (ULI): Condition and measurements of the other teeth, measurements of the superior arch, head and face. *American Journal of Physical Anthropology*, *46*(2), 231–243.

Lee, R. T. (1999). Arch width and form: A review. *American Journal of Orthodontics and Dentofacial Orthopedics*, *115*(3), 305–313.

Lesot, H., Lisi, S., Peterkova, R., Peterka, M., Mitolo, V., & Ruch, J. (2001). Epigenetic signals during odontoblast differentiation. *Advances in Dental Research*, *15*(1), 8–13.

Lindner, A., & Modeer, T. (1989). Relation between sucking habits and dental characteristics in preschoolchildren with unilateral cross-bite. *European Journal of Oral Sciences*, *97*(3), 278–283.

Lundström, A. (1951). Aetiology of crowding of the teeth (based on studies of twins and on morphological investigation) and its bearing on orthodontic treatment (expansion or extraction). *European Orthodontic Society*, 176–190.

Matalova, E., Fleischmannova, J., Sharpe, P. T., & Tucker, A. S. (2008). Tooth agenesis: From molecular genetics to molecular dentistry. *Journal of Dental Research*, *87*(7), 617–623.

Miller, E., & Martin, N. (1995). Analysis of the effect of hormones on opposite-sex twin attitudes. *Acta Geneticae Medicae et Gemellologiae: Twin Research*, *44*(1), 41–52.

Mitchell, M. (2009). *Complexity: A Guided Tour.* Oxford University Press, Oxford.

Mitsiadis, T. A., & Smith, M. M. (2006). How do genes make teeth to order through development? *Journal of Experimental Zoology Part B: Molecular and Developmental Evolution*, *306*(3), 177–182.

Mobley, C. C., & Reifsnider, E. (2005). Pregnancy, child nutrition, and oral health. In R. Touger-Decker, D. A. Sirois, & C. C. Mobley (Eds.), *Nutrition and Oral Medicine. Nutrition and Health.* Humana Press, Totowa, NJ. https://doi.org/10.1385/1-59259-831-5:017

Modeer, T., Odenrtck, L., & Lindner, A. (1982). Sucking habits and their relation to posterior cross-bite in 4-year-old children. *European Journal of Oral Sciences, 90*(4), 323–328.

Moorrees, C. F., & Reed, R. B. (1954). Biometrics of crowding and spacing of the teeth in the mandible. *American Journal of Physical Anthropology, 12*(1), 77–88.

Mustonen, T., Pispa, J., Mikkola, M. L., Pummila, M., Kangas, A. T., Pakkasjärvi, L., Jaatinen, R., & Thesleff, I. (2003). Stimulation of ectodermal organ development by Ectodysplasin-A1. *Developmental Biology, 259*(1), 123–136.

Nanci, A. (2008). *Ten Cate's Oral Histology: Development, Structure, and Formation.* Mosby, St. Louis, MO.

Nanci, A. (2017). *Ten Cate's Oral Histology-e-Book: Development, Structure, and Function.* Elsevier Health Sciences, St. Louis, MO.

Nelson, T. A. B., Willmot, D. R., Elcock, C., Smith, R. N., Robinson, D. L., & Brook, A. H. (2001). The use of computerized image analysis to measure the form and dimensions of the maxillary dental arches in subjects with hypodontia. *Proceedings of the 12th International Symposium on Dental Morphology.* Sheffield Academic Press, Ltd., Sheffield.

Newman, M. (2018). *Networks.* Oxford University Press, Oxford.

Oeschger, E. S., Kanavakis, G., Cocos, A., Halazonetis, D. J., & Gkantidis, N. (2022). Number of teeth is related to craniofacial morphology in humans. *Biology, 11*(4), 544.

Øgaard, B., Larsson, E., & Lindsten, R. (1994). The effect of sucking habits, cohort, sex, intercanine arch widths, and breast or bottle feeding on posterior crossbite in Norwegian and Swedish 3-year-old children. *American Journal of Orthodontics and Dentofacial Orthopedics, 106*(2), 161–166.

Ohazama, A., Johnson, E. B., Ota, M. S., Choi, H. J., Porntaveetus, T., Oommen, S., Itoh, N., Eto, K., Gritli-Linde, A., & Herz, J. (2008). Lrp4 modulates extracellular integration of cell signaling pathways in development. *PLoS ONE, 3*(12), e4092.

Osborne, J. (1978). Morphogenetic gradients: Fields versus clones. *Development, Function and Evolution of Teeth,* 171–201.

Page, S. E. (2018). *The Model Thinker: What You Need to Know to Make Data Work for You.* Basic Books, New York.

Pagella, P., Neto, E., Jiménez-Rojo, L., Lamghari, M., & Mitsiadis, T. A. (2014). Microfluidics co-culture systems for studying tooth innervation. *Frontiers in Physiology, 5,* 326.

Patel, D., Sassani, S., Farella, M., Ranjitkar, S., Yong, R., Swindells, S., & Brook, A. H. (2018). Variations in dental arch morphology are outcomes of the complex adaptive system associated with the developmental variation of hypodontia. *International Journal of Design and Nature and Ecodynamics, 13,* 107–113.

Peterkova, R., Lesot, H., & Peterka, M. (2006). Phylogenetic memory of developing mammalian dentition. *Journal of Experimental Zoology Part B: Molecular and Developmental Evolution, 306*(3), 234–250.

Peterkova, R., Peterka, M., Viriot, L., & Lesot, H. (2002). Development of the vestigial tooth primordia as part of mouse odontogenesis. *Connective Tissue Research, 43*(2–3), 120–128.

Peters, H., Neubüser, A., Kratochwil, K., & Balling, R. (1998). Pax9-deficient mice lack pharyngeal pouch derivatives and teeth and exhibit craniofacial and limb abnormalities. *Genes and Development, 12*(17), 2735–2747.

Proffit, W. R., Mc Glone, R. E., & Barrett, M. J. (1975). Lip and tongue pressures related to dental arch and oral cavity size in Australian aborigines. *Journal of Dental Research, 54*(6), 1161–1172.

Qiu, M., Bulfone, A., Ghattas, I., Meneses, J. J., Christensen, L., Sharpe, P. T., Presley, R., Pedersen, R. A., & Rubenstein, J. L. (1997). Role of the Dlx homeobox genes in proximodistal patterning of the branchial arches: Mutations of Dlx-1, Dlx-2, and Dlx-1 and-2 alter morphogenesis of proximal skeletal and soft tissue structures derived from the first and second arches. *Developmental Biology*, *185*(2), 165–184.

Raberin, M., Laumon, B., Martin, J.-L., & Brunner, F. (1993). Dimensions and form of dental arches in subjects with normal occlusions. *American Journal of Orthodontics and Dentofacial Orthopedics*, *104*(1), 67–72.

Radlanski, R. J., & Renz, H. (2005). Explainable and critical periods during human dental morphogenesis and their control. *Archives of Oral Biology*, *50*(2), 199–203.

Ribeiro, D. C., Brook, A. H., Hughes, T. E., Sampson, W. J., & Townsend, G. C. (2013). Intrauterine hormone effects on tooth dimensions. *Journal of Dental Research*, *92*(5), 425–431.

Richards, L. C., Townsend, G. C., Brown, T., & Burgess, V. B. (1990). Dental arch morphology in south Australian twins. *Archives of Oral Biology*, *35*(12), 983–989.

Riga, A., Belcastro, M. G., & Moggi-Cecchi, J. (2014). Environmental stress increases variability in the expression of dental cusps. *American Journal of Physical Anthropology*, *153*(3), 397–407.

Roseman, C. C., & Delezene, L. K. (2019). The inhibitory cascade model is not a good predictor of molar size covariation. *Evolutionary Biology*, *46*(3), 229–238.

Rugg-Gunn, A., Al-Mohammadi, S., & Butler, T. (1998). Malnutrition and developmental defects of enamel in 2-to 6-year-old Saudi boys. *Caries Research*, *32*(3), 181–192.

Rushton, M. A. (1947). Effects of radium on the dentition. *American Journal of Orthodontics and Oral Surgery*, *33*(12), B828–B830.

Salazar-Ciudad, I., & Jernvall, J. (2002). A gene network model accounting for development and evolution of mammalian teeth. *Proceedings of the National Academy of Sciences*, *99*(12), 8116–8120.

Salazar-Ciudad, I., & Jernvall, J. (2010). A computational model of teeth and the developmental origins of morphological variation. *Nature*, *464*(7288), 583–586.

Sarkar, L., Cobourne, M., Naylor, S., Smalley, M., Dale, T., & Sharpe, P. T. (2000). Wnt/Shh interactions regulate ectodermal boundary formation during mammalian tooth development. *Proceedings of the National Academy of Sciences*, *97*(9), 4520–4524.

Seow, W. A., & Wan, A. (2000). Research reports clinical: A controlled study of the morphometric changes in the primary dentition of pre-term, very-low-birthweight children. *Journal of Dental Research*, *79*(1), 63–69.

Seow, W. K., Brown, J. P., Tudehope, D. A., & O'Callaghan, M. (1984). Dental defects in the deciduous dentition of premature infants with low birth weight and neonatal rickets. *Pediatric Dentistry*, *6*(2), 88–92.

Sharpe, P. T. (1995). Homeobox genes and orofacial development. *Connective Tissue Research*, *32*(1–4), 17–25.

Sillman, J. (1964). Dimensional changes of the dental arches: Longitudinal study from birth to 25 years. *American Journal of Orthodontics*, *50*(11), 824–842.

Sofaer, J., Chung, C., Niswander, J., & Runck, D. (1971). Developmental interaction, size and agenesis among permanent maxillary incisors. *Human Biology*, 36–45.

Steele-Perkins, G., Butz, K. G., Lyons, G. E., Zeichner-David, M., Kim, H.-J., Cho, M.-I., & Gronostajski, R. M. (2003). Essential role for NFI-C/CTF transcription-replication factor in tooth root development. *Molecular and Cellular Biology*, *23*(3), 1075–1084.

Tabatabaeifar, M., Schlingmann, K. P., Litwin, M., Emre, S., Bakkaloglu, A., Mehls, O., Antignac, C., Schaefer, F., & Weber, S. (2009). Functional analysis of BMP4 mutations identified in pediatric CAKUT patients. *Pediatric Nephrology*, *24*(12), 2361–2368.

Tadros, M., Brook, A. H., Ranjitkar, S., & Townsend, G. C. (2019). Compensatory interactions between developing maxillary anterior teeth in a sample of twins. *Archives of Oral Biology, 97*, 198–207.

Taduran, R. J. O., Ranjitkar, S., Hughes, T., Townsend, G., & Brook, A. H. (2016). Complex systems in human development: Sexual dimorphism in teeth and fingerprints of Australian twins. *International Journal of Design & Nature and Ecodynamics, 11*(4), 676–685.

Tapp, A. L., Maybery, M. T., & Whitehouse, A. J. (2011). Evaluating the twin testosterone transfer hypothesis: A review of the empirical evidence. *Hormones and Behavior, 60*(5), 713–722.

Thesleff, I. (2014). Current understanding of the process of tooth formation: Transfer from the laboratory to the clinic. *Australian Dental Journal, 59*, 48–54.

Thesleff, I., Vaahtokari, A., & Partanen, A. M. (1995). Regulation of organogenesis. Common molecular mechanisms regulating the development of teeth and other organs. *The International Journal of Developmental Biology, 39*(1), 35–50.

Thilander, B. (2009). Dentoalveolar development in subjects with normal occlusion. A longitudinal study between the ages of 5 and 31 years. *The European Journal of Orthodontics, 31*(2), 109–120.

Thurner, S., Hanel, R., & Klimek, P. (2018). *Introduction to the Theory of Complex Systems.* Oxford University Press, Oxford.

Townsend, G. C., & Brook, A. H. (2008). Genetic, epigenetic and environmental influences on dental development. *Ortho Tribune, 3*(4), 4–6.

Townsend, G. C., & Brown, T. (1981). Morphogenetic fields within the dentition. *Australian Orthodontic Journal, 7*(1), 3–12.

Townsend, G. C., Corruccini, R. S., Brown, T., & Richards, L. C. (1988). Genetic and environmental determinants of dental occlusal variation in South Australian twins. *Australian Orthodontic Journal, 10*(4), 231–235.

Townsend, G. C., Hughes, T., Bockmann, M., Smith, R., & Brook, A. H. (2009). How studies of twins can inform our understanding of dental morphology. *Frontiers of Oral Biology, 13*, 136–141.

Townsend, G. C., Yamada, H., & Smith, P. (1990). Expression of the entoconulid (sixth cusp) on mandibular molar teeth of an Australian aboriginal population. *American Journal of Physical Anthropology, 82*(3), 267–274.

Tucker, A. S., Headon, D. J., Schneider, P., Ferguson, B. M., Overbeek, P., Tschopp, J., & Sharpe, P. T. (2000). Edar/Eda interactions regulate enamel knot formation in tooth morphogenesis. *Development, 127*(21), 4691–4700.

Tucker, A. S., Khamis, A. A., & Sharpe, P. T. (1998). Interactions between Bmp-4 and Msx-1 act to restrict gene expression to odontogenic mesenchyme. *Developmental Dynamics: An Official Publication of the American Association of Anatomists, 212*(4), 533–539.

Van Valen, L. (1970). An analysis of developmental fields. *Developmental Biology, 23*(3), 456–477.

Vastardis, H. (2000). The genetics of human tooth agenesis: New discoveries for understanding dental anomalies. *American Journal of Orthodontics and Dentofacial Orthopedics, 117*(6), 650–656.

Waddington, C. H. (1959). Canalization of development and genetic assimilation of acquired characters. *Nature, 183*(4676), 1654–1655.

Warren, J. J., & Bishara, S. E. (2002). Duration of nutritive and nonnutritive sucking behaviors and their effects on the dental arches in the primary dentition. *American Journal of Orthodontics and Dentofacial Orthopedics, 121*(4), 347–356.

Weaver, C. A., Miller, S. F., da Fontoura, C. S., Wehby, G. L., Amendt, B. A., Holton, N. E., Allareddy, V., Southard, T. E., & Uribe, L. M. M. (2017). Candidate gene analyses of 3-dimensional dentoalveolar phenotypes in subjects with malocclusion. *American Journal of Orthodontics and Dentofacial Orthopedics*, *151*(3), 539–558.

Weinstein, S., Haack, D. C., Morris, L. Y., Snyder, B. B., & Attaway, H. E. (1963). On an equilibrium theory of tooth position. *The Angle Orthodontist*, *33*(1), 1–26.

Wells, J. C. (2012). Body composition in infants: Evidence for developmental programming and techniques for measurement. *Reviews in Endocrine and Metabolic Disorders*, *13*(2), 93–101.

Williams, M. A., & Letra, A. (2018). The changing landscape in the genetic etiology of human tooth agenesis. *Genes (Basel)*, *9*(5).

Williams, S. D., Hughes, T. E., Adler, C. J., Brook, A. H., & Townsend, G. C. (2014). Epigenetics: A new frontier in dentistry. *Australian Dental Journal*, *59*(Suppl 1), 23–33.

Woodworth, D. A., Sinclair, P. M., & Alexander, R. G. (1985). Bilateral congenital absence of maxillary lateral incisors: A craniofacial and dental cast analysis. *American Journal of Orthodontics*, *87*(4), 280–293.

Wu, C.-T., & Morris, J. R. (2001). Genes, genetics, and epigenetics: A correspondence. *Science*, *293*(5532), 1103–1105.

Xavier, G. M., Seppala, M., Barrell, W., Birjandi, A. A., Geoghegan, F., & Cobourne, M. T. (2016). Hedgehog receptor function during craniofacial development. *Developmental Biology*, *415*(2), 198–215.

Yamashiro, T., Zheng, L., Shitaku, Y., Saito, M., Tsubakimoto, T., Takada, K., Takano-Yamamoto, T., & Thesleff, I. (2007). Wnt10a regulates dentin sialophosphoprotein mRNA expression and possibly links odontoblast differentiation and tooth morphogenesis. *Differentiation*, *75*(5), 452–462.

Yu, M., Wang, H., Fan, Z., Xie, C., Liu, H., Liu, Y., Han, D., Wong, S. W., & Feng, H. (2019). BMP4 mutations in tooth agenesis and low bone mass. *Archives of Oral Biology*, *103*, 40–46.

Yu, M., Wong, S. W., Han, D., & Cai, T. (2019). Genetic analysis: Wnt and other pathways in nonsyndromic tooth agenesis. *Oral Diseases*, *25*(3), 646–651.

Zadzińska, E., Kozieł, S., Kurek, M., & Spinek, A. (2013). Mother's trauma during pregnancy affects fluctuating asymmetry in offspring's face. *Anthropol Anz*, *70*(4), 427–437.

Zeleny, M. (1987). Management support systems: Towards integrated knowledge management. *Human Systems Management*, *7*(1), 59–70.

Zhao, H., Oka, K., Bringas, P., Kaartinen, V., & Chai, Y. (2008). TGF-β type I receptor Alk5 regulates tooth initiation and mandible patterning in a type II receptor-independent manner. *Developmental Biology*, *320*(1), 19–29.

Index

Note: Page numbers in *italics* indicate a figure and page numbers in **bold** indicate a table on the corresponding page.

F

P

R

T